Asymptotic
Approximations
of Integrals

Asymptotic
Approximations
of Integrals

R. Wong
City University of Hong Kong
Kowloon, Hong Kong

Society for Industrial and Applied Mathematics
Philadelphia

Library of Congress Cataloging-in-Publication Data
Wong, R. (Roderick), 1944-
   Asymptotic approximations of integrals / R. Wong --SIAM ed.
      p. cm. -- (Classics in applied mathematics ; 34)
   Includes bibliographical references and index.
   ISBN 0-89871-497-4
      1. Integrals. 2. Approximation theory. 3. Asymptotic expansions.
      I. Title. II. Series.

QA311 . W65 2001
515'.43--dc21

                                                        2001032023

To
my mother
who gave me
life and education

Contents

III   Mellin Transform Techniques

IV   The Summability Method

V   Elementary Theory of Distributions

VI The Distributional Approach

VII Uniform Asymptotic Expansions

VIII   Double Integrals

IX   Higher Dimensional Integrals

Preface to the Classics Edition

Although my book was written twelve years ago, it is probably still the most up-to-date book in the area of asymptotic expansions of integrals. While trying to buy a copy of this book for one of my students in 1999, I found that it was already out of print. Therefore, I was very pleased when I heard that SIAM had decided to include my book in its Classics series. Here I would like to take this opportunity to thank Professor Robert E. O'Malley, Jr., for his recommendation to reprint it in this series.

Since the publication of this book, significant developments have occurred in the general theory of Asymptotic Expansions. These developments include smoothing of the Stokes phenomenon (M. V. Berry, 1989), uniform exponentially improved asymptotic expansions (F. W. J. Olver, 1991), and hyperasymptotics (M. V. Berry and C. J. Howls, 1990). All these new concepts belong to the area of what is now known as exponential asymptotics. For expositions of these new theories, see the papers by R. B. Paris and A. D. Wood (IMA *Bulletin*, 31 (1995), 21–23; J. P. Boyd (*Acta Appl. Math.*, 56 (1999), 1–98); and Wong (*Special Functions*, NATO ASI Series, M. E. H. Ismail and S. K. Suslov, eds., Kluwer, 2000).

Preface

Asymptotic approximation is an important topic in applied analysis, and its applications permeate many fields in science and engineering such as fluid mechanics, electromagnetism, diffraction theory, and statistics. Although it is an old subject, dating back to the time of Laplace, new methods and new applications continue to appear in various publications. There are now several excellent books on this subject, and, in particular, the one by F. W. J. Olver deserves special mention. However, most of these books were written more than 15 years ago, and Olver's book stresses more the differential equation side of asymptotic theory. There is now a need to provide an up-to-date account of methods used in the other main area of asymptotic theory, namely, asymptotic approximation of integrals. The purpose of this book is precisely to fulfil this need. Many of the results appear for the first time in book form. These include logarithmic singularities, Mellin transform technique for multiple integrals, summability method, distributional approach, uniform asymptotic expansions via a rational transformation, and double integrals with a curve of stationary points. For completeness, classical methods are also discussed in detail.In this sense, the book is self-contained. Furthermore, all results are proved rigorously and accompanied by error bounds whenever possible.

The book presupposes that the reader has a thorough knowledge of advanced calculus and is familiar with the basic theory of complex variables. It can be used either as a text for graduate students in mathematics, physics, and engineering, or as a reference tool for research workers in these, and other, areas. As a text, it is suitable for a two-semester course meeting three hours per week, but it can also be used for a one-semester course. For instance, Chapters I to IV and parts of Chapter VII would constitute such a course. Each chapter is self-contained in order to render it accessible to the casual peruser. Each chapter has an extensive set of exercises, many of which are accompanied by hints for their solution. However, the development of the material in the text does not depend on the exercises, and omission of some or all of them does not destroy the continuity of the presentation. Nevertheless, students are strongly advised to read through the exercises, since some of them provide important extensions of the general theory, while others supply completely new results.

Chapters I and II cover classical techniques in the asymptotic evaluation of integrals. More recent methods are introduced in Chapters III, IV, and VI. In Chapter V, a short introduction to distribution theory is presented, and almost all of the results given in this chapter are used later. Integrals which depend on auxiliary parameters in addition to the asymptotic variable are discussed in Chapter VII. Finally, Chapters VIII and IX are devoted to multidimensional integrals.

A short section titled "Supplementary Notes" appears at the end of each chapter, where additional references can be found. Some of these supply sources of material presented, and some pertain to more recent books or papers on closely related topics. Since mention to the references Erdélyi *et al.* (1953, 1953b), Olver (1974a), and Watson (1944) for properties of special functions is made frequently throughout the book, we have omitted their occurrences from the author index.

The writing of the manuscript began when I was a Killam Research Fellow (1982–1984). The preparation of the book has been facilitated by both the Killam Foundation and the Natural Sciences and Engineering Council of Canada. To these agencies I am most grateful.

I must express my special thanks to Professor F. W. J. Olver, who has read the entire manuscript as well as the page proofs, offered numerous suggestions, and corrected innumerable errors, mathematical as well as linguistic. Without his generous advice and constant encouragement, this book would have never been written. I am indebted to my colleague Professor J. P. McClure, and to Professor Qu Chong-kai of Tsinghau University, both of whom read and commented on the manuscript in its preliminary stages. Thanks are also due to Carol Plumridge for her excellent job of typing, and to my student, Tom Lang, for proofreading various parts of the typescript. Finally, my deep gratitude goes to my wife, Edwina, for reading the page proofs and for her patience and understanding during the preparation of the manuscript.

I

Fundamental Concepts of Asymptotics

1. What Is Asymptotics?

In analysis and applied mathematics, one frequently comes across problems concerning the determination of the behavior of a function as one of its parameters tends to a specific value, or of a sequence as its index tends to infinity. The branch of mathematics that is devoted to the investigation of these types of problems is called *asymptotics*. Thus, for instance, results such as

$$\log n! \sim (n + \tfrac{1}{2}) \log n - n + \tfrac{1}{2} \log 2\pi, \tag{1.1}$$

$$H_n \equiv 1 + \frac{1}{2} + \frac{1}{3} + \cdots + \frac{1}{n} \sim \log n, \tag{1.2}$$

and

$$L_n \equiv \frac{1}{\pi} \int_0^\pi \frac{|\sin(n + \tfrac{1}{2})t|}{\sin \tfrac{1}{2}t} \, dt \sim \frac{4}{\pi^2} \log n, \tag{1.3}$$

are all part of this subject. Equation (1.1) is known as the Stirling formula; the numbers H_n are called the harmonic numbers and often

1

occur in the analysis of algorithms (Greene and Knuth 1981); and the numbers L_n in (1.3) are called the Lebesgue constants in the theory of Fourier series. The twiddle sign $\sim$ is used to mean that the quotient of the left-hand side by the corresponding right-hand side approaches 1 as $n \to \infty$. Formulas such as those in (1.1)–(1.3) are called *asymptotic formulas* or *asymptotic equalities*.

Results in (1.1)–(1.3) are all easy to derive and can be found in books on elementary analysis; see, for example, Rudin (1976). However, on many occasions, information given by an asymptotic formula is insufficient and higher term approximations are required. The situation here is very much akin to the one in which the approximation obtained from the mean-value theorem is not sufficient and the use of Taylor's formula with remainder becomes necessary. Higher-term approximations for $\log n!$, H_n, and L_n are given by

$$\log n! \sim \left(n + \frac{1}{2}\right) \log n - n + \frac{1}{2} \log 2\pi + \sum_{s=0}^{\infty} \frac{B_{2s+2}}{(2s+1)(2s+2)n^{2s+1}}, \quad (1.4)$$

$$H_n = \sum_{k=1}^{n} \frac{1}{k} \sim \log n + \gamma + \frac{1}{2n} - \sum_{s=1}^{\infty} \frac{B_{2s}}{(2s)n^{2s}}, \quad (1.5)$$

and

$$L_n \sim \frac{4}{\pi^2} \left\{ \log(2n+1) + A_0 - \sum_{s=1}^{\infty} \frac{A_s}{(2n+1)^{2s}} \right\}, \quad (1.6)$$

where B_{2s} denotes the Bernoulli numbers defined by

$$\frac{z}{e^z - 1} = \sum_{n=0}^{\infty} B_n \frac{z^n}{n!}, \quad (1.7)$$

and

$$A_0 = 2 \sum_{m=1}^{\infty} \frac{\log m}{4m^2 - 1} + 2 \log 2 + \gamma = 2.441\ldots, \quad (1.8)$$

$$A_s = \frac{(1 - 2^{2s-1})B_{2s}}{s} \left[1 - \sum_{m=1}^{s} \frac{(-1)^{m-1}}{(2m)!} B_{2m} \pi^{2m} \right], \quad s \geq 1, \quad (1.9)$$

γ being the Euler constant. Note that the series (1.4)–(1.6) are all divergent; we have extended the symbol $\sim$ to mean that any partial sum of any one of these is an approximation of the corresponding left-hand side with an error that is of the same order of magnitude as the first neglected term. The results are, in fact, more precise than this; that is, one can show that the remainder due to truncation of any one of these series has the property that it is numerically less than, and has the same sign as, the first term omitted. These results are not easy to obtain, and they form typical examples in books on asymptotics. A proof of (1.4) is outlined in Ex. 19, and a proof of (1.5) is given in Section 6. The result in (1.6) is due to Watson (1930), and his proof is reproduced in Section 6.

The present subject of asymptotics can be divided into three main areas. The first area deals with functions that are expressible in the forms of definite integrals or contour integrals. A typical example in this area is given by the integral

$$I_n = \int_a^b \varphi(x)[f(x)]^n \, dx,$$

where $\varphi(x)$ and $f(x)$ are continuous functions defined on the interval $[a, b]$ and $f(x)$ is positive there. Long ago, Laplace made the observation that the major contribution to the integral I_n should come from the neighborhoods of the points where $f(x)$ attains its greatest value. Furthermore, he showed that if $f(x)$ attains its maximum value only at the point ξ in (a, b) where $f'(\xi) = 0$ and $f''(\xi) < 0$, then

$$I_n \sim \varphi(\xi)[f(\xi)]^{n + 1/2}\left\{\frac{-2\pi}{nf''(\xi)}\right\}^{1/2}. \qquad (1.10)$$

This formula is now known as the Laplace approximation. For an excellent introduction to the topics in this area, we refer to the book by Copson (1965).

The second area in asymptotics is concerned with solutions to differential equations. The best known equation here is probably

$$y'' + [\lambda^2 a(x) + b(x)]y = 0, \qquad (1.11)$$

where λ is a large positive parameter and $a(x) > 0$ in $[x_0, x_1]$. Liouville and Green, simultaneously and independently, showed that equation (1.11) has two linearly independent solutions, which behave asymptotically like

$$y^{\pm}(x) \sim \frac{1}{a^{1/4}(x)} \exp\left[\pm i\lambda \int a^{1/2}(x) \, dx\right], \qquad \text{as } \lambda \to \infty. \qquad (1.12)$$

Formula (1.12) has been known as the WKB approximation; it is only recently that (1.12) is called, and rightly so, the Liouville–Green approximation. For a definitive work on this area of asymptotics, see Olver (1974a).

The third area in asymptotics is connected with enumeration problems. A typical example is the following: Let d_n denote the number of partitions of an n-element set (e.g., $d_1 = 1$, $d_2 = 2$, $d_3 = 5$, $d_4 = 15$). It is known that the exponential generating function of these numbers is $\exp(e^z - 1)$, i.e.,

$$\exp(e^z - 1) = \sum_{n=0}^{\infty} \frac{d_n}{n!} z^n. \tag{1.13}$$

The problem here is to obtain the asymptotic behavior of d_n as $n \to \infty$. From a formula of Hayman (see Chapter II, Section 7), we have

$$d_n \sim \frac{\exp[n(r_n + r_n^{-1} - 1) - 1]}{\sqrt{r_n + 1}}, \tag{1.14}$$

where r_n is the root of the equation

$$r \exp(r) = n. \tag{1.15}$$

For a survey of the methods in this area of asymptotics, see Bender (1974).

The asymptotics of the first area will occupy a central portion of the present book. Since problems in the third area are often related to problems in the first area, some of the important methods in this area will also be mentioned.

2. Asymptotic Expansions

In 1886, Poincaré introduced the notion of an asymptotic expansion. This concept enables one to manipulate a large class of divergent series in much the same way as convergent power series. Moreover, it enables one to obtain numerical as well as qualitative results for many problems. The divergent series in (1.4)–(1.6) are all special examples of asymptotic expansions. Before giving a precise definition, let us first recall the O- and o-symbols introduced by Landau.

Let Ω be a point set in the complex z-plane, and let z_0 be a limit point of Ω, possibly the point at infinity. Frequently, Ω will be either a sector

$$0 < |z| < \infty, \qquad \theta_0 < \arg z < \theta_1,$$

or the set of nonnegative integers $N = \{0, 1, 2, \ldots\}$.

Let $f(z)$ and $g(z)$ be two functions defined on Ω. We write

$$f(z) = O(g(z)), \qquad \text{as } z \to z_0, \tag{2.1}$$

to mean that there is a constant $K > 0$ and a neighborhood U of z_0 such that $|f(z)| \le K|g(z)|$ for all $z \in \Omega \cap U$. We also write

$$f(z) = o(g(z)), \qquad \text{as } z \to z_0, \tag{2.2}$$

to mean that for every $\varepsilon > 0$, there exists a neighborhood U_ε of z_0 such that $|f(z)| \le \varepsilon|g(z)|$ for all $z \in \Omega \cap U_\varepsilon$. If $f(z)/g(z)$ tends to unity, then we write

$$f(z) \sim g(z), \qquad \text{as } z \to z_0. \tag{2.3}$$

Definition 1 (**Poincaré**). Let $f(z)$ be defined in an unbounded set Ω. A power series $\sum_{n=0}^{\infty} a_n z^{-n}$, convergent or divergent, is called an *asymptotic expansion* of $f(z)$ if, for every fixed integer $N \ge 0$,

$$f(z) = \sum_{n=0}^{N} a_n z^{-n} + O(z^{-(N+1)}), \qquad \text{as } z \to \infty; \tag{2.4}$$

in which case, we write

$$f(z) \sim \sum_{n=0}^{\infty} a_n z^{-n}, \qquad \text{as } z \to \infty. \tag{2.5}$$

An important property of the Poincaré asymptotic expansion is that the expansion, if it exists, is unique. The coefficients a_n are determined by the recurrence relations:

$$a_0 = \lim_{z \to \infty} f(z), \qquad a_m = \lim_{z \to \infty} z^m \left[f(z) - \sum_{n=0}^{m-1} a_n z^{-n} \right]. \tag{2.6}$$

These formulas, coupled with the fact that $\lim_{z \to \infty} z^m[\exp(-b|z|^p)] = 0$, for every nonnegative integer m and all fixed positive numbers b and p, imply that every function $g(z)$, satisfying $g(z) = O(\exp(-b|z|^p))$, as $z \to \infty$ in Ω, has the unique asymptotic expansion

$$g(z) \sim 0, \qquad \text{as } z \to \infty \text{ in } \Omega. \tag{2.7}$$

In turn, it will be true, that two functions $F(z)$ and $G(z)$ such that

$$F(z) = G(z) + g(z), \tag{2.8}$$

$g(z)$ as above, will have the same asymptotic expansion.

Functions satisfying the order relation $g(z) = O(\exp(-b|z|^p))$, as $z \to \infty$ in Ω, are said to be *exponentially small*, and it is usual to replace such functions by zero at any stage in a proof being used to establish the validity of a Poincaré asymptotic expansion.

The following properties follow immediately from the definition.

Theorem 1. *If* $f(z) \sim \sum_{n=0}^{\infty} a_n z^{-n}$ *and* $g(z) \sim \sum_{n=0}^{\infty} b_n z^{-n}$, *as* $z \to \infty$ *in the same unbounded set* Ω, *then* (i)

$$\alpha f(z) + \beta g(z) \sim \sum_{n=0}^{\infty} (\alpha a_n + \beta b_n) z^{-n}, \qquad \text{as } z \to \infty \text{ in } \Omega, \tag{2.9}$$

providing α *and* β *are constants; and* (ii)

$$f(z)\, g(z) \sim \sum_{n=0}^{\infty} c_n z^{-n}, \qquad \text{as } z \to \infty \text{ in } \Omega, \tag{2.10}$$

where $c_n = \sum_{s=0}^{n} a_s b_{n-s}$.

Theorem 2. *If* $f(z)$ *is continuous in the domain* Ω *defined by* $|z| > a$, $\theta_0 \le \arg z \le \theta_1$, *and if* $f(z) \sim \sum_{n=0}^{\infty} a_n z^{-n}$, *as* $z \to \infty$ *in* Ω, *then*

$$\int_z^{\infty} \left[f(t) - a_0 - \frac{a_1}{t} \right] dt \sim \sum_{n=1}^{\infty} \frac{a_{n+1}}{n} z^{-n}, \qquad \text{as } z \to \infty \text{ in } \Omega, \tag{2.11}$$

where the path of integration is the straight line joining z *to* ∞ *with a fixed argument.*

The problem concerning the differentiation of an asymptotic expansion is much more difficult. In general, termwise differentiation of an asymptotic expansion does not necessarily give an asymptotic expansion. For example, the function $f(z) = e^{-z} \cos e^z$ has the asymptotic expansion $f(z) \sim 0 + 0z^{-1} + 0z^{-2} + \cdots$, as $z \to \infty$ in $|\arg z| \le \pi/2 - \Delta$, $\Delta > 0$; but the termwise differentiated series is not the asymptotic expansion of the derivative $f'(z) = -\sin e^z - e^{-z} \cos e^z$. The following results are nevertheless true.

Theorem 3. *Let Ω be the domain defined by $|z| > R$, $\theta_0 < \arg z < \theta_1$, and assume that $f(z) \sim \sum_{n=0}^{\infty} a_n z^{-n}$, as $z \to \infty$ in Ω. If $f(z)$ has a continuous derivative $f'(z)$, and if $f'(z)$ possesses an asymptotic expansion as $z \to \infty$ in Ω, then*

$$f'(z) \sim - \sum_{n=1}^{\infty} n a_n z^{-n-1}, \qquad as \ z \to \infty \ in \ \Omega. \tag{2.12}$$

Proof. By hypothesis, there exists a sequence $\{b_n\}$ such that

$$f'(z) \sim \sum_{n=0}^{\infty} b_n z^{-n}, \qquad as \ z \to \infty \ in \ \Omega.$$

Since $f'(z)$ is continuous in Ω, we have

$$
\begin{aligned}
f(z_1) - f(z) &= \int_z^{z_1} f'(\zeta) \, d\zeta \\
&= b_0(z_1 - z) + b_1 \log \frac{z_1}{z} + \int_z^{z_1} \left[f'(\zeta) - b_0 - \frac{b_1}{\zeta} \right] d\zeta,
\end{aligned}
\tag{2.13}
$$

where the path of integration is the straight line joining z to z_1 with fixed argument. As $z_1 \to \infty$, $f(z_1) \to a_0$ and the last integral in (2.13) tends to

$$\int_z^{\infty} \left[f'(\zeta) - b_0 - \frac{b_1}{\zeta} \right] d\zeta$$

which is convergent. Therefore, b_0 and b_1 must be zero, and (2.13) becomes

$$a_0 - f(z) = \int_z^{\infty} \left[f'(\zeta) - b_0 - \frac{b_1}{\zeta} \right] d\zeta.$$

By Theorem 2,

$$f(z) \sim a_0 - \sum_{n=1}^{\infty} \frac{b_{n+1}}{n z^n}, \qquad as \ z \to \infty \ in \ \Omega.$$

The uniqueness of asymptotic expansion then implies that $b_{n+1} = -n a_n$, $n = 1, 2, \ldots$; that is,

$$f'(z) \sim - \sum_{n=2}^{\infty} \frac{(n-1) a_{n-1}}{z^n}, \qquad as \ z \to \infty \ in \ \Omega.$$

This proves the theorem. ∎

Theorem 4.   *Let Ω be the domain given in Theorem 3 and assume that $f(z)$ is analytic in Ω. If*

$$f(z) \sim \sum_{n=0}^{\infty} a_n z^{-n}$$

uniformly in $\arg z$ as $z \to \infty$ in any closed sector contained in Ω, then

$$f'(z) \sim - \sum_{n=2}^{\infty} \frac{(n-1)a_{n-1}}{z^n}$$

also uniformly in $\arg z$ as $z \to \infty$ in any closed sector contained in Ω.

Proof.   Put

$$f(z) = \sum_{n=0}^{m-1} \frac{a_n}{z^n} + \frac{1}{z^m} f_m(z).$$

Then

$$f'(z) = - \sum_{n=2}^{m-1} \frac{(n-1)a_{n-1}}{z^n} + \frac{1}{z^m} \varphi_m(z),$$

where

$$\varphi_m(z) = f'_m(z) - (m-1)a_{m-1} - \frac{m}{z} f_m(z).$$

Recall that Ω is the domain given by $\{z : |z| > R, \ \theta_0 < \arg z < \theta_1\}$, and let Ω'' denote the closed sector $\theta_0'' \le \arg z \le \theta_1''$, contained in Ω. Choose θ_0' and θ_1' so that

$$\theta_0 < \theta_0' < \theta_0'' \le \theta_1'' < \theta_1' < \theta_1.$$

The region $\Omega' = \{z : |z| > R, \ \theta_0' < \arg z < \theta_1'\}$ clearly contains Ω'' and is contained in Ω. By hypothesis, there exists $T \ge R$ and $A_m > 0$ such that $|f_m(z)| \le A_m$ for all $z \in \Omega'$ with $|z| \ge T$. Thus, to prove the theorem, it suffices to show that $f'_m(z) = O(1)$ uniformly as $z \to \infty$ in Ω''. For any ζ in $\theta_0'' \le \arg \zeta \le \theta_1''$, there exists a positive number δ such that the circle $\gamma = \{z : |z - \zeta| = \delta|\zeta|\}$ is contained in $\theta_0' \le \arg \zeta \le \theta_1'$. By Cauchy's theorem,

$$f'_m(\zeta) = \frac{1}{2\pi i} \int_{\gamma} \frac{f_m(z)}{(z-\zeta)^2} \, dz = \frac{1}{2\pi} \int_0^{2\pi} \frac{f_m(\zeta + \delta|\zeta|e^{i\theta})}{\delta|\zeta|e^{i\theta}} \, d\theta.$$

Hence

$$|f'_m(\zeta)| \le \frac{A_m}{\delta|\zeta|} \le \frac{A_m}{\delta T}.$$

This completes the proof of Theorem 4. ■

Theorem 5. *Let $f(z)$ be an analytic function in $\Omega = \{z : |z| \ge R\}$ and suppose that*

$$f(z) \sim \sum_{n=0}^{\infty} \frac{a_n}{z^n}, \qquad as\ z \to \infty\ in\ \Omega. \tag{2.14}$$

Then the asymptotic series is convergent and its sum is equal to $f(z)$ for all sufficiently large values of z.

Proof. The Laurent series of $f(z)$ is

$$f(z) = \sum_{-\infty}^{\infty} c_n z^n,$$

for $|z| \ge R_1 > R$, where the coefficient c_n is given by

$$c_n = \frac{1}{2\pi i} \int_{\Gamma} \frac{f(z)}{z^{n+1}}\, dz,$$

Γ being any circle $|z| = \rho > R_1$. Since $f(z) \to a_0$ as $|z| \to \infty$, there exists a constant $M > 0$ such that $|f(z)| \le M$ for $|z| \ge R_1$. For $n > 0$, we have $|c_n| \le M/\rho^n$. Letting $\rho \to \infty$ gives $c_n = 0$ for all $n > 0$. Therefore

$$f(z) = \sum_{n=0}^{\infty} c_{-n} z^{-n}$$

for $|z| \ge R_1$. Since convergent series are asymptotic series, by the uniqueness theorem, $c_{-n} = a_n$. Therefore, $f(z) = \sum_{n=0}^{\infty} a_n z^{-n}$ for all $|z| \ge R_1$. ■

The above results are well-known, and can be found in many standard books on asymptotics; see, for example, Copson (1965), Erdélyi (1956), and Olver (1974a). We include them here merely for completeness.

3. Generalized Asymptotic Expansions

Within the framework provided by Poincaré, the determination of the asymptotic behavior of functions either defined by integrals or as solutions to differential equations has been the subject of intensive study. However, it is easily seen that the Poincaré definition of an asymptotic expansion is unnecessarily restrictive. For instance, the expansions

$$\frac{x^{2/3} + x^{1/2}}{x - 1} = \frac{1}{x^{1/3}} + \frac{1}{x^{1/2}} + \frac{1}{x^{4/3}} + \frac{1}{x^{3/2}} + \cdots, \qquad |x| > 1$$

and

$$\frac{1 + \log \log x}{\log x - 1} = \frac{\log \log x}{\log x} + \frac{1}{\log x} + \frac{\log \log x}{\log^2 x} + \frac{1}{\log^2 x} + \cdots, \qquad x > e,$$

have many of the properties commonly associated with asymptotic expansions but are excluded by the Poincaré definition. In this section, we shall describe a generalization, which was first introduced by Schmidt (1937) and later extended by Erdélyi (1961).

Definition 2. Let $\{\varphi_n(z)\}$ be a sequence of functions defined in a common set Ω, and let z_0 be a limit point of Ω. We say that $\{\varphi_n\}$ is an *asymptotic sequence as $z \to z_0$ in Ω* if, for all $n \geq 0$,

$$\varphi_{n+1}(z) = o(\varphi_n(z)), \qquad \text{as } z \to z_0. \tag{3.1}$$

The notation $\{\varphi_n(z)\}$ will be used throughout to denote an asymptotic sequence, and when ambiguity must be avoided, the limit point z_0 will be specified in some way.

Definition 3. Let $f(z)$ and $f_n(z)$, $n = 0, 1, 2, \ldots$, be functions defined in Ω. The formal series $\sum f_n(z)$ is called a *generalized asymptotic expansion* of $f(z)$ *with respect to* the asymptotic sequence $\{\varphi_n(z)\}$, as $z \to z_0$, if

$$f(z) = \sum_{n=0}^{N} f_n(z) + o(\varphi_N(z)) \tag{3.2}$$

for every fixed $N \geq 0$. In this case, we write

$$f(z) \sim \sum_{n=0}^{\infty} f_n(z); \quad \{\varphi_n\}, \quad \text{as } z \to z_0. \tag{3.3}$$

When $f_n(z) = a_n \varphi_n(z)$, a_n a fixed complex number, for every n, then the above expansion is said to be of *Poincaré type*. Furthermore, if the expansion is of this type, and $\varphi_n(z) = (\xi(z))^{\lambda_n}$, λ_n a fixed complex number, the expansion is said to be of *power series form*.

Definition 4. Two functions $f(z)$ and $g(z)$ defined in some neighborhood of z_0 are said to be *asymptotically equal*, written

$$f(z) \approx g(z); \quad \{\varphi_n\}, \quad \text{as } z \to z_0, \tag{3.4}$$

if

$$f(z) = g(z) + o(\varphi_n(z)), \quad \text{as } z \to z_0, \tag{3.5}$$

for every $n \geq 0$.

Two functions having the same asymptotic expansion are asymptotically equal, and the converse is also true.

Even this degree of generality is not sufficient to describe the asymptotic behavior of many of the known functions of mathematics. The form

$$f(z) \sim g_1(z)\left[\sum_{n=0}^{\infty} f_n^{(1)}(z); \quad \{\varphi_n^{(1)}\}\right] + g_2(z)\left[\sum_{n=0}^{\infty} f_n^{(2)}(z); \quad \{\varphi_n^{(2)}\}\right] + \cdots, \tag{3.6}$$

as $z \to z_0$, with the meaning

$$f(z) = g_1(z)\left[\sum_{n=0}^{N_1} f_n^{(1)}(z) + o(\varphi_{N_1}^{(1)})\right] + g_2(z)\left[\sum_{n=0}^{N_2} f_n^{(2)}(z) + o(\varphi_{N_2}^{(2)})\right] + \cdots, \tag{3.7}$$

as $z \to z_0$, where $N_1, N_2, \ldots$ are arbitrary fixed integers, must often be used to give asymptotic information for many of the higher transcendental functions. Expansions of the form (3.7) are called *compound asymptotic expansions*.

It is easy to give examples of functions that have generalized asymptotic expansions but are not of Poincaré type. Let Ω be the set of real numbers $x \geq c > 1$. In Ω, the function defined by the uniformly convergent series

$$f(x) = \sum_{n=0}^{\infty} \frac{\sin(n+1)\pi x}{x^n} \tag{3.8}$$

satisfies, for every fixed integer N,

$$f(x) = \sum_{n=0}^{N} \frac{\sin(n+1)\pi x}{x^n} + o(x^{-N}). \tag{3.9}$$

Hence (3.8) is certainly an asymptotic expansion with respect to the asymptotic sequence $\{x^{-n}\}$. However, it is not of Poincaré type since the sequence $\varphi_n(x) = \sin(n+1)\pi x / x^n$ does not satisfy $\varphi_{n+1} = o(\varphi_n)$, as $x \to \infty$. For instance, $\varphi_1 = \sin 2\pi x / x = 0$, when $x = \frac{1}{2}m$, where m is any nonnegative integer, but $\varphi_2(\frac{1}{2}m) = \sin \frac{3}{2}\pi m / (\frac{1}{2}m)^2 \neq 0$, when m is an odd integer; thus $\varphi_2 \neq o(\varphi_1)$ as $x \to \infty$.

A conspicuous feature of the theory of generalized asymptotic expansions is the lack of uniqueness. That is, there is no analogue of formula (2.6) for constructing successive terms. In fact, a function possessing a generalized asymptotic expansion with respect to an asymptotic sequence can possess infinitely many such expansions with respect to the same sequence, and a generalized asymptotic expansion of a function can also be a generalized asymptotic expansion with respect to infinitely many different asymptotic sequences. To illustrate these points, we give the example:

$$\frac{1}{1+z} \sim \sum_{n=0}^{\infty} \frac{(-1)^n}{z^{n+1}}; \quad \{z^{-n-1}\}, \tag{3.10}$$

$$\frac{1}{1+z} \sim \sum_{n=0}^{\infty} (-1)^n \frac{1+e^{-z}}{z^{n+1}}; \quad \{z^{-n-1}\}, \tag{3.11}$$

$$\frac{1}{1+z} \sim \sum_{n=0}^{\infty} \frac{(-1)^n}{z^{n+1}}; \quad \{z^{-(n+1)/2}\}, \tag{3.12}$$

as $z \to \infty$ in $|\arg z| \leq \pi/2 - \Delta, \Delta > 0$.

Another criticism of the theory of generalized asymptotic expansions is that its definition allows expansions that are useless either numerically or analytically. An example of such a result is supplied by Riekstiņš (1966):

$$\frac{\sin x}{x} \sim \sum_{s=1}^{\infty} \frac{s!\, e^{-(s+1)x/2s}}{(\log x)^s}; \qquad \{(\log x)^{-s}\}, \tag{3.13}$$

as $x \to \infty$. Similarly, it is true that

$$\frac{\cos x}{x} \sim \sum_{s=1}^{\infty} \frac{s!\, e^{-(s+1)x/2s}}{(\log x)^s}; \qquad \{(\log x)^{-s}\}, \tag{3.14}$$

as pointed out by Olver (1980, p. 197).

In view of the chaotic conditions that seem to have been introduced by the consideration of generalized asymptotic expansions against the tidy theory that can be given for asymptotic expansions of power series form, the usefulness of the generalized theory is called into question. There indeed exists a viewpoint among asymptotic analysts which suggests that we should dispense with the formal generalized definition and simply use truncated expansions accompanied by appropriate O-estimates (or o-estimates) for the error terms; see Olver (1980). Of course, there also exists the viewpoint that the definition of generalized expansions opens up new possibilities, leads to a considerable flexibility of method, and greatly extends the class of functions whose asymptotic behavior can be obtained; see Wyman (1963).

In the present book, we shall take a neutral stand; that is, we shall accept the definition of generalized asymptotic expansions, but will use it only as a notational device for the result given in (3.2). We find this definition convenient to use, as it clearly represents the result obtained and, in many instances, it considerably simplifies some complicated expressions. However, we do not think it is worthwhile to develop a theory based on this definition. Also, we believe it is a wrong attitude to prescribe an asymptotic sequence $\{\varphi_n\}$ on an *a priori* basis and then investigate procedures for producing asymptotic expansions $\sum f_n$ for a specific function with respect to $\{\varphi_n\}$. We prefer to first produce, by any procedure, a useful asymptotic expansion $\sum f_n$ for a given function f and then ask for the determination of an asymptotic sequence $\{\varphi_n\}$ for which (3.2) is valid. Finally, we emphasize the importance to consider, at the end of any investigation, whether or not the result obtained is worth having. For instance, results such as (3.13) and (3.14) are obviously not worth having, and results like (3.11) and (3.12) also can not be considered superior to that given in (3.10).

4. Integration by Parts

A simple yet powerful technique for deriving asymptotic expansions of definite integrals is the method of integration by parts. Each integration produces a term in the expansion, and the error term is given explicitly as an integral. This idea can be made clear by considering the following examples.

Example 1. *The Exponential Integral.*
This integral is defined by

$$\mathrm{Ei}(z) = \int_{-\infty}^{z} \frac{e^t}{t}\, dt, \qquad |\arg(-z)| < \pi, \tag{4.1}$$

where the integration can be taken along any path C in the complex t-plane with a cut along the positive real axis. A possible choice of C is the infinite line segment

$$-\infty < \mathrm{Re}\, t < \mathrm{Re}\, z, \qquad \mathrm{Im}\, t = \mathrm{Im}\, z, \tag{4.2}$$

joining $t = -\infty$ to $t = z$ and parallel to the real axis. Repeated integration by parts gives

$$\mathrm{Ei}(z) = \frac{e^z}{z} + \int_{-\infty}^{z} \frac{e^t}{t^2}\, dt = \frac{e^z}{z} + \frac{e^z}{z^2} + 2\int_{-\infty}^{z} \frac{e^t}{t^3}\, dt$$

$$= \frac{e^z}{z}\left[\sum_{k=0}^{n} \frac{k!}{z^k} + \varepsilon_n(z) \right], \tag{4.3}$$

where

$$\varepsilon_n(z) = (n+1)!\, ze^{-z} \int_{-\infty}^{z} \frac{e^t}{t^{n+2}}\, dt, \qquad |\arg(-z)| < \pi. \tag{4.4}$$

To estimate the remainder $\varepsilon_n(z)$, we take C to be the line segment given in (4.2). Suppose $|\arg(-z)| \le \pi - \delta$, where δ is any positive number in $(0, \pi)$, and write $z = x + iy$. Along the segment $t = \tau + iy$, $-\infty < \tau \le x$, we have $|e^t| = e^\tau$, $|t| \ge |z| \sin \delta$. Hence

$$|\varepsilon_n(z)| \le \frac{(n+1)!}{|z|^{n+1}(\sin \delta)^{n+2}}. \tag{4.5}$$

Accordingly, as $z \to \infty$ in $|\arg(-z)| \le \pi - \delta$,

$$\text{Ei}(z) \sim \frac{e^z}{z} \sum_{k=0}^{\infty} \frac{k!}{z^k}. \tag{4.6}$$

If $\text{Re } z \le 0$, i.e., $|\arg(-z)| \le \pi/2$, then we may take $\delta = \pi/2$ and (4.5) becomes

$$|\varepsilon_n(z)| \le \frac{(n+1)!}{|z|^{n+1}}. \tag{4.7}$$

In this case, the error committed does not exceed the first neglected term in absolute value.

In this example, we are fortunate to have a sharp numerical upper bound for the absolute value of the remainder term. In general, it is often very difficult to find such an upper bound, and, even in cases where an upper bound can be obtained, it is often so weak that it is quite useless for numerical calculation. This does not mean that the asymptotic expansion obtained is of no value, since for many purposes, particularly in applied mathematics, the asymptotic expansions obtained give much more information than we need to know. In fact, the knowledge of the first one or two terms of these expansions is usually sufficient for our purpose. However, to obtain results that are absolutely reliable for numerical computations, it is necessary to construct a sharp upper bound for the remainder term. Thus, if it is possible to obtain such an upper bound, then we must not neglect to do so.

Example 2. *Fourier Integrals.*

A second type of integral to which the method of integration by parts can be applied is the Fourier integral

$$F(x) = \int_a^b f(t)e^{ixt}\, dt, \tag{4.8}$$

where (a, b) is a real finite interval and $f(t)$ is an N-times continuously differentiable function in $[a, b]$. By successive integration by parts, we have

$$F(x) = \sum_{n=0}^{N-1} \left(\frac{i}{x}\right)^{n+1} [e^{iax}f^{(n)}(a) - e^{ibx}f^{(n)}(b)] + \varepsilon_N(x), \tag{4.9}$$

where

$$\varepsilon_N(x) = \left(\frac{i}{x}\right)^N \int_a^b f^{(N)}(t)e^{ixt}\, dt. \tag{4.10}$$

The Riemann–Lebesgue lemma then gives $\varepsilon_N(x) = o(x^{-N})$ as $x \to +\infty$, thus showing that equation (4.9) is the asymptotic expansion of $F(x)$ as far as terms of order x^{-N}. If N is infinite, then (4.8) can be written as

$$F(x) \sim \sum_{n=0}^{\infty} \left(\frac{i}{x}\right)^{n+1} \{e^{iax}f^{(n)}(a) - e^{ibx}f^{(n)}(b)\};\quad \{x^{-n-1}\},\quad \text{as } x \to +\infty. \tag{4.11}$$

A simple upper bound for the error term $\varepsilon_N(x)$ is given by

$$|\varepsilon_N(x)| \le x^{-N} \int_a^b |f^{(N)}(t)|\, dt; \tag{4.12}$$

but this estimate falls short of the actual result $\varepsilon_N(x) = O(x^{-N-1})$. To improve this estimate, we integrate (4.10) by parts one more time and arrive at

$$\varepsilon_N(x) = \left(\frac{i}{x}\right)^{N+1} \{e^{iax}f^{(N)}(a) - e^{ibx}f^{(N)}(b)\}$$

$$+ \left(\frac{i}{x}\right)^{N+1} \int_a^b f^{(N+1)}(t)e^{ixt}\, dt, \tag{4.13}$$

which in turn gives

$$|\varepsilon_N(x)| \le x^{-N-1}\left[|f^{(N)}(a)| + |f^{(N)}(b)| + \int_a^b |f^{(N+1)}(t)|\, dt\right]. \tag{4.14}$$

The foregoing results can easily be extended to semi-infinite intervals. For instance, if $f(t)$ is infinitely differentiable on $[a, \infty)$ and

$$f^{(s)}(t) = O(t^{-1-\varepsilon}),\quad \text{as } t \to \infty, \tag{4.15}$$

for some $\varepsilon > 0$ and for every $s \ge 0$, then by letting $b \to \infty$ in (4.11) we establish

$$\int_a^{\infty} f(t)e^{ixt}\, dt \sim e^{iax} \sum_{n=0}^{\infty} f^{(n)}(a)\left(\frac{i}{x}\right)^{n+1},\quad \text{as } x \to +\infty. \tag{4.16}$$

The above technique of integration by parts needs some modification when there are algebraic singularities at the end points. As a simple illustration, let us consider the infinite Fourier integral

$$F(x) = \int_0^\infty t^{\alpha - 1} f(t) e^{ixt} \, dt, \tag{4.17}$$

where $0 < \alpha < 1$. Put

$$g_0(t) = t^{\alpha - 1} e^{ixt} \tag{4.18}$$

and define

$$g_1(t) = -\int_t^{t + i\infty} \tau^{\alpha - 1} e^{ix\tau} \, d\tau, \tag{4.19}$$

where the path of integration is the vertical line $\tau = t + iy$, $y \geq 0$. It is easily shown that $g_1'(t) = g_0(t) = t^{\alpha - 1} e^{ixt}$ and

$$g_1(0) = -\int_0^{i\infty} \tau^{\alpha - 1} e^{ix\tau} \, d\tau = -e^{i\pi\alpha/2} \Gamma(\alpha) x^{-\alpha}. \tag{4.20}$$

We assume that condition (4.15) again holds (with $\varepsilon = 0$). Upon integration by parts, we have from (4.17)

$$F(x) = f(0) e^{i\pi\alpha/2} \Gamma(\alpha) x^{-\alpha} - \int_0^\infty f'(t) \, g_1(t) \, dt. \tag{4.21}$$

To continue this process, we define

$$g_{n+1}(t) = -\int_t^{t + i\infty} g_n(\tau) \, d\tau = \frac{(-1)^{n+1}}{n!} \int_t^{t + i\infty} (z - t)^n z^{\alpha - 1} e^{ixz} \, dz, \tag{4.22}$$

and observe that

$$\frac{d}{dt} g_{n+1}(t) = g_n(t) \tag{4.23}$$

and

$$g_{n+1}(0) = (-1)^{n+1} \frac{\Gamma(n + \alpha)}{n!} e^{i\pi(n + \alpha)/2} x^{-n - \alpha}. \tag{4.24}$$

Repeated integration by parts then gives

$$F(x) = \sum_{n=0}^{N-1} \frac{f^{(n)}(0)}{n!} e^{i\pi(n + \alpha)/2} \Gamma(n + \alpha) x^{-n - \alpha} + R_N, \tag{4.25}$$

where

$$R_N = (-1)^N \int_0^\infty f^{(N)}(t) \, g_N(t) \, dt. \qquad (4.26)$$

To find a bound for R_N, we must first estimate $g_N(t)$. Note that along the path of integration in (4.22), we have

$$|g_{n+1}(t)| \leq \frac{t^{\alpha-1}}{n!} \int_0^\infty y^n e^{-xy} \, dy = t^{\alpha-1} x^{-n-1} \qquad (4.27)$$

for $t > 0$, $x > 0$, $n = 0, 1, 2, \ldots$. From (4.26), it now follows that

$$|R_N| \leq x^{-N} \int_0^\infty t^{\alpha-1} |f^{(N)}(t)| \, dt. \qquad (4.28)$$

The clever technique given in this example is due to Erdélyi (1955). Although clever, the field of its application is rather limited. A more powerful method for treating oscillatory integrals is discussed in Chapter IV.

Example 3. Cases of Failure.

The above examples clearly illustrate the basic idea behind the method of integration by parts. However, in applying this method, one often has the tendency to think that there is no need for the estimation of the remainder, since each integration yields a term in the asymptotic expansion and hence repeating this procedure infinitely many times should produce the entire expansion. This thought is incorrect, however, as we shall see in the present example. Let

$$S(x) = \int_0^\infty \frac{1}{\sqrt[3]{1 + t(x + t)}} \, dt. \qquad (4.29)$$

Repeated integration by parts gives

$$
\begin{aligned}
S(x) &= -\frac{3}{2}\frac{1}{x} + \frac{3 \cdot 1}{2} \int_0^\infty \frac{(1 + t)^{2/3}}{(x + t)^2} \, dt \\
&= -\frac{3}{2}\frac{1}{x} - \frac{3^2 \cdot 1}{2 \cdot 5}\frac{1}{x^2} + \frac{3^2 \cdot 2!}{2 \cdot 5} \int_0^\infty \frac{(1 + t)^{5/3}}{(x + t)^3} \, dt \qquad (4.30) \\
&= -\sum_{n=1}^{N-1} \frac{3^n (n-1)!}{2 \cdot 5 \cdots (3n-1)} \frac{1}{x^n} + \delta_N(x),
\end{aligned}
$$

where

$$\delta_N(x) = \frac{3^{N-1}(N-1)!}{2 \cdot 5 \cdots (3N-4)} \int_0^\infty \frac{(1+t)^{N-4/3}}{(x+t)^N} \, dt. \tag{4.31}$$

Thus, one might naturally conclude that

$$S(x) \sim - \sum_{n=1}^\infty \frac{3^n(n-1)!}{2 \cdot 5 \cdots (3n-1)} x^{-n}, \tag{4.32}$$

as $x \to +\infty$, which is obviously false, since the integral $S(x)$ is positive, whereas the terms in its series expansion are all negative. Also, since $t \leq 1 + t \leq x + t$ for $x \geq 1$, we have

$$\int_0^\infty \frac{t^{N-4/3}}{(x+t)^N} \, dt \leq \int_0^\infty \frac{(1+t)^{N-4/3}}{(x+t)^N} \, dt \leq \frac{3}{x^{1/3}}. \tag{4.33}$$

In view of the well-known identity (Olver 1974a, p. 38)

$$\int_0^\infty \frac{t^{\alpha-1}}{(1+t)^{\alpha+\beta}} \, dt = \frac{\Gamma(\alpha)\,\Gamma(\beta)}{\Gamma(\alpha+\beta)}, \tag{4.34}$$

the integral on the extreme left of (4.33) can be explicitly evaluated to be $\Gamma(N - \frac{1}{3})\,\Gamma(\frac{1}{3})/\Gamma(N)x^{1/3}$. Therefore the remainder $\delta_N(x)$ is of higher order of magnitude than any of the terms in the series (4.30), violating the definition of an asymptotic expansion. The correct asymptotic expansion of $S(x)$ is given in Chapter VI. For a similar example involving generalized asymptotic expansions, see Soni (1978).

5. Watson's Lemma

Another simple and useful technique for deriving asymptotic expansion of an integral is that of termwise integration of a series expansion of the integrand. This method is particularly easy to apply, when the integrand is exponentially decaying. To illustrate, we consider the integral

$$L(x) = \int_0^\infty \frac{t^{\lambda-1}}{1+t} e^{-xt} \, dt, \tag{5.1}$$

where $\lambda > 0$. Since

$$\frac{1}{1+t} = \sum_{n=0}^{N-1} (-1)^n t^n + (-1)^N \frac{t^N}{1+t}, \tag{5.2}$$

termwise integration gives

$$L(x) = \sum_{n=0}^{N-1} (-1)^n \frac{\Gamma(n+\lambda)}{x^{n+\lambda}} + R_N(x), \tag{5.3}$$

where

$$R_N(x) = (-1)^N \int_0^\infty \frac{t^{N+\lambda-1}}{1+t} e^{-xt} \, dt. \tag{5.4}$$

Thus

$$|R_N(x)| \le \frac{\Gamma(N+\lambda)}{x^{N+\lambda}}, \tag{5.5}$$

and

$$L(x) \sim \sum_{n=0}^{\infty} (-1)^n \frac{\Gamma(n+\lambda)}{x^{n+\lambda}}, \qquad \text{as } x \to \infty. \tag{5.6}$$

This is a special case of a much more general result due to Watson (1918), now known as Watson's lemma. In view of its importance, Watson's statement and proof of the result are reproduced below.

Watson's Lemma.   *If* (i) $f(t)$ *is analytic when* $|t| \le a + \delta$, *where* $a > 0$, $\delta > 0$, *except at a branch-point at the origin, and*

$$f(t) = \sum_{m=1}^{\infty} a_m t^{m/r-1} \tag{5.7}$$

when $|t| \le a$, *r being positive;* (ii) $|f(t)| < Ke^{bt}$, *where* K *and* b *are independent of* t, *when* t *is positive and* $t \ge a$; (iii) $|\arg z| \le \pi/2 - \Delta$, *where* $\Delta > 0$; *and* (iv) $|z|$ *is sufficiently large; then there exists a complete asymptotic expansion given by the formula*

$$F(z) = \int_0^\infty f(t)e^{-zt} \, dt \sim \sum_{m=1}^{\infty} a_m \Gamma\left(\frac{m}{r}\right) z^{-m/r}. \tag{5.8}$$

Proof.   If M is any fixed integer, we have

$$\left| f(t) - \sum_{m=1}^{M-1} a_m t^{m/r-1} \right| < K_1 t^{M/r-1} e^{bt} \tag{5.9}$$

throughout the range of integration, where K_1 is some number independent of t. Hence

$$F(z) = \sum_{m=1}^{M-1} a_m \int_0^\infty t^{m/r-1} e^{-zt}\, dt + R_M,$$

where

$$|R_M| < K_1 \int_0^\infty t^{M/r-1} e^{bt} |e^{-zt}|\, dt < K_1 \Gamma\!\left(\frac{M}{r}\right)(\mathrm{Re}\ z - b)^{-M/r}, \quad (5.10)$$

provided that $\mathrm{Re}\ z > b$, which is the case when $|z|$ is sufficiently large. Since $(\mathrm{Re}\ z - b)^{-1} = O(1/z)$ for the range of values of z under consideration, we have

$$F(z) = \sum_{m=1}^{M-1} a_m \Gamma\!\left(\frac{m}{r}\right) z^{-m/r} + O(z^{-M/r}), \quad (5.11)$$

and so the integral possesses the complete asymptotic expansion, which is of Poincaré type. ∎

Here it seems appropriate to make the following remark. In (5.11), we have only an order estimate, and not a numerical bound, for the remainder R_M, since the constant K_1, although it exists, is not explicitly known. In order to construct a computable bound for R_M, one must first replace K_1 in (5.10) by a numerical value. The best value of K_1 is given by

$$K_1 = \sup_{(0,\infty)}\left\{ \left| f(t) - \sum_{m=1}^{M-1} a_m t^{m/r-1} \right| t^{1-M/r} e^{-bt} \right\}, \quad (5.12)$$

which, however, may be difficult to obtain even in specific examples. Thus it is possible in some circumstances that one has an asymptotic expansion without a numerical error bound.

Among the known procedures, Watson's lemma is certainly one of the most frequently used methods for finding asymptotic expansions. However, its conditions and path of integration are known to be more restrictive than necessary. In order to generalize, we consider the integral

$$F(z) = \int_0^{\infty e^{i\gamma}} f(t) e^{-zt}\, dt, \quad (5.13)$$

where γ is a fixed real number, and the path of integration is the straight line joining $t = 0$ to $t = \infty e^{i\gamma}$. The following result is given in Wyman (1964, p. 249).

Generalized Watson's Lemma.   *Suppose that the integral in (5.13) exists for some fixed $z = z_0$, and that as $t \to 0$ along $\arg t = \gamma$,*

$$f(t) \sim \sum_{n=0}^{\infty} a_n t^{\lambda_n - 1}, \tag{5.14}$$

where $\operatorname{Re} \lambda_0 > 0$ and $\operatorname{Re} \lambda_{n+1} > \operatorname{Re} \lambda_n$. Then

$$F(z) \sim \sum_{n=0}^{\infty} a_n \Gamma(\lambda_n) z^{-\lambda_n}, \tag{5.15}$$

as $z \to \infty$ in $|\arg(ze^{i\gamma})| \leq \pi/2 - \Delta$, for any real number Δ in the interval $0 < \Delta \leq \pi/2$.

Proof.   The general properties of Laplace integrals (Widder 1941, p. 37) ensure that $F(z)$ exists as long as $\operatorname{Re}(ze^{i\gamma}) > \operatorname{Re}(z_0 e^{i\gamma})$. Put

$$g(t) = \int_c^t f(\tau) e^{-z_0 \tau} \, d\tau. \tag{5.16}$$

Then, by hypothesis, $|g(t)| \leq M < \infty$ for all t on the line segment joining $t = c$ to $t = \infty e^{i\gamma}$. Thus, for any arbitrary fixed choice of $t = c = |c|e^{i\gamma}$, $0 < |c| < \infty$, it is true that

$$\int_c^{\infty e^{i\gamma}} f(t) e^{-zt} \, dt = (z - z_0) \int_c^{\infty e^{i\gamma}} g(t) e^{-(z-z_0)t} \, dt. \tag{5.17}$$

Furthermore, the right-hand side is dominated by

$$\frac{M|z - z_0|}{\operatorname{Re}[(z - z_0)e^{i\gamma}]} \exp\{-\operatorname{Re}[(z - z_0)e^{i\gamma}]|c|\}. \tag{5.18}$$

Since

$$\operatorname{Re}[(z - z_0)e^{i\gamma}] = \operatorname{Re}(ze^{i\gamma}) - \operatorname{Re}(z_0 e^{i\gamma}) \tag{5.19}$$
$$\geq |z| \cos(\arg(ze^{i\gamma})) - |z_0| \geq |z| \sin \Delta - |z_0|,$$

we have

$$\int_c^{\infty e^{i\gamma}} f(t) e^{-zt} \, dt = O(e^{-\delta|z||c|}), \tag{5.20}$$

for $|z| > 2|z_0| \csc \Delta$, where $\delta = \frac{1}{2} \sin \Delta$. Hence

$$\int_c^{\infty e^{i\gamma}} f(t) e^{-zt} \, dt \approx 0; \qquad \{z^{-\lambda_n}\}, \tag{5.21}$$

as $z \to \infty$ in $|\arg(ze^{i\gamma})| \leq \pi/2 - \Delta$. This, of course, implies

$$F(z) \approx \int_0^c f(t) e^{-zt} \, dt; \qquad \{z^{-\lambda_n}\}, \tag{5.22}$$

uniformly in $\arg z$, as $z \to \infty$ in $|\arg(ze^{i\gamma})| \leq \pi/2 - \Delta$.

For every fixed integer $N \geq 1$, there exist positive numbers K_N and r_N such that

$$f(t) = \sum_{n=0}^{N-1} a_n t^{\lambda_n - 1} + R_N \qquad \text{for} \quad |t| \leq r_N \quad \text{and} \quad \arg t = \gamma, \tag{5.23}$$

where

$$|R_N| \leq K_N |t^{\lambda_N - 1}|. \tag{5.24}$$

Hence

$$\int_0^c f(t) e^{-zt} \, dt = \sum_{n=0}^{N-1} a_n \int_0^c t^{\lambda_n - 1} e^{-zt} \, dt + S_N \tag{5.25}$$

with $|c| \leq r_N$ and

$$S_N = \int_0^c R_N e^{-zt} \, dt. \tag{5.26}$$

From (5.20) it follows that

$$\int_0^c t^{\lambda_n - 1} e^{-zt} \, dt = \int_0^{\infty e^{i\gamma}} t^{\lambda_n - 1} e^{-zt} \, dt + O(e^{-\delta|c||z|})$$
$$= \Gamma(\lambda_n) z^{-\lambda_n} + O(e^{-\delta|c||z|}). \tag{5.27}$$

Furthermore, (5.24) and (5.26) give

$$|S_N| \leq K_N \int_0^c |t^{\lambda_N - 1} e^{-zt} \, dt| \leq K_N \int_0^{\infty e^{i\gamma}} |t^{\lambda_N - 1} e^{-zt} \, dt|$$
$$= O(z^{-\lambda_N}), \qquad \text{as } z \to \infty \text{ in } |\arg(ze^{i\gamma})| \leq \pi/2 - \Delta. \tag{5.28}$$

These results couple together to give

$$F(z) \sim \sum_{n=0}^{\infty} a_n \Gamma(\lambda_n) z^{-\lambda_n}, \tag{5.29}$$

uniformly in $\arg z$, as $z \to \infty$ in $|\arg(ze^{i\gamma})| \leq \pi/2 - \Delta$. $\blacksquare$

The above analysis again gives only an asymptotic expansion without an error bound. Although it is possible to modify the argument so that a numerical bound can be constructed for the error associated with the above expansion (see for instance, Olver (1974a, p. 114)), we shall not pursue this possibility, and prefer to illustrate it with some specific examples.

Example 4. *The Bessel Function $J_\nu(x)$.*

When Re $\nu > -\frac{1}{2}$, the Bessel function of the first kind $J_\nu(x)$ has the integral representation

$$
\begin{aligned}
J_\nu(x) &= \frac{(x/2)^\nu}{\Gamma(\tfrac{1}{2})\Gamma(\nu + \tfrac{1}{2})} \int_{-1}^{1} (1 - t^2)^{\nu - 1/2} \cos xt \, dt \\
&= \frac{(x/2)^\nu}{\Gamma(\tfrac{1}{2})\Gamma(\nu + \tfrac{1}{2})} \int_{-1}^{1} (1 - t^2)^{\nu - 1/2} e^{ixt} \, dt.
\end{aligned}
\tag{5.30}
$$

By Cauchy's theorem, the last integral can be written as

$$
\int_{-1}^{-1 + i\infty} (1 - t^2)^{\nu - 1/2} e^{ixt} \, dt - \int_{1}^{1 + i\infty} (1 - t^2)^{\nu - 1/2} e^{ixt} \, dt,
$$

where the paths of integration are the vertical lines through $t = -1$ and $t = +1$. In the first integral, we put $t = -1 + i2\tau$, and in the second integral, we put $t = 1 + i2\tau$. The resulting expression is

$$
\int_{-1}^{1} (1 - t^2)^{\nu - 1/2} e^{ixt} \, dt = 2^{2\nu}\{ e^{-i(x - \pi\nu/2 - \pi/4)} I_-(x) + e^{i(x - \pi\nu/2 - \pi/4)} I_+(x) \},
\tag{5.31}
$$

where

$$
I_\pm(x) = \int_0^\infty \tau^{\nu - 1/2}(1 \pm i\tau)^{\nu - 1/2} e^{-2x\tau} \, d\tau.
\tag{5.32}
$$

Although this result is established when x is real and positive, by analytic continuation it also holds for complex x as long as Re $x > 0$. Application of the generalized Watson's lemma to (5.32) with $\lambda_n = n + \nu + \frac{1}{2}$ and z replaced by $2x$ yields

$$
I_\pm(x) \sim \sum_{n=0}^{\infty} (-1)^n \frac{(\tfrac{1}{2} - \nu)_n}{n!} (\pm i)^n \Gamma(n + \nu + \tfrac{1}{2})(2x)^{-n - \nu - 1/2}, \tag{5.33}
$$

as $x \to \infty$ in $|\arg x| \leq \pi/2 - \Delta$, where

$$(\lambda)_0 = 1, \qquad (\lambda)_k = \frac{\Gamma(\lambda + k)}{\Gamma(\lambda)} = \lambda(\lambda + 1) \cdots (\lambda + k - 1). \qquad (5.34)$$

Substituting (5.33) in (5.31) and replacing x by z in the resulting expansion, we obtain

$$J_\nu(z) \sim \left(\frac{2}{\pi z}\right)^{1/2} \Bigg[\cos(z - \pi\nu/2 - \pi/4) \sum_{k=0}^{\infty} (-1)^k \frac{(\nu, 2k)}{(2z)^{2k}} \\ - \sin(z - \pi\nu/2 - \pi/4) \sum_{k=0}^{\infty} (-1)^k \frac{(\nu, 2k+1)}{(2z)^{2k+1}} \Bigg] \qquad (5.35)$$

as $z \to \infty$ in $|\arg z| \leq \pi/2 - \Delta$. Here we have introduced the notation:

$(\nu, 0) = 1$ and

$$(\nu, k) = \frac{(-1)^k}{k!} (\tfrac{1}{2} - \nu)_k (\tfrac{1}{2} + \nu)_k = \frac{(4\nu^2 - 1)(4\nu^2 - 3^2) \cdots [4\nu^2 - (2k-1)^2]}{2^{2k} \cdot k!}.$$
$$(5.36)$$

To bound the error terms associated with the expansion (5.33), we assume that ν is real. From Taylor's theorem, we have

$$(1 \pm i\tau)^{\nu - 1/2} = \sum_{k=0}^{n-1} (-1)^k \frac{(\tfrac{1}{2} - \nu)_k}{k!} (\pm i\tau)^k + r_n(\tau), \qquad (5.37)$$

where

$$r_n(\tau) = (-1)^n \frac{(\tfrac{1}{2} - \nu)_n}{(n-1)!} (\pm i\tau)^n \int_0^1 (1-t)^{n-1}(1 \pm i\tau t)^{\nu - 1/2 - n} \, dt. \qquad (5.38)$$

Choose $n > \nu - \tfrac{1}{2}$ so that $|1 \pm i\tau t|^{\nu - n - 1/2} \leq 1$ and

$$|r_n(\tau)| \leq \frac{|(\tfrac{1}{2} - \nu)_n|}{n!} \tau^n. \qquad (5.39)$$

From this it follows that the n-th error term of (5.33) does not exceed

$$\frac{|(\tfrac{1}{2} - \nu)_n|}{n!} \frac{\Gamma(\nu + n + \tfrac{1}{2})}{[2|x| \cos(\arg x)]^{\nu + n + 1/2}}.$$

If x is positive, then this implies that the n-th error term of (5.33) is bounded in absolute value by the first neglected term of the expansion. A combination of this result together with (5.35) gives

$$J_\nu(x) = \sqrt{\frac{2}{\pi x}} \sum_{k=0}^{n-1} (-1)^k \frac{(\nu, k)}{2^k} \frac{\cos[x - (\nu + \frac{1}{2} + k)\pi/2]}{x^k} + \varepsilon_n(x), \quad (5.40)$$

where

$$|\varepsilon_n(x)| \le \sqrt{\frac{2}{\pi|x|}} \frac{|(\frac{1}{2} - \nu)_n|}{n!} \frac{\Gamma(\nu + n + \frac{1}{2})}{|\Gamma(\nu + \frac{1}{2})|} \frac{[\sec(\arg x)]^{\nu + n + 1/2}}{(2|x|)^n}. \quad (5.41)$$

If x is positive, then (5.41) simplifies to

$$|\varepsilon_n(x)| \le \sqrt{\frac{2}{\pi x}} \frac{|(\nu, n)|}{2^n} x^{-n}. \quad (5.42)$$

The sector of validity of (5.35) can be extended to $|\arg z| \le \pi - \Delta$ by rotating the path of integration in (5.32) through an angle γ, where $-\pi/2 < \gamma < \pi/2$. To do this, we note that for $x > 0$, we have by Cauchy's theorem

$$I_\pm(x) = \int_0^{\infty e^{i\gamma}} \tau^{\nu - 1/2}(1 \pm i\tau)^{\nu - 1/2} e^{-2x\tau} \, d\tau. \quad (5.43)$$

The integral on the right-hand side actually defines an analytic function in $-\pi/2 - \gamma < \arg x < \pi/2 - \gamma$. Thus, by varying γ in the interval $(-\pi/2, \pi/2)$, (5.43) provides an analytic continuation for $I_\pm(x)$ in the larger sector $|\arg x| \le \pi - \Delta$. We shall denote the analytic continuation again by $I_\pm(x)$. With this understanding, the identity in (5.31) now holds for $|\arg x| \le \pi - \Delta$. Applying the generalized Watson's lemma to (5.43) then shows that (5.33) is also valid for $-\pi/2 - \gamma + \Delta \le \arg x \le \pi/2 - \gamma - \Delta$. This, upon varying γ from $-\pi/2$ to $\pi/2$, would imply that (5.35) holds in the sector $|\arg z| \le \pi - \Delta$.

Example 5. Consider the integral

$$F(x) = \int_0^x \frac{e^t - 1}{t} \, dt, \qquad x > 0, \quad (5.44)$$

which, as we shall see later, is related to the exponential integral $\mathrm{Ei}(x)$ discussed in Example 1. The problem of finding the asymptotic expansion of the integral was proposed by a statistician. There are several ways of deriving this result, and the most elementary one is probably to use l'Hôpital's rule. For instance, by this rule, we have

$$\lim_{x \to \infty} xe^{-x}F(x) = 1,$$

$$\lim_{x \to \infty} x^2 e^{-x} \left[F(x) - \frac{e^x}{x} \right] = 1,$$

$$\vdots$$

$$\lim_{x \to \infty} x^{N+1} e^{-x} \left[F(x) - e^x \sum_{n=0}^{N-1} \frac{n!}{x^{n+1}} \right] = N!,$$

$$\vdots$$

Thus it follows that

$$F(x) \sim e^x \sum_{n=0}^{\infty} \frac{n!}{x^{n+1}}, \qquad \text{as } x \to +\infty. \tag{5.45}$$

The drawback of this approach is that there are no error bounds for the expansion (5.45).

An alternative method is to apply the generalized Watson's lemma, and this can be done by making the substitution $t = x(1 - v)$, followed by an integration by parts. The result is

$$F(x) = -xe^x \int_0^1 e^{-xv} \log(1 - v) \, dv \sim e^x \sum_{n=0}^{\infty} \frac{n!}{x^{n+1}}. \tag{5.46}$$

To bound the error, we may split the integral in (5.46) at $v = \frac{1}{2}$ and let

$$\delta(x) = -\int_{1/2}^1 e^{-xv} \log(1 - v) \, dv. \tag{5.47}$$

It is easily shown that

$$0 \le \delta(x) \le \tfrac{1}{2}(1 + \log 2)e^{-x/2}. \tag{5.48}$$

By writing

$$-\log(1 - v) = \sum_{n=1}^{N-1} \frac{1}{n} v^n + R_N(v) \tag{5.49}$$

with

$$R_N(v) = \frac{v^N}{N} \frac{1}{(1 - \xi)^N}, \qquad 0 < \xi < v, \tag{5.50}$$

we also have

$$-\int_0^{1/2} e^{-xv} \log(1-v)\,dv = \sum_{n=1}^{N-1} \frac{(n-1)!}{x^{n+1}} - \varepsilon_{N,1}(x) + \varepsilon_{N,2}(x), \quad (5.51)$$

where

$$\varepsilon_{N,1}(x) = \sum_{n=1}^{N-1} \frac{1}{n} \Gamma(n+1, x/2)x^{-n-1} \qquad (5.52)$$

and

$$\varepsilon_{N,2}(x) = \int_0^{1/2} e^{-xv} R_N(v)\,dv, \qquad (5.53)$$

$\Gamma(\alpha, x)$ being the complementary incomplete Gamma function defined by

$$\Gamma(\alpha, x) = \int_x^\infty e^{-t} t^{\alpha-1}\,dt. \qquad (5.54)$$

On making the substitution $t = x(1 + \tau)$, the last integral can be written as

$$\Gamma(\alpha, x) = e^{-x}x^\alpha \int_0^\infty e^{-x\tau}(1+\tau)^{\alpha-1}\,d\tau.$$

Since $1 + \tau \le e^\tau$, it follows that

$$\Gamma(\alpha, x) \le \frac{e^{-x}x^\alpha}{x - \alpha + 1} \qquad (5.55)$$

if $\alpha > 1$ and $x > \alpha - 1$. Therefore

$$0 \le \varepsilon_{N,1}(x) \le \frac{e^{-x/2}}{x - 2N + 2} \sum_{n=1}^{N-1} \frac{1}{2^n n} \le e^{-x/2} \log 2 \qquad (5.56)$$

if $x > 2N - 1$. To estimate $\varepsilon_{N,2}(x)$, we insert (5.50) into (5.53). The result is

$$0 \le \varepsilon_{N,2}(x) \le 2^N \frac{(N-1)!}{x^{N+1}}. \qquad (5.57)$$

A combination of (5.46), (5.47), and (5.51) then yields

$$F(x) = xe^x \left[\sum_{n=1}^{N-1} \frac{(n-1)!}{x^{n+1}} - \varepsilon_{N,1}(x) + \varepsilon_{N,2}(x) + \delta(x) \right], \qquad (5.58)$$

where the error terms on the right-hand side have the estimates given in (5.48), (5.56), and (5.57).

Although bounds have now been obtained for the errors associated with the expansion (5.45), there still remains the question of how good are these bounds. In the following, we shall present yet another method, which will lead to the more precise result

$$F(x) = -\log x + \gamma + e^x \left[\sum_{s=0}^{n-1} \frac{s!}{x^{s+1}} - \rho_n(x) \right],$$ (5.59)

where

$$\rho_n(x) = \text{Re}\left[i \cdot n! \int_0^\infty \frac{e^{i\tau}}{(x + i\tau)^{n+1}} \, d\tau \right].$$ (5.60)

An integration by parts coupled with a simple estimation gives

$$|\rho_n(x)| \le \frac{n!}{x^{n+1}}\left[1 + \frac{\sqrt{\pi}\,\Gamma(\frac{1}{2}(n+1)+1)}{\Gamma(\frac{1}{2}(n+1)+\frac{1}{2})} \right].$$ (5.61)

A comparison of (5.58) and (5.59) shows immediately that the latter result is definitely superior to the former, even without numerical computation.

To derive (5.59), we first recall the exponential integral Ei(z) discussed in Example 1. Since (4.1) does not define Ei(z) for $z = x > 0$, we extend the definition of this integral by defining

$$\text{Ei}(x) = \int_{-\infty}^x \frac{e^t}{t} \, dt \qquad \text{for } x > 0,$$ (5.62)

with the integral being understood in the sense of a Cauchy principal value. The connection between $F(x)$ and Ei(x) is found by writing

$$F(x) = \int_0^1 \frac{e^t - 1}{t} \, dt - \int_{-\infty}^1 \frac{e^t}{t} \, dt - \log x + \text{Ei}(x).$$ (5.63)

Note that the difference of the first two integrals is a constant and can be shown to be equal to the negative of the Euler constant (Olver 1974a, p. 40)

$$\gamma = \int_0^1 \frac{1 - e^{-t}}{t} \, dt - \int_1^\infty \frac{e^{-t}}{t} \, dt.$$ (5.64)

Thus we have

$$F(x) = -\gamma - \log x + \text{Ei}(x).$$ (5.65)

To derive the asymptotic expansion of $\text{Ei}(x)$ for $x > 0$, we apply Cauchy's theorem to obtain the contour integral representation

$$\text{Ei}(x) = -\frac{1}{2}\left[\int_x^{x+i\infty}\frac{e^t}{t}\,dt + \int_x^{x-i\infty}\frac{e^t}{t}\,dt\right]. \tag{5.66}$$

By partial integration, it follows that

$$\text{Ei}(x) = e^x\left[\sum_{s=0}^{n-1}\frac{s!}{x^{s+1}} - \rho_n(x)\right], \tag{5.67}$$

where $\rho_n(x)$ is as given in (5.60). Coupling (5.65) and (5.67) yields the desired result (5.59). The expansion (5.67) can be found in Olver (1974a, p. 227, Ex. 13.5).

It is often believed that reasonably good estimates can be obtained for the error terms associated with the asymptotic expansions by retracing the asymptotic methods employed and tightening up all the inequalities involved. We hope the above example has shown the contrary; that is, to construct realistic error bounds, quite different analyses are often needed. In fact, as we shall see in later chapters, it is from these new methods that advances of the present subject were made.

We shall conclude this section with a discussion of a converse to Watson's lemma. When

$$F(z) = \int_0^\infty f(t)e^{-zt}\,dt, \tag{5.68}$$

the conditions under which the inverse

$$f(t) = \frac{1}{2\pi i}\int_{\sigma-i\infty}^{\sigma+i\infty}F(z)e^{tz}\,dz \tag{5.69}$$

exists are well known; see, for example, Widder (1941, p. 66). If $F(z)$ has a convergent expansion of the form

$$F(z) = \sum_{n=0}^\infty a_n\Gamma(\lambda_n)z^{-\lambda_n} \tag{5.70}$$

which can be integrated term by term to give

$$f(t) = \sum_{n=0}^\infty a_n\frac{\Gamma(\lambda_n)}{2\pi i}\int_{\sigma-i\infty}^{\sigma+i\infty}z^{-\lambda_n}e^{tz}\,dz = \sum_{n=0}^\infty a_n t^{\lambda_n-1}, \tag{5.71}$$

then the behavior as $t \to 0^+$ would be established. However, in practice, the expansion in (5.70) is often not convergent or the termwise integration procedure in (5.71) is not valid. The following result shows that if (5.70) holds as an asymptotic expansion, then so does (5.71).

Converse to Watson's Lemma.   *Let $f(t)$ be a continuous function in $(0, \infty)$, $f(t) = 0$ for $t < 0$, and $e^{-ct}f(t) \in L^1(0, \infty)$. Let $F(z)$ be the Laplace transform of $f(t)$ as given in (5.68). If*

$$F(z) \sim \sum_{n=0}^{\infty} a_n \Gamma(\lambda_n) z^{-\lambda_n} \tag{5.72}$$

as $z \to \infty$ uniformly in $|\arg(z - c)| \le \pi/2$, where $\mathrm{Re}\ \lambda_n \uparrow +\infty$ as $n \to \infty$, then as $t \to 0^+$

$$f(t) \sim \sum_{n=0}^{\infty} a_n t^{\lambda_n - 1}. \tag{5.73}$$

Proof.   Under the given hypotheses, $F(z)$ is analytic in $\mathrm{Re}\ z > c$ and continuous in $\mathrm{Re}\ z \ge c$, and the inversion formula in (5.69) holds for any $\sigma \ge c$. Put

$$F(z) = \sum_{n=0}^{N-1} a_n \Gamma(\lambda_n) z^{-\lambda_n} + F_N(z).$$

It follows from (5.72) that there exists a constant K_N such that $|F_N(z)| \le K_N |z^{-\lambda_N}|$ for all z in $\mathrm{Re}\ z \ge c$. The inversion formula gives

$$f(t) = \sum_{n=0}^{N-1} a_n t^{\lambda_n - 1} + f_N(t) \qquad \text{with} \quad f_N(t) = \frac{1}{2\pi i} \int_{\sigma - i\infty}^{\sigma + i\infty} F_N(z) e^{tz}\, dz.$$

For any fixed t in $(0, 1)$, we may choose $\sigma = c/t$ and put $z = (c/t)(1 + i\tau)$. If $\lambda_N = \alpha_N + i\beta_N$ and $\alpha_N > 1$, then

$$|F_N(z)| \le K_N \left(\frac{t}{c}\right)^{\alpha_N} (1 + \tau^2)^{-\alpha_N/2} e^{\pi|\beta_N|/2}.$$

Thus

$$|f_N(t)| \le \frac{K_N}{2\pi} \left(\frac{t}{c}\right)^{\alpha_N - 1} \int_{-\infty}^{\infty} (1 + \tau^2)^{-\alpha_N/2} e^{\pi|\beta_N|/2 + c}\, d\tau$$

$$= O(t^{\lambda_N - 1}), \qquad \text{as } t \to 0^+. \qquad \blacksquare$$

The above argument is given in E. J. Watson (1981, p. 89).

6. The Euler–Maclaurin Summation Formula

In the study of analytic number theory, one often encounters approximation problems concerning sums of the form $\sum_{k \le n} f(k)$. The best known tool for this type of problem is probably the Euler–Maclaurin summation formula, which estimates the sum $f(0) + f(1) + \cdots + f(n)$ by the integral $\int_0^n f(t)\, dt$, with a correction term expressible as a finite series involving the values of $f(t)$ and its odd derivatives at $t = 0$ and $t = n$, together with an explicit error term.

To develop this summation formula, we need a set of polynomials, $B_0(x), B_1(x), B_2(x), \ldots$, defined by the expansion

$$\frac{te^{xt}}{e^t - 1} = \sum_{s=0}^{\infty} B_s(x) \frac{t^s}{s!}, \qquad |t| < 2\pi, \tag{6.1}$$

for all values of x. These are known as the *Bernoulli polynomials*. By differentiation, it is easily seen that

$$B_s'(x) = sB_{s-1}(x), \qquad s = 1, 2, \ldots. \tag{6.2}$$

The quantities $B_s(0)$ are called the *Bernoulli numbers* and are simply denoted by B_s. Thus

$$\frac{t}{e^t - 1} = \sum_{s=0}^{\infty} B_s \frac{t^s}{s!}, \qquad |t| < 2\pi. \tag{6.3}$$

Since

$$\frac{t}{e^t - 1} = -t + \frac{-t}{e^{-t} - 1},$$

all Bernoulli numbers of odd suffix vanish, except for B_1, i.e., $B_3 = B_5 = \cdots = 0$. Furthermore, since the generating function has simple poles at $t = \pm 2\pi k i$, $k = 1, 2, \ldots$, by enlarging the simple closed contour $\gamma: |t| = r < 2\pi$ in

$$\frac{B_{2s}}{(2s)!} = \frac{1}{2\pi i} \int_{\gamma} \frac{1}{e^t - 1}\, t^{-2s}\, dt,$$

we obtain from the Cauchy residue theorem

$$\frac{B_{2s}}{(2s)!} = \frac{2(-1)^{s-1}}{(2\pi)^{2s}} \sum_{k=1}^{\infty} \frac{1}{k^{2s}}, \tag{6.4}$$

which is of course also an asymptotic expansion of B_{2s} as $s \to \infty$.

From the defining equations (6.1) and (6.3), we have

$$e^{xt} \sum_{s=0}^{\infty} B_s \frac{t^s}{s!} = \sum_{s=0}^{\infty} B_s(x) \frac{t^s}{s!}.$$

Equating coefficients gives

$$B_s(x) = \sum_{j=0}^{s} \binom{s}{j} B_{s-j} x^j. \tag{6.5}$$

Thus the Bernoulli polynomials can be constructed from the Bernoulli numbers.

Since

$$\frac{te^{xt}}{e^t - 1} = \frac{(-t)e^{(1-x)(-t)}}{e^{-t} - 1},$$

it follows from (6.1) that

$$B_s(1 - x) = (-1)^s B_s(x), \tag{6.6}$$

and hence

$$B_s(1) = (-1)^s B_s(0) = (-1)^s B_s. \tag{6.7}$$

Putting $x = 1$ in (6.5) gives

$$B_{s-1} = -\frac{1}{s} \sum_{j=0}^{s-2} \binom{s}{j} B_j, \qquad s = 2, 3, \ldots . \tag{6.8}$$

This is a recurrence formula, and since $B_0 = 1$ we have, in particular,

$$B_1 = -\frac{1}{2}, \qquad B_2 = \frac{1}{6}, \qquad B_4 = -\frac{1}{30}. \tag{6.9}$$

From (6.5) we also have

$$B_0(x) = 1, \qquad B_1(x) = x - \frac{1}{2}, \qquad B_2(x) = x^2 - x + \frac{1}{6},$$

$$B_3(x) = x^3 - \frac{3}{2}x^2 + \frac{1}{2}x, \qquad B_4(x) = x^4 - 2x^3 + x^2 - \frac{1}{30}. \tag{6.10}$$

Lemma 1. (i) *For $s > 0$, $B_{2s+1}(x)$ vanishes when $x = 0, \frac{1}{2}$, and 1, and vanishes at no other point inside $[0, 1]$.* (ii) *$B_{2s+2}(x) - B_{2s+2}$ vanishes at $x = 0$ and 1, and is of constant sign in $(0, 1)$.* (iii) *The extreme value of $B_{2s+2}(x) - B_{2s+2}$ in $(0, 1)$ occurs at $x = \frac{1}{2}$.*

Proof. The first statement in (i) follows immediately from (6.6) and (6.7). Furthermore, (6.6) shows that if $B_{2s+1}(x)$ vanishes at any point inside $(0, 1)$ other than $x = \frac{1}{2}$, then it must vanish at least at two such points. By Rolle's theorem, its derivative $B'_{2s+1}(x) = (2s+1)B_{2s}(x)$ vanishes at least four times in $(0, 1)$. By similar reasoning, $B_{2s-1}(x)$ must vanish at least at two points inside $(0, 1)$ other than $x = \frac{1}{2}$. The argument can now be repeated, and shows that $B_{2s-3}(x), \ldots, B_3(x)$ all have the same property. This is impossible since $B_3(x) = x(x - \frac{1}{2})(x - 1)$ has only one zero inside $(0, 1)$, which occurs at $x = \frac{1}{2}$. Therefore, the only zeros of $B_{2s+1}(x)$ in $[0, 1]$ are at $x = 0, \frac{1}{2}$, and 1.

The first statement in (ii) also follows from (6.7), and the second statement in (ii) is proved by an argument entirely similar to the one given above.

To prove (iii), we only need observe that the extreme value of $B_{2s}(x) - B_{2s}$ can occur only at its critical point $x = \frac{1}{2}$. This completes the proof of the lemma. ∎

Let $[x]$ denote the largest integer less than or equal to x and define

$$s!\, P_s(x) = B_s(x - [x]). \tag{6.11}$$

The functions $P_s(x)$ are periodic functions with period 1, and agree with $(1/s!)B_s(x)$ in the interval $[0, 1)$. The graphs of $P_1(x)$, $P_2(x)$, and $P_3(x)$ are shown in Figures 1.1–1.3.

Lemma 2. *For $m = 1, 2, \ldots$, we have*

$$P_{m+1}(x) = \int P_m(x)\, dx, \tag{6.12}$$

$$P_{2m}(2x) = 2^{2m-1}[P_{2m}(x) + P_{2m}(x + \tfrac{1}{2})], \tag{6.13}$$

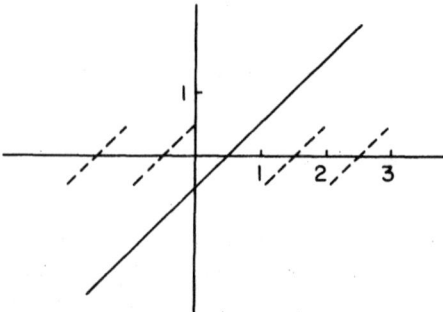

Fig. 1.1 $B_1(x)$ —— and $P_1(x)$ - - - -.

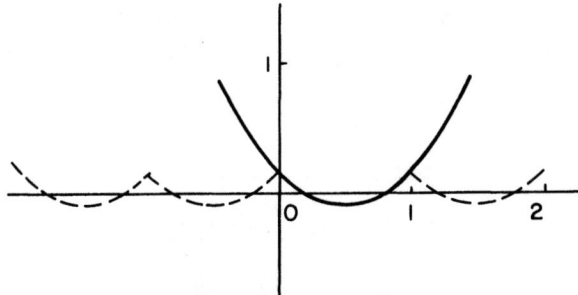

Fig. 1.2 $B_2(x)$ —— and $2!\,P_2(x)$ ---.

and

$$(-1)^{m+1}[P_{2m}(x) - P_{2m}(\tfrac{1}{2})] \geq 0. \qquad (6.14)$$

Proof. Equation (6.12) follows immediately from (6.2). To prove (6.13), we use the identity

$$\frac{te^{(x+1/2)t}}{e^t - 1} = 2\,\frac{(t/2)e^{xt}}{e^{t/2} - 1} - \frac{te^{xt}}{e^t - 1},$$

which in turn gives

$$B_s(x + \tfrac{1}{2}) = 2^{1-s}B_s(2x) - B_s(x). \qquad (6.15)$$

Equation (6.13) is a consequence of (6.11) and (6.15).

Now consider the function $(-1)^{m+1}B_{2m}(x)$. By (6.4) and Lemma 1 (ii), this function attains the same positive value $(-1)^{m+1}B_{2m}$ at $x = 0$ and

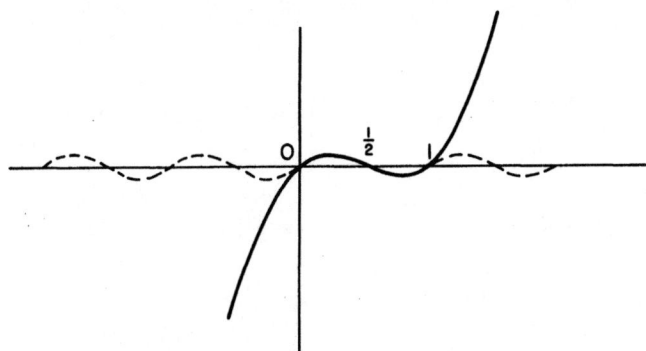

Fig. 1.3 $B_3(x)$ —— and $3!\,P_3(x)$ -----.

$x = 1$. Furthermore, by Lemma 1 (iii), its extreme value in $[0, 1]$ can occur only at its critical point $x = \frac{1}{2}$. From (6.15), we have

$$B_{2m}(\tfrac{1}{2}) = (2^{1-2m} - 1)B_{2m}. \tag{6.16}$$

Thus $(-1)^{m+1}B_{2m}(\tfrac{1}{2})$ is negative and

$$(-1)^{m+1}B_{2m}(x) \geq (-1)^{m+1}B_{2m}(\tfrac{1}{2}). \tag{6.17}$$

The inequality in (6.14) now follows from (6.17) and the fact that $P_{2m}(x)$ is a periodic function with period 1 and agrees with $B_{2m}(x)/(2m)!$ in $[0, 1]$.   ∎

Theorem 6.   *Let $f(t)$ be a real- or complex-valued function defined on $0 \leq t < \infty$. If $f^{(2m)}(t)$ is absolutely integrable on $(0, \infty)$ then, for $n = 1$, $2, \ldots$,*

$$f(0) + \cdots + f(n) = \int_0^n f(x)\,dx + \frac{1}{2}[f(0) + f(n)]$$
$$+ \sum_{s=1}^{m-1} \frac{B_{2s}}{(2s)!}[f^{(2s-1)}(n) - f^{(2s-1)}(0)] + R_m(n), \tag{6.18}$$

where the remainder is given by

$$R_m(n) = \int_0^n \frac{B_{2m} - B_{2m}(x - [x])}{(2m)!} f^{(2m)}(x)\,dx \tag{6.19}$$

and satisfies

$$|R_m(n)| \leq (2 - 2^{1-2m}) \frac{|B_{2m}|}{(2m)!} \int_0^n |f^{(2m)}(x)|\,dx. \tag{6.20}$$

Proof.   Since $\lim_{x \to 0^+} P_1(x) = -\frac{1}{2}$ and $\lim_{x \to 1^-} P_1(x) = +\frac{1}{2}$, integration by parts gives

$$\int_j^{j+1} P_1(x)f'(x)\,dx = \frac{1}{2}[f(j+1) + f(j)] - \int_j^{j+1} f(x)\,dx.$$

Summing up from $j = 0$ to $j = n - 1$, we have

$$\frac{1}{2}f(0) + f(1) + \cdots + \frac{1}{2}f(n) = \int_0^n f(x)\,dx + \int_0^n P_1(x)f'(x)\,dx,$$

or equivalently

$$f(0) + f(1) + \cdots + f(n) = \int_0^n f(x)\,dx + \frac{1}{2}[f(0) + f(n)]$$

$$+ \int_0^n P_1(x)f'(x)\,dx. \tag{6.21}$$

Now observe that for $s = 2, 3, \ldots$, we have from (6.7)

$$\lim_{x \to (j+1)^-} P_s(x) = \lim_{x \to j^+} P_s(x) = \frac{1}{s!}B_s.$$

Thus by partial integrations

$$\int_j^{j+1} P_1(x)f'(x)\,dx = \frac{1}{2!}B_2[f'(j+1) - f'(j)] - \int_j^{j+1} P_2(x)f''(x)\,dx$$

$$= \frac{1}{2!}B_2[f'(j+1) - f'(j)] - \frac{1}{3!}B_3[f''(j+1) - f''(j)]$$

$$+ \int_j^{j+1} P_3(x)f'''(x)\,dx.$$

Since $B_3 = B_5 = \cdots = 0$, continuing the process we arrive at

$$\int_j^{j+1} P_1(x)f'(x)\,dx = \sum_{s=1}^m \frac{B_{2s}}{(2s)!}[f^{(2s-1)}(j+1) - f^{(2s-1)}(j)]$$

$$- \int_j^{j+1} P_{2m}(x)f^{(2m)}(x)\,dx.$$

In view of (6.11), summation gives

$$\int_0^n P_1(x)f'(x)\,dx = \sum_{s=1}^{m-1} \frac{B_{2s}}{(2s)!}[f^{(2s-1)}(n) - f^{(2s-1)}(0)] + R_m(n), \tag{6.22}$$

where $R_m(n)$ is as given in (6.19). The desired result (6.18) now follows from (6.21) and (6.22).

To estimate the remainder, we note that Lemma 1 (iii) implies $|B_{2m}(x) - B_{2m}| \le |B_{2m}(\frac{1}{2}) - B_{2m}|$. Hence by (6.16),

$$|B_{2m}(x) - B_{2m}| \le (2 - 2^{1-2m})|B_{2m}|. \tag{6.23}$$

This completes the proof of the theorem. ∎

Corollary 1.   *Assume the hypotheses of Theorem 6 hold. (i) If $f^{(2m)}(t)$ does not change sign in $(0, n)$, then $R_m(n)$ is bounded in absolute value by $2 - 2^{1-2m}$ times the first neglected term in (6.18), and has the same sign. (ii) If $f^{(2m)}(x)$ and $f^{(2m+2)}(x)$ have the same constant sign in $(0, n)$, then $R_m(n)$ is bounded in absolute value by, and has the same sign as, the first neglected term in (6.18).*

Proof.   From (6.16), we know that the sign of $B_{2m} - B_{2m}(\tfrac{1}{2})$ is the same as the sign of B_{2m}. By Lemma 1 (ii) and (iii), we conclude that the same is true for $B_{2m} - B_{2m}(x)$ in $[0, 1]$. Statement (i) is now an immediate consequence of (6.19) and (6.20).

The result in (ii) follows from the so-called *error test*, which states that if consecutive error terms associated with a series expansion have opposite signs, then each error term is numerically less than, and has the same sign as, the first neglected term of the series. This result is easy to prove, and is given in Steffensen (1950, p. 4) and Olver (1974a, p. 68). Since $f^{(2m)}(x)$ and $f^{(2m+2)}(x)$ have the same sign in $(0, n)$, and since $B_{2m} - B_{2m}(x)$ is of the same sign as B_{2m} in $[0, 1]$, the remainders $R_m(n)$ and $R_{m+1}(n)$ obviously have opposite signs. The corollary is thus proved. ∎

Example 6.   *(The Harmonic Numbers)*

These numbers are defined in (1.2). In this example, we shall derive the asymptotic expansion (1.5). Let $f(x) = (1 + x)^{-1}$ and apply (6.18) with n replaced by $n - 1$. The result is

$$1 + \frac{1}{2} + \cdots + \frac{1}{n} = \log n + \frac{1}{2}\left(1 + \frac{1}{n}\right)$$

$$+ \sum_{s=1}^{m-1} \frac{B_{2s}}{(2s)!}\left[-\frac{(2s-1)!}{n^{2s}} + (2s-1)!\right] + R_m(n-1), \tag{6.24}$$

where

$$R_m(n-1) = \int_0^{n-1} \frac{B_{2m} - B_{2m}(x - [x])}{(1+x)^{2m+1}}\, dx. \tag{6.25}$$

Now fix m and let $n \to +\infty$. Since $|B_{2m} - B_{2m}(x - [x])|$ is bounded by $(2 - 2^{1-2m})|B_{2m}|$, the last integral converges if we replace n by $+\infty$. This establishes the existence of the limit

$$\gamma \equiv \lim_{n\to\infty}\left(\sum_{k=1}^{n} \frac{1}{k} - \log n\right) = \frac{1}{2} + \sum_{s=1}^{m-1} \frac{B_{2s}}{2s} + E_m, \tag{6.26}$$

where

$$E_m = \int_0^\infty \frac{B_{2m} - B_{2m}(x - [x])}{(1+x)^{2m+1}} \, dx. \tag{6.27}$$

The number γ is the well-known *Euler constant*; see Olver (1974a, p. 34). Since E_m and E_{m+1} have opposite signs, by the error test E_m is of the same sign as, and is numerically less than, the first neglected term in (6.26). Thus, the expansion (6.26) can be used to calculate the number γ numerically. Inserting (6.26) in (6.24) gives

$$\sum_{k=1}^n \frac{1}{k} = \log n + \gamma + \frac{1}{2n} - \sum_{s=1}^{m-1} \frac{B_{2s}}{2s} \, n^{-2s} + \rho_m(n) \tag{6.28}$$

with

$$\rho_m(n) = -\int_{n-1}^\infty \frac{B_{2m} - B_{2m}(x - [x])}{(1+x)^{2m+1}} \, dx. \tag{6.29}$$

The error test then implies

$$0 \le (-1)^m \rho_m(n) \le \frac{|B_{2m}|}{2m} \, n^{-2m}, \tag{6.30}$$

i.e., the error in (6.28) is bounded in absolute value by the first neglected term and has the same sign.

Example 7. Observe that for any function defined on $[0, \infty)$, we have

$$\sum_{j=1}^n f(2j-1) = \sum_{j=0}^{2n} f(j) - \sum_{j=0}^n f(2j). \tag{6.31}$$

Thus, Example 6 can be used to give

$$\sum_{j=0}^{n-1} \frac{1}{2j+1} \sim \frac{1}{2} \log n + \log 2 + \frac{\gamma}{2} - \sum_{s=1}^\infty \frac{B_{2s}}{2s} \frac{(1 - 2^{2s-1})}{(2n)^{2s}}, \tag{6.32}$$

as $n \to \infty$. However, the error analysis for the expansion (6.28) no longer applies to the expansion (6.32). (All one can conclude at the moment is that the remainder is bounded in absolute value by twice the first neglected term.) To obtain a similar result for (6.32), we also observe the identity

$$\frac{1}{2} \int_0^{2n} f(x) \, dx = \int_0^{2n} f(x) \, dx - \int_0^n f(2x) \, dx. \tag{6.33}$$

A combination of (6.13), (6.18), (6.31), and (6.33) gives

$$\sum_{j=1}^{n} f(2j - 1) = \frac{1}{2} \int_{0}^{2n} f(x)\,dx + \sum_{s=1}^{m-1} \frac{B_{2s}}{(2s)!} (1 - 2^{2s-1})$$

$$\times [f^{(2s-1)}(2n) - f^{(2s-1)}(0)] + r_m(n), \qquad (6.34)$$

where

$$r_m(n) = -2^{2m} \int_{0}^{n} f^{(2m)}(2x)[P_{2m}(x + \tfrac{1}{2}) - P_{2m}(\tfrac{1}{2})]\,dx. \qquad (6.35)$$

From (6.14) and the error test, it also follows that if $f^{(2m)}(x)$ and $f^{(2m+2)}(x)$ have the same constant sign in $(0, 2n)$, then $r_m(n)$ is bounded in absolute value by, and has the same sign as, the first neglected term in (6.34). We now apply (6.34)–(6.35) to the function $f(x) = 1/(x + 2)$. In the resulting expression we replace n by $n - 1$ and use the result

$$\lim_{n \to \infty} \left(\sum_{j=0}^{n-1} \frac{1}{2j + 1} - \frac{1}{2} \log n \right) = \log 2 + \frac{\gamma}{2}$$

obtained from (6.32). What we then have is the identity

$$\sum_{j=0}^{n-1} \frac{1}{2j + 1} = \frac{1}{2} \log n + \log 2 + \frac{\gamma}{2} - \sum_{s=1}^{m-1} \frac{B_{2s}}{2s} \frac{(1 - 2^{2s-1})}{(2n)^{2s}} + \delta_m(n), \qquad (6.36)$$

where

$$\delta_m(n) = 2^{2m}(2m)! \int_{n-1}^{\infty} \frac{P_{2m}(x + \tfrac{1}{2}) - P_{2m}(\tfrac{1}{2})}{(2x + 2)^{2m+1}}\,dx. \qquad (6.37)$$

In view of (6.14) and the error test, we have

$$0 \le (-1)^{m+1} \delta_m(n) \le \frac{|B_{2m}|}{2m} \frac{(2^{2m-1} - 1)}{(2n)^{2m}}. \qquad (6.38)$$

Example 8. The Lebesgue constants L_n are defined by the integral in (1.3), or equivalently,

$$L_n = \frac{2}{\pi} \int_{0}^{\pi/2} \frac{|\sin(2n + 1)t|}{\sin t}\,dt. \qquad (6.39)$$

To derive their asymptotic expansion (1.6), we first note the Fourier cosine series expansion

$$|\sin x| = \frac{2}{\pi} - \frac{2}{\pi} \sum_{m=1}^{\infty} \frac{2}{4m^2 - 1} \cos 2mx; \qquad (6.40)$$

see Tolstov (1962, p. 26). With $x = 0$, (6.40) gives

$$\sum_{m=1}^{\infty} \frac{2}{4m^2 - 1} = 1. \tag{6.41}$$

Since $\cos 2mx = 1 - 2\sin^2 mx$, we have by coupling (6.40) and (6.41)

$$|\sin x| = \frac{8}{\pi} \sum_{m=1}^{\infty} \frac{\sin^2 mx}{4m^2 - 1}. \tag{6.42}$$

Writing $(2n + 1)t$ for x and integrating term by term, we obtain

$$L_n = \frac{16}{\pi^2} \sum_{m=1}^{\infty} \frac{1}{4m^2 - 1} \int_0^{\pi/2} \frac{\sin^2(2n+1)mt}{\sin t} \, dt. \tag{6.43}$$

The integrals in (6.43) can be evaluated by using the identity

$$\sin x + \sin 3x + \cdots + \sin(2k - 1)x = \frac{\sin^2 kx}{\sin x}, \tag{6.44}$$

which can be obtained by summing the series $e^{ix} + e^{i3x} + \cdots + e^{i(2k-1)x}$ and then taking the imaginary part. From (6.44), it follows that

$$\int_0^{\pi/2} \frac{\sin^2 kx}{\sin x} \, dx = 1 + \frac{1}{3} + \frac{1}{5} + \cdots + \frac{1}{2k - 1},$$

$$L_n = \frac{16}{\pi^2} \sum_{m=1}^{\infty} \frac{1}{4m^2 - 1} \left[1 + \frac{1}{3} + \cdots + \frac{1}{2m(2n+1) - 1} \right]. \tag{6.45}$$

The above ingenious argument is due to Szegö (1921). Watson's derivation of (1.6) is based on the formula (6.45), and proceeds as follows. First replace m by p in (6.36), and then insert the resulting expression in (6.45). Thus

$$L_n = \frac{4}{\pi^2} \left\{ \log(2n + 1) + A_0 - \sum_{s=1}^{p-1} \frac{A_s}{(2n+1)^{2s}} + \varepsilon_p(n) \right\}, \tag{6.46}$$

where A_0 is given in (1.8),

$$A_s = 2\frac{B_{2s}}{s}(1 - 2^{2s-1}) \sum_{m=1}^{\infty} \frac{1}{(4m^2 - 1)(2m)^{2s}}, \tag{6.47}$$

and

$$\varepsilon_p(n) = 4 \sum_{m=1}^{\infty} \frac{1}{4m^2 - 1} \delta_p(m(2n+1)). \tag{6.48}$$

In view of the identity

$$\sum_{m=1}^{\infty} \frac{1}{(4m^2-1)(2m)^{2s}} = \sum_{m=1}^{\infty} \left\{ \frac{1}{4m^2-1} - \frac{1}{(2m)^2} - \frac{1}{(2m)^4} - \cdots - \frac{1}{(2m)^{2s}} \right\}$$

$$= \frac{1}{2} \left[1 - \sum_{r=1}^{s} \frac{(-1)^{r-1}}{(2r)!} B_{2r} \pi^{2r} \right], \qquad (6.49)$$

equation (6.47) agrees with equation (1.9). The second equality in (6.49) follows from (6.4). To estimate the remainder $\varepsilon_p(n)$, we apply (6.38) to (6.48). The result is

$$0 \le (-1)^{p+1} \varepsilon_p(n)$$

$$\le \frac{|B_{2p}|}{p} (2^{2p-1} - 1) \left[1 - \sum_{r=1}^{p} \frac{(-1)^{r-1}}{(2r)!} B_{2r} \pi^{2r} \right] \frac{1}{(2n+1)^{2p}},$$

or equivalently

$$0 \le (-1)^{p+1} \varepsilon_p(n) \le \frac{|A_p|}{(2n+1)^{2p}}. \qquad (6.50)$$

Here we have again made use of (6.49). This completes the proof of the result stated in (1.6).

Exercises

1. The complementary error function is defined by

$$\text{erfc } z = \frac{2}{\sqrt{\pi}} \int_z^{\infty} e^{-t^2} \, dt.$$

Show that

$$\text{erfc } z = \frac{e^{-z^2}}{\sqrt{\pi}\, z} \left[1 + \sum_{k=1}^{n} (-1)^k \frac{1 \cdot 3 \cdots (2k-1)}{(2z^2)^k} + r_n(z) \right],$$

where

$$r_n(z) = (-1)^{n+1} \frac{1 \cdot 3 \cdots (2n+1)}{2^n} z e^{z^2} \int_z^{\infty} \frac{e^{-t^2}}{t^{2n+2}} \, dt.$$

Also show that for $|\arg z| \le \pi/2 - \delta$,

$$|r_n(z)| \le \frac{1 \cdot 3 \cdots (2n+1)}{(2|z|^2)^{n+1} \sin \delta}.$$

2. The logarithmic integral is defined by

$$\mathrm{li}(z) = \int_0^z \frac{dt}{\log t}, \qquad |\arg z| < \pi, \quad |\arg(1 - z)| < \pi,$$

where the integral is taken along any path belonging to the plane
with two cuts along the segments $(-\infty, 0]$ and $[1, \infty)$ of the real
axis. Show that

$$\mathrm{li}(z) = \frac{z}{\log z} \left[\sum_{k=0}^n \frac{k!}{(\log z)^k} + r_n(z) \right], \qquad \text{where } |r_n(z)| \le \frac{(n+1)!}{|\log z|^{n+1}}$$

for $|z| < 1$ and $|\arg z| \le \pi - \delta$.

3. The complementary incomplete gamma function $\Gamma(\alpha, x)$ is defined
by

$$\Gamma(\alpha, x) = \int_x^\infty e^{-t} t^{\alpha - 1} \, dt.$$

Show that when α and x are positive,

$$\Gamma(\alpha, x) \sim e^{-x} x^{\alpha - 1} \sum_{s=0}^\infty \frac{(\alpha - 1)(\alpha - 2) \cdots (\alpha - s)}{x^s},$$

as $x \to +\infty$. Furthermore, show that the nth error term is bounded
in absolute value by the $(n + 1)$th term of the series and has the
same sign, when $n \ge \alpha - 1$. In particular, deduce that $\Gamma(\alpha, x) \le$
$e^{-x} x^{\alpha - 1}$, $\alpha \le 1$, $x > 0$.

4. Show that for $\mathrm{Re}\, z > 0$, $\Gamma(\alpha, z)$ can be written as

$$\Gamma(\alpha, z) = e^{-z} z^\alpha \int_0^\infty e^{-zt} (1 + t)^{\alpha - 1} \, dt.$$

Prove that for fixed α

$$\Gamma(\alpha, z) = e^{-z} z^{\alpha - 1} \left\{ \sum_{s=0}^{n-1} \frac{(\alpha - 1)(\alpha - 2) \cdots (\alpha - s)}{z^s} + \varepsilon_n(z) \right\},$$

$n = 0, 1, 2, \ldots$, where $\varepsilon_n(z) = O(z^{-n})$ as $z \to \infty$ in the sector
$|\arg z| \le \pi - \delta < \pi$. Also prove that for $n \ge \alpha - 1$ and $|\arg z| \le \pi/2$,

$$|\varepsilon_n(z)| \le \frac{|(\alpha - 1)(\alpha - 2) \cdots (\alpha - n)|}{|z|^n}.$$

5. Let

$$L(x) = \int_0^\infty t^\alpha f(t) e^{-xt}\, dt,$$

where Re $\alpha > -1$. Assume that $f(t)$ is analytic in a neighborhood of the origin, and that the integral exists for all $x > x_0 \geq 0$. Define

$$f_0(t) = f(t), \qquad f_{n+1}(t) = t^{-\alpha}\frac{d}{dt}\{[f_n(t) - f_n(0)]t^\alpha\},$$

and prove that each $f_n(t)$ is analytic in the neighborhood of the origin. Verify that

$$\int_0^\infty t^\alpha f_n(t) e^{-xt}\, dt = f_n(0)\frac{\Gamma(\alpha+1)}{x^{\alpha+1}} + \frac{1}{x}\int_0^\infty t^\alpha f_{n+1}(t) e^{-xt}\, dt,$$

and deduce that

$$L(x) = \frac{\Gamma(\alpha+1)}{x^{\alpha+1}}\left[f_0(0) + \frac{f_1(0)}{x} + \cdots + \frac{f_{n-1}(0)}{x^{n-1}}\right]$$

$$+ x^{-n}\int_0^\infty t^\alpha f_n(t) e^{-xt}\, dt.$$

Show that this expansion has the same coefficients as those obtained by using Watson's lemma.

6. Consider

$$M(x) = \int_0^\infty f(t) e^{-xt^2}\, dt,$$

where $f(t)$ is analytic in a neighborhood of $t = 0$. Define

$$f_0(t) = f(t), \qquad f_{n+1}(t) = \frac{d}{dt}\left\{\frac{f_n(t) - f_n(0)}{2t}\right\},$$

and show that

$$\int_0^\infty f_n(t) e^{-xt^2}\, dt = \frac{1}{2}\left(\frac{\pi}{x}\right)^{1/2} f_n(0) + \frac{f_n'(0)}{2x} + x^{-1}\int_0^\infty f_{n+1}(t) e^{-xt^2}\, dt.$$

Derive the expansion

$$M(x) = \frac{1}{2}\left(\frac{\pi}{x}\right)^{1/2}\left[f_0(0) + \frac{f_1(0)}{x} + \cdots + \frac{f_n(0)}{x^n}\right]$$

$$+ \frac{1}{2x}\left[f_0'(0) + \frac{f_1'(0)}{x} + \cdots + \frac{f_n'(0)}{x^n}\right]$$

$$+ x^{-n-1}\int_0^\infty f_{n+1}(t) e^{-xt^2}\, dt,$$

and verify that the coefficients are the same as those obtained by termwise integration.

7. Let

$$I(x) = \int_0^\infty t^{\alpha-1}f(t)e^{-xt}\,dt, \qquad \text{Re } \alpha > 0,$$

and define

$$f_0(t) = f(t), \qquad f_{k+1}(t) = \frac{d}{dt}\left\{\frac{f_k(t) - f_k((\alpha+k)/x)}{t - (\alpha+k)/x}\right\}.$$

Derive the expansion

$$I(x) = \frac{\Gamma(\alpha)}{x^\alpha}\sum_{k=0}^{n-1}\frac{(\alpha)_k}{x^{2k}}f_k\left(\frac{\alpha+k}{x}\right) + \delta_n(x),$$

where

$$\delta_n(x) = x^{-n}\int_0^\infty t^{\alpha+n-1}f_n(t)e^{-xt}\,dt.$$

Assume that $f(t)$ is analytic for Re $t > 0$ and that $|f^{(i)}(t)| \le Me^{\mu|t|}$, $i = 0, 1, \ldots, 2n$, M and μ being nonnegative constants. Show by induction that $|f_k^{(i)}(t)| \le M_k e^{\mu|t|}$ for $i = 0, 1, \ldots, 2n-2k$ and $k = 0, 1, \ldots, n$. From this deduce that $\delta_n(x) = O(x^{-2n-\text{Re}\,\alpha})$ as $x \to +\infty$.

8. Let

$$I(x) = \int_0^1 t^\alpha J_0(xt)\,dt, \qquad -1 < \alpha < \tfrac{1}{2},$$

where $J_m(t)$ is the Bessel function of the first kind of order m. Use the identity $(t^m J_m(t))' = t^m J_{m-1}(t)$ to show that

$$I(x) = \sum_{k=1}^n 2^{k-1}\frac{\Gamma(k-\tfrac{1}{2}-\tfrac{1}{2}\alpha)}{\Gamma(\tfrac{1}{2}-\tfrac{1}{2}\alpha)}x^{-k}J_k(x)$$

$$+ 2^n\frac{\Gamma(n+\tfrac{1}{2}-\tfrac{1}{2}\alpha)}{\Gamma(\tfrac{1}{2}-\tfrac{1}{2}\alpha)}x^{-n}\int_0^1 t^{\alpha-n}J_n(xt)\,dt.$$

This suggests that the integral $I(x)$ may have the generalized asymptotic expansion

$$I(x) \sim \sum_{k=1}^\infty 2^{k-1}\frac{\Gamma(k-\tfrac{1}{2}-\tfrac{1}{2}\alpha)}{\Gamma(\tfrac{1}{2}-\tfrac{1}{2}\alpha)}x^{-k}J_k(x); \qquad \{x^{-k}\}, \qquad \text{as } x \to \infty.$$

Now, use the identity (Erdélyi *et al.* 1953b, p. 49)

$$\int_0^\infty t^{\rho-1} J_\mu(xt)\, dt = 2^{\rho-1} x^{-\rho} \frac{\Gamma(\frac{1}{2}\mu + \frac{1}{2}\rho)}{\Gamma(1 + \frac{1}{2}\mu - \frac{1}{2}\rho)},$$

$-\mathrm{Re}\,\mu < \mathrm{Re}\,\rho < \frac{3}{2}$, to show that as $x \to \infty$,

$$x^{-n} \int_0^1 t^{\alpha-n} J_n(xt)\, dt \sim x^{-\alpha-1} 2^{\alpha-n} \frac{\Gamma(\frac{1}{2} + \frac{1}{2}\alpha)}{\Gamma(n + \frac{1}{2} - \frac{1}{2}\alpha)},$$

thus proving the above generalized asymptotic expansion invalid. With the aid of the same identity, derive the expansion

$$I(x) = 2^\alpha \frac{\Gamma(\frac{1}{2} + \frac{1}{2}\alpha)}{\Gamma(\frac{1}{2} - \frac{1}{2}\alpha)} x^{-\alpha-1}$$

$$+ \sum_{k=1}^n 2^{k-1} \frac{\Gamma(k - \frac{1}{2} - \frac{1}{2}\alpha)}{\Gamma(\frac{1}{2} - \frac{1}{2}\alpha)} x^{-k} J_k(x) + R_n,$$

where

$$R_n = -2^n \frac{\Gamma(n + \frac{1}{2} - \frac{1}{2}\alpha)}{\Gamma(\frac{1}{2} - \frac{1}{2}\alpha)} x^{-n} \int_1^\infty t^{\alpha-n} J_n(xt)\, dt.$$

Show that as $x \to \infty$, $R_n = O(x^{-n-1/2})$.

9. The confluent hypergeometric function of the second kind $\Psi(\alpha, \gamma, z)$ has the integral representation

$$\Psi(\alpha, \gamma, z) = \frac{1}{\Gamma(\alpha)} \int_0^\infty e^{-zt} t^{\alpha-1} (1 + t)^{\gamma-\alpha-1}\, dt,$$

when $\mathrm{Re}\,\alpha > 0$ and $\mathrm{Re}\,z > 0$. Derive the asymptotic expansion

$$\Psi(\alpha, \gamma, z) = z^{-\alpha} \sum_{k=0}^n \left[\frac{(-1)^k (\alpha)_k (1 + \alpha - \gamma)_k}{k!} z^{-k} + O(|z|^{-n-1}) \right],$$

as $|z| \to \infty$ in $|\arg z| \le \pi - \delta < \pi$. Use the identity

$$\Psi(\alpha, \gamma, z) = z\Psi(\alpha + 1, \gamma + 1, z) + (1 + \alpha - \gamma)\Psi(\alpha + 1, \gamma, z)$$

to remove the condition $\mathrm{Re}\,\alpha > 0$.

10. The Airy function $\mathrm{Ai}(z)$ has the integral representation

$$\mathrm{Ai}(z) = \frac{1}{\pi} \exp\left(-\frac{2}{3} z^{3/2}\right) \int_0^\infty e^{-z^{1/2} t^2} \cos\left(\frac{1}{3} t^3\right) dt$$

for $|\arg z| < \pi$. Prove that

$$\text{Ai}(z) \sim \frac{1}{2\pi z^{1/4}} \exp\left(-\frac{2}{3} z^{3/2}\right) \sum_{n=0}^{\infty} \frac{\Gamma(3n + \frac{1}{2})}{3^{2n}(2n!)} \frac{(-1)^n}{z^{3n/2}},$$

as $|z| \to \infty$ in $|\arg z| < \pi$.

11. The quotient of two Gamma functions $\Gamma(z + a)/\Gamma(z + b)$ is expressible in the form

$$\frac{\Gamma(z + a)}{\Gamma(z + b)} = \frac{1}{\Gamma(b - a)} \int_0^{\infty} f(t)e^{-zt}\, dt,$$

where $f(t) = e^{-at}(1 - e^{-t})^{b-a-1}$ as long as $\text{Re}(b - a) > 0$ and $\text{Re}(z + a) > 0$. Derive the asymptotic expansion

$$\frac{\Gamma(z + a)}{\Gamma(z + b)} \sim \sum_{n=0}^{\infty} \frac{(-1)^n}{n!} B_n^{(a-b+1)}(a) \frac{\Gamma(b - a + n)}{\Gamma(b - a)} \frac{1}{z^{b-a+n}},$$

as $|z| \to \infty$ in $|\arg z| < \pi - \delta < \pi$. $B_n^{(l)}(\alpha)$ is the generalized Bernoulli polynomial defined by

$$\frac{t^l e^{\alpha t}}{(e^t - 1)^l} = \sum_{n=0}^{\infty} B_n^{(l)}(\alpha) \frac{t^n}{n!}, \qquad |t| < 2\pi.$$

12. The function

$$\mu(z, \beta, \alpha) = \frac{1}{\Gamma(\beta + 1)} \int_0^{\infty} \frac{z^{\alpha + t} t^{\beta}}{\Gamma(\alpha + t + 1)}\, dt, \qquad \text{Re } \beta > -1,$$

occurs in connection with operational calculus and integral equations. Show that

$$\mu(z, \beta, \alpha) \sim z^{\alpha} \sum_{n=0}^{\infty} \frac{\text{Rg}^{(n)}(\alpha + 1)}{n!} \frac{(\beta + 1)_n}{(-\log z)^{\beta + n + 1}},$$

as $z \to 0$ in an unrestricted manner, where $(\alpha)_n$ is the Pochhammer notation, $(\alpha)_0 = 1$ and $(\alpha)_n = \alpha(\alpha + 1)\cdots(\alpha + n - 1)$, $n = 1, 2, \ldots$, and $\text{Rg}(x) \equiv 1/\Gamma(x)$.

13. The Legendre function of the second kind, $Q_\nu^{-\mu}(\cosh z)$, has the integral representation (Erdélyi et al. 1953a, p. 155)

$$Q_\nu^{-\mu}(\cosh z) = \frac{\sqrt{\frac{1}{2}\pi}\, \exp(-\pi i \mu)}{\Gamma(\frac{1}{2} + \mu)(\sinh z)^{\mu}}$$

$$\times \int_z^{\infty} e^{-(\nu + 1/2)t}(\cosh t - \cosh z)^{\mu - 1/2}\, dt.$$

where $z > 0$, $\text{Re}(v - \mu + 1) > 0$, and $\text{Re}\,\mu > -\frac{1}{2}$. Use the substitution $t = u + z$ to show that the above equation can be written as

$$Q_v^{-\mu}(\cosh z) = A \int_0^\infty u^{\mu - 1/2} e^{-(v + 1/2)u} f(u)\, du,$$

where

$$A = \frac{\sqrt{\tfrac{1}{2}\pi}\,\exp[-\pi i \mu - (v + \tfrac{1}{2})z]}{\Gamma(\tfrac{1}{2} + \mu)(\sinh z)^{1/2}}$$

and

$$f(u) = \left[\frac{\cosh(u + z) - \cosh z}{u \sinh z}\right]^{\mu - 1/2}.$$

Now derive the asymptotic expansion

$$Q_v^{-\mu}(\cosh z) \sim \sqrt{\frac{\tfrac{1}{2}z}{\sinh z}}\; \frac{e^{-\pi i \mu - (v + \tfrac{1}{2})z}}{\Gamma(\tfrac{1}{2} + \mu)} \sum_{s=0}^\infty \frac{c_s \Gamma(s + \mu + \tfrac{1}{2})}{(v + \tfrac{1}{2})^{s + \mu + 1/2}},$$

as $v \to \infty$ in $|\arg v| \le \pi/2 - \delta < \pi/2$, where c_s is the Maclaurin coefficient of $f(u)$. Calculate c_0 and c_1.

14. (Barnes' lemma.) Consider the loop integral

$$I(z) = \frac{i}{2\pi} \int_C (-t)^{\beta - 1} f(t) e^{-zt}\, dt,$$

where C is the Gamma-function contour shown in Figure 1.4, which encloses the origin and embraces a ray P from the origin along which $\text{Re}(zt)$ is positive. Assume that (i) the function $f(t)$ is continuous on, and analytic within, the contour, (ii) the integral is convergent, and (iii) $f(t)$ admits the convergent expansion $f(t) = \sum_{n=0}^\infty c_n(-t)^n$ for $|t| < l$. Show that

$$I(z) \sim \sum_{n=0}^\infty \frac{c_n}{\Gamma(1 - \beta - n)z^{\beta + n}},$$

uniformly in $\arg z$, as $z \to \infty$ in the sector $|\arg(ze^{iy})| \le \pi/2 - \Delta$, $\Delta > 0$. (Hint: Use the result (Erdélyi *et al.* 1953a, p. 14)

$$\frac{i}{2\pi} \int_C (-t)^{\alpha - 1} e^{-zt}\, dt = \frac{1}{\Gamma(1 - \alpha)z^\alpha},$$

and pattern the proof after that of Watson's lemma).

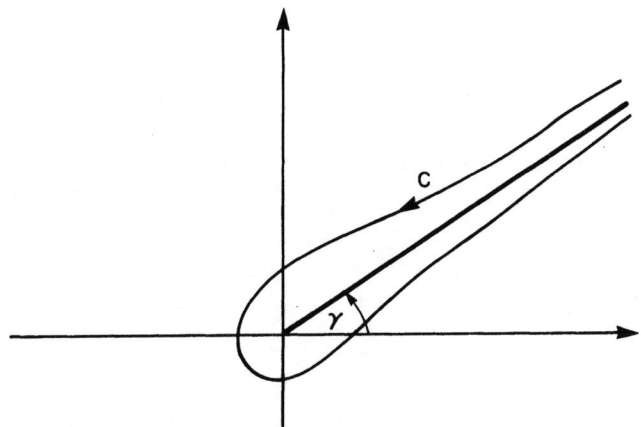

Fig. 1.4 Loop contour C, $-\pi + \gamma \le \arg(-t) \le \pi + \gamma$.

15. Consider the loop integral

$$G(z) = \int_{\infty}^{(0+)} f(t)e^{-zt}\, dt,$$

where the contour is shown in Figure 1.5 and

$$f(t) = e^{-at}(1 - e^{-t})^{b-a-1}.$$

Show that if $\mathrm{Re}(b - a) > 0$, then the size of the circle can be shrunk to zero, and

$$G(z) = [e^{2\pi i(b-a)} - 1] \int_{0}^{\infty} f(t)e^{-zt}\, dt.$$

From this and Ex. 11, conclude that

$$G(z) = [e^{2\pi i(b-a)} - 1]\Gamma(b-a)\frac{\Gamma(z+a)}{\Gamma(z+b)},$$

and that the condition "$\mathrm{Re}(b-a) > 0$" there can be removed.

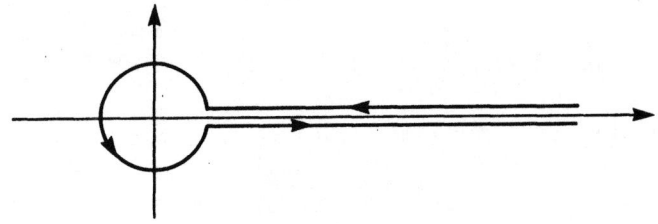

Fig. 1.5

16. Consider the loop integral

$$\bar{\mu}(z, \beta, \alpha) = \Gamma(-\beta) \frac{i}{2\pi} \int_\infty^{(0+)} \frac{(-t)^\beta z^{\alpha + t}}{\Gamma(\alpha + t + 1)} \, dt,$$

where the contour is as shown in Figure 1.5. Show that if $-1 <$ Re $\beta < 0$, then $\bar{\mu}(z, \beta, \alpha) = \mu(z, \beta, \alpha)$, where $\mu(z, \beta, \alpha)$ is the function defined in Ex. 12. Furthermore, show that with $\bar{\mu}(z, \beta, \alpha)$ being the analytic continuation of $\mu(z, \beta, \alpha)$ for Re $\beta \le -1$, the condition "Re $\beta > -1$" in that exercise can be removed.

17. Let

$$I(x) = \int_0^\infty f(t)K_0(xt) \, dt,$$

where $K_0(t)$ denotes the modified Bessel function of order 0. Assume that

$$f(t) = \sum_{s=0}^{n-1} a_s t^{s + \alpha - 1} + f_n(t),$$

where $\alpha > 0$ and $|f_n(t)| \le M_n t^{n + \alpha - 1} e^{\sigma t}$, $0 < t < \infty$. Use the results

$$\int_0^\infty t^{\mu - 1} K_0(t) \, dt = 2^{\mu - 2} \Gamma^2\left(\frac{\mu}{2}\right), \qquad \mu > 0,$$

and

$$0 \le K_0(t) \le \sqrt{\frac{\pi}{2t}} \, e^{-t}, \qquad t > 0,$$

to derive the asymptotic expansion

$$I(x) = \frac{1}{4} \sum_{s=0}^{n-1} \Gamma^2\left(\frac{s + \alpha}{2}\right) \frac{a_s}{(\frac{1}{2}x)^{s + \alpha}} + \delta_n(x).$$

Show that the remainder term satisfies

$$|\delta_n(x)| \le \sqrt{\frac{\pi}{2x}} \, M_n \Gamma(n + \alpha - \tfrac{1}{2})(x - \sigma)^{-n - \alpha + 1/2}.$$

18. Obtain a similar result for the integral

$$I(x) = \int_0^\infty f(t) \, \mathrm{Ai}(xt) \, dt,$$

where Ai(x) is the Airy function which has the following proper-ties:

$$\int_0^\infty t^{\mu-1}\, \mathrm{Ai}(t)\, dt = \frac{3^{2\mu/3-7/6}}{2\pi}\, \Gamma\!\left(\frac{\mu}{3}\right)\Gamma\!\left(\frac{\mu+1}{3}\right), \qquad \mu > 0,$$

and

$$0 \le \mathrm{Ai}(t) \le \frac{1}{2\sqrt{\pi}}\, t^{-1/4}\exp\!\left(-\frac{2}{3}t^{3/2}\right), \qquad 0 < t < \infty.$$

19. Verify the identity

$$\sum_{l=0}^{n}\log(z+l) = \left(z+n+\frac{1}{2}\right)\log(z+n) - \left(z-\frac{1}{2}\right)\log z - n$$

$$+ \sum_{s=1}^{m-1}\frac{B_{2s}}{(2s)(2s-1)}\left[\frac{1}{(z+n)^{2s-1}} - \frac{1}{z^{2s-1}}\right]$$

$$-\frac{1}{2m}\int_0^n \frac{B_{2m}-B_{2m}(x-[x])}{(z+x)^{2m}}\, dx,$$

and deduce from this the special case

$$\log n! = \left(n+\frac{1}{2}\right)\log n - (n-1) - \frac{1}{2}\int_0^{n-1}\frac{B_2-B_2(x-[x])}{(1+x)^2}\, dx.$$

Now use the Stirling formula (Rudin 1976, p. 194) $n! \sim n^n e^{-n}\sqrt{2\pi n}$ to show

$$1 - \frac{1}{2}\int_0^\infty \frac{B_2-B_2(x-[x])}{(1+x)^2}\, dx = \frac{1}{2}\log 2\pi.$$

Finally, from Euler's limit formula (Olver 1974a, p. 33)

$$\Gamma(z) = \lim_{n\to\infty}\frac{n!\, n^z}{z(z+1)(z+2)\cdots(z+n)},$$

derive the Stirling series

$$\log \Gamma(z) = \left(z-\frac{1}{2}\right)\log z - z + \frac{1}{2}\log 2\pi$$

$$+ \sum_{s=1}^{m-1}\frac{B_{2s}}{(2s)(2s+1)z^{2s+1}} + R_m(z),$$

where

$$R_m(z) = \frac{1}{2m} \int_0^\infty \frac{B_{2m} - B_{2m}(x - [x])}{(z + x)^{2m}}\, dx.$$

Verify also that for z real and positive,

$$R_m(z) = \frac{B_{2m}}{2m(2m - 1)} \frac{\theta_m}{z^{2m - 1}}, \qquad \text{where } \theta_m \text{ is a number in the inter-}$$

val $(0, 1)$.

20. For any $m = 1, \dots, [\tfrac{1}{2}k] + 1$, show that

$$\sum_{j=0}^{n-1} (2j + 1)^k = \frac{(2n)^{k+1}}{k+1} \sum_{s=0}^{m-1} \binom{k+1}{2s}(1 - 2^{2s-1}) \frac{B_{2s}}{(2n)^{2s}} + \varepsilon_m^{(k)}(n),$$

where $\varepsilon_m^{(k)}(n) = 0$ if $m = [\tfrac{1}{2}k] + 1$, and

$$0 \le (-1)^m \varepsilon_m^{(k)}(n)$$

$$\le \frac{(2n)^{k+1}}{k+1} \binom{k+1}{2m}(2^{2m-1} - 1) \frac{|B_{2m}|}{(2n)^{2m}} \qquad \text{if } m < [\tfrac{1}{2}k] + 1.$$

21. Let α be any real number, not -1 or a positive integer. Show that as $n \to \infty$,

$$\sum_{j=0}^{n-1} (2j + 1)^\alpha - (1 - 2^\alpha)\, \zeta(-\alpha) \sim \frac{(2n)^{\alpha+1}}{\alpha+1} \sum_{s=0}^{\infty} \binom{\alpha+1}{s}(1 - 2^{2s-1}) \frac{B_{2s}}{(2n)^{2s}},$$

where $\zeta(z)$ is the Riemann Zeta function. Also show that if the expansion is truncated at the term $s = m - 1$, where $m > \tfrac{1}{2}(\alpha + 1)$, then the remainder is bounded in absolute value by the next term and has the same sign.

22. Let

$$\Lambda_n = \frac{1}{n} \sum_{j=0}^{n-1} \cot \frac{(2j + 1)\pi}{4n}.$$

From the expansion

$$\cot z = \frac{1}{z} - \sum_{r=1}^{\infty} 2^{2r}|B_{2r}| \frac{z^{2r-1}}{(2r)!}, \qquad |z| < \pi,$$

derive the asymptotic series

$$\Lambda_n = \frac{2}{\pi} \log n + A_0 + \frac{8}{\pi} \sum_{s=1}^{m-1} \frac{(-1)^{s+1} A_s}{(2n)^{2s}} + \rho_m(n),$$

where

$$A_0 = \frac{4}{\pi} \log 2 + \frac{2}{\pi} \gamma - \frac{1}{\pi} \sum_{r=1}^{\infty} \frac{\pi^{2r}|B_{2r}|}{r(2r)!},$$

$$(-1)^{s+1} A_s = \frac{B_{2s}}{2s} \frac{(2^{2s-1}-1)}{2}$$

$$+ \frac{B_{2s}}{(2s)!} \frac{2^{2s-1}-1}{2} \sum_{r=s+1}^{\infty} \frac{|B_{2r}|}{(2r-2s)!} \frac{\pi^{2r}}{2r},$$

and

$$\rho_m(n) = \frac{4}{\pi} \left[\delta_m(n) - \sum_{r=m+1}^{\infty} \frac{|B_{2r}|}{(2r)!} \left(\frac{\pi}{2n} \right)^{2r} \varepsilon_m^{(2r-1)}(n) \right],$$

$\delta_m(n)$ and $\varepsilon_m^{(2r-1)}(n)$ being the same as given in (6.38) and Ex. 20, respectively. Now, use the identities

$$\log \sin z = \log z - \sum_{r=1}^{\infty} \frac{2^{2r-1}|B_{2r}|}{r(2r)!} z^{2r}, \qquad 0 < |z| < \pi,$$

$$\tan z = \sum_{r=1}^{\infty} \frac{2^{2r}(2^{2r}-1)}{(2r)!} |B_{2r}| z^{2r-1}, \qquad |z| < \frac{\pi}{2},$$

to show that

$$A_0 = \frac{2}{\pi} \left(\gamma + \log \frac{8}{\pi} \right) \qquad \text{and} \qquad A_s = (2^{2s-1}-1)^2 \frac{\pi^{2s}}{2s} \frac{B_{2s}^2}{(2s)!}.$$

Finally, verify the inequalities

$$0 \le (-1)^{m+1} \rho_m(n) \le \frac{8}{\pi} \frac{(2^{2m-1}-1)^2}{(2m)!} \frac{\pi^{2m}}{2m} \frac{B_{2m}^2}{(2n)^{2m}}.$$

23. Let

$$\lambda_n = \frac{2}{n} \sum_{l=1}^{n-1} \csc \frac{l\pi}{n}.$$

Show that

$$\frac{\pi}{4} \lambda_n - \log \frac{2n}{\pi} \sim \gamma + \sum_{s=1}^{\infty} (-\pi^2)^s \frac{(2^{2s-1}-1)}{s(2s)!} \frac{B_{2s}^2}{n^{2s}}, \qquad \text{as } n \to \infty.$$

Supplementary Notes

§1.   In addition to Copson (1965) and Olver (1974a) mentioned in this
section, specialized books dealing with asymptotics include Erdélyi (1956), de Bruijn (1970), Hsu (1958), Evgrafov (1961), Jeffreys
(1962), Berg (1968), Sirovich (1971), Dingle (1973), Lauwerier
(1974), Riekstiņš (1974, 1977, 1981), Bleistein and Handelsman
(1975a), Fedoryuk (1977), and Murray (1984).

§2.   The material in this section is classical, and the presentation here
was certainly influenced by that in Copson (1965, §4).

§3.   Another criticism of generalized asymptotic expansions is that
there is no longer a calculus to manipulate them. For basic
properties of these expansions, we refer to Erdélyi and Wyman
(1963). For applications of these expansions, see the last reference
and also Wimp (1980).

§4.   Example 2 is also given in Erdélyi (1955).

§5.   When $v = 0$, the expansion (5.40) and its associated error bound
(5.42) are given in Szegö (1967, p. 193).

§6.   For an alternative solution to the problem in Example 8, see
Hardy (1942).

Exercises.   The integration-by-parts techniques in Exs. 5 and 6 were
given in Rosser (1955). For Ex. 7, see Franklin and Friedman (1957).
Exercise 8 is based on the result in Soni (1978). The asymptotic
expansion of the ratio of two gamma functions in Exs. 11 and 15 was
obtained by Tricomi and Erdélyi (1951). A related expansion had been
given by Fields (1966). Recently, Frenzen (1987) has established computable error bounds for Fields' expansion when the parameters are all
real. For the behavior of $\mu(z, \alpha, \beta)$ in Ex. 12, as $z \to \infty$, see Wyman and
Wong (1969). The result in Ex. 14 was given by Barnes (1906), but it is
better known as Watson's lemma for loop integrals. The number Λ_n in
Ex. 22 is known as the Lebesgue constant associated with the polynomial interpolation at the zeros of the Chebyshev polynomials. The
result in this exercise was given by Shivakumar and Wong (1982).
Exercise 23 is taken from Henrici (1974).

II

Classical Procedures

1. Laplace's Method

In the study of probabilities in large numbers of trials one often encounters integrals of the form

$$I_n = \int_a^b \varphi(x)[f(x)]^n \, dx, \qquad (1.1)$$

where $\varphi(x)$ and $f(x)$ are continuous functions defined on the finite or infinite interval $[a, b]$, and $f(x)$ is positive there. For instance, if X_1, $X_2, \ldots$ is a sequence of independent identically distributed random variables with a common density $f(x)$, then the density $f_n(x)$ of the sum $S_n = X_1 + \cdots + X_n$ is given by the n-fold convolution $f * \cdots * f$. If $\varphi(\zeta)$ denotes the characteristic function of $f(x)$, then the characteristic function of f_n is $[\varphi(\zeta)]^n$; see Feller (1966, p. 474). Thus, by Fourier inversion, we have

$$f_n(x) = \frac{1}{2\pi} \int_{-\infty}^{\infty} e^{-i\zeta x}[\varphi(\zeta)]^n \, d\zeta. \qquad (1.2)$$

If $f(x)$ is the uniform density

$$f(x) = \begin{cases} \dfrac{1}{2a} & |x| < a \\ 0 & \text{otherwise,} \end{cases}$$

then (1.2) becomes

$$f_n(x) = \frac{1}{2\pi} \int_{-\infty}^{\infty} \cos \zeta x \left[\frac{\sin a\zeta}{a\zeta} \right]^n d\zeta; \tag{1.3}$$

cf. Henrici (1974, p. 471). If $f(x)$ is the bilateral exponential density $f(x) = \frac{1}{2}e^{-|x|}$, $-\infty < x < \infty$, then (1.2) becomes

$$f_n(x) = \frac{1}{2\pi} \int_{-\infty}^{\infty} \cos \zeta x \left[\frac{1}{1 + \zeta^2} \right]^n d\zeta. \tag{1.4}$$

These integrals are, of course, exactly of the form I_n.

Long ago, Laplace made the observation that the major contribution to the integral I_n should come from the neighborhoods of the points where $f(x)$ attains its greatest value. For instance, if $f(x)$ has its maximum value only at the point ξ in (a, b) where $f'(\xi) = 0$ and $f''(\xi) < 0$, then Laplace's result is

$$I_n \sim \varphi(\xi)[f(\xi)]^{n + 1/2} \left[\frac{-2\pi}{nf''(\xi)} \right]^{1/2} \tag{1.5}$$

as $n \to \infty$. An alternative form of this approximation is obtained by putting $f(x) = e^{h(x)}$; that is, if $h(x)$ attains its maximum only at $x = \xi$ where $h'(\xi) = 0$ and $h''(\xi) < 0$, then

$$\int_a^b \varphi(x)e^{nh(x)} \, dx \sim \varphi(\xi)e^{nh(\xi)} \left[\frac{-2\pi}{nh''(\xi)} \right]^{1/2} \tag{1.6}$$

as $n \to \infty$. Formulas (1.5) and (1.6) are now known as *Laplace's approximations* or *Laplace's formulas*.

A heuristic argument for these formulas may proceed as follows. First, replace $\varphi(x)$ and $h(x)$ by the leading terms in their Taylor series expansions at $x = \xi$, and then extend the integration limits to $-\infty$ and ∞. Thus,

$$\int_a^b \varphi(x)e^{nh(x)} \, dx \approx \int_a^b \varphi(\xi)e^{n[h(\xi) + (x - \xi)^2 h''(\xi)/2]} \, dx$$

$$\approx \varphi(\xi)e^{nh(\xi)} \int_{-\infty}^{\infty} e^{n(x - \xi)^2 h''(\xi)/2} \, dx \tag{1.7}$$

$$= \varphi(\xi)e^{nh(\xi)} \left[\frac{-2\pi}{nh''(\xi)} \right]^{1/2}.$$

To give a rigorous proof of the Laplace approximation, we consider the equivalent integral

$$I(\lambda) = \int_a^b \varphi(x)e^{-\lambda h(x)} \, dx, \tag{1.8}$$

where λ is a large positive parameter. Our method is simply to reduce this integral to a Laplace transform and then apply Watson's lemma. First, we observe that by subdividing the range of integration at the minima and maxima of $h(x)$, and by reversing the sign of x whenever necessary, we may assume, without loss of generality, that $h(x)$ has only one minimum in $[a, b]$ which occurs at $x = a$. Next, we assume that

$$h(x) \sim h(a) + \sum_{s=0}^{\infty} a_s(x - a)^{s + \mu}, \tag{1.9}$$

and

$$\varphi(x) \sim \sum_{s=0}^{\infty} b_s(x - a)^{s + \alpha - 1}, \tag{1.10}$$

as $x \to a^+$, and that the expansion (1.9) can be termwise differentiated, that is,

$$h'(x) \sim \sum_{s=0}^{\infty} a_s(s + \mu)(x - a)^{s + \mu - 1} \tag{1.11}$$

as $x \to a^+$. Here μ is a positive constant, and α can be real or complex as long as Re $\alpha > 0$. Also, we suppose, without loss of generality, that $a_0 \neq 0$ and $b_0 \neq 0$. The following result is an extension of Laplace's formula.

Theorem 1.   *For the integral*

$$I(\lambda) = \int_a^b \varphi(x)e^{-\lambda h(x)}\, dx, \tag{1.12}$$

we assume that (i) $h(x) > h(a)$ *for all* $x \in (a, b)$, *and for every* $\delta > 0$ *the infimum of* $h(x) - h(a)$ *in* $[a + \delta, b)$ *is positive;* (ii) $h'(x)$ *and* $\varphi(x)$ *are continuous in a neighborhood of* $x = a$, *except possibly at* a; (iii) *the expansions* (1.9), (1.10) *and* (1.11) *hold; and* (iv) *the integral* $I(\lambda)$ *converges absolutely for all sufficiently large* λ. *Then*

$$I(\lambda) \sim e^{-\lambda h(a)} \sum_{s=0}^{\infty} \Gamma\left(\frac{s + \alpha}{\mu}\right) \frac{c_s}{\lambda^{(s+\alpha)/\mu}}, \tag{1.13}$$

as $\lambda \to +\infty$, *where the coefficients* c_s *are expressible in terms of* a_s *and* b_s.

The first three coefficients are explicitly given by

$$c_0 = \frac{b_0}{\mu a_0^{\alpha/\mu}}, \qquad c_1 = \left\{\frac{b_1}{\mu} - \frac{(\alpha + 1)a_1 b_0}{\mu^2 a_0}\right\} \frac{1}{a_0^{(\alpha+1)/\mu}}, \tag{1.14}$$

and

$$c_2 = \left[\frac{b_2}{\mu} - \frac{(\alpha + 2)a_1 b_1}{\mu^2 a_0} + \{(\alpha + \mu + 2)a_1^2 - 2\mu a_0 a_2\} \frac{(\alpha + 2)b_0}{2\mu^3 a_0^2}\right] \frac{1}{a_0^{(\alpha+2)/\mu}}. \tag{1.15}$$

Proof.   Conditions (ii) and (iii) imply the existence of a number c in (a, b) such that $h'(x)$ and $\varphi(x)$ are continuous in $(a, c]$, and $h'(x)$ is also positive there. Put $T = h(c) - h(a)$, and introduce the new variable

$$t = h(x) - h(a). \tag{1.16}$$

Since $h(x)$ is increasing in (a, c), we may write

$$e^{\lambda h(a)} \int_a^c \varphi(x)e^{-\lambda h(x)}\, dx = \int_0^T f(t)e^{-\lambda t}\, dt \tag{1.17}$$

with $f(t)$ being the continuous function in $(0, T]$ given by

$$f(t) = \varphi(x)\frac{dx}{dt} = \frac{\varphi(x)}{h'(x)}. \tag{1.18}$$

Now substitute (1.9) into equation (1.16). Reversion of the resulting expression yields an asymptotic expansion of the form

$$x - a \sim \sum_{s=1}^{\infty} \alpha_s t^{s/\mu}, \qquad \text{as } t \to 0^+ ; \qquad (1.19)$$

see Exercise 1. The first three coefficients are given by

$$\alpha_1 = \frac{1}{a_0^{1/\mu}}, \qquad \alpha_2 = -\frac{a_1}{\mu a_0^{1+2/\mu}}, \qquad (1.20)$$

and

$$\alpha_3 = \frac{(\mu+3)a_1^2 - 2\mu a_0 a_2}{2\mu^2 a_0^{2+3/\mu}}. \qquad (1.21)$$

Substitution of (1.19) in (1.18) then gives

$$f(t) \sim \sum_{s=0}^{\infty} c_s t^{(s+\alpha-\mu)/\mu}, \qquad \text{as } t \to 0^+. \qquad (1.22)$$

To the integral on the right-hand side of (1.17) we can now apply Watson's lemma. Thus

$$\int_a^c \varphi(x) e^{-\lambda h(x)} \, dx \sim e^{-\lambda h(a)} \sum_{s=0}^{\infty} \Gamma\left(\frac{s+\alpha}{\mu}\right) \frac{c_s}{\lambda^{(s+\alpha)/\mu}}, \qquad (1.23)$$

as $\lambda \to +\infty$.

To complete the proof, we must also estimate the integral on the remaining range (c, b). Let λ_0 be a value of λ for which $I(\lambda)$ is absolutely convergent, and set

$$\varepsilon = \inf_{c \le x < b} \{h(x) - h(a)\}. \qquad (1.24)$$

By condition (i), ε is positive. Now we compute, for $\lambda \ge \lambda_0$,

$$\lambda[h(x) - h(a)] = (\lambda - \lambda_0)[h(x) - h(a)] + \lambda_0[h(x) - h(a)]$$
$$\ge (\lambda - \lambda_0)\varepsilon + \lambda_0[h(x) - h(a)]$$

and

$$\left| e^{\lambda h(a)} \int_c^b \varphi(x) e^{-\lambda h(x)} \, dx \right| \le M e^{-\varepsilon \lambda}, \qquad (1.25)$$

where M is a constant, namely,

$$M = e^{\lambda_0[\varepsilon + h(a)]} \int_c^b |\varphi(x)| e^{-\lambda_0 h(x)} \, dx.$$

The desired result (1.13) now follows from (1.23) and (1.25).   ∎

Example 1.   The Gamma function has the well-known integral representation

$$\Gamma(\lambda + 1) = \int_0^\infty u^\lambda e^{-u} \, du, \qquad \lambda > 0. \tag{1.26}$$

The integrand has a maximum at $u = \lambda$ and its maximum value is $e^{-\lambda} \lambda^\lambda$. Since the location of the maximum is not fixed, we make the substitution $u = \lambda(1 + x)$. Then (1.26) takes the form

$$\Gamma(\lambda + 1) = e^{-\lambda} \lambda^{\lambda + 1} \int_{-1}^\infty [(1 + x) e^{-x}]^\lambda \, dx = e^{-\lambda} \lambda^{\lambda + 1} \int_{-1}^\infty e^{-\lambda h(x)} \, dx,$$
$$\tag{1.27}$$

where

$$h(x) = x - \log(1 + x). \tag{1.28}$$

Splitting the interval of integration at $x = 0$, we have

$$e^\lambda \lambda^{-\lambda - 1} \Gamma(\lambda + 1) = \int_0^\infty e^{-\lambda h(x)} \, dx + \int_0^1 e^{-\lambda h(-x)} \, dx. \tag{1.29}$$

To each of the integrals on the right-hand side, we can now apply Theorem 1. Since

$$h(x) = \frac{1}{2} x^2 - \frac{1}{3} x^3 + \frac{1}{4} x^4 - \cdots, \qquad x \in [0, 1),$$

we have, in the notations of (1.9) and (1.10), $\mu = 2$, $\alpha = 1$, $b_0 = 1$, $b_1 = b_2 = \cdots = 0$, and $a_s = (-1)^s/(s + 2)$. Thus, by (1.14) and (1.15), $c_0 = 1/\sqrt{2}$, $c_1 = 2/3$, and $c_2 = \sqrt{2}/12$. Expansion (1.13) then gives

$$\int_0^\infty e^{-\lambda h(x)} \, dx \sim \sqrt{\frac{\pi}{2}} \frac{1}{\lambda^{1/2}} + \frac{2}{3} \frac{1}{\lambda} + \frac{1}{12} \sqrt{\frac{\pi}{2}} \frac{1}{\lambda^{3/2}} + \cdots, \tag{1.30}$$

as $\lambda \to +\infty$. Similarly,

$$\int_0^1 e^{-\lambda h(-x)} \, dx \sim \sqrt{\frac{\pi}{2}} \frac{1}{\lambda^{1/2}} - \frac{2}{3} \frac{1}{\lambda} + \frac{1}{12} \sqrt{\frac{\pi}{2}} \frac{1}{\lambda^{3/2}} - \cdots, \tag{1.31}$$

as $\lambda \to +\infty$. A combination of the last two expansions, together with (1.29) and the identity $\Gamma(\lambda + 1) = \lambda\Gamma(\lambda)$, gives

$$\Gamma(\lambda) \sim e^{-\lambda}\lambda^{\lambda}\sqrt{\frac{2\pi}{\lambda}}\left[1 + \frac{1}{12\lambda} + \cdots\right], \qquad \lambda \to +\infty. \qquad (1.32)$$

To obtain higher order approximations, we repeat some of the arguments used in the proof of Theorem 1. Equation (1.16) suggests that in (1.30) we set

$$t = h(x) = \frac{1}{2}x^2 - \frac{1}{3}x^3 + \frac{1}{4}x^4 - \cdots. \qquad (1.33)$$

Reversion of this expansion gives

$$x = \sum_{s=1}^{\infty} \alpha_s t^{s/2}. \qquad (1.34)$$

An explicit expression for the coefficient α_s can be obtained from Lagrange's inversion formula (see Copson (1935, p. 125)), and a recurrence relation will be given later. From (1.18) and (1.22), it follows that

$$f(t) = \frac{dx}{dt} = \sum_{s=0}^{\infty} c_s t^{(s-1)/2}, \qquad (1.35)$$

where

$$c_s = \frac{s+1}{2}\alpha_{s+1}. \qquad (1.36)$$

By Watson's lemma, we have

$$\int_0^{\infty} e^{-\lambda h(x)}\,dx = \int_0^{\infty} e^{-\lambda t}\frac{dx}{dt}\,dt \sim \sum_{s=0}^{\infty} \Gamma\left(\frac{s+1}{2}\right)\frac{c_s}{\lambda^{(s+1)/2}}. \qquad (1.37)$$

In exactly the same manner, we set

$$t = h(-x) = \frac{1}{2}(-x)^2 - \frac{1}{3}(-x)^3 + \frac{1}{4}(-x)^4 - \cdots$$

in (1.31), and obtain, by reversion,

$$-x = \sum_{s=1}^{\infty} (-1)^s \alpha_s t^{s/2}.$$

Thus

$$\frac{dx}{dt} = \sum_{s=0}^{\infty} (-1)^s c_s t^{(s-1)/2},$$

and

$$\int_0^1 e^{-\lambda h(-x)} dx = \int_0^{\infty} e^{-\lambda t} \frac{dx}{dt} dt \sim \sum_{s=0}^{\infty} (-1)^s \Gamma\left(\frac{s+1}{2}\right) \frac{c_s}{\lambda^{(s+1)/2}}. \quad (1.38)$$

A combination of (1.29), (1.37), and (1.38) gives

$$\Gamma(\lambda) \sim e^{-\lambda} \lambda^{\lambda} \sum_{s=0}^{\infty} \Gamma(s + \tfrac{1}{2}) \frac{2c_{2s}}{\lambda^{s+1/2}}. \quad (1.39)$$

To determine the coefficients, we return to (1.33), i.e., $t = x - \log(1 + x)$. By differentiation, we have

$$x \frac{dx}{dt} = (1 + x). \quad (1.40)$$

Substitution of (1.34) in (1.40) yields $\alpha_1 = \sqrt{2}$ and the recurrence relation

$$\frac{s+2}{\sqrt{2}} \alpha_{s+1} = \alpha_s - \sum_{k=0}^{s-2} \frac{k+2}{2} \alpha_{k+2} \alpha_{s-k}, \quad (1.41)$$

$s = 1, 2, \ldots$. As usual, empty sums are understood to be zero. From (1.41), it follows readily that

$$\alpha_2 = \frac{2}{3}, \ \alpha_3 = \frac{\sqrt{2}}{18}, \ \alpha_4 = -\frac{2}{135}, \ \alpha_5 = \frac{\sqrt{2}}{1080}, \ \text{and} \ \ c_2 = \frac{\sqrt{2}}{12}, \ c_4 = \frac{\sqrt{2}}{432}.$$

Thus, we have finally

$$\Gamma(\lambda) \sim e^{-\lambda} \lambda^{\lambda} \left(\frac{2\pi}{\lambda}\right)^{1/2} \left[1 + \frac{1}{12\lambda} + \frac{1}{288\lambda^2} + \cdots \right]. \quad (1.42)$$

The leading term in this expansion gives the Stirling formula quoted in Chapter I, Ex. 19.

Example 2.   Consider the integral

$$H(\rho) = \int_0^{\infty} \tau^{\gamma-1} \exp(-\rho\tau - \alpha\tau^{-\beta}) d\tau, \quad (1.43)$$

where $\alpha > 0$, $\beta > 0$, and γ is arbitrary. Although this integral is not in the form to which Laplace's method can be applied directly, it can be transformed into such a form by the simple substitution

$$\tau = \left(\frac{\alpha\beta}{\rho}\right)^{1/(\beta+1)} x. \qquad (1.44)$$

Upon making this substitution, we have

$$H(\rho) = \left(\frac{\alpha\beta}{\rho}\right)^{\gamma/(\beta+1)} \int_0^\infty x^{\gamma-1} \exp\left[-\lambda\left(x + \frac{1}{\beta}x^{-\beta}\right)\right] dx, \qquad (1.45)$$

where

$$\lambda = (\alpha\beta\rho^\beta)^{1/(\beta+1)}. \qquad (1.46)$$

The minimum value of $x + (1/\beta)x^{-\beta}$ in $(0, \infty)$ is $1 + 1/\beta$ and occurs at $x = 1$. To apply Theorem 1, we split the last integral at $x = 1$ and write

$$H(\rho) = \left(\frac{\alpha\beta}{\rho}\right)^{\gamma/(\beta+1)} [I_1(\lambda) + I_2(\lambda)], \qquad (1.47)$$

with $I_1(\lambda)$ and $I_2(\lambda)$ corresponding, respectively, to the intervals $(0, 1)$ and $(1, \infty)$. By Theorem 1, with $a = 1$, $b = \infty$, $\varphi(x) = x^{\gamma-1}$, and $h(x) = x + \beta^{-1}x^{-\beta}$, we have

$$I_2(\lambda) \sim e^{-\lambda(1+\beta^{-1})} \sum_{s=0}^\infty \Gamma\left(\frac{s+1}{2}\right) \frac{c_{2,s}}{\lambda^{(s+1)/2}}, \qquad (1.48)$$

where

$$c_{2,0} = \frac{1}{\sqrt{2(\beta+1)}}, \qquad c_{2,1} = \frac{[\gamma - 1 + \frac{1}{3}(\beta+2)]}{\beta+1},$$

and

$$c_{2,2} = \frac{[(\gamma-1)(\gamma+\beta) + \frac{1}{12}(\beta+2)(2\beta+1)]}{\sqrt{2(\beta+1)^3}}.$$

Similarly, the integral $I_1(\lambda)$ can be written as

$$I_1(\lambda) = \int_{-1}^0 (-x)^{\gamma-1} \exp\left\{-\lambda\left[-x + \frac{1}{\beta}(-x)^{-\beta}\right]\right\} dx. \qquad (1.49)$$

With $\varphi(x) = (-x)^{y-1}$ and $h(x) = -x + (-x)^{-\beta}/\beta$, Theorem 1 gives

$$I_1(\lambda) \sim e^{-\lambda(1+1/\beta)} \sum_{s=0}^{\infty} \Gamma\left(\frac{s+1}{2}\right) \frac{c_{1,s}}{\lambda^{(s+1)/2}}, \tag{1.50}$$

where $c_{1,0} = c_{2,0}$, $c_{1,1} = -c_{2,1}$, and $c_{1,2} = c_{2,2}$. Adding (1.48) and (1.50) together yields

$$H(\rho) \sim \left(\frac{\alpha\beta}{\rho}\right)^{y/(\beta+1)} e^{-\lambda(1+1/\beta)} \sum_{s=0}^{\infty} \Gamma\left(\frac{s+1}{2}\right) \frac{c_s}{\lambda^{(s+1)/2}}, \tag{1.51}$$

where $c_s = c_{1,s} + c_{2,s}$. In particular, we obtain

$$H(\rho) \sim \left(\frac{\alpha\beta}{\rho}\right)^{y/(\beta+1)} e^{-\lambda(1+1/\beta)} \sqrt{\frac{2\pi}{\beta+1}}\, \lambda^{-1/2}. \tag{1.52}$$

The above result was given by Erdélyi (1961).

We now show that the coefficients c_s in (1.51) all vanish when s is odd, and that $c_s = 2c_{2,s}$ when s is even. To this end, we consider the equation

$$t = h(x) - h(1) = x + \frac{1}{\beta}x^{-\beta} - 1 - \frac{1}{\beta}$$

$$= \frac{\beta+1}{2}(x-1)^2 - \frac{(\beta+1)(\beta+2)}{6}(x-1)^3 + \cdots, \qquad 1 < x < 2. \tag{1.53}$$

By Lagrange's inversion formula (see Copson (1935, p. 125)),

$$x - 1 = \sum_{s=1}^{\infty} \alpha_{2,s} t^{s/2},$$

where

$$\alpha_{2,s} = \frac{1}{s!}\left[\frac{d^{s-1}}{dx^{s-1}}\{\varphi(x)\}^s\right]_{x=1}$$

and

$$\left[x + \frac{1}{\beta}x^{-\beta} - 1 - \frac{1}{\beta}\right]^{1/2} = \frac{x-1}{\varphi(x)}.$$

Similarly, inversion of the equation

$$t = h(-x) - h(1) = -x + \frac{1}{\beta}(-x)^{-\beta} - 1 - \frac{1}{\beta}$$

$$= \frac{\beta+1}{2}(-x-1)^2 - \frac{(\beta+1)(\beta+2)}{6}(-x-1)^3 + \cdots, \qquad -1 < x < 0, \tag{1.54}$$

gives

$$-x - 1 = \sum_{s=1}^{\infty} \alpha_{2,s}(-\sqrt{t})^s.$$

Put

$$\sigma_{\pm}(\sqrt{t}) = \sum_{s=1}^{\infty} \alpha_{2,s}(\pm\sqrt{t})^s$$

and

$$F_{\pm}(\sqrt{t}) = [1 + \sigma_{\pm}(\sqrt{t})]^{\gamma-1} \frac{d}{dt} \sigma_{\pm}(\sqrt{t}). \tag{1.55}$$

Then $x = 1 + \sigma_{+}(\sqrt{t})$ for $1 < x < 2$, and $-x = 1 + \sigma_{-}(\sqrt{t})$ for $-1 < x < 0$. Since $\sigma_{-}(\sqrt{t}) = \sigma_{+}(-\sqrt{t})$, we also have

$$F_{-}(\sqrt{t}) = F_{+}(-\sqrt{t}). \tag{1.56}$$

We now make t in (1.53) and (1.54) the variable of integration in (1.45). The integrals $I_1(\lambda)$ and $I_2(\lambda)$ then become

$$I_1(\lambda) = -e^{-\lambda(1 + 1/\beta)} \int_0^{\infty} f_1(t)e^{-\lambda t}\, dt$$

and

$$I_2(\lambda) = e^{-\lambda(1 + 1/\beta)} \int_0^{\infty} f_2(t)e^{-\lambda t}\, dt,$$

where

$$f_2(t) = x^{\gamma-1}\frac{dx}{dt}, \quad x > 1, \quad \text{and} \quad f_1(t) = (-x)^{\gamma-1}\frac{d(-x)}{dt}, \quad -1 < x < 0.$$

From (1.47), it follows that

$$H(\rho) = \left(\frac{\alpha\beta}{\rho}\right)^{\gamma/(\beta+1)} e^{-\lambda(1 + 1/\beta)} \int_0^{\infty} f(t)e^{-\lambda t}\, dt, \tag{1.57}$$

where

$$f(t) = f_2(t) - f_1(t). \tag{1.58}$$

On account of (1.55), it is clear that for small positive values of t,

$$f_1(t) = F_{-}(\sqrt{t}) \quad \text{and} \quad f_2(t) = F_{+}(\sqrt{t}). \tag{1.59}$$

Since $x^{\gamma-1}$ is analytic in a neighborhood of $x = 1$, it is also clear that there exists a positive number δ such that for $0 < t < \delta$, $F_+(\sqrt{t})$ has a convergent expansion of the form $F_+(\sqrt{t}) = \sum_{s=0}^{\infty} c_{2,s}(\sqrt{t})^{s-1}$. Equation (1.56) then implies $F_-(\sqrt{t}) = \sum_{s=0}^{\infty} c_{2,s}(-\sqrt{t})^{s-1}$. Coupling (1.58) and (1.59) together gives

$$f(t) = F_+(\sqrt{t}) - F_-(\sqrt{t}) = 2 \sum_{n=0}^{\infty} c_{2,2n} t^{n-1/2} \equiv \sum_{n=0}^{\infty} c_{2n} t^{n-1/2}.$$

By Watson's lemma,

$$H(\rho) \sim \left(\frac{\alpha\beta}{\rho}\right)^{\gamma/(\beta+1)} e^{-\lambda(1+1/\beta)} \sum_{n=0}^{\infty} \Gamma(n + \tfrac{1}{2}) \frac{c_{2n}}{\lambda^{n+1/2}}.$$

This proves our claim that the odd-indexed coefficients in (1.51) all vanish. The vanishing of the odd coefficients is not peculiar to the expansion of the function $H(\rho)$; for example, see also the expansion (1.39). This phenomenon, in fact, occurs whenever the function $h(x)$ in (1.12) has a simple minimum at an interior point of the interval of integration; cf. Olver (1974a, p. 127, Theorem 7.1).

2.   Logarithmic Singularities

In some problems, we encounter integrals of the form given in (1.1) which do not satisfy the conditions necessary for the validity of the approximation (1.5). For instance, in a study of probability theory, there arose the integral

$$I(n) = \int_{-\infty}^{\infty} xe^{-x^2} \left[\frac{1 + \Theta(x)}{2}\right]^n dx, \tag{2.1}$$

where

$$\Theta(x) = \frac{2}{\sqrt{\pi}} \int_0^x e^{-u^2} du. \tag{2.2}$$

Tricomi (1933) showed that

$$I(n) \sim \frac{\sqrt{\pi \log n}}{n}, \qquad \text{as } n \to \infty. \tag{2.3}$$

Note that the function $\frac{1}{2}[1 + \Theta(x)]$ increases monotonically from 0 to 1 as x varies from $-\infty$ to $+\infty$. Thus, the greatest value of this function is not attained at a finite point but at infinity, and the approximation (1.5) does not apply. However, the substitution

$$e^{-t} = \frac{1 + \Theta(x)}{2} \tag{2.4}$$

places $I(n)$ in the form

$$I(n) = \sqrt{\pi} \int_0^\infty x(t) e^{-(n+1)t} \, dt. \tag{2.5}$$

By monotonicity, the inverse $x = x(t)$ is uniquely defined and decreases from $+\infty$ to $-\infty$ as t increases from 0 to ∞. Since e^{-t} has its greatest value at the finite end point $t = 0$, one naturally attempts to apply the result in (1.5). In fact, since the integral in (2.5) is a Laplace transform, it is even more natural to use Watson's lemma directly. This will work provided that $x(t)$ has only an algebraic singularity at $t = 0$. The following lemma shows that $x(t)$ has a singularity of a quite different type.

Lemma 1. *For small positive values of t, the real root of equation (2.4) is given by*

$$x(t) = \sqrt{-\log t} - \frac{\log(-\log t)}{4\sqrt{-\log t}} + O\!\left(\frac{1}{\sqrt{-\log t}}\right). \tag{2.6}$$

Proof. By integration by parts, it can easily be shown that

$$\Theta(x) \sim 1 - \frac{e^{-x^2}}{\sqrt{\pi} x}\left[1 - \frac{1}{2x^2} + \frac{3}{4x^4} - \cdots\right], \tag{2.7}$$

as $x \to +\infty$. Equation (2.4) can hence be written as

$$\frac{e^{-x^2}}{2\sqrt{\pi} x}\left[1 + O\!\left(\frac{1}{x^2}\right)\right] = 1 - e^{-t}, \tag{2.8}$$

or

$$\frac{e^{-x^2}}{2\sqrt{\pi} x}\left[1 + O\!\left(\frac{1}{x^2}\right)\right] = t[1 + O(t)], \tag{2.9}$$

as $x \to +\infty$ and $t \to 0^+$. The last equation gives

$$-x^2 - \log 2\sqrt{\pi} - \log x + O\left(\frac{1}{x^2}\right) = \log t + O(t). \qquad (2.10)$$

Taking the dominant terms on both sides, we get

$$x^2 \approx -\log t, \qquad \text{as } t \to 0^+, \qquad (2.11)$$

and hence

$$x \sim \sqrt{-\log t}, \qquad \text{as } t \to 0^+. \qquad (2.12)$$

This is the first approximation to the solution of (2.4). To improve this result, we set

$$x^2 = -\log t + \varepsilon(t). \qquad (2.13)$$

Note that by (2.11), we have

$$\frac{\varepsilon(t)}{-\log t} \to 0, \qquad \text{as } t \to 0^+. \qquad (2.14)$$

Inserting (2.13) in (2.10) yields

$$\log t - \varepsilon(t) - \log 2\sqrt{\pi} - \frac{1}{2}\log(-\log t) - \frac{1}{2}\log\left(1 + \frac{\varepsilon(t)}{-\log t}\right)$$
$$+ O\left(\frac{1}{-\log t + \varepsilon(t)}\right) = \log t + O(t).$$

Hence, in view of (2.14),

$$\varepsilon(t) = -\frac{1}{2}\log(-\log t) - \log 2\sqrt{\pi} + o(1), \qquad (2.15)$$

as $t \to 0^+$. Coupling (2.13) and (2.15), we obtain

$$x^2 = (-\log t)\left[1 - \frac{1}{2}\frac{\log(-\log t)}{-\log t} + O\left(\frac{1}{-\log t}\right)\right]$$

and

$$x = \sqrt{-\log t} - \frac{\log(-\log t)}{4\sqrt{-\log t}} + O\left(\frac{1}{\sqrt{-\log t}}\right)$$

as $t \to 0^+$, thus proving (2.6). ∎

Substituting (2.6) in (2.5) suggests that there is a need to study the asymptotic behavior of Laplace transforms whose kernel has a logarithmic singularity at the origin. To begin, we recall the well-known identity

$$\int_0^{\infty e^{i\gamma}} t^{\lambda-1} e^{-zt} \, dt = \Gamma(\lambda) z^{-\lambda}, \tag{2.16}$$

where Re $\lambda > 0$ and $|\arg(ze^{i\gamma})| < \pi/2$. When this identity is differentiated m times with respect to λ, the result is

$$\int_0^{\infty e^{i\gamma}} t^{\lambda-1} (\log t)^m e^{-zt} \, dt = \frac{d^m}{d\lambda^m} [\Gamma(\lambda) z^{-\lambda}], \tag{2.17}$$

or equivalently,

$$\int_0^{\infty e^{i\gamma}} t^{\lambda-1} (-\log t)^m e^{-zt} \, dt = (\log z)^m z^{-\lambda} \sum_{r=0}^{m} (-1)^r \binom{m}{r} \Gamma^{(r)}(\lambda)(\log z)^{-r}. \tag{2.18}$$

From (2.18), one might reasonably conjecture that for any complex number μ,

$$\int_0^{\infty e^{i\gamma}} t^{\lambda-1} (-\log t)^\mu e^{-zt} \, dt = (\log z)^\mu z^{-\lambda} \sum_{r=0}^{\infty} (-1)^r \binom{\mu}{r} \Gamma^{(r)}(\lambda)(\log z)^{-r}. \tag{2.19}$$

However, as shown in Ex. 5, this series diverges for all finite values of $\log z$, and the conjecture is false. Instead, with appropriate conditions,

$$\int_0^{\infty e^{i\gamma}} t^{\lambda-1} (-\log t)^\mu e^{-zt} \, dt \sim (\log z)^\mu z^{-\lambda} \sum_{r=0}^{\infty} (-1)^r \binom{\mu}{r} \Gamma^{(r)}(\lambda)(\log z)^{-r}, \tag{2.20}$$

as $|z| \to \infty$. We now prove this result.

By (5.20) of Chapter I,

$$\int_c^{\infty e^{i\gamma}} t^{\lambda-1} (-\log t)^\mu e^{-zt} \, dt = O(\exp(-\varepsilon|c||z|)) \tag{2.21}$$

for any point $t = c = |c|e^{i\gamma}$, where $\varepsilon = \frac{1}{2} \sin \Delta$ and $z \to \infty$ in the sector $|\arg(ze^{i\gamma})| \le \pi/2 - \Delta < \pi/2$. If $\gamma = 0$ and $0 < c \le 1$, then the condition

Re $\mu > -1$ must be added. Anticipating the final result (2.20), it suffices to consider only the integral

$$L(\lambda, \mu, z) = \int_0^c t^{\lambda - 1}(-\log t)^\mu e^{-zt}\, dt, \qquad (2.22)$$

where $c = |c|e^{i\gamma}$, $0 < |c| < 1$, and the path of integration is the straight line joining $t = 0$ to $t = c$.

Theorem 2. *For any complex numbers λ and μ with Re $\lambda > 0$, the integral (2.22) possesses the asymptotic expansion*

$$L(\lambda, \mu, z) \sim z^{-\lambda}(\log z)^\mu \sum_{r=0}^\infty (-1)^r \binom{\mu}{r}\Gamma^{(r)}(\lambda)(\log z)^{-r}, \qquad (2.23)$$

uniformly in arg z, *as* $z \to \infty$ *in* $|\arg(ze^{i\gamma})| \le \pi/2 - \Delta$.

Proof.   The substitution $u = zt$ gives

$$
\begin{aligned}
L(\lambda, \mu, z) &= z^{-\lambda}\int_0^{cz} u^{\lambda - 1}(\log z - \log u)^\mu e^{-u}\, du \\
&= z^{-\lambda}(\log z)^\mu \int_0^{cz} u^{\lambda - 1}\left(1 - \frac{\log u}{\log z}\right)^\mu e^{-u}\, du.
\end{aligned}
\qquad (2.24)
$$

Let N be an arbitrary integer such that $N + 1 \ge \mu$. Then, by using the binomial expansion, it is easily seen that

$$\left(1 - \frac{\log u}{\log z}\right)^\mu = \sum_{n=0}^N (-1)^n \binom{\mu}{n}\frac{(\log u)^n}{(\log z)^n} + R_N \qquad (2.25)$$

for all points on the path of integration, where

$$|R_N| \le K_N\left|\frac{(\log u)^{N+1}}{(\log z)^{N+1}}\right| \qquad (2.26)$$

for some fixed $K_N > 0$. Inserting (2.25) in (2.24), we obtain

$$L(\lambda, \mu, z) = z^{-\lambda}(\log z)^\mu$$
$$\times \left[\sum_{n=0}^N \binom{\mu}{n}(-\log z)^{-n}\int_0^{cz} u^{\lambda - 1}(\log u)^n e^{-u}\, du + r_N\right],$$

where

$$r_N = \int_0^{cz} u^{\lambda - 1}e^{-u}R_N\, du.$$

Let $\tau = \arg(ze^{i\gamma})$. By taking $z = 1$ and $\mu = n$, and replacing c and γ by cz and τ, respectively, in (2.21), we have

$$\int_{cz}^{\infty e^{it}} u^{\lambda-1}(\log u)^n e^{-u}\, du = O(\exp(-\varepsilon|c||z|)).$$

Thus, it follows that

$$L(\lambda, \mu, z) = z^{-\lambda}(\log z)^{\mu}\left[\sum_{n=0}^{N}(-1)^n\binom{\mu}{n}\Gamma^{(n)}(\lambda)(\log z)^{-n}\right.$$

$$\left. + O(\exp(-\varepsilon|c||z|)) + r_N\right],$$

as $z \to \infty$ in $|\arg(ze^{i\gamma})| \leq \pi/2 - \Delta$. Furthermore,

$$|r_N| \leq K_N|\log z|^{-(N+1)}\int_0^{cz}|u^{\lambda-1}(\log u)^{N+1}e^{-u}\, du|$$

$$\leq K_N|\log z|^{-(N+1)}\int_0^{\infty e^{it}}|u^{\lambda-1}(\log u)^{N+1}e^{-u}\, du|. \tag{2.27}$$

It is trivial to show that the integral in (2.27) exists and is bounded in $\arg z$. Hence

$$L(\lambda, \mu, z) = z^{-\lambda}(\log z)^{\mu}\left[\sum_{n=0}^{N}(-1)^n\binom{\mu}{n}\Gamma^{(n)}(\lambda)(\log z)^{-n} + O((\log z)^{-N-1})\right],$$

as $z \to \infty$ in $|\arg(ze^{i\gamma})| \leq \pi/2 - \Delta$. Since the order relation is independent of $\arg z$, this proves the required result (2.23). ∎

In a similar manner, one can show, for example, that the integral

$$F(z) = \int_0^c t^{\lambda-1}\log(-\log t)e^{-zt}\, dt, \tag{2.28}$$

where Re $\lambda > 0$ and $0 < c < 1$, has the asymptotic expansion

$$z^{\lambda}F(z) - \Gamma(\lambda)\log\log z \sim -\sum_{n=1}^{\infty}\frac{1}{n}\Gamma^{(n)}(\lambda)(\log z)^{-n}, \tag{2.29}$$

as $z \to \infty$, and that the integral

$$G(z) = \int_0^c \log(-\log t)(-\log t)^{-1/2}e^{-zt}\, dt, \tag{2.30}$$

where $0 < c < 1$, has the asymptotic expansion

$$G(z) \sim \frac{\log \log z}{z\sqrt{\log z}} \sum_{n=0}^{\infty} (-1)^n \binom{-1/2}{n} \Gamma^{(n)}(1)(\log z)^{-n}$$

$$+ \frac{1}{z\sqrt{\log z}} \sum_{n=0}^{\infty} c_n \Gamma^{(n)}(1)(\log z)^{-n}, \tag{2.31}$$

as $z \to \infty$, where the constants c_n are given by

$$c_n = -\sum_{r=0}^{n-1} \binom{-\frac{1}{2}}{r} \frac{(-1)^r}{n-r}; \tag{2.32}$$

see Exercise 6.

Returning to the integral $I(n)$ given in (2.1), we observe that the function $\Theta(x)$ in (2.2) is odd and hence from (2.7)

$$\Theta(x) \sim -1 - \frac{e^{-x^2}}{\sqrt{\pi} x} \left(1 - \frac{1}{2x^2} + \frac{3}{4x^4} - \cdots \right)$$

as $x \to -\infty$. This coupled with (2.4) gives

$$e^{-t} = \frac{1 + \Theta(x)}{2} = -\frac{e^{-x^2}}{2\sqrt{\pi} x} [1 + o(1)],$$

which obviously implies that a constant K exists such that $|x(t)| \le K\sqrt{t}$ for sufficiently large t. This is sufficient to show that $\int_c^{\infty} x(t) e^{-(n+1)t}\, dt$ is exponentially small for any $0 < c < 1$. Therefore

$$I(n) \approx \sqrt{\pi} \int_0^c x(t) e^{-(n+1)t}\, dt. \tag{2.33}$$

The asymptotic sequence involved in writing (2.33) must be such that terms which are exponentially small can be neglected. Inserting (2.6) in (2.33) yields

$$I(n) = \sqrt{\pi} \int_0^c (-\log t)^{1/2} e^{-(n+1)t}\, dt - \frac{\sqrt{\pi}}{4} \int_0^c \frac{\log(-\log t)}{\sqrt{-\log t}} e^{-(n+1)t}\, dt$$

$$+ O\!\left(\frac{1}{(n+1)\sqrt{\log(n+1)}} \right) \tag{2.34}$$

as $n \to \infty$, the estimate of the remainder being obtained by (2.23). A combination of (2.23), (2.31), and (2.34) gives

$$I(n) = \frac{\sqrt{\pi} \log(n + 1)}{n + 1} - \frac{\sqrt{\pi}}{4} \frac{\log \log(n + 1)}{(n + 1)\sqrt{\log(n + 1)}}$$
$$+ O\left(\frac{1}{(n + 1)\sqrt{\log(n + 1)}}\right), \tag{2.35}$$

which of course agrees with (2.3).

An integral somewhat similar to (2.1) is

$$J(k) = \int_{-\infty}^{\infty} [\Phi(x) + 1 - \Phi(x + L)]^{k-1} \, d\Phi(x), \tag{2.36}$$

where $L \geq 0$ and

$$\Phi(x) = \frac{1}{\sqrt{2\pi}} \int_{-\infty}^{x} e^{-u^2/2} \, du. \tag{2.37}$$

This integral arose in a comparison of individual means in the analysis of variance. By methods similar to those employed above, it can be shown that when L is positive,

$$J(k) \sim \frac{1}{k} + \frac{2e^{-L^2/2}}{ke^{L\sqrt{2\log k}}}, \qquad \text{as } k \to \infty; \tag{2.38}$$

see McClure and Wong (1986).

Logarithmic singularities appear not only in Laplace-type integrals, they also occur in Fourier integrals. For example, in a study of absolute summability of some series related to Fourier series (see Ray (1970)), it was shown that

$$\int_{0}^{\pi} \left(\log \frac{2\pi}{t}\right)^{\beta} \frac{\sin nt}{t} \, dt \sim \frac{\pi}{2} (\log n)^{\beta} \tag{2.39}$$

for all β, and that

$$\int_{0}^{\pi} \left(\log \frac{2\pi}{t}\right)^{\beta} \cos nt \, dt \sim \frac{\pi}{2} \frac{\beta(\log n)^{\beta-1}}{n} \tag{2.40}$$

for all β except $\beta = 0$. Both of these integrals are special cases of the finite Fourier integral

$$I(\lambda, \mu, x) = \int_{0}^{a} t^{\lambda-1}(-\log t)^{\mu} e^{ixt} \, dt \tag{2.41}$$

where Re $\lambda > 0$, μ is complex, $0 < a < 1$, and x is positive; see Example 3 below.

To derive the asymptotic expansion of $I(\lambda, \mu, x)$, we first consider the modified integral

$$J(\lambda, \mu, x) = \int_0^c t^{\lambda-1}(-\log t)^\mu e^{ixt}\, dt, \qquad (2.42)$$

where $|c| < 1$, arg $c = \gamma$ and $0 < \gamma < \pi/2$. Since $J(\lambda, \mu, x) = L(\lambda, \mu, -ix)$, where $L(\lambda, \mu, z)$ is defined in (2.22), we have from Theorem 2

$$J(\lambda, \mu, x) \sim \frac{e^{\lambda \pi i/2}}{x^\lambda} \sum_{r=0}^\infty (-1)^r \binom{\mu}{r} \Gamma^{(r)}(\lambda) \left(\log x - \frac{\pi}{2} i\right)^{\mu-r}, \qquad (2.43)$$

as $x \to +\infty$ through positive real values. By expanding $(\log x - \pi i/2)^{\mu-r}$ in powers of $(\log x)^{-1}$, we obtain

$$J(\lambda, \mu, x) \sim \frac{e^{\lambda \pi i/2}}{x^\lambda} \sum_{r=0}^\infty c_r(\lambda, \mu)(\log x)^{\mu-r} \qquad (2.44)$$

as $x \to +\infty$, where the coefficients $c_r(\lambda, \mu)$ are given by

$$c_r(\lambda, \mu) = (-1)^r \binom{\mu}{r} \sum_{k=0}^r \binom{r}{k} \Gamma^{(k)}(\lambda) \left(\frac{\pi i}{2}\right)^{r-k}. \qquad (2.45)$$

Theorem 3. *Let $g_s(\lambda, \mu)$ denote the s-th derivative of $t^{\lambda-1}(-\log t)^\mu$ at $t = a$ and let $c_r(\lambda, \mu)$ be the constants given in (2.45). Then the integral (2.41) has the asymptotic expansion*

$$I(\lambda, \mu, x) \sim \frac{e^{\lambda \pi i/2}}{x^\lambda} \sum_{r=0}^\infty c_r(\lambda, \mu)(\log x)^{\mu-r} + e^{ixa} \sum_{s=0}^\infty (-1)^s g_s(\lambda, \mu)(ix)^{-s-1}, \qquad (2.46)$$

as $x \to +\infty$ through positive real values.

Proof. By Cauchy's theorem, we have

$$I(\lambda, \mu, x) = \left(\int_0^c - \int_a^c\right) t^{\lambda-1}(-\log t)^\mu e^{ixt}\, dt,$$

where the paths of integration are indicated by the solid lines in Figure 2.1. In the second integral we replace t by $a - t$. This, together with (2.42), gives

$$I(\lambda, \mu, x) = J(\lambda, \mu, x) + e^{ixa} \int_0^{a-c} G_{\lambda\mu}(t) e^{-ixt}\, dt, \qquad (2.47)$$

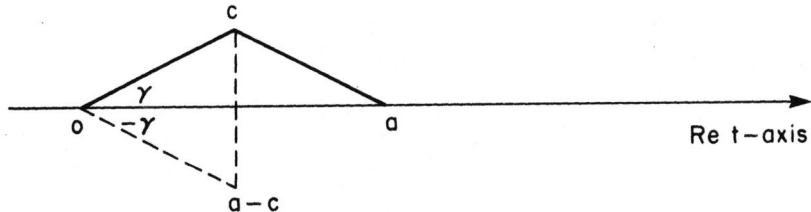

Fig. 2.1

where $G_{\lambda\mu}(t) = (a - t)^{\lambda - 1}[-\log(a - t)]^{\mu}$. Expanding $G_{\lambda\mu}(t)$ in the Maclaurin series

$$G_{\lambda\mu}(t) = \sum_{s=0}^{\infty} \frac{(-1)^s}{s!} g_s(\lambda, \mu)t^s,$$

and applying Watson's lemma, we obtain

$$\int_0^{a-c} G_{\lambda\mu}(t)e^{-ixt}\, dt \sim \sum_{s=0}^{\infty} (-1)^s g_s(\lambda, \mu)(ix)^{-s-1}, \qquad (2.48)$$

as $x \to \infty$ in the sector $-\pi + \gamma + \Delta \leq \arg x \leq \gamma - \Delta$, $0 < \Delta < \gamma$. In particular, (2.48) holds as $x \to +\infty$ through positive real values. The desired result (2.46) now follows upon inserting (2.44) and (2.48) in (2.47). ∎

Example 3. Consider the integral

$$S(n) = \int_0^\pi \left(\log \frac{2\pi}{t}\right)^\beta \frac{\sin nt}{t}\, dt$$

given in (2.39). Putting $t = 2\pi\tau$ gives

$$S(n) = \int_0^{1/2} (-\log \tau)^\beta \frac{\sin 2\pi n\tau}{\tau}\, d\tau.$$

By an integration by parts, we have

$$\frac{\beta + 1}{2\pi n} S(n) = \int_0^{1/2} (-\log \tau)^{\beta + 1} \cos 2\pi n\tau\, d\tau,$$

which is the real part of the integral $\int_0^{1/2}(-\log \tau)^{\beta + 1}e^{i2\pi n\tau}\, d\tau$. To this integral, we can apply Theorem 3 with $a = \frac{1}{2}$, $\lambda = 1$, $\mu = \beta + 1$, and

$x = 2\pi n$. The first few terms in its asymptotic expansion are given by

$$\frac{i}{2\pi n}\left[(\log 2\pi n)^{\beta + 1} - (\beta + 1)\left(\frac{\pi i}{2} - \gamma\right)(\log 2\pi n)^{\beta} + O((\log 2\pi n)^{\beta - 1})\right]$$

$$+ e^{i\pi n}\left[-(\log 2)^{\beta + 1}\frac{i}{2\pi n} + O(n^{-2})\right].$$

Taking only the real part of this expansion, we get

$$\int_0^{1/2}(-\log \tau)^{\beta + 1}\cos 2\pi n\tau \, d\tau = \frac{\beta + 1}{4n}(\log 2\pi n)^{\beta} + O((\log 2\pi n)^{\beta - 1}),$$

which of course implies the result (2.39). The asymptotic formula (2.40) can be proved in a similar manner.

3. The Principle of Stationary Phase

According to Watson (1944, p. 229–230), the principle of stationary phase was first introduced explicitly by Lord Kelvin (1887), although the essence of this principle can be found in the earlier works of Cauchy, Stokes, and Riemann. Kelvin's original problem was to find the asymptotic behavior of the integral

$$u(x) = \frac{1}{2\pi}\int_0^{\infty}\cos\{m[x - tf(m)]\} \, dm, \qquad \text{as } x \to \infty,$$

where all variables are real. However, it is now more common to state the principle in terms of the integral

$$I(x) = \int_a^b g(t)e^{ixf(t)} \, dt, \tag{3.1}$$

where a, b, and the function $f(t)$ are real, and x is a large real parameter.

The underlying principle of stationary phase is the assertion that the major asymptotic contribution to the integral (3.1) comes from points where the *phase function* $f(t)$ is *stationary*, i.e., where $f'(t)$ vanishes. Suppose that $f(t)$ has only one such point, say c, in (a, b), and that $f''(t)$

3. The Principle of Stationary Phase

is positive there, i.e., $f(c)$ is a simple minimum. Then, heuristically, we have

$$I(x) \approx \int_{c-\varepsilon}^{c+\varepsilon} g(t)e^{ixf(t)} \, dt \qquad (\varepsilon > 0)$$

$$\approx g(c) \int_{c-\varepsilon}^{c+\varepsilon} e^{ix[f(c) + f''(c)(t-c)^2/2]} \, dt$$

$$\approx g(c)e^{ixf(c)} \int_{-\infty}^{\infty} e^{ixf''(c)(t-c)^2/2} \, dt$$

$$= g(c) \sqrt{\frac{2\pi}{xf''(c)}} \, e^{i[xf(c) + \pi/4]}. \tag{3.2}$$

Similarly, if $f''(c) < 0$, i.e., $f(c)$ is a maximum, then

$$I(x) \sim g(c) \sqrt{\frac{2\pi}{-xf''(c)}} \, e^{i[xf(c) - \pi/4]}. \tag{3.3}$$

A rigorous proof of (3.2) has been given by Watson, but his discussion is rather restricted in scope. In particular, his argument does not seem capable of producing the higher-order terms in the asymptotic expansion. Considerably more satisfactory results have subsequently been obtained by van der Corput (1934, 1936), Erdélyi (1955), and Olver (1974b). The argument which we shall present is due to Erdélyi.

First we assume that $f(t)$ has only a finite number of stationary points in the interval under consideration, i.e., $f'(t)$ has only a finite number of zeros there. Next, by subdividing the interval, we may assume that in each subinterval there is at most one stationary point, and that it is located at one of the endpoints. Without loss of generality, we may suppose that this point is the left endpoint. By changing the variable of integration from t to $-t$, or by changing $f(t)$ to $-f(t)$ and x to $-x$, we may also suppose that $f(t)$ is strictly increasing. Thus we shall consider the integral (3.1), in which $f(t)$ and $g(t)$ satisfy the following conditions:

(a) $f'(t) > 0$ in (a, b), and $f(t)$ has the form

$$f(t) = f(a) + (t-a)^\rho f_1(t), \, f_1(a) \neq 0,$$

where $\rho \geq 1$ and $f_1(t)$ is infinitely differentiable in $[a, b]$.

(b) $g(t)$ is of the form

$$g(t) = (t - a)^{\lambda - 1} g_1(t),$$

where $0 < \lambda \le 1$ and $g_1(t)$ is also infinitely differentiable in $[a, b]$.

Our strategy is simply to reduce the integral $I(\lambda)$ in (3.1) to a Fourier-type integral, and then apply the integration-by-parts technique discussed in Chapter I, Section 4. To this end, we introduce the variable u defined by

$$u^\rho = f(t) - f(a). \qquad (3.4)$$

Since $f(t)$ is strictly increasing in $[a, b]$, there is a one-to-one relationship between t and u. Also, in consequence of condition (a), t is an infinitely differentiable function of u in $0 \le u \le B$, where $B^\rho = f(b) - f(a)$. In terms of the variable u, $I(x)$ becomes

$$I(x) = e^{ixf(a)} \int_0^B u^{\lambda - 1} h(u) e^{ixu^\rho} \, du, \qquad (3.5)$$

where

$$h(u) = u^{1 - \lambda} g(t) \frac{dt}{du} = \left(\frac{t - a}{u}\right)^{\lambda - 1} g_1(t) \frac{dt}{du}. \qquad (3.6)$$

The smoothness conditions in (a) and (b) imply that $h(u)$ is a C^∞-function on $[0, B]$.

To proceed further, we extend $h(u)$ to a C^∞-function on $[0, \infty)$ with

$$h(u) \equiv 0 \quad \text{in a neighborhood of infinity}, \qquad (3.7)$$

and write

$$I(x) = e^{ixf(a)}\left(\int_0^\infty - \int_B^\infty\right) u^{\lambda - 1} h(u) e^{ixu^\rho} \, du \equiv e^{ixf(a)}[I_1(\lambda) - I_2(\lambda)]. \qquad (3.8)$$

The explicit extension of $h(u)$ on $[B, \infty)$ is not required; a construction of this extension is indicated in Rudin (1974, p. 418, Exs. 11 and 12). To apply successive integration by parts, we set

$$k_0(u) = u^{\lambda - 1} e^{ixu^\rho} \qquad (3.9)$$

and define inductively

$$k_{n+1}(u) = -\int_u^{u + \infty e^{i\pi/2\rho}} k_n(z) \, dz, \qquad n = 0, 1, 2, \ldots, \qquad (3.10)$$

where the path of integration is the ray $\arg(z - u) = \pi/2\rho$; cf. Chapter I, equations (4.21)–(4.23). By interchanging the order of integration, it can be verified that

$$k_{n+1}(u) = \frac{(-1)^{n+1}}{n!} \int_u^{u + \infty e^{i\pi/2\rho}} (z - u)^n z^{\lambda - 1} e^{ixz^\rho} \, dz, \qquad (3.11)$$

from which it follows that

$$k_{n+1}(0) = \frac{(-1)^{n+1}}{n!} \int_0^{\infty e^{i\pi/2\rho}} z^{n+\lambda-1} e^{ixz^\rho} \, dz$$

$$= \frac{(-1)^{n+1}}{n! \, \rho} \Gamma\left(\frac{n + \lambda}{\rho}\right) e^{i\pi(n+\lambda)/2\rho} x^{-(n+\lambda)/\rho}. \qquad (3.12)$$

Since $dk_{n+1}/du = k_n(u)$, partial integration gives

$$I_1(x) = \sum_{n=0}^{N-1} (-1)^{n+1} h^{(n)}(0) k_{n+1}(0) + R_N^{(1)}(x)$$

$$= \sum_{n=0}^{N-1} \frac{1}{n! \, \rho} \Gamma\left(\frac{n + \lambda}{\rho}\right) h^{(n)}(0) e^{i\pi(n+\lambda)/2\rho} x^{-(n+\lambda)/\rho} + R_N^{(1)}(x) \qquad (3.13)$$

for any $N \geq 1$, the contributions from the upper limit of integration all vanishing in view of (3.7). The remainder $R_N^{(1)}(x)$ in (3.13) is given by

$$R_N^{(1)}(x) = (-1)^N \int_0^\infty h^{(N)}(u) k_N(u) \, du. \qquad (3.14)$$

The coefficients $h^{(n)}(0)$ in (3.13) can be expressed in terms of the original functions $f(t)$ and $g(t)$ via (3.4) and (3.6). In particular, we have

$$h(0) = g_1(a) f_1(a)^{-\lambda/\rho}. \qquad (3.15)$$

To estimate $R_N^{(1)}(x)$, we first note that a further integration by parts gives

$$R_N^{(1)}(x) = (-1)^{N+1} \left[h^{(N)}(0) k_{N+1}(0) + \int_0^\infty h^{(N+1)}(u) k_{N+1}(u) \, du \right].$$

Hence, by (3.12),

$$|R_N^{(1)}(x)| \leq \frac{1}{N! \, \rho} \Gamma\left(\frac{N + \lambda}{\rho}\right) |h^{(N)}(0)| x^{-(N+\lambda)/\rho}$$

$$+ \int_0^\infty |h^{(N+1)}(u)| \, |k_{N+1}(u)| \, du. \qquad (3.16)$$

We next observe that along the path of integration in (3.11), we have $z = u + \zeta e^{i\pi/2\rho}$, where $u \geq 0$ and $\zeta \geq 0$. Clearly $|z|^{\lambda-1} \leq u^{\lambda-1}$ for $\lambda \leq 1$, and

$$\rho \int_0^u (\eta + \zeta e^{i\pi/2\rho})^{\rho-1} \, d\eta = z^\rho - i\zeta^\rho.$$

Since the imaginary part of the integrand is positive, it follows that $\mathrm{Re}(ixz^\rho) + x\zeta^\rho \leq 0$, which in turn gives $|\exp(ixz^\rho)| \leq \exp(-x\zeta^\rho)$. Consequently,

$$|k_{N+1}(u)| \leq \frac{u^{\lambda-1}}{N!} \int_0^\infty \zeta^N e^{-x\zeta^\rho} \, d\zeta = \frac{1}{N! \, \rho} \Gamma\left(\frac{N+1}{\rho}\right) u^{\lambda-1} x^{-(N+1)/\rho}. \quad (3.17)$$

Inserting (3.17) in (3.16), we obtain

$$|R_N^{(1)}(x)| \leq \frac{1}{N! \, \rho} \left[\Gamma\left(\frac{N+\lambda}{\rho}\right) |h^{(N)}(0)| x^{-(N+\lambda)/\rho} \right.$$
$$\left. + \Gamma\left(\frac{N+1}{\rho}\right) \left\{ \int_0^\infty u^{\lambda-1} |h^{(N+1)}(u)| \, du \right\} x^{-(N+1)/\rho} \right].$$

Since $0 < \lambda \leq 1$, this demonstrates that

$$R_N^{(1)}(x) = O(x^{-(N+\lambda)/\rho}) \qquad \text{for } x > 0. \quad (3.18)$$

We now turn to the second integral

$$I_2(x) = \int_B^\infty u^{\lambda-1} h(u) e^{ixu^\rho} \, du$$

in (3.8). The analysis for this integral is considerably simpler, since here the singular point $u = 0$ is outside the range of integration. The substitution $u^\rho = v$ gives

$$I_2(x) = \frac{1}{\rho} \int_{B^\rho}^\infty v^{\lambda/\rho-1} h(v^{1/\rho}) e^{ixv} \, dv.$$

Since the integrand is infinitely differentiable, by repeated integration by parts we get

$$I_2(x) = \frac{1}{\rho} e^{ixB^\rho} \sum_{n=0}^{M-1} h_1^{(n)}(B^\rho) \left(\frac{i}{x}\right)^{n+1} + R_M^{(2)}(x), \quad (3.19)$$

for any $M \geq 1$, where

$$h_1(v) = v^{\lambda/\rho-1} h(v^{1/\rho}) \quad (3.20)$$

and

$$R_M^{(2)}(x) = \frac{1}{\rho}\left(\frac{i}{x}\right)^M \int_{B^\rho}^{\infty} h_1^{(M)}(v)e^{ixv}\, dv, \tag{3.21}$$

the contribution from the upper limit of integration again vanishing in view of (3.7); cf. Chapter I, Eq. (4.16). By the Riemann–Lebesgue lemma, $R_M^{(2)}(x) = o(x^{-M})$ as $x \to \infty$. Hence, a combination of (3.8), (3.13) and (3.19) gives

$$I(x) \sim e^{ixf(a)} \sum_{n=0}^{\infty} \frac{1}{n!\,\rho}\,\Gamma\!\left(\frac{n+\lambda}{\rho}\right)h^{(n)}(0)e^{i\pi(n+\lambda)/2\rho}x^{-(n+\lambda)/\rho}$$
$$- e^{ixf(b)} \sum_{n=0}^{\infty} \frac{1}{\rho}\,h_1^{(n)}(B^\rho)\left(\frac{i}{x}\right)^{n+1}, \tag{3.22}$$

as $x \to +\infty$. The coefficients $(1/\rho)h_1^{(n)}(B^\rho)$ can be expressed in terms of the original functions $f(t)$ and $g(t)$. Indeed, from (3.4), (3.6), and (3.20), we have

$$\frac{1}{\rho}\,h_1^{(n)}(B^\rho) = \left\{\frac{1}{f'(t)}\frac{d}{dt}\right\}^n \frac{g(t)}{f'(t)}\bigg|_{t=b}; \tag{3.23}$$

cf. Olver (1974b, Eq. (2.3)).

If x is a negative parameter, or if $f'(t)$ is negative, then the asymptotic expansion of $I(x)$ can be obtained by taking complex conjugates or, equivalently, by changing the sign of i throughout.

To illustrate how the stationary phase approximation (3.2) follows from (3.22), we first split the integral (3.1) at $t = c$. Thus

$$I(x) = \int_a^c g(t)e^{ixf(t)}\, dt + \int_c^b g(t)e^{ixf(t)}\, dt.$$

We then replace t by $-t$ in the first integral on the right-hand side. By using (3.22), we find that each integral on the right is asymptotically equal to half of the quantity

$$g(c)\sqrt{\frac{2\pi}{xf''(c)}}\,e^{i[xf(c)+\pi/4]}.$$

This of course establishes (3.2).

Example 4. The function $F(x)$ defined by $F(x) = \int_0^1 e^{ixu^2}\, du$ is exactly of the form given in (3.5) with $f(a) = 0$, $\lambda = B = 1$, $\rho = 2$, and $h(u) = 1$.

Hence we may apply the result in (3.22) directly with $f(b) = 1$. Since $h^{(n)}(0) = 0$ for $n = 1, 2, \ldots$, and since $h_1(v) = v^{-1/2}$ according to (3.20), it follows immediately that

$$F(x) \sim \frac{1}{2}\sqrt{\frac{\pi}{x}}\, e^{i\pi/4} - \frac{i}{2}e^{ix}\sum_{n=0}^{\infty}(-i)^n\,\frac{\Gamma(n+\frac{1}{2})}{\Gamma(\frac{1}{2})}\,\frac{1}{x^{n+1}},$$

as $x \to +\infty$. For an alternative derivation, based on the method of steepest descent (Sec. 4), we refer to Bender and Orszag (1978, pp. 283–285). In a similar manner, we can also show that the integral $G(x) = \int_0^1 e^{ixt^3}\,dt$ has the asymptotic expansion

$$G(x) \sim \Gamma(\tfrac{4}{3})e^{i\pi/6}x^{-1/3} - e^{ix}\sum_{n=0}^{\infty}\frac{\Gamma(n+\frac{2}{3})}{\Gamma(-\frac{1}{3})}(ix)^{-n-1},$$

as $x \to +\infty$; cf. Erdélyi (1956, p. 56).

Example 5.   Consider the integral

$$H(x) = \frac{1}{\pi}\int_0^{\pi} e^{ix(t-\sin t)}\,dt,$$

which can be expressed in terms of the Anger function $\mathbf{J}_x(x)$ and the Weber function $\mathbf{E}_x(x)$; see Watson (1944, p. 308). In our notations, $a = 0$, $b = \pi$, $g(t) \equiv 1$, and $f(t) = t - \sin t$; hence, $\lambda = 1$, $g_1(t) = 1$, $\rho = 3$, and $f_1(t) = (t - \sin t)/t^3$. By definition (3.6),

$$h^{(n)}(u) = \frac{d^{n+1}t}{du^{n+1}}, \qquad n = 0, 1, 2, \ldots,$$

where

$$u^3 = t - \sin t = \frac{t^3}{3!} - \frac{t^5}{5!} + \frac{t^7}{7!} - \cdots.$$

Reverting this series, we obtain $h(0) = 6^{1/3}$, $h''(0) = \frac{1}{10}6$, $h^{(4)}(0) = \frac{6}{70}6^{5/3}$, and $h^{(1)}(0) = h^{(3)}(0) = h^{(5)}(0) = 0$; see Abramowitz and Stegun (1964, p. 16). By (3.23), we also have

$$\frac{1}{\rho}h_1^{(n)}(B^\rho) = \left(\frac{1}{1-\cos t}\frac{d}{dt}\right)^n\frac{1}{1-\cos t}\bigg|_{t=\pi}.$$

Straightforward computation gives

$$\frac{1}{\rho}h_1(B^\rho) = \frac{1}{2}, \qquad \frac{1}{\rho}h_1''(B^\rho) = \frac{1}{2^4}, \qquad \frac{1}{\rho}h_1^{(4)}(B^\rho) = -\frac{1}{2^4}.$$

From (3.22), it follows that

$$H(x) \sim \left[\frac{1}{3}\Gamma\left(\frac{1}{3}\right)e^{i\pi/6}\left(\frac{6}{x}\right)^{1/3} + \frac{i}{60}\left(\frac{6}{x}\right) + \frac{1}{840}\Gamma\left(\frac{5}{3}\right)\left(\frac{6}{x}\right)^{5/3} + \cdots\right]$$

$$+ ie^{inx}\left[-\frac{1}{2x} + \frac{1}{16x^3} - \frac{1}{16x^5} + \cdots\right].$$

This result was given by Olver (1974b), using the summability method (Chapter IV).

Example 6. The study of linear dispersive wave theory begins with the simple harmonic waves given by

$$u(x, t) = ae^{i(\kappa x - \omega t)}, \tag{3.24}$$

where t is time, a is the amplitude, κ is the wave number, and ω is the frequency. The *phase velocity* is the velocity dx/dt for which the phase function $\kappa x - \omega t$ is constant, i.e., $dx/dt = \omega/\kappa$. The term *dispersive* is referred to situations in which ω is a function of κ.

A continuous superposition of (3.24) yields

$$u(x, t) = \int_{-\infty}^{\infty} a(\kappa)e^{i\{\kappa x - \omega(\kappa)t\}} \, d\kappa. \tag{3.25}$$

To investigate the behavior of $u(x, t)$ for large positive t and fixed $\xi = x/t$, we apply the principle of stationary phase. The stationary points $\kappa = k$ are determined from $\omega'(k) = x/t$. The quantity $\omega'(k)$ is referred to as the *group velocity*. If there is only one such point, then by (3.2) and (3.3),

$$u(x, t) \sim a(k)\sqrt{\frac{2\pi}{t|\omega''(k)|}} \, e^{ikx - i\omega(k)t - i(\pi/4)\,\mathrm{sgn}\,\omega''(k)}, \tag{3.26}$$

as $t \to +\infty$. If there are more than one stationary points, then the asymptotic formula of $u(x, t)$ is obtained by adding up terms of the form (3.26).

To provide a concrete example of (3.25), we consider the Cauchy problem for the *Klein–Gordon* equation

$$u_{tt} - \gamma^2 u_{xx} + c^2 u = 0, \qquad t > 0, \quad -\infty < x < \infty, \qquad (3.27)$$

with specified initial data $u(x, 0)$ and $u_t(x, 0)$. Let $U(\kappa, t)$ denote the Fourier transform of $u(x, t)$, i.e.,

$$U(\kappa, t) = \frac{1}{\sqrt{2\pi}} \int_{-\infty}^{\infty} e^{i\kappa x} u(x, t) \, dx.$$

The transformed equation is given by

$$\frac{\partial^2 U}{\partial t^2} + (\gamma^2 \kappa^2 + c^2) U(\kappa, t) = 0,$$

and its solution is

$$U(\kappa, t) = a_+(\kappa) \exp(i\sqrt{\gamma^2\kappa^2 + c^2}\, t) + a_-(\kappa) \exp(-i\sqrt{\gamma^2\kappa^2 + c^2}\, t),$$

where $a_\pm(\kappa)$ are determined by the initial data. Taking the Fourier inverse transform, we obtain

$$
\begin{aligned}
u(x, t) = {}& \frac{1}{\sqrt{2\pi}} \int_{-\infty}^{\infty} a_+(\kappa) \exp\{i[\omega_+(\kappa)t - x\kappa]\} \, d\kappa \\
& + \frac{1}{\sqrt{2\pi}} \int_{-\infty}^{\infty} a_-(\kappa) \exp\{i[\omega_-(\kappa)t - x\kappa]\} \, d\kappa,
\end{aligned}
\qquad (3.28)
$$

where $\omega_\pm(\kappa) = \pm\sqrt{\gamma^2\kappa^2 + c^2}$. Both integrals in (3.28) are of the form (3.25), and hence to each of them we can apply the asymptotic formula (3.26).

4.   Method of Steepest Descents

This is a method for deriving asymptotic expansions of contour integrals of the form

$$I(\lambda) = \int_C g(z) e^{\lambda f(z)} \, dz, \qquad (4.1)$$

where $f(z)$ and $g(z)$ are analytic functions and λ is again a large positive parameter. It was introduced by Debye (1909) in a paper concerning Bessel functions of large order. There are now so many references to this method in the literature that, by name, it is probably the best known procedure for finding asymptotic behavior of integrals. However, most descriptions of this procedure are so sketchy that it is probably also the most difficult to understand among all the classical methods. Debye's basic idea is to deform the contour C into a new path of integration C' so that the following conditions hold:

(i) C' passes through one or more zeros of $f'(z)$;

(ii) the imaginary part of $f(z)$ is constant on C'.

To obtain a geometrical interpretation of the new path of integration C', we write

$$f(z) = u(x, y) + iv(x, y), \qquad (4.2)$$

where $z = x + iy$ and u and v are real. If u is treated as a third axis orthogonal to both x and y, then the equation $u = u(x, y)$ defines a surface S in the (x, y, u) space. Suppose that $z_0 = x_0 + iy_0$ is a zero of $f'(z)$. Then, by the Cauchy–Riemann equations, $f'(z) = u_x - iu_y$. Thus $f'(z_0) = 0$ implies

$$u_x(x_0, y_0) = u_y(x_0, y_0) = 0, \qquad (4.3)$$

i.e., (x_0, y_0) is a critical point of $u(x, y)$. Since u is a harmonic function, $u(x, y)$ cannot have a maximum or a minimum at an interior point. Therefore, (x_0, y_0) must be a saddle point of $u(x, y)$, see Figure 2.2. For this reason, we call z_0 a *saddle point* of $f(z)$.

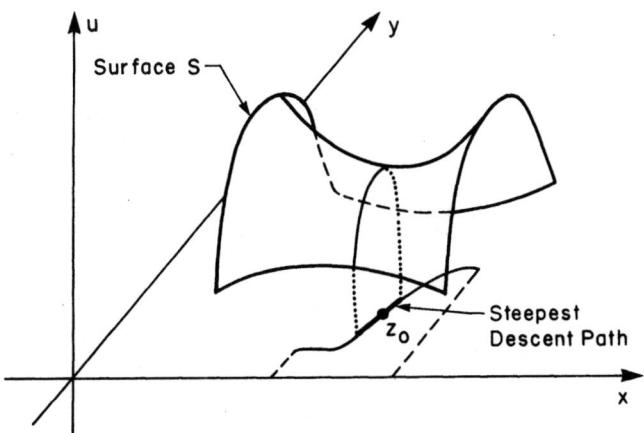

Fig. 2.2 Steepest descent path.

The term "steepest descent" stems from the condition that $v(x, y) =$ constant on the new path C'; see condition (ii) above. To show this, we let $x = x(s)$, $y = y(s)$ be a set of parametric equations of a differentiable curve γ. Then

$$x = x(s), \qquad y = y(s), \qquad u = u(s) = u(x(s), y(s)) \tag{4.4}$$

form a set of parametric equations of a curve on the surface S. For simplicity, we suppose that the parameter s is the arc length of the curve $x = x(s)$, $y = y(s)$ in the (x, y) plane. Thus

$$\left(\frac{dx}{ds}\right)^2 + \left(\frac{dy}{ds}\right)^2 = 1, \tag{4.5}$$

and we may define θ by

$$\cos\theta = \frac{dx}{ds}, \qquad \sin\theta = \frac{dy}{ds}. \tag{4.6}$$

The change of u along γ is represented by

$$\frac{du}{ds} = \frac{\partial u}{\partial x}\frac{dx}{ds} + \frac{\partial u}{\partial y}\frac{dy}{ds} = u_x \cos\theta + u_y \sin\theta. \tag{4.7}$$

The steepness of the curve (4.4) on S is measured by the angle α formed by the u-axis and that curve. Using the formula for the angle between two lines, we have from (4.4)

$$\cos\alpha = \frac{\dfrac{du}{ds}}{\sqrt{1 + \left(\dfrac{du}{ds}\right)^2}}. \tag{4.8}$$

If we consider all differentiable curves through a point on S, then

$$\frac{d}{d\theta}\cos\alpha = \frac{1}{\left[1 + \left(\dfrac{du}{ds}\right)^2\right]^{3/2}}(-u_x \sin\theta + u_y \cos\theta).$$

Hence $d(\cos\alpha)/d\theta = 0$ when

$$-u_x \sin\theta + u_y \cos\theta = 0.$$

By the Cauchy–Riemann equations, the last equation is equivalent to

$$\frac{dv}{ds} = v_x \cos\theta + v_y \sin\theta = 0. \tag{4.9}$$

At values for which $d(\cos \alpha)/d\theta = 0$, we also have

$$\frac{d^2}{d\theta^2} \cos \alpha = - \frac{1}{\left[1 + \left(\dfrac{du}{ds}\right)^2\right]^{3/2}} \frac{du}{ds}.$$

It is now clear that $\cos \alpha$ will have an absolute maximum when $dv/ds = 0$ and $du/ds > 0$, and $\cos \alpha$ will have an absolute minimum when $dv/ds = 0$ and $du/ds < 0$. When $du/ds = 0$, neither of these conclusions need be valid.

If we vary the point (x, y, u), then a curve $v(x, y) = $ constant $(dv/ds = 0)$ will have the property that, as long as we strike no singularities of u or points at which $du/ds = 0$, the tangent vector of the curve $x = x(s)$, $y = y(s)$, $u = u(s)$ will have maximum or minimum inclination with the u-axis. These lead to curves of steepest ascent or steepest descent on the surface S. It is from this property that Debye's method derives its name.

The shape of the surface S can also be represented on the (x, y) plane by drawing the level curves on which u is constant. The domains where $u(x, y) > u(x_0, y_0)$ are called *hills*, and those where $u(x, y) < u(x_0, y_0)$ are called *valleys*. The level curve which passes through the saddle point, i.e., $u(x, y) = u(x_0, y_0)$, separates the nearby part of S into hills and valleys. Suppose that z_0 is a saddle point of *order* $m - 1$, $m \geq 2$, i.e.,

$$f'(z_0) = f''(z_0) = \cdots = f^{(m-1)}(z_0) = 0$$

but $f^{(m)}(z_0) = ae^{i\varphi}$, $a > 0$, and φ real. If $z = z_0 + re^{i\theta}$, then

$$f(z) = f(z_0) + \frac{r^m}{m!} ae^{i(m\theta + \varphi)} + \cdots.$$

Thus, near the saddle point z_0 the level curves and steepest paths are roughly the same as

$$u(x, y) = u(x_0, y_0) + \frac{r^m}{m!} a \cos(m\theta + \varphi)$$

$$v(x, y) = v(x_0, y_0) + \frac{r^m}{m!} a \sin(m\theta + \varphi).$$

Consequently, we can easily determine the directions of the curves $u = $ constant and $v = $ constant as they emanate from the point (x_0, y_0).

The directions of the level curves $u = $ constant are given by the solutions of the equation $\cos(m\theta + \varphi) = 0$, i.e.,

$$\theta = -\frac{\varphi}{m} + \left(k + \frac{1}{2}\right)\frac{\pi}{m}, \qquad k = 0, 1, \ldots, 2m - 1.$$

Analogously, the directions of the steepest paths satisfy $\sin(m\theta + \varphi) = 0$, i.e.,

$$\theta = -\frac{\varphi}{m} + \frac{k\pi}{m}, \qquad k = 0, 1, \ldots, 2m - 1.$$

Therefore there are $2m$ steepest directions from z_0, m directions of steepest ascent and m directions of steepest descent. Near z_0 the level curves $u = u(x_0, y_0)$ divide S into m valleys below the saddle point and m hills above the saddle point. The paths of integration are in the valleys. Figure 2.3 indicates the hills and valleys in the two simplest cases $m = 2$ and $m = 3$ with $\varphi = 0$.

We now suppose that the original path of integration can be deformed into a steepest path $v(x, y) = $ constant $ = \operatorname{Im} f(z_0)$. On this path, we have

$$f(z) = f(z_0) - \tau, \tag{4.10}$$

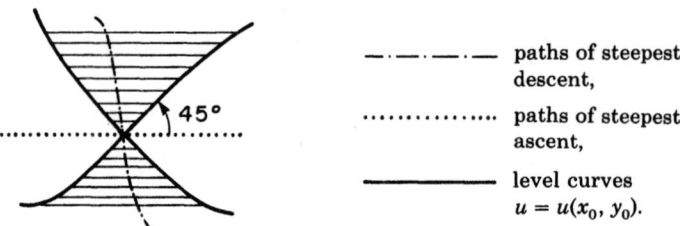

— · — · — paths of steepest descent,

············· paths of steepest ascent,

———————— level curves $u = u(x_0, y_0)$.

(a)   Saddle point of order 1.

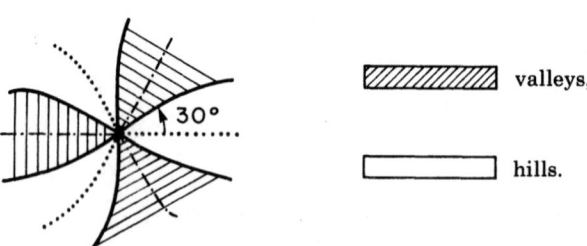

⧄⧄⧄⧄⧄⧄ valleys,

☐ hills.

(b)   Saddle point of order 2.

Fig. 2.3   Behaviour near a saddle point.

where τ is real and is either monotonically increasing or monotonically decreasing. The integrand in (4.1) becomes $g(z)e^{\lambda f(z_0) - \lambda\tau}$. On a path where $\tau \to -\infty$, the integral may be divergent. For this reason we choose paths on which τ is positive and increasing. These are the paths of steepest descent from saddle points. We thus suppose that the original path of integration can be deformed into an equivalent path consisting of paths of steepest descent through some of the saddle points. The saddle points involved in this deformation are called the *relevant saddle points*. Furthermore, if it is possible to express the original integral as a sum of integrals along paths of steepest descent from saddle points, then our problem is reduced to finding the asymptotic behavior of integrals of the form

$$e^{\lambda f(z_0)} \int_0^T g(z) \frac{dz}{d\tau} e^{-\lambda\tau} \, d\tau, \tag{4.11}$$

where T is a positive real number. In most cases, T is at $+\infty$, unless, for example, the integration path strikes another saddle point. The full contribution to the asymptotic expansion of the original integral (4.1) can be obtained by adding the contributions from all relevant saddle points. However, only the highest relevant saddle points, i.e. saddle points at which Re $f(z_0)$ is greatest, are really needed for the expansion. (Note that sometimes there may be more than one dominant relevant saddle point.)

To the integral in (4.11), we can apply Watson's lemma. To do this, we first require a series expansion for $g(z) \, dz/d\tau$ in ascending powers of τ. If z_0 is a saddle point of order $m - 1$, then

$$f(z) = f(z_0) - (z - z_0)^m f_1(z) = f(z_0) - \sum_{n=0}^{\infty} a_n(z - z_0)^{n+m}, \tag{4.12}$$

where $a_0 \neq 0$. Comparing this with (4.10), we have

$$\tau = (z - z_0)^m f_1(z). \tag{4.13}$$

Put $v^m = \tau$ and let ω denote the angle between the real axis and the tangent line to the steepest descent path at z_0, i.e.,

$$\omega = \lim\{\arg(z - z_0)\}, \quad z \to z_0 \quad \text{along the steepest descent path.} \tag{4.14}$$

Then, for small $|z - z_0|$, (4.12) and (4.13) give

$$v = a_0^{1/m}(z - z_0)\left\{1 + \frac{a_1}{ma_0}(z - z_0) + \cdots\right\}, \tag{4.15}$$

where

$$a_0^{1/m} = \exp\left\{\frac{1}{m}\left[\log|a_0| + i \arg a_0\right]\right\} \tag{4.16}$$

and $\arg a_0$ is chosen to satisfy

$$|\arg \lambda + m\omega + \arg a_0| < \frac{\pi}{2}. \tag{4.17}$$

(The asymptotic variable λ in (4.1) need not be positive, and here we shall allow it to be complex.) From (4.15), it follows that v is a single-valued analytic function of z in a neighborhood of z_0, and dv/dz does not vanish at z_0. By the inverse function theorem (Copson 1935, §6.22),

$$z - z_0 = \sum_{s=1}^{\infty} \alpha_s v^s = \sum_{s=1}^{\infty} \alpha_s \tau^{s/m}, \tag{4.18}$$

where

$$\alpha_1 = \frac{1}{a_0^{1/m}}, \qquad \alpha_2 = -\frac{a_1}{ma_0^{1+2/m}}, \qquad \alpha_3 = \frac{(m+3)a_1^2 - 2ma_0a_2}{2m^2a_0^{2+3/m}}. \tag{4.19}$$

Furthermore, for small τ, $g(z)\,dz/d\tau$ has a convergent expansion of the form

$$g(z)\frac{dz}{d\tau} = \sum_{s=0}^{\infty} c_s \tau^{(s+1-m)/m}, \tag{4.20}$$

where

$$c_0 = \frac{b_0}{ma_0^{1/m}}, \qquad c_1 = \left\{\frac{b_1}{m} - \frac{2a_1b_0}{m^2a_0}\right\}\frac{1}{a_0^{2/m}},$$

$$c_2 = \left\{\frac{b_2}{m} - \frac{3a_1b_1}{m^2a_0} + [(m+3)a_1^2 - 2ma_0a_2]\frac{3b_0}{2m^3a_0^2}\right\}\frac{1}{a_0^{3/m}}, \tag{4.21}$$

and $b_s = g^{(s)}(z_0)/s!$; cf. Eqs. (1.14) and (1.15). The asymptotic expansion of the integral in (4.11) can now be obtained directly from Watson's lemma, i.e., by inserting (4.20) in (4.11) and integrating term-by-term.

Example 7. The Airy integral given by

$$\mathrm{Ai}(z) = \frac{1}{2\pi i}\int_L \exp\left(\frac{1}{3}t^3 - zt\right)dt, \tag{4.22}$$

where L is any contour which begins at infinity in the sector $-\pi/2 <$ arg $t < -\pi/6$ and ends at infinity in the sector $\pi/6 <$ arg $t < \pi/2$, plays an important role in "uniform asymptotic expansions" given later. For this reason, we shall give a complete discussion of the asymptotic behavior of this function and a summary of some of the analytical properties that it possesses.

From (4.22), it can be verified that Ai(z) has the Maclaurin expansion

$$\text{Ai}(z) = 3^{-2/3} \left[\sum_{n=0}^{\infty} \frac{z^{3n}}{3^{2n} n! \, \Gamma(n + \frac{2}{3})} - 3^{-2/3} \sum_{n=0}^{\infty} \frac{z^{3n+1}}{3^{2n} n! \, \Gamma(n + \frac{4}{3})} \right]. \quad (4.23)$$

Thus, Ai(z) is an entire function. Furthermore, by integration by parts it can be verified that Ai(z) is a solution of

$$\frac{d^2 w}{dz^2} - zw = 0. \quad (4.24)$$

Clearly Ai(ωz) and Ai($\omega^2 z$) are also solutions of this differential equation, where $\omega = \exp(2\pi i/3)$. With the aid of Cauchy's theorem, we can show from (4.22) that the three solutions are connected by the relation

$$\text{Ai}(z) + \omega \, \text{Ai}(\omega z) + \omega^2 \, \text{Ai}(\omega^2 z) = 0. \quad (4.25)$$

Instead of Ai(ωz) and Ai($\omega^2 z$), the function

$$\text{Bi}(z) = i\omega^2 \, \text{Ai}(\omega^2 z) - i\omega \, \text{Ai}(\omega z) \quad (4.26)$$

is often used as the second solution. It has the advantage of being real when z is real.

To put the integral (4.22) in the form of (4.1), we suppose for a moment that z is real and positive, and make the substitution $t = z^{1/2} u$. This yields

$$\text{Ai}(z) = \frac{z^{1/2}}{2\pi i} \int_L e^{\lambda f(u)} \, du, \quad (4.27)$$

where $\lambda = z^{3/2}$ and $f(u) = \frac{1}{3} u^3 - u$. By analytic continuation, it follows that (4.27) is in fact valid for $-\pi/3 <$ arg $z < \pi/3$. The saddle points of the integrand in (4.27) are at $u = \pm 1$. Putting $u = s + it$, we have

$$f(u) = \frac{1}{3} s^3 - st^2 - s + i(s^2 t - t - \frac{1}{3} t^3). \quad (4.28)$$

Since Im $f(\pm 1) = 0$, if $f(u) = U + iV$ then $V = 0$ implies either $t = 0$ or $t^2 - 3s^2 + 3 = 0$. We must now find the directions in which U decreases on the curves. On $t = 0$, we have $U = \frac{1}{3}s^3 - s$, which has a local minimum at $s = 1$ and a local maximum at $s = -1$. Thus, near $s = 1$, $t = 0$ is a steepest ascent curve. The other equation $t^2 - 3s^2 + 3 = 0$ represents two branches of a hyperbola with the asymptotes $t = \pm\sqrt{3}s$. On this hyperbola, $U = -\frac{8}{3}s^3 + 2s$, which decreases for $s > 1$. From Figure 2.4, it is clear that the branch of the hyperbola on the right half-plane is our desired path of steepest descent through the saddle point $u = 1$. (In Figure 2.4, arrows indicate the direction in which U decreases.)

We may now deform the original contour L into this path of steepest descent, and write

$$\text{Ai}(z) = \frac{z^{1/2}}{2\pi i}\left(\int_1^{\infty e^{\pi i/3}} - \int_1^{\infty e^{-\pi i/3}}\right)e^{\lambda f(u)}\,du. \qquad (4.29)$$

In both integrals above, $f(u) - f(1)$ is real and has a maximum at $u = 1$. Also, $f(u) - f(1)$ is decreasing as u moves away from $u = 1$. Put

$$\tau = f(1) - f(u) = -\frac{1}{3}u^3 + u - \frac{2}{3} = -(u-1)^2 - \frac{1}{3}(u-1)^3. \qquad (4.30)$$

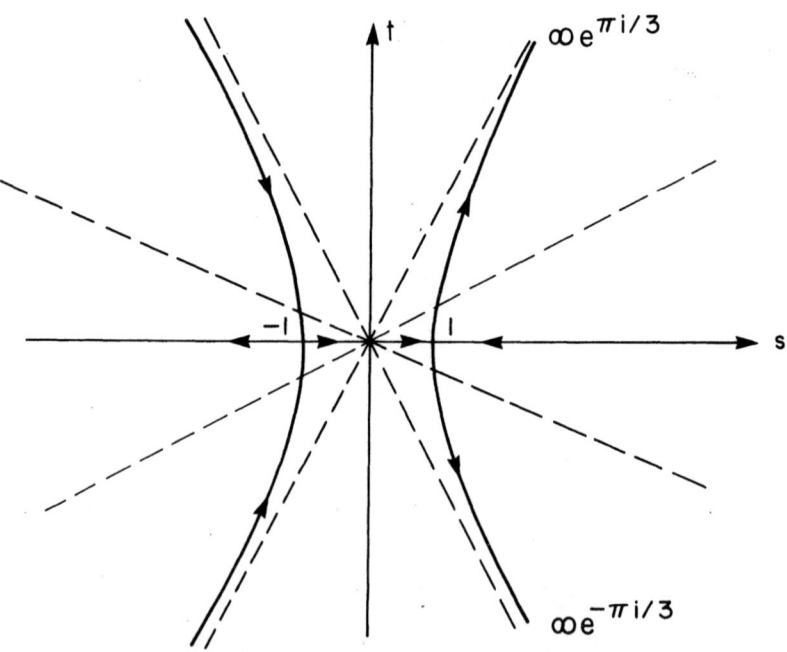

Fig. 2.4 Steepest paths for Ai(z).

On our steepest descent path, τ is real; cf. Eq. (4.10). From (4.30), we have

$$\pm i\tau^{1/2} = (u - 1)[1 + \tfrac{1}{3}(u - 1)]^{1/2}, \qquad (4.31)$$

where $[\ldots]^{1/2}$ is that branch which reduces to 1 at $u = 1$. By Lagrange's formula for the reversion of series (see Copson (1935, §6.23)),

$$u^{\pm} = 1 + \sum_{n=1}^{\infty} \alpha_n (\pm i\tau^{1/2})^n, \qquad (4.32)$$

where

$$\alpha_n = \frac{1}{n!}\frac{d^{n-1}}{du^{n-1}}\left[1 + \frac{1}{3}(u - 1)\right]^{-n/2}\Bigg|_{u=1}. \qquad (4.33)$$

Since u^+ enters the first quadrant for increasing τ, we must take u^+ for the first integral in (4.29). Similarly, we take u^- for the second integral there. Equation (4.29) then becomes

$$\text{Ai}(z) = \frac{z^{1/2}}{2\pi i}e^{-2\lambda/3}\int_0^\infty \left(\frac{du^+}{d\tau} - \frac{du^-}{d\tau}\right)e^{-\lambda\tau}\,d\tau. \qquad (4.34)$$

Since $n\alpha_n$ is the coefficient of $(u - 1)^{n-1}$ in the Taylor expansion of $[1 + \tfrac{1}{3}(u - 1)]^{-n/2}$ at $u = 1$, it is easily shown that

$$\alpha_n = \frac{(-1)^{n-1}\Gamma(\tfrac{3}{2}n - 1)}{n!\,\Gamma(\tfrac{1}{2}n)3^{n-1}}. \qquad (4.35)$$

From this, it follows that

$$\frac{du^+}{d\tau} - \frac{du^-}{d\tau} = 2i\sum_{m=0}^{\infty}\frac{(-1)^m\Gamma(3m + \tfrac{1}{2})(m + \tfrac{1}{2})}{(2m + 1)!\,\Gamma(m + \tfrac{1}{2})9^m}\tau^{m-1/2}. \qquad (4.36)$$

By Watson's lemma,

$$\text{Ai}(z) \sim \frac{1}{2\pi z^{1/4}}\exp\left(-\frac{2}{3}z^{3/2}\right)\sum_{m=0}^{\infty}\frac{(-1)^m\Gamma(3m + \tfrac{1}{2})}{(2m)!\,9^m}z^{-3m/2}, \qquad (4.37)$$

as $z \to \infty$, uniformly in arg z for $|\arg z| \leq \pi/3 - \Delta, \Delta > 0$.

This example is purely illustrative, since we have already indicated in Chapter I, Exercise 10, a much simpler method by which the expansion (4.37) can be derived for the wider range of validity $|\arg z| \leq \pi - \Delta, \Delta > 0$. To obtain an expansion in a sector that includes arg $z = \pi$, one can use (4.25). Indeed, if $\pi/3 < \arg z < 5\pi/3$, then

$\omega = e^{2\pi i/3}$ implies $-\pi/3 < \arg(\omega^{-1}z) < \pi$ and $-\pi < \arg(\omega^{-2}z) < \pi/3$. From (4.25), we have $\mathrm{Ai}(\omega^{-2}z) + \omega\,\mathrm{Ai}(\omega^{-1}z) + \omega^2\,\mathrm{Ai}(z) = 0$, which implies $\mathrm{Ai}(z) = -\omega^{-2}\,\mathrm{Ai}(\omega^{-2}z) - \omega^{-1}\,\mathrm{Ai}(\omega^{-1}z)$. Hence

$$
\begin{aligned}
\mathrm{Ai}(z) \sim {}& \frac{1}{2\pi z^{1/4}} \exp\!\left(-\frac{2}{3}z^{3/2}\right) \sum_{m=0}^{\infty} \frac{(-1)^m \Gamma(3m + \frac{1}{2})}{(2m)!\, 9^m}\, z^{-3m/2} \\
& + \frac{i}{2\pi z^{1/4}} \exp\!\left(\frac{2}{3}z^{3/2}\right) \sum_{m=0}^{\infty} \frac{\Gamma(3m + \frac{1}{2})}{(2m)!\, 9^m}\, z^{-3m/2},
\end{aligned}
\tag{4.38}
$$

an expansion that is uniformly valid in $\pi/3 + \Delta \le \arg z \le 5\pi/3 - \Delta$.

Observe that both expansions in (4.37) and (4.38) are valid in $\pi/3 + \Delta \le \arg z \le \pi - \Delta$, where Δ is positive but can be arbitrarily small. However, there is no inconsistency, since the first term on the right-hand side of (4.38) is *dominant* (exponentially increasing) and the second term is *recessive* (exponentially decreasing) in the sector $\pi/3 < \arg z < \pi$. The roles of these two terms are interchanged in the sector $-\pi/3 < \arg z < \pi/3$. The ray $\arg z = \pi/3$ is called a *Stokes line*. As $z \to \infty$ along this ray, the exponential factors in both terms are bounded, and bounded away from zero. Note also that there is a discontinuous change in the coefficients of the asymptotic expansion of $\mathrm{Ai}(z)$ when $\arg z$ changes continuously across $\arg z = \pi/3$; see equation (4.38). This circumstance was first discovered by Stokes, and is now known as *Stokes' phenomenon*.

Example 8.   The Hankel function $H_\nu^{(1)}(z)$ has the integral representation

$$
H_\nu^{(1)}(z) = \frac{1}{\pi i} \int_{-\infty}^{\infty + \pi i} e^{z \sinh t - \nu t}\, dt.
\tag{4.39}
$$

If $z = \nu a$, where a is a positive constant, then

$$
H_\nu^{(1)}(\nu a) = \frac{1}{\pi i} \int_{-\infty}^{\infty + \pi i} e^{\nu f(t)}\, dt,
\tag{4.40}
$$

where

$$
f(t) = a \sinh t - t.
\tag{4.41}
$$

Similarly,

$$
H_\nu^{(2)}(\nu a) = -\frac{1}{\pi i} \int_{-\infty}^{\infty - \pi i} e^{\nu f(t)}\, dt;
\tag{4.42}
$$

see Olver (1974a, p. 240).

With v regarded as the large parameter, the saddle points of the integrand in (4.40) and (4.42) are the roots of the equation

$$a \cosh t = 1. \tag{4.43}$$

It is easily seen that in the region $-\pi < \operatorname{Im} t < \pi$, there are two real saddle points if $0 < a < 1$, two purely imaginary saddle points if $a > 1$, and exactly one saddle point, at $t = 0$, if $a = 1$. Write $t = u + iv$. Then

$$f(t) = (a \sinh u \cos v - u) + i(a \cosh u \sin v - v). \tag{4.44}$$

Case (i): $0 < a < 1$. Let $\alpha > 0$ be defined by $\cosh \alpha = a^{-1}$. The saddle points in $|\operatorname{Im} t| < \pi$ are located at $t = \pm\alpha$. Since $f(\pm\alpha) = \pm(\tanh \alpha - \alpha)$ is real, the steepest curves through the saddle points are given by

$$\cosh u \sin v - v \cosh \alpha = 0.$$

They consist of the real axis $v = 0$ and the two curves passing through $t = \pm\alpha$ and approaching the horizontal lines $v = +\pi$ and $v = -\pi$ from below and from above, respectively; see Figure 2.5 (arrows indicate direction of descent). It is clear that the path of integration in (4.40) can be deformed into the curve ABCD, and that the curve ABCE is equivalent to the integration path in (4.42). On these two curves, the maximum of $f(t)$ occurs at $t = -\alpha$, and its value is given by $\alpha - \tanh \alpha$. If

$$\tau = \alpha - \tanh \alpha - a \sinh t + t, \tag{4.45}$$

then τ is real on the curves ABCD and ABCE. Furthermore, as t travels along the contour from $-\infty$ to $\infty + i\pi$, τ decreases from $+\infty$ to 0 and

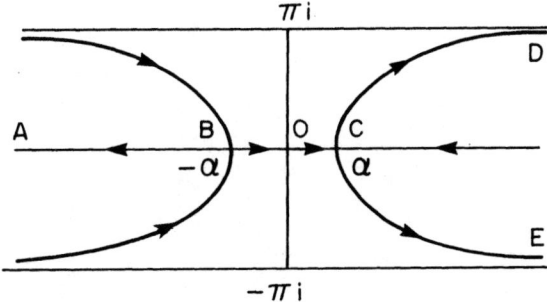

Fig. 2.5 Debye's contours.

then increases to $+\infty$. Thus, for each positive τ, there are two values of t, which we call t_1 and t_2; t_1 corresponds to the portion of the contour along the negative real-axis from $-\infty$ to $-\alpha$, and t_2 corresponds to the portion from $t = -\alpha$ to $\infty + i\pi$. In terms of τ, the integral (4.40) becomes

$$H_\nu^{(1)}(\nu a) = e^{\nu(\alpha - \tanh \alpha)} \frac{1}{\pi i} \int_0^\infty e^{-\nu\tau} \left(\frac{dt_2}{d\tau} - \frac{dt_1}{d\tau} \right) d\tau. \qquad (4.46)$$

We now discuss the expansions of t_1 and t_2 in powers of τ. From (4.45), we have

$$\tau = \frac{1}{2} \tanh \alpha \, (t + \alpha)^2 \left[1 - \frac{2}{3!} \coth \alpha \, (t + \alpha) + \frac{2}{4!} (t + \alpha)^2 - \cdots \right],$$

and

$$\tau^{1/2} = \pm \sqrt{\frac{1}{2} \tanh \alpha} \, (t + \alpha) \left[1 - \frac{2}{3!} \coth \alpha \, (t + \alpha) + \frac{2}{4!} (t + \alpha)^2 - \cdots \right]^{1/2},$$

$$(4.47)$$

where the sign $+$ or $-$ depends on whether $\mathrm{Re}\, t > -\alpha$ or $\mathrm{Re}\, t < -\alpha$. By Lagrange's inversion theorem,

$$t_1 + \alpha = \sum_{m=0}^\infty (-1)^{m+1} \frac{a_m}{m+1} \tau^{(m+1)/2},$$

$$t_2 + \alpha = \sum_{m=0}^\infty \frac{a_m}{m+1} \tau^{(m+1)/2} \qquad\qquad (4.48)$$

for sufficiently small values of τ, where

$$m! \, a_m = \left(\frac{2}{\tanh \alpha} \right)^{(m+1)/2} \frac{d^m}{dt^m} \left[1 - \frac{2}{3!} \coth \alpha \, (t + \alpha) + \cdots \right]^{-(m+1)/2} \Bigg|_{t=-\alpha}.$$

$$(4.49)$$

Now

$$\frac{dt_2}{d\tau} - \frac{dt_1}{d\tau} = \sum_{n=0}^\infty a_{2n} \tau^{n-1/2},$$

and, since $d\tau/dt = -a \cosh t + 1$, $dt_2/d\tau - dt_1/d\tau = O(1)$, as $\tau \to +\infty$. Hence, by Watson's lemma,

$$H_\nu^{(1)}(\nu a) \sim -\frac{i}{\pi} e^{\nu(\alpha - \tanh \alpha)} \sum_{n=0}^\infty a_{2n} \frac{\Gamma(n + \frac{1}{2})}{\nu^{n+1/2}}, \qquad \text{as } \nu \to +\infty. \quad (4.50)$$

In a similar manner, we have

$$H_\nu^{(2)}(va) \sim \frac{i}{\pi} e^{\nu(\alpha - \tanh \alpha)} \sum_{n=0}^{\infty} a_{2n} \frac{\Gamma(n + \frac{1}{2})}{\nu^{n+1/2}}, \qquad \text{as } \nu \to +\infty. \qquad (4.51)$$

In fact, since $H_\nu^{(1)}(x) = \overline{H_\nu^{(2)}(\bar{x})}$, (4.51) follows immediately from (4.50).

The asymptotic expansion of $J_\nu(va)$, however, cannot be obtained from (4.50) and (4.51), even though we have the connection formula

$$J_\nu(x) = \tfrac{1}{2}[H_\nu^{(1)}(x) + H_\nu^{(2)}(x)]. \qquad (4.52)$$

Inserting (4.50) and (4.51) into (4.52) yields only a null series. This indicates that $J_\nu(x)$ is exponentially small in comparison to the leading terms in (4.50) and (4.51). To derive the asymptotic expansion of $J_\nu(va)$, we use Schläfli's integral representation (see Olver (1974a, p. 58))

$$J_\nu(va) = \frac{1}{2\pi i} \int_{\infty - \pi i}^{\infty + \pi i} e^{\nu f(t)} \, dt, \qquad (4.53)$$

where $f(t)$ is as given in (4.41). From the above discussion, it is clear that the relevant saddle point is at $t = \alpha$ and the steepest descent path is the curve ECD in Figure 2.5. The maximum value of $f(t)$ is $\tanh \alpha - \alpha$ and occurs at $t = \alpha$. Put

$$\tau = \tanh \alpha - \alpha - a \sinh t + t. \qquad (4.54)$$

It is readily verified that τ is real along the curve ECD, and that, as t travels along this curve from E to D, τ decreases from $+\infty$ to 0, and then increases to $+\infty$. Thus, for each positive τ, there are two values of t, which we call t_1 and t_2, such that $\operatorname{Im} t_1 > 0$ and $\operatorname{Im} t_2 < 0$. In terms of τ, (4.53) can be written as

$$J_\nu(va) = e^{\nu(\tanh \alpha - \alpha)} \frac{1}{2\pi i} \int_0^\infty e^{-\nu \tau} \left(\frac{dt_1}{d\tau} - \frac{dt_2}{d\tau} \right) d\tau. \qquad (4.55)$$

We now consider the expansions of t_1 and t_2 in powers of τ. From (4.54), we have

$$\tau = -\frac{1}{2} \tanh \alpha \, (t - \alpha)^2 - \frac{1}{3!}(t - \alpha)^3 - \frac{1}{4!} \tanh \alpha \, (t - \alpha)^4 - \cdots$$

and hence

$$\tau^{1/2} = \pm i \sqrt{\frac{1}{2} \tanh \alpha} \, (t - \alpha) \left[1 + \frac{2}{3!} \coth \alpha \, (t - \alpha) + \cdots \right]^{1/2}.$$

Reversion of the last expansion gives

$$t_1 - \alpha = \sum_{m=0}^{\infty} i^{m+1} \frac{b_m}{m+1} \tau^{(m+1)/2},$$

$$t_2 - \alpha = \sum_{m=0}^{\infty} (-i)^{m+1} \frac{b_m}{m+1} \tau^{(m+1)/2},$$

where

$$m! \, b_m = \left(\frac{2}{\tanh \alpha}\right)^{(m+1)/2} \frac{d^m}{dt^m} \left[1 + \frac{2}{3!} \coth \alpha \, (t-\alpha) + \cdots\right]^{-(m+1)/2} \Bigg|_{t=\alpha}$$

Consequently,   (4.56)

$$\frac{dt_1}{d\tau} - \frac{dt_2}{d\tau} = i \sum_{n=0}^{\infty} (-1)^n b_{2n} \tau^{n-1/2}.$$

By application of Watson's lemma, we now derive the first of Debye's results:

$$J_\nu(\nu a) \sim \frac{1}{2\pi} e^{\nu(\tanh \alpha - \alpha)} \sum_{n=0}^{\infty} (-1)^n b_{2n} \frac{\Gamma(n+\frac{1}{2})}{\nu^{n+1/2}}, \qquad \text{as } \nu \to +\infty. \quad (4.57)$$

Case (ii): $a > 1$. Let $\cos \alpha = a^{-1}$, $0 < \alpha < \pi/2$. The relevant saddle points are $t = \pm i\alpha$. For the integral associated with $H_\nu^{(1)}(\nu a)$, we need consider only the saddle point $t = i\alpha$. The steepest curves in this case are given by

$$a \cosh u \sin v - v = a \sin \alpha - \alpha,$$

or equivalently,

$$\cosh u = \frac{\sin \alpha + (v - \alpha) \cos \alpha}{\sin v}. \qquad (4.58)$$

For $v \in (0, \pi)$, the minimum of $\sin \alpha + (v - \alpha) \cos \alpha - \sin v$ is zero, occurring at $v = \alpha$. Thus,

$$\sin \alpha + (v - \alpha) \cos \alpha \geq \sin v, \qquad 0 < v < \pi.$$

From this, it follows that for each v in $(0, \pi)$, there are two real values of u satisfying (4.58). These two values are equal but opposite in sign, and coincide only when $v = \alpha$. Furthermore, they are infinite when $v = 0$ or $v = \pi$. The curves given by (4.58) are shown in the upper half of Figure

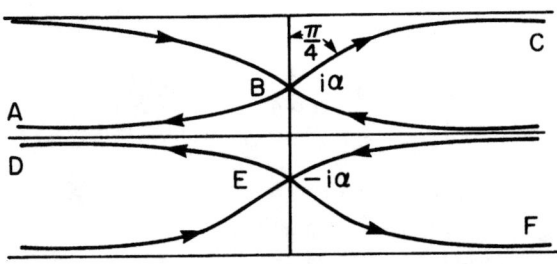

Fig. 2.6

2.6 (arrows indicating directions of descent). If

$$\tau = i(a \sin \alpha - \alpha) - (a \sinh t - t), \qquad (4.59)$$

then τ is real on the curve ABC, and, as t travels along this curve from A to C, τ decreases from $+\infty$ to 0 and then increases to $+\infty$. Thus, corresponding to each positive value of τ, there are two values of t, say t_1 and t_2, satisfying (4.59) with Re $t_1 > 0$ and Re $t_2 < 0$. In terms of τ, (4.40) becomes

$$H_\nu^{(1)}(va) = e^{i\nu(a \sin \alpha - \alpha)} \frac{1}{\pi i} \int_0^\infty e^{-\nu\tau}\left(\frac{dt_1}{d\tau} - \frac{dt_2}{d\tau}\right) d\tau. \qquad (4.60)$$

To obtain series expansions of t_1 and t_2 in terms of τ, we return to (4.59) and observe that for t near $i\alpha$

$$\tau = -\frac{i}{2}\tan\alpha\,(t - i\alpha)^2\left[1 + \frac{2}{3! \, i}\cot\alpha\,(t - i\alpha) + \frac{2}{4!}(t - i\alpha)^2 + \cdots\right].$$

For the descent from B to C, the angle ω defined in (4.14) is $\pi/4$, and for the descent from B to A, $\omega = -3\pi/4$. Consequently, the correct choice of square root is

$$\tau^{1/2} = \pm e^{-i\pi/4}\sqrt{\frac{1}{2}\tan\alpha}\,(t - i\alpha)\left[1 + \frac{2}{3! \, i}\cot\alpha\,(t - i\alpha) + \cdots\right]^{1/2}, \qquad (4.61)$$

where the sign $+$ or $-$ depends on whether Re $t > 0$ or Re $t < 0$; cf. (4.17). Reverting the series gives

$$t_1 - i\alpha = \sum_{m=0}^\infty \frac{c_m}{m+1}\tau^{(m+1)/2},$$

$$t_2 - i\alpha = \sum_{m=0}^\infty (-1)^{m+1}\frac{c_m}{m+1}\tau^{(m+1)/2}, \qquad (4.62)$$

where

$$c_m = \left(\frac{2i}{\tan \alpha}\right)^{(m+1)/2} C_m \qquad (4.63)$$

and

$$m! \, C_m = \frac{d^m}{dt^m}\left[1 + \frac{2}{3! \, i}\cot \alpha \, (t - i\alpha) + \cdots\right]^{-(m+1)/2}\Bigg|_{t = i\alpha}. \qquad (4.64)$$

Proceeding now as in case (i), we find

$$H_\nu^{(1)}(\nu a) \sim e^{i\nu(a\sin\alpha - \alpha) - i\pi/4}\sqrt{\frac{2}{\nu\pi \tan \alpha}}\sum_{n=0}^{\infty} C_{2n}\frac{\Gamma(n + \frac{1}{2})}{\Gamma(\frac{1}{2})}\left(\frac{2i}{\nu \tan \alpha}\right)^n, \qquad (4.65)$$

as $\nu \to +\infty$.

For the integral associated with $H_\nu^{(2)}(\nu a)$, the saddle point is at $t = -i\alpha$ and the steepest descent path is the curve DEF in Figure 2.6. In exactly the same manner, one can show that

$$H_\nu^{(2)}(\nu a) \sim e^{-i\nu(a\sin\alpha - \alpha) + i\pi/4}\sqrt{\frac{2}{\nu\pi \tan \alpha}}\sum_{n=0}^{\infty} C_{2n}\frac{\Gamma(n + \frac{1}{2})}{\Gamma(\frac{1}{2})}\left(\frac{-2i}{\nu \tan \alpha}\right)^n,$$

$$(4.66)$$

as $\nu \to +\infty$. This result could, of course, also be deduced from the relation $H_\nu^{(2)}(\nu a) = \overline{H_\nu^{(1)}(\nu a)}$.

Combination of (4.65) with (4.66) gives the compound expansion

$$\begin{aligned}J_\nu(\nu a) \sim \left(\frac{2}{\nu\pi \tan \alpha}\right)^{1/2}\Bigg[&\cos\left(\nu \tan \alpha - \nu\alpha - \frac{\pi}{4}\right) \\ &\times \sum_{m=0}^{\infty}(-1)^m\frac{\Gamma(2m + \frac{1}{2})}{\Gamma(\frac{1}{2})}\frac{C_{4m}}{(\frac{1}{2}\nu \tan \alpha)^{2m}} \\ &- \sin\left(\nu \tan \alpha - \nu\alpha - \frac{\pi}{4}\right) \\ &\times \sum_{m=0}^{\infty}(-1)^m\frac{\Gamma(2m + \frac{3}{2})}{\Gamma(\frac{1}{2})}\frac{C_{4m+2}}{(\frac{1}{2}\nu \tan \alpha)^{2m+1}}\Bigg],\end{aligned} \qquad (4.67)$$

as $\nu \to +\infty$.

Case (iii): $a = 1$.   From (4.41), it is evident that here we have only one saddle point in the domain of interest, which occurs at $t = 0$ and is of order 2, i.e., $f'(0) = f''(0) = 0$. Since $f(0) = 0$, the steepest curves through $t = 0$ are given by

$$\cosh u \sin v - v = 0; \tag{4.68}$$

cf. (4.44). They consist of the real axis $v = 0$ and the two curves shown in Figure 2.7 with directions of descent indicated by arrows. It is obvious that the curve ABC is the steepest descent path for $H_\nu^{(1)}(v)$ and the curve ABD is for $H_\nu^{(2)}(v)$. On the curve ABC, $\sinh t - t$ is real and negative. Put

$$\tau = t - \sinh t. \tag{4.69}$$

Then, corresponding to each positive τ, there are two values of t, called t_1 and t_2, such that t_1 is a complex number with a positive real part and t_2 is a negative real number. In terms of τ, we have

$$H_\nu^{(1)}(v) = \frac{1}{\pi i} \int_0^\infty e^{-v\tau} \left(\frac{dt_1}{d\tau} - \frac{dt_2}{d\tau} \right) d\tau. \tag{4.70}$$

The ascending series of t_1 and t_2 in powers of τ can be obtained as before. From (4.69), we have

$$-\tau = t^3 \left[\frac{1}{3!} + \sum_{n=1}^\infty \frac{t^{2n}}{(2n+3)!} \right],$$

and hence

$$\omega \tau^{1/3} = t \left[\frac{1}{3!} + \sum_{n=1}^\infty \frac{t^{2n}}{(2n+3)!} \right]^{1/3}, \tag{4.71}$$

where ω is a complex cube root of -1 and $[\ldots]^{1/3}$ is positive at the origin. For t_1, we choose $\omega = e^{i\pi/3}$, and for t_2, we choose $\omega = -1$.

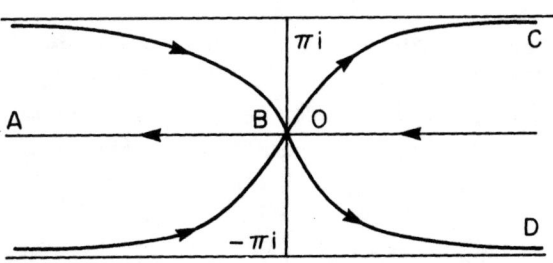

Fig. 2.7

Reverting the series in (4.71) gives

$$t = \sum_{n=0}^{\infty} \frac{d_n}{n+1} (\omega \tau^{1/3})^{n+1}, \tag{4.72}$$

where

$$n! \, d_n = \frac{d^n}{dt^n} \left[\frac{1}{3!} + \frac{t^2}{5!} + \frac{t^4}{7!} + \cdots \right]^{-(n+1)/3} \Bigg|_{t=0}. \tag{4.73}$$

Consequently, for sufficiently small τ,

$$\frac{dt_1}{d\tau} = \frac{1}{3} \sum_{n=0}^{\infty} d_n e^{i(n+1)\pi/3} \tau^{(n-2)/3}, \qquad \frac{dt_2}{d\tau} = \frac{1}{3} \sum_{n=0}^{\infty} d_n e^{i(n+1)\pi} \tau^{(n-2)/3},$$

and

$$\frac{dt_1}{d\tau} - \frac{dt_2}{d\tau} = -\frac{2i}{3} \sum_{n=0}^{\infty} d_n e^{i(n+1)2\pi/3} \sin\left[\frac{(n+1)\pi}{3} \right] \tau^{(n-2)/3}.$$

For large positive τ, it is clear from (4.69) that e^{t_1} and e^{t_2} are approximately equal to $-\tau$. Hence $dt/d\tau = 1/(1 - \cosh t)$ is bounded. By Watson's lemma,

$$H_\nu^{(1)}(\nu) \sim -\frac{2}{3\pi} \sum_{n=0}^{\infty} d_n e^{i(n+1)2\pi/3} \sin\left[\frac{(n+1)\pi}{3} \right] \frac{\Gamma\left(\dfrac{n+1}{3}\right)}{\nu^{(n+1)/3}}. \tag{4.74}$$

Taking the complex conjugate, we obtain

$$H_\nu^{(2)}(\nu) \sim -\frac{2}{3\pi} \sum_{n=0}^{\infty} d_n e^{-i(n+1)2\pi/3} \sin\left[\frac{(n+1)\pi}{3} \right] \frac{\Gamma\left(\dfrac{n+1}{3}\right)}{\nu^{(n+1)/3}}. \tag{4.75}$$

The asymptotic expansion of $J_\nu(\nu)$ can be obtained as the mean of the results in (4.74) and (4.75). Note that when $n + 1 = 3k$, $k = 1, 2, \ldots$, we have $\sin[(n+1)\pi/3] = 0$. Furthermore, when $n + 1 = 3k + 1$ or $3k + 2$, $k = 0, 1, 2, \ldots$, we have $\cos[2(n+1)\pi/3] = -\frac{1}{2}$. Hence, it follows that

$$J_\nu(\nu) \sim \frac{1}{3\pi} \sum_{n=0}^{\infty} d_n \sin\left[\frac{(n+1)\pi}{3} \right] \frac{\Gamma\left(\dfrac{n+1}{3}\right)}{\nu^{(n+1)/3}}; \tag{4.76}$$

cf. (4.52). From (4.69), it is evident that t is an odd function of $\tau^{1/3}$. Thus, the odd-indexed coefficients d_{2n+1} in (4.72) all vanish. The leading coefficient d_0 is given by

$$d_0 = (3!)^{1/3}; \qquad (4.77)$$

cf. (4.73). Since $d_1 = d_3 = 0$ and $\sin \pi = 0$, the first two nonzero terms in (4.76) are

$$J_\nu(\nu) = \frac{\Gamma(\frac{1}{3})}{2^{2/3} \cdot 3^{1/6} \pi \nu^{1/3}} - \frac{3^{7/6}\Gamma(\frac{5}{3})}{2^{1/3} \cdot 140 \pi \nu^{5/3}} + O\left(\frac{1}{\nu^{7/3}}\right). \qquad (4.78)$$

5. Perron's Method

In §4 we have already presented a method, namely, the method of steepest descents, to derive asymptotic expansions for integrals of the form

$$I(\lambda) = \int_C g(z)e^{\lambda f(z)} \, dz. \qquad (5.1)$$

However, in many specific cases, it is often found impractical to construct paths of steepest descent. Consequently, methods have been developed to replace these paths by simpler ones. Such methods have been referred to as the *saddle-point methods*. Another difficulty with the method of steepest descent is the inversion problem of finding $z(\tau)$ from the equation $\tau = f(z_0) - f(z)$ in (4.10), z_0 being a saddle point; see Copson (1965, pp. 91–92).

In this section, we shall describe a method due to Perron (1917), which avoids the construction of the steepest descent paths and the inversion problem mentioned above. This does not necessarily mean that Perron's method is superior to that of steepest descent. In some instances, one method may have a distinct advantage over the other, whereas in other instances, the converse may be true (see the Supplementary Notes at the end of this chapter). Our presentation of Perron's method follows closely that given by Wyman (1964).

Throughout this section, we shall consider λ to be a complex variable, and C to be a continuous curve in the z-plane beginning at $z = a$ and ending at $z = b$, where b may be finite or infinite. Instead of requiring C to be a steepest descent path from $z = a$, we shall impose conditions that only require C to be restricted to the point set for which

$$\mathrm{Re}\{\lambda f(z)\} < \mathrm{Re}\{\lambda f(a)\} \tag{5.2}$$

for each $z \neq a$ on C, i.e., $\mathrm{Re}\{\lambda f(z)\}$ attains its maximum at $z = a$. Note that (5.2) implies

$$|\arg\{\lambda f(a) - \lambda f(z)\}| < \frac{\pi}{2}, \tag{5.3}$$

whereas the steepest descent method requires $\arg\{\lambda f(a) - \lambda f(z)\} = 0$. To ensure that (5.2) holds for all λ varying in some unbounded point set and for all z on the curve C, we impose the following conditions which slightly strengthen (5.2):

(C_1)   The path C lies in the sector

$$|\arg\{\lambda f(a) - \lambda f(z)\}| \leq \frac{\pi}{2} - \delta, \tag{5.4}$$

where δ is a fixed positive number.

(C_2)   For each point $c \neq a$ on C, there exists a fixed number Δ, depending on c, such that $|f(z) - f(a)| \geq \Delta > 0$ for all z on the portion of C joining $z = c$ to $z = b$.

The above two conditions are completely analogous to the criteria used in Laplace's method for real integrals. In addition to these, we assume that the following conditions also hold:

(C_3)   In the neighborhood of a,

$$f(z) = f(a) - \mu(z - a)^p[1 - \varphi(z)], \tag{5.5}$$

where $\mu \neq 0$ is a complex number, p is real and positive and $\varphi(z)$ is analytic at $z = a$ with $\varphi(a) = 0$.

Note that since p need not be a positive integer, $f(z)$ is not necessarily analytic at a. To make $(z - a)^p$ single-valued, we introduce a cut in the z-plane from $z = a$ to infinity along some convenient radial line $\arg(z - a) = \gamma$.

(C_4)   The curve C may touch or coincide with one of the sides of the cut, but must not cross this cut. Furthermore, there exists a point $c' \neq a$ on C such that for any c on C with $|ac| < |ac'|$, the portion of C from $z = a$ to $z = c$ can be deformed into the radial line $\arg(z - a) = \arg(c - a)$.

Let β denote the angle of slope of C at a, i.e.,

$$\beta = \lim\{\arg(z - a)\}, \qquad z \to a \text{ along C.} \tag{5.6}$$

Observe that in order for (5.4) to hold, the point set through which $\lambda \to \infty$ must be contained in the sector

$$(4k - 1)\frac{\pi}{2} + \delta \le \arg \lambda + \arg \mu + p\beta \le (4k + 1)\frac{\pi}{2} - \delta \tag{5.7}$$

for some fixed δ in $0 < \delta \le \pi/2$, where k is a fixed integer.

Theorem 4. *Consider the integral $I(\lambda)$ in (5.1), and assume that it exists absolutely for every fixed λ satisfying the inequalities in (5.7). Furthermore, assume that in a neighborhood of a,*

$$g(z) = (z - a)^{\nu - 1}\psi(z), \tag{5.8}$$

where ν is a fixed complex number, $\mathrm{Re}\ \nu > 0$, and $\psi(z)$ is analytic at $z = a$ with $\psi(a) \neq 0$. Then, under the conditions (C_1) to (C_4), we have

$$I(\lambda) \sim \exp\{\lambda f(a)\} \sum_{s=0}^{\infty} a_s \exp\left\{2\pi k i\frac{(\nu + s)}{p}\right\}\lambda^{-(\nu + s)/p}, \tag{5.9}$$

as $|\lambda| \to \infty$, uniformly with respect to $\arg \lambda$ confined to the sector (5.7). The coefficients a_s are given by

$$a_s = \frac{\Gamma\left(\dfrac{\nu + s}{p}\right)}{p\mu^{(\nu + s)/p}s!}\frac{d^s}{dz^s}\left\{\psi(z)\left[\frac{\mu(z - a)^p}{f(a) - f(z)}\right]^{(\nu + s)/p}\right\}_{z = a}. \tag{5.10}$$

Proof. Since $\psi(z)\exp\{w\varphi(z)\}$ is, for each fixed w, analytic at $z = a$, we have

$$\psi(z)\exp\{w\varphi(z)\} = \sum_{s=0}^{\infty} P_s(w)(z - a)^s, |z - a| < \rho, \tag{5.11}$$

for sufficiently small ρ, where $P_s(w)$ is a polynomial in w whose degree does not exceed s. Furthermore,

$$P_s(w) = \frac{1}{s!}\frac{d^s}{dz^s}\left[\psi(z)\exp\{w\varphi(z)\}\right]_{z = a}. \tag{5.12}$$

In view of (5.5), (5.12) can be written as

$$e^{-w}P_s(w) = \frac{1}{s!}\frac{d^s}{dz^s}\left[\psi(z)\exp\left\{w\frac{f(z)-f(a)}{\mu(z-a)^p}\right\}\right]_{z=a}. \tag{5.13}$$

By Cauchy's inequality,

$$|P_s(w)| \le \frac{M(r)}{r^s} \qquad \text{for } 0 < r < \rho,$$

where

$$M(r) = \max_{|z-a|=r}|\psi(z)\exp\{w\varphi(z)\}|.$$

Recall that $\varphi(z)$ vanishes at $z = a$. Hence, for any $0 < r < \rho$,

$$|P_s(w)| \le \frac{K_1 \exp\{K_2|w|r\}}{r^s},$$

where K_1 and K_2 are fixed numbers independent of r and w. Therefore, we may write, for any fixed integer $N > 0$,

$$\psi(z)\exp\{w\varphi(z)\} = \sum_{s=0}^{N} P_s(w)(z-a)^s + R_N, \tag{5.14}$$

where, for $|z - a| \le r < \rho$,

$$|R_N| \le K_1'|z-a|^{N+1}\exp(K_2|w|r), \tag{5.15}$$

K_1' being a fixed number which depends on N but is independent of z and w.

In (5.14), we now put $w = \lambda\mu(z-a)^p$. From (5.5) and (5.8), it follows that

$$g(z)\exp\{\lambda f(z) - \lambda f(a)\}$$
$$= (z-a)^{\nu-1}\psi(z)\exp\{-w + w\varphi(z)\}$$
$$= \sum_{s=0}^{N} e^{-w}P_s(w)(z-a)^{s+\nu-1} + e^{-w}(z-a)^{\nu-1}R_N \tag{5.16}$$

and, for $|z - a| \le r < \rho$,

$$|e^{-w}(z-a)^{\nu-1}R_N| \le K_1'|(z-a)^{N+\nu}|\exp(-\operatorname{Re} w + K_2|w|r). \tag{5.17}$$

Returning to the integral $I(\lambda)$ in (5.1), we write

$$I(\lambda) = \left(\int_a^c + \int_c^b\right) g(z)\, \exp\{\lambda f(z)\}\, dz, \qquad (5.18)$$

where c is a point on the path of integration C. We choose c sufficiently close to a so that the portion of C from $z = a$ to $z = c$ can be deformed into the radial line $\arg(z - a) = \arg(c - a)$; see condition (C_4). Further-more, we require $|c - a| < r$ so that the expansion (5.16)–(5.17) holds for all z on the new path of integration from $z = a$ to $z = c$. Inserting (5.16) in the first integral on the right of (5.18) gives

$$\int_a^c g(z)\, \exp\{\lambda f(z)\}\, dz = \exp\{\lambda f(a)\}\left[\sum_{s=0}^N I_s(\lambda) + E_N(\lambda)\right], \qquad (5.19)$$

where

$$I_s(\lambda) = \int_a^c e^{-w} P_s(w)(z - a)^{s + v - 1}\, dz \qquad (5.20)$$

and

$$E_N(\lambda) = \int_a^c e^{-w}(z - a)^{v-1} R_N\, dz. \qquad (5.21)$$

By choosing c closer to a if necessary, we also have from (5.7)

$$(4k - 1)\frac{\pi}{2} + \delta_1 \le \arg\lambda + \arg\mu + p\,\arg(c - a) \le (4k + 1)\frac{\pi}{2} - \delta_1$$

for some $0 < \delta_1 < \delta$. This of course implies that on the path of integra-tion, which is now the straight line $\arg(z - a) = \arg(c - a)$,

$$(4k - 1)\frac{\pi}{2} + \delta_1 \le \arg w \le (4k + 1)\frac{\pi}{2} - \delta_1, \qquad (5.22)$$

and therefore $\operatorname{Re} w \ge |w| \sin\delta_1$. Thus we obtain from (5.17)

$$|e^{-w}(z - a)^{v-1} R_N| \le K_1'|(z - a)^{N + v}|\, \exp(-|w|\sin\delta_1 + K_2|w|r)$$

for the points on the path of integration. As long as we choose $r < (\sin\delta_1)/K_2$, there will exist a constant $K_2' > 0$ such that for all z with $\arg(z - a) = \arg(c - a)$ and $|z - a| \le r < \rho$,

$$|e^{-w}(z - a)^{v-1} R_N| \le K_1'|(z - a)^{N + v}|\, \exp(-K_2'|w|),$$

and therefore,

$$|E_N(\lambda)| \le K_1' \int_a^c |(z-a)^{N+v}| \exp(-K_2'|\lambda\mu(z-a)^p|)\,|dz|. \quad (5.23)$$

If we let $\tau = \arg(c-a)$ and $z - a = \rho e^{i\tau}$ in the integral in (5.23), then

$$|E_N(\lambda)| \le K_1' \int_0^\infty |\rho^{v+N}| \exp(-K_2'|\lambda||\mu|\rho^p)\,d\rho = O(\lambda^{-(v+N+1)/p}). \quad (5.24)$$

We now turn our attention to the integrals $I_s(\lambda)$ in (5.20). First we observe that for each s, we have

$$\int_c^{a+\infty e^{i\tau}} e^{-w} P_s(w)(z-a)^{s+v-1}\,dz = O(\exp(-\varepsilon|\lambda|)) \quad (5.25)$$

for some $\varepsilon > 0$, where the O-term depends on s but is independent of λ. This can be shown by either expressing the integral on the left-hand side of (5.25) in terms of the complementary incomplete gamma function defined in Chapter I, Exercises 3 and 4, or directly by using an argument similar to that for (5.31) below. Next we note, from (5.22), that $w = \lambda\mu(z-a)^p$ gives

$$z - a = w^{1/p} \exp\left(\frac{2k\pi i}{p}\right)\mu^{-1/p}\lambda^{-1/p}$$

and

$$(z-a)^{s+v-1} = w^{(s+v-1)/p} \exp\left(2k\pi i \frac{s+v-1}{p}\right)\mu^{-(s+v-1)/p}\lambda^{-(s+v-1)/p}.$$

Hence, making w the variable of integration, we obtain

$$\begin{aligned} I_s(\lambda) = \frac{1}{p}\,\mu^{-(s+v)/p}\lambda^{-(s+v)/p}\exp\left(2k\pi i\,\frac{s+v}{p}\right) \\ \times \int_0^{\infty e^{i\tau'}} e^{-w} P_s(w) w^{(s+v)/p-1}\,dw + O(\exp(-\varepsilon|\lambda|)), \end{aligned} \quad (5.26)$$

where, for fixed λ, $|\tau'| < \pi/2$. The path of integration in (5.26) can be deformed into the positive real axis. Thus

$$\begin{aligned} I_s(\lambda) = \frac{1}{p}\,\mu^{-(s+v)/p}\lambda^{-(s+v)/p}\exp\left(2k\pi i\,\frac{s+v}{p}\right) \\ \times \int_0^\infty e^{-w} P_s(w) w^{(s+v)/p-1}\,dw + O(\exp(-\varepsilon|\lambda|)). \end{aligned} \quad (5.27)$$

We now insert (5.13) in (5.27), and then interchange the order of integration and differentiation. This leads to

$$I_s(\lambda) = a_s \exp\left(2k\pi i\, \frac{s+v}{p} \right)\lambda^{-(s+v)/p} + O(\exp(-\varepsilon|\lambda|)). \qquad (5.28)$$

A combination of (5.19), (5.24) and (5.28) yields

$$\int_a^c g(z) \exp\{\lambda f(z)\}\, dz = \exp\{\lambda f(a)\}$$

$$\times \left[\sum_{s=0}^{N} a_s e^{2k\pi(s+v)i/p}\lambda^{-(s+v)/p} + O(\lambda^{-(N+v+1)/p}) \right].$$

$$(5.29)$$

The exponentially small terms from $I_s(\lambda)$, $s = 0, 1, \ldots, N$, can all be included in the O-term in (5.29).

It remains to consider only the second integral on the right-hand side of (5.18). Choose a point λ_0 satisfying the inequalities in (5.7) and write

$$\int_c^b g(z) \exp\{\lambda f(z)\}\, dz = \int_c^b g(z) \exp\{\lambda_0 f(z)\} \exp\{(\lambda - \lambda_0)f(z)\}\, dz.$$

Then

$$\left| \int_c^b g(z) \exp\{\lambda f(z)\}\, dz \right| \le \max|\exp\{(\lambda - \lambda_0)f(z)\}| \int_c^b |g(z) \exp\{\lambda_0 f(z)\}\, dz|,$$

where the maximum is taken over all points on the portion of C joining $z = c$ to $z = b$. Since we have assumed that the integral on the right exists, we must have

$$\left| \int_c^b g(z) \exp\{\lambda f(z)\}\, dz \right| \le K \max|\exp\{(\lambda - \lambda_0)f(z)\}|$$

for some constant $K > 0$, from which it follows that

$$\left| \int_c^b g(z) \exp\{\lambda f(z)\}\, dz \right|$$

$$\le K'|\exp\{\lambda f(a)\}|\max|\exp\{-(\lambda - \lambda_0)[f(a) - f(z)]\}|, \quad (5.30)$$

where $K' = K \exp\{-\lambda_0 f(a)\}$. For convenience, we put $p(z) = f(a) - f(z)$. Then

$$\operatorname{Re}\{(\lambda - \lambda_0)p(z)\} = |p(z)|\{|\lambda| \cos[\arg(\lambda p(z))] - |\lambda_0| \cos[\arg(\lambda_0 p(z))]\}$$

$$= |\lambda|\,|p(z)|\{\cos[\arg(\lambda p(z))] - o(1)\},$$

as $\lambda \to \infty$. Hence, by conditions (C_1) and (C_2),

$$\text{Re}\{(\lambda - \lambda_0)p(z)\} \geq |\lambda|\Delta[\sin \delta - o(1)],$$

as $\lambda \to \infty$ in the sector (5.7), and consequently,

$$\text{Re}\{(\lambda - \lambda_0)[f(a) - f(z)]\} \geq \varepsilon_1|\lambda| \qquad \text{for some } \varepsilon_1 > 0,$$

uniformly in arg λ for all λ satisfying (5.7) and uniformly in z for all z on C from $z = c$ to $z = b$. This together with (5.30) implies that

$$\left| \int_c^b g(z) \exp\{\lambda f(z)\} \, dz \right| = O\{\exp[\lambda f(a) - \varepsilon_1|\lambda|]\}. \qquad (5.31)$$

The final result now follows from (5.18), (5.29), and (5.31). ∎

Example 9.   To illustrate the above result, we reconsider the gamma function studied in §1, but now allow the argument to be complex. Our starting point is again the integral representation

$$\Gamma(\lambda + 1) = \int_0^\infty u^\lambda e^{-u} \, du, \qquad \text{Re } \lambda > 0.$$

If we consider λ real and positive, then, as in (1.27), the substitution $u = \lambda(1 + t)$ gives

$$\Gamma(\lambda) = e^{-\lambda}\lambda^\lambda \int_{-1}^\infty e^{\lambda f(t)} \, dt, \qquad (5.32)$$

where

$$f(t) = \log(1 + t) - t. \qquad (5.33)$$

By analytic continuation, (5.32) continues to hold as long as Re $\lambda > 0$. Following (1.29), we write

$$e^\lambda \lambda^{-\lambda} \Gamma(\lambda) = \int_0^\infty e^{\lambda f(t)} \, dt + \int_0^1 e^{\lambda f(-t)} \, dt. \qquad (5.34)$$

Instead of applying Theorem 1, we now use Theorem 4 on each of the integrals on the right-hand side. In terms of the notation of (5.1), we have $f(t) = \log(1 + t) - t$ for the first integral and $f(t) = \log(1 - t) + t$ for the second integral on the right-hand side of (5.34). This yields $a = 0$, $\mu = \frac{1}{2}$, $p = 2$, $\nu = 1$, and $\psi(t) \equiv 1$ in (5.5) and (5.8). We wish to first derive the asymptotic expansion of $\Gamma(\lambda)$ in the sector $|\arg \lambda| \leq \pi/2 - \delta$.

In order for the sector given in (5.7), i.e.,

$$(4k - 1)\frac{\pi}{2} + \delta \leq \arg \lambda + \arg \mu + p\beta \leq (4k + 1)\frac{\pi}{2} - \delta,$$

to match $-\pi/2 + \delta \leq \arg \lambda \leq \pi/2 - \delta$ with $\beta = 0$, we must take $k = 0$. Consequently, from (5.9) and (5.10), it follows that

$$\int_0^\infty e^{\lambda f(t)} \, dt \sim \sum_{s=b}^\infty a_s \lambda^{-(s+1)/2}$$

and

$$\int_0^1 e^{\lambda f(-t)} \, dt \sim \sum_{s=0}^\infty (-1)^s a_s \lambda^{-(s+1)/2},$$

as $\lambda \to \infty$, uniformly in $|\arg \lambda| \leq \pi/2 - \delta$, where

$$a_s = \frac{1}{2s!} \Gamma\left(\frac{s+1}{2}\right) \frac{d^s}{dt^s} \left\{ \left[\frac{t^2}{t - \log(1+t)} \right]^{(s+1)/2} \right\}_{t=0}.$$

Adding the two expansions together gives

$$\Gamma(\lambda) \sim e^{-\lambda} \lambda^{\lambda - 1/2} \sum_{s=0}^\infty \frac{b_s}{\lambda^s}, \tag{5.35}$$

where

$$b_s = \frac{1}{(2s)!} \Gamma\left(s + \frac{1}{2}\right) \frac{d^{2s}}{dt^{2s}} \left\{ \left[\frac{t^2}{t - \log(1+t)} \right]^{(s+1)/2} \right\}_{t=0}. \tag{5.36}$$

The expansion in (5.35) holds uniformly when $|\arg \lambda| \leq \pi/2 - \delta$, and, as expected, it agrees with the expansion given in (1.39) when λ is real. It should be noted that although here we have given an explicit expression for the coefficients in the expansion, the same expression, i.e., (5.36), can be obtained by applying the Lagrange inversion formula (Copson 1935, p. 125) to (1.33). Thus, the present example is purely illustrative, and the same result (5.35) with complex argument could have been easily derived by the method in §1.

The expansion (5.35) is actually uniformly valid in the region $|\arg \lambda| \leq \pi - \delta$. For an extension of the domain of validity to this larger region, we refer to the outline given in Ex. 21.

Example 10.   Consider the integral representation of the Bessel function

$$J_\alpha(z) = \frac{1}{2\pi i} \int_L t^{-\alpha-1} \exp\left\{\frac{1}{2} z(t - t^{-1})\right\} dt, \qquad (5.37)$$

where L is the loop contour shown in Figure 2.8. We shall first restrict the approach of z to infinity to the sector $\delta \le \arg z \le \pi - \delta, \delta > 0$. For z in this range, we choose $\pi/2 - \delta \le \Delta < \pi/2$ so that the integral in (5.37) is absolutely convergent for every fixed α. Since the function $\exp\{\frac{1}{2}z(t - t^{-1})\}$ is single-valued in the entire t-plane, we need discuss the behavior of this function only on the unit circle and on the line $\arg t = \pi - \Delta, |t| \ge 1$. On this line, we have $t = -re^{-i\Delta}$ and $r \ge 1$. Hence, if we write $z = |z|e^{i\varphi}$, then

$$\left| \exp\left\{\frac{1}{2}z(t - t^{-1})\right\} \right| = \exp\left\{-\frac{1}{2}|z|\left[r\cos(\varphi - \Delta) - \frac{\cos(\varphi + \Delta)}{r}\right]\right\}.$$

Since

$$r\cos(\varphi - \Delta) - \frac{\cos(\varphi + \Delta)}{r} = \left(r - \frac{1}{r}\right)\cos(\varphi - \Delta) + \frac{2}{r}\sin\varphi\sin\Delta \ge 0,$$

it follows that $|\exp\{\frac{1}{2}z(t - t^{-1})\}| \le 1$ on both straight-line portions of the contour L.

On the unit circle $t = e^{i\theta}$, we have $|\exp\{\frac{1}{2}z(t - t^{-1})\}| = \exp\{-|z|\sin\varphi \sin\theta\}$. Therefore, for the entire path of integration, there exists a

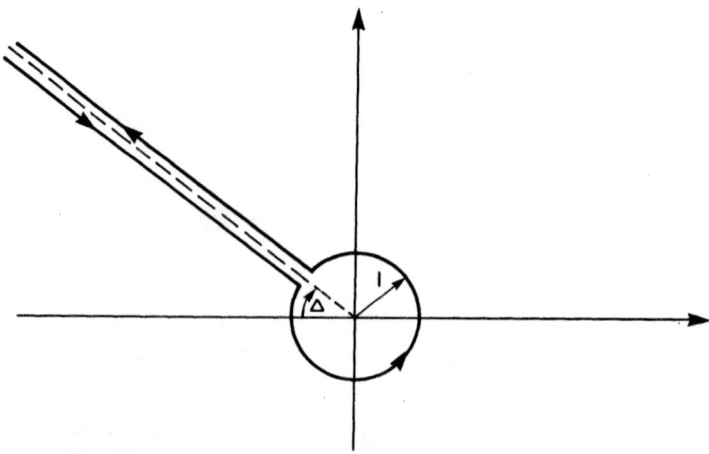

Fig. 2.8   Contour L, $0 < \Delta < \pi/2$, $-\pi - \Delta \le \arg t \le \pi - \Delta$.

unique absolute maximum, which occurs at $\theta = -\pi/2$. The maximum
value is $\exp\{|z| \sin \varphi\}$, and is greater than $\exp\{|z| \sin \delta\}$. If we make the
substitution $t' = t + i$, then the maximum point is translated to the
origin. For convenience, we shall simply write t for t'. Thus

$$
J_\alpha(z) = \frac{1}{2\pi i} e^{-iz} \int_{L'} (t-i)^{-\alpha-1} \exp\left\{\frac{1}{2} z\left(t - i - \frac{1}{t-i} + 2i\right)\right\} dt
$$

$$
= \frac{1}{2\pi i} \exp\left\{-i\left(z - \alpha\frac{\pi}{2} - \frac{\pi}{2}\right)\right\} \int_{L'} (1+it)^{-\alpha-1} \exp\left\{\frac{izt^2}{2(1+it)}\right\} dt,
$$

(5.38)

L' being the image of L under the translation. To apply Theorem 4, we
first divide the contour L' into two pieces L'_1 and L'_2. L'_1 is the portion of
L' which travels along the lower edge of the cut $\arg(t-i) = \pi - \Delta$ from
$t - i = -\infty e^{-i\Delta}$ to the point $t = 0$ on the circular part of L', and L'_2 is
the remaining portion of L'. Let the last integral in (5.38) along L'_i,
$i = 1, 2$, be denoted by $J^*_{\alpha,i}(z)$, that is,

$$
J^*_{\alpha,i}(z) = \int_{L'_i} (1+it)^{-\alpha-1} \exp\left\{\frac{izt^2}{2(1+it)}\right\} dt.
$$

In the notation of (5.5) and (5.8), we have

$$
f(t) = \frac{it^2}{2(1+it)}, \quad a = 0, \quad \mu = -\frac{i}{2}, \quad p = 2,
$$

$$
v = 1, \quad \psi(t) = (1+it)^{-\alpha-1}.
$$

Furthermore, since the tangent line to L' at the origin is the real axis,
the angle β, defined by (5.6), is equal to $-\pi$ for $J^*_{\alpha,1}(z)$ and 0 for $J^*_{\alpha,2}(z)$,
respectively. From (5.9) and (5.10), it follows that

$$
J^*_{\alpha,1}(z) \sim -\sum_{s=0}^{\infty} a_s \exp\{k(s+1)\pi i\} z^{-(s+1)/2},
$$

(5.39)

as $|z| \to \infty$ uniformly in the sector $2(k+1)\pi + \delta \le \arg z \le (2k+3)\pi - \delta$, and

$$
J^*_{\alpha,2}(z) \sim \sum_{s=0}^{\infty} a_s \exp\{k'(s+1)\pi i\} z^{-(s+1)/2},
$$

(5.40)

as $|z| \to \infty$ uniformly in $2k'\pi + \delta \le \arg z \le (2k'+1)\pi - \delta$, where

$$
a_s = \frac{\Gamma\left(\frac{s+1}{2}\right)}{2(-\frac{1}{2}i)^{(s+1)/2}s!} \left[\frac{d^s}{dt^s} (1+it)^{(s+1)/2-\alpha-1}\right]_{t=0}.
$$

(5.41)

For the domains of validity of these expansions to include $\delta \leq \arg z \leq \pi - \delta$, we must take $k = -1$ in (5.39) and $k' = 0$ in (5.40). Adding (5.39) to (5.40) gives

$$J_\alpha(z) \sim \frac{1}{2\pi i} \exp\left\{-i\left(z - \alpha\frac{\pi}{2} - \frac{\pi}{2}\right)\right\} \sum_{s=0}^{\infty} 2a_{2s} z^{-s-1/2},$$

as $|z| \to \infty$ uniformly in $\delta \leq \arg z \leq \pi - \delta$. By using the duplication formula $(2s)! = 2^{2s} s! \, \Gamma(s + \tfrac{1}{2})/\sqrt{\pi}$, it can be written as

$$J_\alpha(z) \sim \frac{\exp\left\{-i\left(z - \alpha\dfrac{\pi}{2} - \dfrac{\pi}{4}\right)\right\}}{\sqrt{2\pi z}} \sum_{k=0}^{\infty} \frac{(-i)^k \Gamma(k + \tfrac{1}{2} - \alpha)}{2^k k! \, \Gamma(\tfrac{1}{2} - k - \alpha)} z^{-k}, \quad (5.42)$$

uniformly in $\delta \leq \arg z \leq \pi - \delta$. Furthermore, since

$$\exp\left\{i\left(z - \alpha\frac{\pi}{2} - \frac{\pi}{4}\right)\right\}$$

is asymptotically negligible in the range $\delta \leq \arg z \leq \pi - \delta$, we can also write (5.42) in the form

$$
\begin{aligned}
J_\alpha(z) \sim {} & \sqrt{\frac{2}{\pi z}} \cos\left(z - \frac{\pi\alpha}{2} - \frac{\pi}{4}\right) \sum_{k=0}^{\infty} \frac{(-1)^k \, \Gamma(2k + \tfrac{1}{2} - \alpha)}{2^{2k}(2k)! \, \Gamma(-2k + \tfrac{1}{2} - \alpha)} z^{-2k} \\
& - \sqrt{\frac{2}{\pi z}} \sin\left(z - \frac{\pi\alpha}{2} - \frac{\pi}{4}\right) \\
& \times \sum_{k=0}^{\infty} \frac{(-1)^k \, \Gamma(2k + \tfrac{3}{2} - \alpha)}{2^{2k+1}(2k + 1)! \, \Gamma(-2k - \tfrac{1}{2} - \alpha)} z^{-2k-1},
\end{aligned} \quad (5.43)
$$

as $|z| \to \infty$ uniformly in $\delta \leq \arg z \leq \pi - \delta$.

To show that the expansion (5.43) actually holds in the range $|\arg z| \leq \pi - \delta$, we must also discuss the sector $0 \leq \arg z \leq \delta$. Note that when $\arg z = 0$, $|\exp\{\tfrac{1}{2}z(t - t^{-1})\}| \equiv 1$ for all points on the unit circle. Hence, condition (5.2) does not hold, and Theorem 4 cannot be applied. However, only an infinitesimal deformation is required to make Theorem 4 applicable, and the shape of the contour L which we shall choose is shown in Figure 2.9. In this figure, AB is a straight line segment passing through the critical point $t = -i$ and perpendicular to OB, and ED is its mirror image. It is easily seen that $|OB| = \cos \Delta$, $|CB| = \sin \Delta$, $|AC| = \sin \Delta/\cos 2\Delta$ and $|OA| = \cos \Delta/\cos 2\Delta$. Also, we

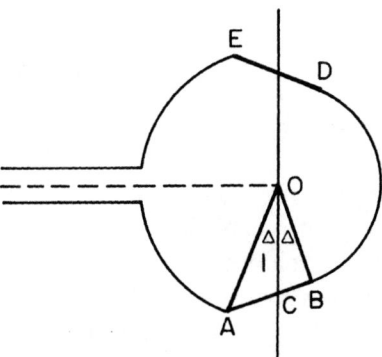

Fig. 2.9 Contour L.

have angle OCA equal to $\pi/2 + \Delta$, and angle OAC equal to $\pi/2 - 2\Delta$. The curves BD and AE are portions of the circles $|t| = |OB|$ and $|t| = |OA|$, respectively.

As before, the integral is absolutely convergent for each fixed z in $|\arg z| \le \delta$, and we need only consider the behavior of $|\exp\{\frac{1}{2}z(t - t^{-1})\}|$ on L. On the straight line portion $t = -r$, $r \ge \cos\Delta/\cos 2\Delta$, we have

$$\left|\exp\left\{\frac{1}{2}z(t - t^{-1})\right\}\right| = \exp\left\{-\frac{1}{2}|z|\cos\varphi\left(r - \frac{1}{r}\right)\right\},$$

where $\varphi = \arg z$. Since $\cos\varphi \ge \cos\delta$ and

$$r - \frac{1}{r} \ge \frac{(\cos^2\Delta - \cos^2 2\Delta)}{(\cos\Delta)(\cos 2\Delta)},$$

there exists a positive ε_1 such that

$$|\exp\{\tfrac{1}{2}z(t - t^{-1})\}| \le \exp(-\varepsilon_1|z|).$$

For the circular portion of L from A to E, $|t| = |OA| = \cos\Delta/\cos 2\Delta = 1 + \frac{3}{2}\Delta^2 + O(\Delta^4)$ and $\pi/2 + \Delta \le \theta \le 3\pi/2 - \Delta$, where $\theta = \arg t$. Since both δ and Δ may be chosen arbitrarily small, we shall take $\delta = \Delta^4$ so that $\sin\varphi = O(\Delta^4)$. Simple calculation gives

$$|\exp\{\tfrac{1}{2}z(t - t^{-1})\}| = \exp\{\tfrac{1}{2}|z|[\cos(\varphi + \theta) - \cos(\varphi - \theta)$$
$$+ \tfrac{3}{2}\Delta^2(\cos(\varphi + \theta) + \cos(\varphi - \theta)) + O(\Delta^4)]\}, \quad \text{as } \Delta \to 0^+.$$

Since

$$\cos(\varphi + \theta) - \cos(\varphi - \theta) = -2\sin\varphi\sin\theta = O(\Delta^4)$$

and

$$\cos(\varphi + \theta) + \cos(\varphi - \theta) = 2 \cos \varphi \cos \theta \le -2 \sin \Delta + O(\Delta^9)$$

as $\Delta \to 0^+$, it follows that $\exp\{\frac{1}{2}|z|(t - t^{-1})\} = O(\exp(-\varepsilon_2|z|))$ for some $\varepsilon_2 > 0$.

Exactly the same result holds for the portion of L from B to D. Hence, only the infinitesimal portions of the path L through the critical points $t = \pm i$ remain to be considered. The asymptotic contribution from these portions can be obtained in exactly the same manner as was done for the range $\delta \le \arg z \le \pi - \delta$. Indeed, the asymptotic expansion from $t = i$ is as given in (5.42), and the asymptotic expansion from $t = -i$ is given by

$$J_\alpha(z) \sim \frac{\exp\left\{i\left(z - \alpha\frac{\pi}{2} - \frac{\pi}{4}\right)\right\}}{\sqrt{2\pi z}} \sum_{k=0}^{\infty} \frac{i^k \Gamma(k + \frac{1}{2} - \alpha)}{2^k k! \, \Gamma(\frac{1}{2} - k - \alpha)} z^{-k}.$$

Upon adding the results, we again have (5.43), an expansion that has now been proved to be uniformly valid for $0 \le \arg z \le \pi - \delta$. From the connection formula $J_\alpha(z) = \overline{J_{\bar{\alpha}}(\bar{z})}$, one immediately extends the domain of uniform validity to $|\arg z| \le \pi - \delta$.

As with Example 9, the present example is purely illustrative, since the same result has already been derived in Chapter I, Section 5, by an entirely different method.

6.   Darboux's Method

The problem of obtaining the asymptotic behavior for the coefficients a_n of the Maclaurin expansion

$$f(z) = \sum_{n=0}^{\infty} a_n z^n \tag{6.1}$$

arises in many instances. The main sources for these problems are number theory and combinatorics.

The natural starting point is the Cauchy integral formula

$$a_n = \frac{1}{2\pi i} \int_C f(z) z^{-n-1} \, dz, \tag{6.2}$$

where C is any simple closed contour which encloses $z = 0$ and lies within the domain of analyticity. One of the methods used to determine the asymptotic form of the integral (6.2) for large values of n is to expand the contour C as large as possible and to apply the residue calculus. An example of this procedure is given in Chapter I, Section 6, concerning the Bernoulli numbers. This procedure, however, fails when $f(z)$ has singularities that are not poles. As far as we are aware, Darboux was the first to attack general problems of this nature. He showed that when $f(z)$ has only a finite number of singularities on its circle of convergence, all of which are algebraic in nature, an asymptotic expansion could be obtained for a_n as $n \to \infty$. The details of his result are the contents of this section.

Let $f(z)$ be an analytic function within the circle $|z| < R$, where $0 < R < \infty$. Since R is finite, we may without loss of generality assume that $R = 1$. For simplicity, we shall also assume that on the circle $|z| = 1$, $f(z)$ has exactly one singularity. The result for a finite number of singularities can be obtained by adding the contribution from each; see Theorem 5 below. Furthermore, if the singularity occurs at $z = e^{i\theta}$, then the substitution $z = e^{i\theta}z'$ locates the singularity at $z' = 1$. We therefore assume that $f(z)$ has a singularity at $z = 1$, and that in a neighborhood of $z = 1$, $f(z)$ is of the form

$$f(z) = (1 - z)^\alpha g(z), \tag{6.3}$$

where $g(z)$ is an analytic function at $z = 1$. Near $z = 1$, we write

$$g(z) = \sum_{r=0}^{\infty} c_r (1 - z)^r. \tag{6.4}$$

The m-th *Darboux approximant* of $f(z)$ is defined by

$$f_m(z) = \sum_{r=0}^{m} c_r (1 - z)^{\alpha + r}. \tag{6.5}$$

Since $f_m(z)$ is analytic within $|z| < 1$, it has a Maclaurin expansion of the form

$$f_m(z) = \sum_{n=0}^{\infty} b_{mn} z^n. \tag{6.6}$$

A simple calculation gives

$$b_{mn} = \frac{f_m^{(n)}(0)}{n!} = \sum_{r=0}^{m} c_r \binom{\alpha + r}{n} (-1)^n. \tag{6.7}$$

Let a_n denote the Maclaurin coefficients of $f(z)$ as given in (6.1). By Cauchy's theorem

$$a_n - b_{mn} = \frac{1}{2\pi i} \int_C [f(z) - f_m(z)]z^{-n-1}\, dz, \qquad (6.8)$$

where C is any contour which encircles $z = 0$ and lies within $|z| < 1$. We put

$$\varepsilon_m(z) = f(z) - f_m(z) \qquad (6.9)$$

and

$$\delta_m(n) = \frac{1}{2\pi i} \int_C \varepsilon_m(z)z^{-n-1}\, dz. \qquad (6.10)$$

In view of (6.7), the result in (6.8) can now be rewritten as

$$a_n = \sum_{r=0}^{m} c_r \binom{\alpha + r}{n} (-1)^n + \delta_m(n). \qquad (6.11)$$

We claim that

$$\delta_m(n) = o(n^{-\alpha - m - 1}) \qquad \text{as } n \to \infty. \qquad (6.12)$$

Integrating by parts N times, we obtain from (6.10)

$$\delta_m(n) = \frac{1}{2\pi i} \frac{(n - N)!}{n!} \int_C \varepsilon_m^{(N)}(z)z^{-(n-N+1)}\, dz. \qquad (6.13)$$

Since

$$\varepsilon_m(z) = c_{m+1}(1 - z)^{\alpha + m + 1} + c_{m+2}(1 - z)^{\alpha + m + 2} + \cdots$$

in a neighborhood of $z = 1$, we have

$$\varepsilon_m^{(N)}(z) = O((1 - z)^{\alpha + m + 1 - N}) \qquad (6.14)$$

as $z \to 1$. As long as N satisfies

$$\operatorname{Re} \alpha + m + 1 \le N < \operatorname{Re} \alpha + m + 2, \qquad (6.15)$$

the contour C can be expanded so that (6.13) becomes

$$\delta_m(n) = \frac{1}{2\pi} \frac{(n - N)!}{n!} \int_0^{2\pi} \varepsilon_m^{(N)}(e^{i\theta})e^{-i(n-N)\theta}\, d\theta. \qquad (6.16)$$

Since the last integral is absolutely integrable, the Riemann–Lebesgue lemma gives

$$\delta_m(n) = o\left(\frac{(n-N)!}{n!}\right) = o(n^{-N}), \qquad \text{as } n \to \infty, \qquad (6.17)$$

which of course implies (6.12), by virtue of (6.15). In the notation of generalized asymptotic expansions, the result (6.11)–(6.12) can be written as

$$(-1)^n a_n \sim \sum_{r=0}^{\infty} c_r \binom{\alpha+r}{n}; \qquad \{n^{-\alpha-r-1}\}, \qquad \text{as } n \to \infty. \qquad (6.18)$$

It is interesting to note that the above derivation also gives an error bound for the expansion (6.18). Indeed, from (6.16) it follows that

$$|\delta_m(n)| \le \frac{(n-N)!}{2\pi n!} \int_0^{2\pi} |\varepsilon_m^{(N)}(e^{i\theta})|\, d\theta \le \frac{1}{2\pi}\left(\frac{e}{n}\right)^N \int_0^{2\pi} |\varepsilon_m^{(N)}(e^{i\theta})|\, d\theta. \qquad (6.19)$$

If $\varepsilon_m^{(N+1)}(z)$ is absolutely integrable on $|z| = 1$, then another bound for $\delta_m(n)$ can be obtained by applying a further integration by parts to (6.13). This is given by

$$|\delta_m(n)| \le \frac{(n-N-1)!}{2\pi n!} \int_0^{2\pi} |\varepsilon_m^{(N+1)}(e^{i\theta})|\, d\theta. \qquad (6.20)$$

Although the latter bound has the advantage of being $O(n^{-N-1})$, the finiteness of the integral in (6.20) is not guaranteed by our conditions.

We now proceed to consider a more general situation formulated by Szegö (1967, p. 207). Assume that $f(z)$ has a finite number of singularities $\alpha_1, \alpha_2, \ldots, \alpha_l$, $(\alpha_i \ne \alpha_j, i \ne j)$ on its circle of convergence $|z| = R$. In the neighborhood of α_k, it has an expansion of the form

$$f(z) = \sum_{r=0}^{\infty} c_r^{(k)}\left(1 - \frac{z}{\alpha_k}\right)^{a_k + rb_k}, \qquad k = 1, \ldots, l, \qquad (6.21)$$

where $b_k > 0$.

Theorem 5. *With the above assumptions, the Maclaurin coefficient of $f(z)$ has the generalized asymptotic expansion*

$$(-1)^n a_n \sim \sum_{r=0}^{\infty} \sum_{k=1}^{l} c_r^{(k)}\binom{a_k + rb_k}{n}\alpha_k^{-n}; \qquad \{n^{-Q_r}\}, \qquad (6.22)$$

as $n \to \infty$, where Q_r is the positive integer given by

$$Q_r = \min_{1 \le k \le l} \{\operatorname{Re} a_k + rb_k + 1\}. \tag{6.23}$$

The derivation of this result is virtually the same as that of (6.18).

Example 11. The following problem has been discussed by Robinson (1951). "Let there be n straight lines in a plane, no three of which meet at a point. Determine the number g_n of groups of n of their points of intersection such that no three of the points of the group be on one of the straight lines."

Robinson showed that g_n satisfies the recurrence relation

$$g_{n+1} = ng_n + \binom{n}{2}g_{n-2}, \qquad n \ge 3,$$

where $g_1 = g_2 = 0$ and $g_3 = 1$. If we let $u_n = 2g_n/(n-1)!$, then $u_1 = u_2 = 0$, $u_3 = 1$, and $u_{n+3} = u_{n+2} + u_n/2n$, $n \ge 1$. Robinson also proved that the number

$$b = \lim_{n \to \infty} \frac{u_n^2}{n}$$

exists and conjectured that this might represent a new geometrical constant. In what follows, we shall use the method of Darboux to show that

$$b = \lim_{n \to \infty} \frac{u_n^2}{n} = \frac{4}{\pi} e^{-3/2}; \tag{6.24}$$

thus Robinson's conjecture is incorrect.

If we define

$$f(z) = 2 \sum_{n=3}^{\infty} \frac{g_n}{n!} z^n = \sum_{n=3}^{\infty} \frac{u_n}{n} z^n,$$

then $f(z)$ satisfies the differential equation

$$(1-z)f'(z) - \tfrac{1}{2}z^2 f(z) = z^2.$$

Obviously, $f(z) = -2$ is a particular solution, and therefore

$$f(z) = -2 + \frac{A}{\sqrt{1-z}} \exp\left(-\frac{z^2}{4} - \frac{z}{2}\right)$$

is the general solution. Since $f(0) = 0$, we must take $A = 2$. Thus

$$f(z) = -2 + \frac{2}{\sqrt{1-z}} \exp\left(-\frac{z^2}{4} - \frac{z}{2}\right).$$

In the neighborhood of $z = 1$,

$$f(z) = -2 + \frac{2}{\sqrt{1-z}} \exp\left[-\frac{3}{4} + (1-z) - \frac{1}{4}(1-z)^2\right]$$

$$= -2 + 2e^{-3/4} \sum_{k=0}^{\infty} \frac{1}{2^k k!} H_k(1)(1-z)^{k-1/2},$$

where $H_k(x)$ is the Hermite polynomial (Erdélyi *et al.* 1953b, p. 194). From (6.18), it immediately follows that

$$\frac{u_n}{n} \sim 2e^{-3/4} \sum_{k=0}^{\infty} \frac{H_k(1)}{2^k k!} \frac{\Gamma(n-k+\frac{1}{2})}{n!\,\Gamma(\frac{1}{2}-k)}; \qquad \{n^{-k-1/2}\}.$$

In view of the result

$$\frac{\Gamma(n+\alpha)}{\Gamma(n+\beta)} = n^{\alpha-\beta}\left[1 + \frac{(\alpha-\beta)(\alpha+\beta-1)}{2n} + O(n^{-2})\right]$$

given in Chapter I, Exs. 11 and 15, we have

$$\frac{u_n}{n} = \frac{2e^{-3/4}}{\sqrt{\pi}\,n^{1/2}}\left[1 - \frac{5}{8n} + O(n^{-2})\right],$$

for which the desired limit (6.24) now follows. The derivation presented here is due to Wyman (1960).

Example 12. Let S be a subset of the set of all positive integers, and let $a_n(S)$ denote the number of elements in the symmetric group S_n on n elements whose cycle lengths belong to the set S. Then it is known that the exponential generating function for $a_n(S)$ is

$$\prod_{k\in S} e^{z^k/k} = (1-z)^{-1}\prod_{k\notin S} e^{-z^k/k}. \tag{6.25}$$

If the complement of S is finite, then (6.18) applies and we have

$$a_n(S) \sim n!\exp\left(-\sum_{k\notin S}\frac{1}{k}\right), \qquad \text{as } n\to\infty.$$

This example is taken from Bender (1974, p. 499).

If the set S is finite then the generating function given in (6.25) is entire, in which case Darboux's method fails. We shall study this situation in the following section.

7.   A Formula of Hayman

The method of Darboux requires the Maclaurin expansion

$$f(z) = \sum_{n=0}^{\infty} a_n z^n \tag{7.1}$$

to have a finite radius of convergence. If $f(z)$ is an entire function, then the series (7.1) has an infinite radius of convergence and Darboux's method does not apply. In 1956, Hayman has given an asymptotic formula for a_n, which is applicable to a large class of entire functions including

$$e^z, \qquad \exp\left(z + \frac{1}{p} z^p\right), \qquad \exp(e^z - 1), \qquad \exp(ze^z). \tag{7.2}$$

The Maclaurin coefficients of each of the generating functions in (7.2) have an important mathematical meaning.

To describe Hayman's result, we use the integral representation

$$a_n = \frac{1}{2\pi r^n} \int_{-\pi}^{\pi} f(re^{i\theta}) e^{-in\theta} \, d\theta, \qquad r > 0, \tag{7.3}$$

and introduce the functions

$$a(r) = r \frac{d}{dr} \log f(r) = r \frac{f'(r)}{f(r)}, \tag{7.4}$$

$$b(r) = \left(r \frac{d}{dr}\right)^2 \log f(r) = ra'(r). \tag{7.5}$$

If the function $\log f(re^{i\theta}) - in\theta$ is expanded in a Maclaurin series about $\theta = 0$, then we have

$$\log f(re^{i\theta}) - in\theta = \log f(r) + i\theta(a(r) - n) - \tfrac{1}{2}\theta^2 b(r) + \cdots. \tag{7.6}$$

Hayman's method is to choose r so that the linear term vanishes, and to obtain conditions on $f(z)$ which are sufficient to ensure that only an infinitesimal interval $|\theta| < \delta$ contributes to the asymptotic expansion of a_n. Furthermore, the quadratic approximation (7.6) is sufficient to determine the first term of this expansion. Thus, returning to the integral in (7.3), Hayman's conditions will imply

$$2\pi r^n a_n \sim \int_{-\delta}^{\delta} f(r) e^{-\theta^2 b(r)/2}\, d\theta, \tag{7.7}$$

and ultimately

$$a_n \sim \frac{f(r)}{2\pi r^n} \int_{-\infty}^{\infty} e^{-\theta^2 b(r)/2}\, d\theta = \frac{f(r)}{r^n \sqrt{2\pi b(r)}}, \tag{7.8}$$

where r is a function of n, implicitly defined by $a(r) = n$.

To establish the approximation (7.8), we assume that $f(z)$ satisfies the following conditions:

(F_1) *there exists a finite number $R_0 > 0$ such that $f(r) > 0$ for $R_0 < r < \infty$;*

(F_2) $\lim\limits_{r \to \infty} b(r) = +\infty$;

(F_3) *for some function $\delta(r)$ defined in the interval $R_0 < r < \infty$ and satisfying $0 < \delta(r) < \pi$, we have*

$$f(re^{i\theta}) \sim f(r) e^{i\theta a(r) - \theta^2 b(r)/2}, \text{ as } r \to \infty, \tag{7.9}$$

uniformly in $|\theta| \le \delta(r)$, and

$$f(re^{i\theta}) = o\!\left(\frac{f(r)}{\sqrt{b(r)}} \right), \quad \text{as } r \to \infty, \tag{7.10}$$

uniformly in $\delta(r) \le |\theta| \le \pi$.

Hayman called these functions *admissible functions*.

The existence of the function $\delta(r)$ in condition (F_3) requires some comment. If we extend the quadratic approximation in (7.6) to a full expansion, we obtain

$$\log f(re^{i\theta}) = \log f(r) + \sum_{\nu=1}^{\infty} \alpha_\nu(r) \frac{(i\theta)^\nu}{\nu!}, \tag{7.11}$$

where $\alpha_1(r) = a(r)$, $\alpha_2(r) = b(r)$, and $\alpha_\nu(r) = r\alpha_{\nu-1}'(r)$ for $\nu > 1$. Hence

$$f(re^{i\theta}) = f(r) \exp\!\left\{ \sum_{\nu=1}^{\infty} \alpha_\nu(r) \frac{(i\theta)^\nu}{\nu!} \right\}.$$

The validity of (7.9) and (7.10) then requires $\delta(r)$ to satisfy

$$\sum_{v=3}^{\infty} \alpha_v(r) \frac{(i\theta)^v}{v!} \to 0, \qquad \text{as } r \to \infty, \tag{7.12}$$

uniformly in $|\theta| \le \delta(r)$, and

$$\delta^2(r)b(r) \to \infty, \qquad \text{as } r \to \infty. \tag{7.13}$$

These two limiting conditions may be used to suggest the appropriate choice of the function $\delta(r)$.

Lemma 2. *Let $f(z) = \sum_{n=0}^{\infty} a_n z^n$ be an entire function with the above conditions and define $a_n = 0$ for $n < 0$. Then, as $r \to \infty$, we have uniformly in all integers n*

$$a_n r^n = \frac{f(r)}{\sqrt{2\pi b(r)}} \left[\exp\left\{ -\frac{[a(r) - n]^2}{2b(r)} \right\} + o(1) \right]. \tag{7.14}$$

Proof. By Cauchy's formula, we can write $a_n r^n = I_1 + I_2$, for all integers n, where

$$I_1 = \frac{1}{2\pi} \int_{-\delta}^{\delta} f(re^{i\theta}) e^{-in\theta} \, d\theta \qquad \text{and} \qquad I_2 = \frac{1}{2\pi} \int_{\delta}^{2\pi-\delta} f(re^{i\theta}) e^{-in\theta} \, d\theta.$$

By condition (F$_3$), we have $I_2 = o[f(r)/\sqrt{b(r)}]$ as $r \to \infty$, uniformly in n, and

$$
\begin{aligned}
I_1 &= \frac{f(r)}{2\pi} \int_{-\delta}^{\delta} [1 + o(1)] e^{i[a(r) - n]\theta - \theta^2 b(r)/2} \, d\theta \\
&= \frac{f(r)}{2\pi} \left[\int_{-\delta}^{\delta} e^{i[a(r) - n]\theta - b(r)\theta^2/2} \, d\theta + o\left(\int_{-\infty}^{\infty} e^{-b(r)\theta^2/2} \, d\theta \right) \right] \\
&= \frac{f(r)}{2\pi} \left[\int_{-\delta}^{\delta} e^{i[a(r) - n]\theta - b(r)\theta^2/2} \, d\theta + o(1/\sqrt{b(r)}) \right]
\end{aligned}
$$

as $r \to \infty$, also uniformly in n. At the boundary point $\theta = \delta$, (7.9) and (7.10) together imply

$$\left| \frac{f(re^{i\theta})}{f(r)} \right| \sim e^{-b(r)\delta^2/2} = o\left(\frac{1}{\sqrt{b(r)}} \right), \qquad \text{as } r \to \infty,$$

from which it follows that

$$\delta^2 b(r) \to \infty, \qquad \text{as } r \to \infty. \tag{7.15}$$

If we put $c = [a(r) - n]\sqrt{2/b(r)}$ and $y = \theta\sqrt{b(r)/2}$, we have

$$I_1 = \frac{f(r)}{\pi\sqrt{2b(r)}}\left[\int_{-h}^{h} e^{-y^2 + icy}\, dy + o(1)\right], \qquad \text{where} \quad h(r) = \delta(r)\sqrt{\frac{b(r)}{2}}.$$

In view of (7.15), we obtain

$$I_1 = \frac{f(r)}{\pi\sqrt{2b(r)}}\left[\int_{-\infty}^{\infty} e^{-y^2 + icy}\, dy + o(1)\right].$$

A simple calculation shows

$$\int_{-\infty}^{\infty} e^{-y^2 + icy}\, dy = e^{-c^2/4}\int_{-\infty}^{\infty} e^{-(y - ic/2)^2}\, dy$$

$$= e^{-c^2/4}\int_{-\infty}^{\infty} e^{-y^2}\, dy = \sqrt{\pi}\,e^{-c^2/4}.$$

Thus we have as $r \to \infty$, uniformly in n,

$$I_1 = \frac{f(r)}{\sqrt{2\pi b(r)}}\left[\exp\left\{-\frac{[a(r) - n]^2}{2b(r)}\right\} + o(1)\right]$$

and

$$a_n r^n = \frac{f(r)}{\sqrt{2\pi b(r)}}\left[\exp\left\{-\frac{[a(r) - n]^2}{2b(r)}\right\} + o(1)\right].$$

This completes the proof of the lemma. ■

Lemma 3. *The function $a(r)$ is positive and increasing in some interval $r_0 \le r < \infty$ and*

$$b(r) = o(a^2(r)), \qquad as\ r \to \infty. \tag{7.16}$$

Proof. Since by condition (F_2), $b(r) = ra'(r) \to +\infty$ as $r \to \infty$, we have $a'(r) > 0$, and so, $a(r)$ must become an increasing function in some interval $\rho < r < \infty$. Hence, either $a(r)$ is finally positive, or $a(r)$ remains bounded by a non-positive constant as $r \to \infty$. To exclude the latter possibility, we use Lemma 2 with $n = -1$. Since $a_{-1} = 0$, (7.14) gives

$$\exp\left\{-\frac{[a(r) + 1]^2}{2b(r)}\right\} = o(1),$$

and hence

$$\lim_{r \to \infty} \frac{[a(r) + 1]^2}{2b(r)} = +\infty, \tag{7.17}$$

which of course implies $a(r) \to \infty$ as $r \to \infty$. Therefore, $a(r)$ is a positive increasing function in some interval $r_0 \le r < \infty$. The result in (7.16) follows immediately from (7.17). ∎

Theorem 6.   *Let r_n be the unique solution of $a(r) = n$ in the interval $r_0 < r < \infty$. Then*

$$a_n \sim \frac{f(r_n)}{r_n^n \sqrt{2\pi b(r_n)}}, \qquad \text{as } n \to \infty. \tag{7.18}$$

Proof.   Let $n > a(r_0)$. Since $a(r)$ is an increasing function and $\lim_{r \to \infty} a(r) = \infty$, the equation $a(r) = n$ must have a unique solution $r = r_n$ in $r_0 < r_n < \infty$. Furthermore, r_n increases with n and $r_n \to \infty$ as $n \to \infty$. The result (7.18) now follows from Lemma 2. ∎

To facilitate the use of the above theorem, we also state another result of Hayman, which enables one to tell in specific examples, almost at a glance, whether or not the generating function is admissible. However, since the proof of this result is outside the subject of "asymptotics", it will not be included here.

Theorem 7.   *Let $f(z)$ and $g(z)$ be admissible functions, i.e., functions satisfying conditions (F_1), (F_2), and (F_3), and let $P(z)$ be a polynomial with real coefficients.*

(i)   *$e^{f(z)}$ and $f(z)g(z)$ are admissible.*

(ii)   *$f(z) + P(z)$ is admissible, and if the leading coefficient of $P(z)$ is positive, then $f(z)P(z)$ and $P(f(z))$ are admissible.*

(iii)   *If the coefficients a_n of the power series for $e^{P(z)}$ are positive real numbers for all sufficiently large n, then $e^{P(z)}$ is admissible.*

From this result, it is almost immediate that all the generating functions given in (7.2) are admissible.

Example 13. Let d_n denote the number of partitions of an n-element set. Then it is known that the exponential generating function for d_n is $\exp(e^z - 1)$, i.e.,

$$\exp(e^z - 1) = \sum_{n=0}^{\infty} d_n \frac{z^n}{n!}. \qquad (7.19)$$

In this case we have $f(r) = \exp(e^r - 1)$, $a(r) = re^r$, and $b(r) = (r^2 + r)e^r$. Hence, by Hayman's formula (7.18),

$$d_n \sim \frac{n! \exp(e^{r_n} - 1)}{r_n^n \sqrt{2\pi(r_n + 1)r_n \exp(r_n)}}, \qquad (7.20)$$

where r_n is defined by

$$r_n \exp(r_n) = n. \qquad (7.21)$$

From (7.21), we also have $e^{r_n} = n/r_n$ and $r_n^n = n^n e^{-nr_n}$. This implies

$$d_n \sim \frac{n! \exp[n(r_n + r_n^{-1}) - 1]}{n^{n+1/2}\sqrt{2\pi(r_n + 1)}}. \qquad (7.22)$$

We now use Stirling's formula for $n!$ to obtain from (7.22)

$$d_n \sim \frac{\exp[n(r_n + r_n^{-1} - 1) - 1]}{\sqrt{r_n + 1}}. \qquad (7.23)$$

It would be natural to express r_n, using (7.21), asymptotically in terms of n and then to obtain an asymptotic expansion for d_n entirely in terms of n. However, this procedure is not satisfactory as we shall now see. Let us write (7.21) in the form $r_n = \log n - \log r_n$. Starting with the approximation $r_n = \log n$ and iterating, we obtain

$$r_n = \log n - \log(\log n) + \frac{\log(\log n)}{\log n} + \cdots. \qquad (7.24)$$

Further terms contain higher powers of $\log n$ in the denominators. However, in (7.23) we have a term of the form nr_n. Hence, it is clear from (7.24) that none of the terms in the asymptotic expansion of nr_n can be dropped. For this reason it is better to retain (7.23) as our final result.

Exercises

1. Suppose that

$$h(x) \sim h(a) + \sum_{s=0}^{\infty} a_s(x-a)^{s+\mu}, \qquad \text{as } x \to a^+,$$

and that $t = h(x) - h(a)$. Show that there exists an asymptotic representation of the form

$$x - a \sim \sum_{s=1}^{\infty} \alpha_s t^{s/\mu}, \qquad \text{as } t \to 0^+,$$

and calculate the coefficients α_1, α_2, and α_3; cf. (1.20) and (1.21).

2. Find the first two terms in the asymptotic expansion of each of the following integrals as $x \to +\infty$.

    (a) $\displaystyle\int_0^{\pi/2} e^{-x\sin^2 t}\, dt,$

    (b) $\displaystyle\frac{1}{\pi}\int_0^{\pi} e^{x\cos\theta}\cos n\theta\, d\theta,$ \qquad modified Bessel function $I_n(x)$,

    (c) $\displaystyle\int_0^{\pi/2} e^{-x\sec t}\, dt,$

    (d) $\displaystyle\int_0^{\infty} t^n e^{-t^2 - x/t}\, dt.$

3. Show that, as $x \to +\infty$,

$$\int_0^{\infty} t^{-t} e^{xt}\, dt \sim \sqrt{2\pi}\, e^{(x-1)/2 + e^{x-1}}.$$

(Hint: Note that the maximum value of the integrand occurs at $t = e^{x-1}$. Therefore, to make the location of the maximum independent of x, introduce a new integration variable $u = te^{1-x}$ and use $y = e^{x-1}$ as the asymptotic variable.)

4. Consider the integral

$$F(\lambda) = \int_{c/\lambda}^{\infty} \left(t - \frac{c}{\lambda}\right)^{\nu-1} f(t) \exp\left\{-\lambda\left(t + \frac{1}{t}\right)\right\} dt,$$

where ν and c are positive numbers, and $f(t)$ is a continuous function in $(0, \infty)$ satisfying the following conditions: (i) $f(t) =$

$O(t^{-\delta})$, as $t \to 0^+$, for some $\delta \geq 0$; (ii) $f(t) = O(e^{\rho t})$, as $t \to \infty$, for some $\rho > 0$; and (iii) $f(t)$ is analytic in a neighborhood of $t = 1$. Assume that $\lambda > 2c$, and split the interval of integration at $t = \frac{1}{2}$ and $t = 1$ so that $F(\lambda) = E(\lambda) + F_1(\lambda) + F_2(\lambda)$, where $E(\lambda)$, $F_1(\lambda)$, and $F_2(\lambda)$ correspond, respectively, to the intervals $(c/\lambda, 1/2)$, $(1/2, 1)$, and $(1, \infty)$. Show that $E(\lambda) = O(\lambda^\delta e^{-\lambda^2/c})$. In the integrals $F_1(\lambda)$ and $F_2(\lambda)$, make the change of variables $t + 1/t - 2 = \tau$. Show that

$$F_1(\lambda) + F_2(\lambda) = e^{-2\lambda} \int_0^\infty f(\tau; \lambda) e^{-\lambda \tau} \, d\tau$$

for some $f(\tau; \lambda)$. Furthermore, prove that in a neighborhood of $\tau = 0$,

$$f(\tau; \lambda) = \sum_{n=0}^\infty c_{n,v}(\lambda) \tau^{n-1/2},$$

where each $c_{n,v}(\lambda)$ is a bounded continuous function of λ in $(0, \infty)$ and is bounded away from zero for all large values of λ. In particular, $c_{0,v}(\lambda) = (1 - c/\lambda)^{v-1} f(1)$. Finally, derive the generalized asymptotic expansion

$$F(\lambda) \sim e^{-2\lambda} \left[\sum_{n=0}^\infty c_{n,v}(\lambda) \Gamma(n + \tfrac{1}{2}) \lambda^{-n-1/2}; \quad \{\lambda^{-n-1/2}\} \right],$$

as $\lambda \to +\infty$ (Frenzen and Wong, 1985).

5. Prove that the series

$$\sum_{r=0}^\infty (-1)^r \binom{\mu}{r} \Gamma^{(r)}(\lambda) (\log z)^{-r}$$

is divergent for all finite values of $\log z$.

6. Prove the two asymptotic results given in (2.29) and (2.31).

7. Show that the integral

$$I(x) = \int_0^1 (-\log t)^{-1/2} e^{ixt} \, dt$$

has the asymptotic expansion

$$I(x) = e^{i(x - \pi/4)} \left(\frac{\pi}{x}\right)^{1/2} + \frac{e^{i\pi/2}}{x(\log x)^{1/2}} + O\left\{\frac{1}{x(\log x)^{3/2}}\right\} \qquad \text{as } x \to +\infty.$$

8.   Find the first two terms in the asymptotic expansion of

$$I(x) = \int_0^1 \frac{e^{ixt}}{(1 - \log t)^\mu} \, dt, \qquad \text{as } x \to +\infty.$$

(Hint: Replace t by et and use $\lambda = xe$ as the asymptotic variable.)

9.   Find the leading term in the asymptotic expansion of each of the following functions as $x \to +\infty$.

(a) $\displaystyle\int_0^\pi \cos\{x(\theta - \sin \theta)\} \, d\theta,$

(b) $\displaystyle\int_a^\infty \frac{1}{t} \exp\left\{i\left(t + \frac{x}{t}\right)\right\} \, dt, \qquad a > 0,$

(c) $\displaystyle\int_a^\infty \frac{1}{t} \exp\left\{i\left(t - \frac{x}{t}\right)\right\} \, dt, \qquad a > 0.$

10. Is the integral

$$I(x) = \int_0^\infty \exp\{it(\log t - x)\} \, dt$$

convergent for all positive x? Prove that

$$I(x) \sim \sqrt{\frac{2\pi}{e}} \exp\left\{\frac{x}{2} - ie^{x-1} + i\frac{\pi}{4}\right\}, \qquad x \to +\infty.$$

11. Let N be a positive integer. Let $g(t)$ be N times, and $f(t)$ be $N + 1$ times, continuously differentiable with $f'(t) \neq 0$ on $[a, b]$. Show that the oscillatory integral

$$F(x) = \int_a^b g(t) e^{ixf(t)} \, dt$$

has the expansion

$$F(x) = \sum_{n=0}^{N-1} \frac{(-1)^{n+1}}{(ix)^{n+1}} h^{(n)}(A) e^{if(a)x} + \sum_{n=0}^{N-1} \frac{(-1)^n}{(ix)^{n+1}} h^{(n)}(B) e^{if(b)x} + R_N,$$

where $A = f(a)$, $B = f(b)$, $h(u) = g(t)/f'(t)$, $u = f(t)$, and

$$R_N = \frac{(-1)^N}{(ix)^N} \int_a^b h^{(N)}(u) e^{ixu} \, du = o(x^{-N}), \qquad \text{as } x \to +\infty.$$

12. Consider the integral

$$I(x, t) = \int_{-\infty}^{\infty} F(\lambda) \exp\{-tg(\lambda) - ix\lambda\} \, d\lambda,$$

where x and t are real, $F(\lambda)$ is a bounded and absolutely integrable function on $(-\infty, \infty)$ and $g(\lambda)$ is a continuous function satisfying the following conditions: (i) $g(\lambda) = 0$ if and only if $\lambda = 0$, (ii) Re $g(\lambda) \geq 0$, and (iii) as $\lambda \to 0$, $g(\lambda) = i\alpha\lambda + \beta\lambda^2 + O(|\lambda|^3)$, with α, β real, $\beta > 0$. Show that the major contribution to the asymptotic behavior of $I(x, t)$ comes from the neighborhood of $\lambda = 0$, and that as $t \to +\infty$, there exists a small positive δ such that

$$I(x, t) = \int_{-\infty}^{\infty} \exp\{-\beta\lambda^2 t - i\alpha\lambda t - i\lambda x\} F(\lambda) \, d\lambda + O(t^{-(1-\delta)}).$$

From this, deduce the asymptotic formula

$$I(x, t) \sim \sqrt{\frac{\pi}{\beta t}} \, F(0) \exp\left[-\frac{(x + \alpha t)^2}{4\beta t}\right], \qquad \text{as } t \to +\infty.$$

13. Consider the Cauchy problem for the dissipative wave equation (or the telegrapher's equation)

$$u_{tt} - u_{xx} + u_t = 0, \qquad t > 0, \quad -\infty < x < \infty,$$

with smooth initial data $u(x, 0)$ and $u_t(x, 0)$. Let $U(\lambda, t)$ denote the Fourier transform of $u(x, t)$. Assuming that u and u_x tend to zero as $x \to \pm\infty$, show that $U(\lambda, t)$ satisfies the ordinary differential equation

$$\frac{\partial^2 U}{\partial t^2} + \frac{\partial U}{\partial t} + \lambda^2 U = 0, \qquad t > 0,$$

with the initial values $U(\lambda, 0)$ and $U_t(\lambda, 0)$ given in terms of the Fourier transforms of $u(x, 0)$ and $u_t(x, 0)$. Derive the solution

$$U(\lambda, t) = F_+(\lambda) \exp[(-\tfrac{1}{2} + \tfrac{1}{2}\sqrt{1 - 4\lambda^2})t] + F_-(\lambda) \exp[(-\tfrac{1}{2} - \tfrac{1}{2}\sqrt{1 - 4\lambda^2})t],$$

where $F_+(\lambda)$ and $F_-(\lambda)$ are specified in terms of the initial data. Show that $u(x, t) = u_+(x, t) + u_-(x, t)$, where

$$u_\pm(x, t) = \frac{1}{\sqrt{2\pi}} \int_{-\infty}^{\infty} F_\pm(\lambda) \exp[(-\tfrac{1}{2} \pm \tfrac{1}{2}\sqrt{1 - 4\lambda^2})t - ix\lambda] \, d\lambda.$$

With the assumption $\int_{-\infty}^{\infty} |F_{\pm}(\lambda)| \, d\lambda < \infty$, prove that

$$|u_-(x, t)| \le \frac{e^{-t/2}}{\sqrt{2\pi}} \int_{-\infty}^{\infty} |F_-(\lambda)| \, d\lambda,$$

and, using Ex. 12, prove also that

$$u_+(x, t) \sim \frac{F_+(0)}{\sqrt{2t}} e^{-x^2/4t}, \qquad \text{as } t \to \infty.$$

Thus, conclude that

$$u(x, t) \sim \frac{F_+(0)}{\sqrt{2t}} \exp\left(-\frac{x^2}{4t}\right), \qquad \text{as } t \to \infty.$$

14. Let $f(t)$ be a real-valued C^∞-function in $[a, b]$, and $g(t)$ be a C^∞-function in $[a, b]$ which vanishes identically at the two end-points. Suppose that t_0 is an interior point of the interval $[a, b]$, and that $g(t)$ is not identically zero near t_0. Furthermore, suppose that $f'(t_0) = 0$, $f''(t_0) \ne 0$ and $f'(t) \ne 0$ for $t \ne t_0$. Put $\sigma = \text{sgn } f''(t_0)$ and $\varphi(t) = \sqrt{\sigma[f(t) - f(t_0)]}/(t - t_0)$. (Note that $\sqrt{\sigma[f(t) - f(t_0)]}$ is a C^∞-function in $[a, b]$ and is positive for $t > t_0$.) Prove that for $x > 0$ and for any integer $n \ge 1$,

$$\int_a^b g(t)e^{ixf(t)} \, dt = e^{ixf(t_0)} \sum_{j=0}^{n-1} a_j(f, g)x^{-j-1/2} + R_n(x),$$

where

$$a_j(f, g) = \frac{\Gamma(j + \frac{1}{2})}{(2j)!} e^{i\pi\sigma(2j+1)/4}\left\{\left(\frac{d}{dt}\right)^{2j} [\varphi(t)^{-2j-1}g(t)]\right\}_{t=t_0}$$

and $R_n(x) = O(x^{-n})$ as $x \to +\infty$. (Hint: Assume $\sigma > 0$, and put $u^2 = f(t) - f(t_0)$. Apply (3.13) with $\rho = 2$, $\lambda = 1$ and $h(u) = g(t) \, dt/du$, and show that

$$a_j(f, g) = \frac{\Gamma(j + \frac{1}{2})}{(2j)!} e^{i\pi(2j+1)/4}h^{(2j)}(0).$$

Furthermore, prove that if f and g are analytic functions in a neighborhood of t_0, then by Cauchy's formula, we have

$$\frac{1}{n!}h^{(n)}(0) = \frac{1}{2\pi i}\int_C \frac{g(t)}{(t - t_0)^{n+1}}\left(\frac{t - t_0}{u}\right)^{n+1} dt,$$

where C is a positively oriented circle around t_0. From this, deduce that for analytic functions f and g, $h^{(n)}(0)/n!$ is the coefficient of $(t - t_0)^n$ in the expansion of

$$g(t)\left[\frac{t - t_0}{\sqrt{f(t) - f(t_0)}}\right]^{n+1} = \varphi(t)^{-n-1}g(t),$$

i.e.,

$$h^{(n)}(0) = \frac{d^n}{dt^n}\left[\varphi(t)^{-n-1}g(t)\right].$$

Finally, conclude that the last equation holds when f and g are only C^∞-functions.)

15. The Airy integral of negative argument has the integral representation

$$\text{Ai}(-x) = \frac{x^{1/2}}{2\pi i}\int_L \exp\left\{\lambda\left(\frac{u^3}{3} + u\right)\right\} du, \qquad x > 0,$$

where $\lambda = x^{3/2}$ and L is a contour which lies in the half-plane $\text{Re } u > 0$, beginning at $\infty\, e^{-i\pi/3}$ and ending at $\infty\, e^{i\pi/3}$ (see Figure 2.4). Show that by integration by parts, the integral on the right-hand side can be written as

$$-\frac{1}{\lambda}\int_L f_1(u)\exp\left\{\lambda\left(\frac{u^3}{3} + u\right)\right\} du, \qquad \text{where} \quad f_1(u) = \frac{d}{du}\frac{1}{u^2 + 1}.$$

By repeating the process, show that

$$2\pi i x^{-1/2}\,\text{Ai}(-x) = \left(-\frac{1}{\lambda}\right)^n\int_L f_n(u)\exp\left\{\lambda\left(\frac{u^3}{3} + u\right)\right\} du,$$

where $f_n(u)$ is defined inductively by

$$f_n(u) = \frac{d}{du}\left\{\frac{1}{u^2 + 1}f_{n-1}(u)\right\}, \qquad n \geq 2.$$

The last integral is clearly absolutely convergent. Hence, $\text{Ai}(-x) = O(x^{-(3n-1)/2})$ as $x \to \infty$, for any $n \geq 0$, but this contradicts the result in (4.38). Find the flaw in the argument.

16. Let

$$I(\lambda) = \int_0^\infty e^{\lambda[(1 + i)x - x^3]}\, dx.$$

Show that the path of integration can be replaced by the radial
line $z = \rho e^{i\pi/8}$, $\rho > 0$, and that as $\lambda \to +\infty$,

$$I(\lambda) \sim e^{-i\pi/16} 2^{-1/8} 3^{-1/4} \pi^{1/2} \lambda^{-1/2} \exp(2^{7/4} 3^{-3/2} e^{i3\pi/8} \lambda).$$

(Dieudonné 1968, p. 251.)

17. Derive the asymptotic expansion of the integral

$$I(\lambda) = \int_{-\infty}^{\infty} z^{\alpha} e^{-z^2 + 2i\lambda z} \, dz$$

as $\lambda \to \infty$ in $|\arg \lambda| \le \pi/4 - \Delta$, $\Delta > 0$, where α is any number, real or
complex, and the path of integration passes above the origin.

18. Use the method of steepest descent to derive the asymptotic
expansion of the Bessel function

$$J_{\alpha}(x) = \frac{1}{2\pi i} \int_{-\infty}^{(0+)} t^{-\alpha-1} \exp\left\{\frac{x}{2}(t - t^{-1})\right\} dt,$$

as $x \to +\infty$, where α is any complex number, and the loop contour
starts from $-\infty$ on the real axis, passes around the origin in a
counterclockwise direction and returns to $-\infty$; (cf. Example 10).

19. Use the method of steepest descent to show that the Hankel
function $H_{\nu}^{(1)}(x)$, defined in (4.39), has the asymptotic expansion

$$\sqrt{\frac{2}{\pi x}} \exp\left[i\left(x - \frac{\nu}{2}\pi - \frac{\pi}{4}\right)\right]\left\{1 + \frac{(\frac{1}{4} - \nu^2)}{2ix} + \cdots\right\},$$

as $x \to +\infty$, for fixed values of ν.

20. In the theory of the wave pattern due to a concentrated pressure
moving over the free surface of water, there arose the integral

$$I(N, \theta) = \int_{-\infty - i\pi/4}^{\infty - i\pi/4} \cosh^3 u \, e^{Nf(u, \theta)} \, du,$$

where $f(u, \theta) = i(\cos \theta \cosh u - \frac{1}{2} \sin \theta \sinh 2u)$. Physically, N is
the large non-dimensional radius measured from the disturbance,
and $\theta > 0$ is the polar angle which appears as a parameter.

(a) Show that the above integral is absolutely convergent for all
$N > 0$.

(b) Show that there exists a critical value θ_c such that (i) for
$0 < \theta < \theta_c$, there are two real saddle points, (ii) for $\theta = \theta_c$, there is

only one real saddle point which is of order 2, and (iii) for $\theta_c < \theta < \pi/2$, there are two conjugate complex saddle points in $|\text{Im } u| < \pi/2$.

(c) Draw the steepest descent curves of $f(u, \theta)$ for each of the three cases mentioned in (b).

(d) Show that in case (i), we have

$$I(N, \theta) \sim \sqrt{\frac{2\pi}{N}} \frac{1}{(\sin \theta)^{1/2}(\cot^2 \theta - 8)^{1/4}}$$

$$\times \left\{ \exp\left(iN \frac{\sin \theta \cosh^3 \xi_-(\theta)}{\sinh \xi_-(\theta)} + i\frac{\pi}{4} \right) \right.$$

$$\left. + \exp\left(iN \frac{\sin \theta \cosh^3 \xi_+(\theta)}{\sinh \xi_+(\theta)} - i\frac{\pi}{4} \right) \right\},$$

as $N \to +\infty$, where $\xi_+(\theta)$ and $\xi_-(\theta)$ are the two real saddle points of $f(u, \theta)$ and $\xi_+(\theta) > \xi_-(\theta)$. Prove that the error in this approximation is $O(N^{-3/2}(\sin \theta)^{-3/2}(\cot^2 \theta - 8)^{-7/4})$.

(e) Show also that in case (ii), we have

$$I(N, \theta_c) = e^{i\sqrt{3}N/2} \cdot \frac{1}{N^{1/3}} \left[\frac{3^{4/3}}{2^{3/2}} \Gamma\left(\frac{1}{3}\right) - \frac{i}{N^{1/3}} \frac{5 \cdot 3^{7/6}}{2^{5/2}} \Gamma\left(\frac{2}{3}\right) + O\left(\frac{1}{N}\right) \right],$$

as $N \to +\infty$.

21. Consider the integral

$$e^\lambda \lambda^{-\lambda} \Gamma(\lambda) = \int_{-1}^{\infty} e^{\lambda\{\log(1 + t) - t\}} \, dt,$$

which is clearly convergent for Re $\lambda > 0$. Let L be the portion of the negative real axis joining $t = -1$ to $t = 0$ and then the radial line arg $t = \gamma$ joining $t = 0$ to $t = \infty\, e^{i\gamma}$. Show that for any γ in $|\gamma| < \pi/2$, there is always a sector $\alpha \le \text{arg } \lambda \le \beta$ for which we may write

$$e^\lambda \lambda^{-\lambda} \Gamma(\lambda) = \int_L e^{\lambda\{\log(1 + t) - t\}} \, dt.$$

Make the substitution $w = [t - \log(1 + t)]^{1/2}$, where the branch is taken to be real and positive when t is real, and positive, so that

$$e^\lambda \lambda^{-\lambda} \Gamma(\lambda) = \int_{L'} \frac{dt}{dw} e^{-\lambda w^2} \, dw,$$

where L' is the image of L. Show that L' consists of the whole negative real-axis plus a curve joining $w = 0$ to $w = \infty\, e^{i\gamma/2}$. Show also that $dt/dw = 2w(1 + t)/t$, and deduce from it that the singularities of $t = t(w)$ in the finite w-plane occur at $w^2 = 2n\pi i$, and hence along $\arg w = \pi/4$ or $\arg w = -3\pi/4$. Finally, prove that the path of integration L' can be deformed into the straight line through the origin extending from $w = \infty\, e^{i(-\pi + \gamma/2)}$ to $w = \infty\, e^{i\gamma/2}$, and derive the expansion (5.35) for $|\arg(\lambda e^{i\gamma})| \leq \pi/2 - \delta$. From this, conclude that the domain of uniform validity of (5.35) is $|\arg \lambda| \leq \pi - \delta$.

22. Let

$$F(z) = \int_0^1 \exp\left[iz\left(\frac{t^3}{3} + t\right)\right] dt.$$

By deforming the path of integration into one similar to that given in Figure 2.1, show that

$$F(z) \sim \sum_{n=0}^{\infty} \frac{(3n)!}{3^n n!\,(-iz)^{3n+1}} + e^{i4z/3} \sum_{n=0}^{\infty} \frac{a_n}{(iz)^{n+1}},$$

as $z \to \infty$ in the region $|\arg z| \leq \pi - \delta$, $\delta > 0$, where

$$a_n = \frac{1}{2^{n+1}} \left[\frac{d^n}{dt^n} \left\{ \frac{6t}{3t + 1 - (1 - t)^3} \right\}^{n+1}\right]_{t=0}.$$

Also derive an asymptotic expansion for $F(z)$ in the sector $\alpha < \arg z < \beta$, where $\alpha < \pi < \beta$.

23. Consider the function

$$F(z) = \int_0^{\infty} \exp\left[iz\left(\frac{t^3}{3} + t\right)\right] dt,$$

where $0 < \arg z < \pi$. Show that, by rotating the path of integration through an angle ψ, $0 < \psi < \pi/3$, $F(z)$ can be continued analytically to the region $-\pi/3 < \arg z < \pi$. Use Theorem 4 to derive the asymptotic expansion

$$F(z) \sim i \sum_{m=0}^{\infty} \frac{(3m)!}{m!\, 3^m} z^{-2m-1},$$

as $z \to \infty$, uniformly in the sector $-\pi/3 + \delta \leq \arg z \leq \pi - \delta$, $\delta > 0$. Note that this expansion holds, in particular, as $z \to +\infty$ through the positive real values. Also note that, since $F(z) = \overline{F(-z)}$, the domain of validity for the above expansion can be extended to $-\pi/3 + \delta \leq \arg z \leq \pi + \pi/3 - \delta$. (For a derivation of the above expansion based on the steepest descent method, see Erdélyi (1956, p. 44) or Henrici (1974, p. 422).)

24. Let $f(z) = (1 - z)^{\lambda - 1}g(z)$, where λ is a fixed complex number and $g(z)$ is an analytic function in $|z| \leq 1 + \delta$, δ being a fixed positive number. Take the cut in the z-plane along the real-axis from $z = 1$ to $z = +\infty$, and let a_n denote the Maclaurin coefficients of $f(z)$. By Cauchy's theorem,

$$a_n = \frac{1}{2\pi i}\int_C z^{-n-1}f(z)\,dz,$$

where C is the contour shown in Figure 2.10. Let $\delta_n = n^{-1/2} < \delta$, and show that

$$\int_{|z| = 1 + \delta_n} z^{-n-1}f(z)\,dz = O\{\exp(-\varepsilon\sqrt{n})\}, \qquad \text{as } n \to \infty,$$

for some fixed $\varepsilon > 0$. Write

$$\frac{1}{2\pi i}\int_{|z-1| = \delta_n} z^{-n-1}f(z)\,dz = \frac{i}{2\pi}\int_{\gamma_n} (z + 1)^{-n-1}f(z + 1)\,dz,$$

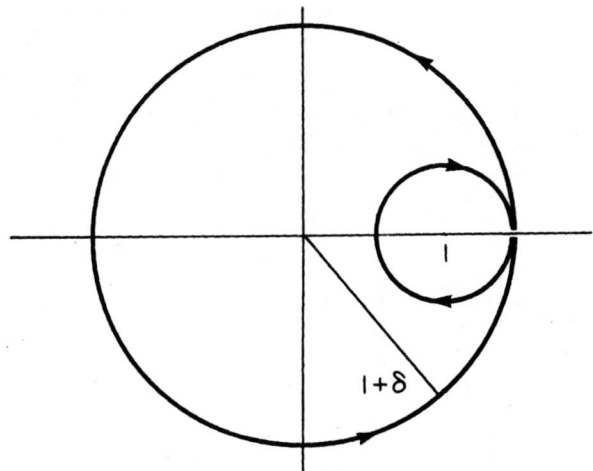

Fig. 2.10 Contour C.

where the path of integration on $|z - 1| = \delta_n$ is oriented in the clockwise direction and γ_n is the contour which traverses on the circle $|z| = \delta_n$ in the counter clockwise direction. Furthermore, write

$$(z + 1)^{-n-1} f(z + 1) = (-z)^{\lambda - 1} G(z; n) e^{-(n+1)z},$$

where $G(z; n) = e^{(n+1)z} (z + 1)^{-n-1} g(z + 1)$.

Show that if we set $w = (n + 1)z$, then $G(z; n)$ has the convergent expansion

$$G(z; n) = \sum_{k=0}^{\infty} P_k(w) z^k, \qquad |z| \le \delta_n,$$

where $P_k(w)$ is a polynomial of degree k. Deduce from this the asymptotic expansion

$$a_n \sim \sum_{k=0}^{\infty} b_k(\lambda)(n + 1)^{-\lambda - k}, \qquad \text{as } n \to \infty,$$

where $b_k(\lambda)$ is some constant. Calculate $b_0(\lambda)$ and $b_1(\lambda)$.

25.   Consider the integral

$$I_n = \int_{a_n}^{1} t^{k_n - 1}(1 - t)^{n - k_n} \, dt,$$

where k_n and a_n are real-valued functions of the integer variable n satisfying $k_n/n = p + O(1/n)$, $n^s |a_n - a| \to 0$ as $n \to \infty$, for each $s = 1, 2, \ldots$, and $0 < p < a < 1$. Put $\lambda(t) = t^p(1 - t)^{1-p}$ and $b = \frac{1}{2}(1 + a)$, and assume that n is sufficiently large that all a_n lie to the right of p and to the left of b. Set

$$I_n = \left(\int_{a_n}^{a} + \int_{a}^{b} + \int_{b}^{1} \right) t^{k_n - 1}(1 - t)^{n - k_n} \, dt = I_{n,1} + I_{n,2} + I_{n,3}.$$

Show that $I_{n,1} = O[\lambda^n(a)n^{-s}]$, $s = 1, 2, \ldots$, and $I_{n,3} = O[\lambda^{n-\zeta}(b)]$ for some $\zeta > 0$. Now write

$$I_{n,2} = \int_{a}^{b} \psi(t; n) e^{-nh(t)} \, dt, \qquad \text{with}$$

$$h(t) = -p \log t - (1 - p) \log(1 - t)$$

and $\psi(t; n) = t^{k_n - np - 1}(1 - t)^{np - k_n}$. Show that the minimum of $h(t)$

in $[a, b]$ occurs at $t = a$. Furthermore, write

$$I_{n,2} = e^{-nh(a)} \int_a^b G(t; n)e^{-nh'(a)(t-a)}\, dt,$$

with $G(t; n) = \psi(t; n)\exp\{-n[\frac{1}{2}h''(a)(t-a)^2 + \cdots]\}$. Prove that

$$G(t; n) = \sum_{j=0}^{m-1} \frac{1}{j!} G^{(j)}(a; n)(t-a)^j + (t-a)^m O(n^{m/2}),$$

where the O-symbol is uniform with respect to n and t for $a \le t \le b$, and consequently that

$$I_{n,2} = e^{-nh(a)}\left\{ \sum_{j=0}^{m-1} G^{(j)}(a; n)[nh'(a)]^{-j-1} + O(n^{-(m+2)/2}) \right\}.$$

Finally, prove that

$$I_n \sim [\lambda(a)]^n \left\{ \frac{c_{1,n}}{n} + \frac{c_{2,n}}{n^2} + \cdots \right\}, \qquad \text{as } n \to \infty,$$

where the coefficients $c_{i,n}$ are bounded functions of n. In particular, show that

$$c_{1,n} = \frac{a}{a-p}\left(\frac{1-a}{a} \right)^{np-k_n+1}$$

and

$$c_{2,n} = \frac{a}{(a-p)^2}\left(\frac{1-a}{a} \right)^{np-k_n+1}(a - np + k_n - 1).$$

(The integral I_n arises in the study of probabilities of large deviation.)

26. Let $X_1, X_2, \ldots$ be a sequence of independent identically distributed random variables with a common density function $f(x)$. Let $\varphi(\zeta)$ be the characteristic function of $f(x)$, i.e., $\varphi(\zeta) = \int_{-\infty}^{\infty} e^{i\zeta x} f(x)\, dx$, and denote the moments of f by $\mu_k = \int_{-\infty}^{\infty} t^k f(t)\, dt$. For simplicity, assume (without loss of generality) that the mean μ_1 is zero and the variance μ_2 is 1. It is well known that the density function $f_n(x)$ of the normalized sum $(X_1 + \cdots + X_n)/\sqrt{n}$ is given by the Fourier inversion formula

$$f_n(x) = \frac{1}{2\pi} \int_{-\infty}^{\infty} e^{-i\zeta x}\left[\varphi\left(\frac{\zeta}{\sqrt{n}} \right) \right]^n d\zeta.$$

Assume that $\int_{-\infty}^{\infty} |\varphi(\zeta)|^p \, d\zeta < \infty$ for some $p \geq 1$, and that the integrals defining the μ_k's are absolutely convergent. Furthermore, assume that $\rho \equiv \int_{-\infty}^{\infty} |x|^3 f(x) \, dx < \infty$. Write $f_n(x) = I_n(x) + J_n(x)$ with

$$I_n(x) = \frac{1}{2\pi} \int_{|\zeta| < \sqrt{n}/\rho} e^{-i\zeta x} \varphi^n\left(\frac{\zeta}{\sqrt{n}}\right) d\zeta,$$

$$J_n(x) = \frac{1}{2\pi} \int_{|\zeta| \geq \sqrt{n}/\rho} e^{-i\zeta x} \varphi^n\left(\frac{\zeta}{\sqrt{n}}\right) d\zeta.$$

Show that

$$|J_n(x)| \leq \frac{\sqrt{n}}{2\pi} q^{n-q} \int_{-\infty}^{\infty} |\varphi(\zeta)|^p \, d\zeta,$$

where

$$0 < \sup_{|\zeta| \geq 1/\rho} |\varphi(\zeta)| = q < 1.$$

Write

$$I_n(x) = \frac{1}{2\pi} \int_{|\zeta| < \sqrt{n}/\rho} e^{n\psi(\zeta/\sqrt{n})} e^{-\zeta^2/2 - ix\zeta} \, d\zeta,$$

where

$$\psi(\zeta) = \log \varphi(\zeta) + \frac{1}{2} \zeta^2.$$

Show that $|\psi(\zeta)| \leq (5/12)\rho|\zeta|^3$ for $|\zeta| < 1/\rho$, and that

$$e^{n\psi(\zeta/\sqrt{n})} = \sum_{k=0}^{N} \frac{1}{k!} \left[n\psi\left(\frac{\zeta}{\sqrt{n}}\right) \right]^k + R_{N+1}(\zeta, n),$$

where

$$|R_{N+1}(\zeta, n)| \leq \left(\frac{e}{N+1}\right)^{N+1} \left(\frac{5}{12}\rho\right)^{N+1} \frac{1}{n^{(N+1)/2}} |\zeta|^{3(N+1)} e^{5\rho|\zeta|^3/12\sqrt{n}}.$$

Note that formally we have

$$e^{n\psi(\zeta/\sqrt{n})} \text{ `` = ''} 1 + \sum_{\nu=1}^{\infty} n^{-\nu/2} Q_\nu(\zeta),$$

where $Q_\nu(\zeta)$ is a polynomial in ζ. Deduce from these results the following well-known *Edgeworth series* in probability theory:

$$f_n(x) = \eta(x) + \eta(x) \sum_{k=3}^{r} n^{-1/2k+1} P_k(x) + o(n^{-1/2r+1}),$$

as $n \to \infty$, uniformly in x, where $P_k(x)$ is a real polynomial depending only on $\mu_1, \ldots, \mu_k$ but not on n or r, and $\eta(x)$ denotes the normal density $\eta(x) = (1/\sqrt{2\pi})e^{-x^2/2}$. (The leading term of Edgeworth series gives the *central limit theorem*: $\lim_{n \to \infty} f_n(x) = \eta(x)$.)

27. The Legendre polynomials are generated by

$$\frac{1}{(1 - 2z \cos\theta + z^2)^{1/2}} = \sum_{n=0}^{\infty} P_n(\cos\theta)z^n, \qquad |z| < 1.$$

Apply Darboux's method to derive the asymptotic expansion

$$P_n(\cos\theta) \sim \left(\frac{2}{\sin\theta}\right)^{1/2} \sum_{v=0}^{\infty} \binom{-\frac{1}{2}}{v}\binom{v - \frac{1}{2}}{n} \frac{\cos\theta_{n,v}}{(2\sin\theta)^v}; \qquad \{n^{-v-1/2}\},$$

as $n \to \infty$, for $0 < \theta < \pi$, where $\theta_{n,v} = (n - v + \frac{1}{2})\theta + (n - \frac{1}{2}v - \frac{1}{4})\pi$.

28. Let $\{a_n\}$ be defined by $a_1 = 0$, $a_2 = 1$, $a_n = n(n-1)a_{n-1} + \frac{1}{2}n(n-1)^2 a_{n-2}$, $n \geq 3$, and let $b_n = 2a_{n+2}/(n+2)!\,(n+1)!$. Prove that $b_0 = b_1 = 1$ and

$$b_n = b_{n-1} + \frac{1}{2n} b_{n-2}, \qquad n \geq 2.$$

Furthermore, prove that the generating function $B(z) = \sum_{n \geq 0} b_n z^n$ satisfies $2B'(z) = 2zB'(z) + (2 + z)B(z)$. Deduce from this the closed-form expression $B(z) = e^{-z/2}(1 - z)^{-3/2}$. Show that

$$b_n = (\pi e n)^{-1/2}[2n + \tfrac{5}{4} + O(n^{-1})]$$

and

$$a_n \sim 2\sqrt{\pi}\, n^{2n+1/2} e^{-2n-1/2}.$$

(This problem arose in counting Hamiltonian cycles for bipartite graphs; see Knuth (1980).)

29.   Let $f(z) = \exp\{(1 - z)^\alpha\}$, where α is a complex number, Re $\alpha > 0$, $\alpha \neq$ positive integer. Show that the $(n + 1)$th Maclaurin coefficient of $f(x)$, denoted by b_n, has the asymptotic expansion

$$b_n \sim \sum_{r=1}^{\infty} \frac{\Gamma(n - \alpha r)}{r! \, n! \, \Gamma(-\alpha r)}; \qquad \{n^{-1-\alpha r}\}, \qquad \text{as } n \to \infty.$$

30.   Let $\{y_n\}$ be a sequence of positive numbers, and suppose that the power series $y(x) = \sum_{n=0}^{\infty} y_n x^n$ has radius of convergence $\rho = 0.3383\ldots$. Suppose also that, about the point $x = \rho$, $y(x)$ has an expansion of the form

$$y(x) = 1 - b(\rho - x)^{1/2} + b_2(\rho - x) + b_3(\rho - x)^{3/2} + \cdots,$$

where $b = 2.6811\ldots$. Show that $y_n \sim (b/2\sqrt{\pi})\rho^{-n+1/2}n^{-3/2}$ as $n \to \infty$. (This problem occurred in a study of rooted unlabelled trees; see Otter (1948).)

31.   Let $A_{n,d}$ denote the number of solutions of $x^d = 1$ in the symmetric group S_n of degree n. It is known that if d is a prime p, then

$$\sum_{n=0}^{\infty} A_{n,p} \frac{x^n}{n!} = \exp\left(z + \frac{z^p}{p}\right).$$

Show that

$$A_{n,2} \sim \frac{1}{\sqrt{2}} \left(\frac{n}{e}\right)^{n/2} \exp(\sqrt{n} - \tfrac{1}{4})$$

and

$$A_{n,p} \sim \frac{1}{\sqrt{p}} \left(\frac{n}{e}\right)^{n(1 - 1/p)} \exp(n^{1/p}) \quad \text{for } p > 2.$$

(Moser and Wyman 1955a.)

32.   Let U_n be the number of idempotent elements in the symmetric semigroup T_n on n elements. It is known that

$$\exp(ze^z) = 1 + \sum_{n=1}^{\infty} U_n \frac{z^n}{n!};$$

cf. Harris and Schoenfeld (1968).

Derive the asymptotic formula

$$U_n \sim \frac{n!}{r_n^n} \sqrt{\frac{r_n + 1}{2\pi n C_n}} \exp\left\{\frac{n}{r_n + 1}\right\},$$

where r_n is defined by $r_n(r_n + 1)\exp(r_n) = n$ and $C_n = r_n^2 + 3r_n + 1$.

33. The Stirling numbers σ_n^m of the second kind are given by

$$\frac{(e^z - 1)^m}{m!} = \sum_{n=m}^{\infty} \sigma_n^m \frac{z^n}{n!}.$$

These numbers play an important role in the calculus of finite differences. Show that

$$\frac{m!}{n!} \sigma_n^m \sim \frac{(e^{r_n} - 1)^m}{r_n^n \sqrt{2\pi m r_n H_n}},$$

where r_n is the solution of

$$\frac{r}{1 - e^{-r}} = \frac{n}{m} \quad \text{and} \quad H_n = \frac{e^{r_n}(e^{r_n} - 1 - r_n)}{(e^{r_n} - 1)^2}.$$

(Moser and Wyman 1958). Furthermore, show that by using Stirling's formula and the approximation $r_n \sim n/m$, we have $\sigma_n^m \sim m^n/m!$ as $n \to \infty$.

Supplementary Notes

§1. A considerable amount of material has been written on Laplace's method. In addition to the books on asymptotics listed in the supplementary notes of Chapter I, we mention Pólya and Szegö (1972, Part II, Chapter 5) and Widder (1941, Chapter 7). A derivation of the infinite asymptotic expansion (1.13) can also be found in Erdélyi (1956). Fulks (1960) has shown that the smoothness condition in Theorem 1 can be removed. For results on Laplace's method for two parameters, we refer to Pederson (1965) and the references given there. Our presentation here is based on Olver (1974a, Chapter 3, §§7 and 8). In fact, he has given a detailed discussion on the construction of error bounds for the asymptotic expansion (1.13). In this regard, we also mention the papers by Olver (1968) and Jones (1972).

   If the parameter β in (1.43) is a negative integer, then the integral in Example 2 has been studied, prior to Erdélyi (1961), by Bakhoom (1933). If the interval of integration $(0, \infty)$ in (1.43) is replaced by a loop contour around the origin, then the behavior of this integral has also been investigated by Wright (1934).

§2. When the upper limit c is a real number in $(0, 1)$, the Laplace integral $L(\lambda, \mu, z)$ in (2.22) was first considered by Erdélyi (1961). When $c = 1$ and μ is positive, the result (2.23) is also given as an exercise in Olver (1974a, p. 325). Our presentation here is essentially based on Wong and Wyman (1972). We choose to allow c to be a complex number in order that we can use Theorem 2 to derive a similar result for the Fourier integral $I(\lambda, \mu, x)$ in (2.41). This approach is taken from Wong and Lin (1978). For other results concerning logarithmic singularities, we mention Riekstiņš (1974), Bleistein (1977), Wong (1977, 1978), and Armstrong and Bleistein (1980).

§3. A drawback of Erdélyi's treatment of stationary phase is that it does not lead to satisfactory error bounds. This is due to the fact that his method depends on the artificial extension of the function $h(u)$ from the finite interval $[0, B]$ to the infinite interval $[0, \infty)$; see (3.7) and (3.8). A simpler and more effective approach, based on Abel summability, has since been given by Olver (1974b). Olver's method not only provides computable bounds for the error term but also can be used to derive asymptotic expansions for a large class of oscillatory integrals. Although Erdélyi's treatment is of limited use in obtaining error bounds, it is still a worth-while technique to know in asymptotics. For this reason, we present his method in the present section, and demonstrate the more powerful method of Olver in Chapter IV (§1 and Ex. 4). For a more detailed discussion of the Klein–Gordon equation (3.27) we refer to Zauderer (1983, pp. 247–251).

§4. The material in this section is classical. The geometrical discussion of the steepest descent method is taken from Wyman (1964); for a different discussion, see Olver (1974a, Chapter 4, §10.3). Examples 7 and 8 are based on Erdélyi (1956, §2.6) and Watson (1944, Chapter VIII), respectively.

   It is known that to obtain the asymptotic expansion of the integral (4.1) there is no need to deform the integration path into paths of steepest descent; see Copson (1965, §36), de Bruijn (1970,

Chapter 5), and also §5 of this chapter. However, as Olver (1970a; 1974a, p. 137) has indicated, these paths are needed if one wishes to obtain the maximum region of validity of the resulting asymptotic expansion in the complex plane, and to construct sharp error bounds.

§5. The major difference between the method of steepest descent and that of Perron is the following: in the first method, we reduce the contour integral (5.1) to a canonical form by introducing a new variable of integration $f(z) - f(a) = -\tau$, where a is a saddle point, and then apply Watson's lemma to the resulting integral. In the second method, we set $f(z) - f(a) = -\mu(z - a)^p + \mu(z - a)^p \varphi(z)$ (see eq. (5.5)), and write the integrand in the form $g(z)e^{\lambda f(z)} = G(\omega, z)e^{-\omega}$, where $\omega = \lambda\mu(z - a)^p$; cf. (5.16). Since the function $G(\omega, z)$ depends on the asymptotic variable λ, Watson's lemma does not apply directly. However, by following the argument used in that lemma, a corresponding asymptotic expansion results. Perron's idea of factorizing the integrand is particularly suitable to situations in which f and g depend on λ; see Exs. 24 and 25, and also Erdélyi and Wyman (1963).

Example 10 is taken from Wyman (1965).

§6. Darboux's method has been extended by Wong and Wyman (1974) to allow the generating function $f(z)$ in (6.1) to have logarithmic-type singularities on its circle of convergence. In a different direction, it has also been extended by Fields (1968), who assumed that the generating function has three algebraic singularities on the circle of convergence, say $|z| = 1$, and that one is fixed at $z = 1$, and the other two are symmetrically located with respect to 1. As the distance between the singularities varies, the two symmetrically located ones may coalesce with each other at $z = -1$, or else may coincide with $z = 1$. Fields obtained generalized asymptotic expansions that are uniform with respect to the distance between the singularities. Fields' results are, however, too complicated to apply in their present form, and need further investigation to see whether simplifications are possible.

§7. Hayman's argument is essentially similar to that given in §5 for Perron's method. In fact, Moser and Wyman (1955a, 1955b, 1956, 1957, 1958) have used the Perron-type argument to derive asymptotic expansions for the Maclaurin coefficients of several specific entire functions. Other authors have used the saddle point method to achieve similar results; see de Bruijn (1970, §6.2) and Olver (1974a, Chapter 9, §3 and Chapter 8, §10.4). We choose to present Hayman's result, since it is easy to apply and is applicable to a large class of entire functions. For other results on this subject, we mention Wyman (1959), Grosswald (1966, 1969), and Harris and Schoenfeld (1968).

Exercises.   Exercise 3 is based on an example in Evgrafov (1961, p. 27) and Dieudonné (1968, p. 109). For a solution to Ex. 5, see Wong (1970). The results in Exs. 7 and 8 are taken from Wong and Lin (1978). The contents of Exs. 12 and 13 are given in Zauderer (1983, Example 5.14). The results in Ex. 20 are due to Peters (1949), but our presentation is based on Ursell (1972). The argument outlined in Ex. 21 is given by Wyman (1965). The alternative treatment of Darboux's method described in Ex. 24 is due to Erdélyi and Wyman (1963). The integral I_n in Ex. 25 occurs in the study of probabilities of large deviations; for a discussion of this problem, see Fu and Wong (1980). The derivation of the Edgeworth series outlined in Ex. 26 is taken from Feller (1966). The generalized asymptotic expansion of the Legendre polynomial $P_n(\cos \theta)$ given in Ex. 27 can be shown to be convergent, but it converges to $2P_n(\cos \theta)$. This interesting example of a paradox in asymptotics is due to Olver (1970b).

III

Mellin Transform Techniques

1. Introduction

The Mellin transform of a locally integrable function $f(t)$ on $(0, \infty)$ is defined by

$$M[f; z] = \int_0^\infty t^{z-1} f(t) \, dt, \tag{1.1}$$

when the integral converges. Its domain of analyticity is usually an infinite strip $a < \operatorname{Re} z < b$. The inversion formula for this transform is given by

$$f(t) = \frac{1}{2\pi i} \int_{c-i\infty}^{c+i\infty} t^{-z} M[f; z] \, dz, \tag{1.2}$$

where $a < c < b$. One of the two convolution integrals studied in the Mellin transform theory takes the form

$$I(x) = \int_0^\infty f(t) h(xt) \, dt. \tag{1.3}$$

Note that many important integral transforms, including the Laplace, Fourier, Hankel, and Stieltjes transforms, can be put in the form of

(1.3). Taking Mellin transforms on both sides of (1.3) gives

$$M[I; z] = M[f; 1 - z]\, M[h; z]. \tag{1.4}$$

Conditions for (1.1), (1.2), and (1.4) are summarized in §2. If $M[f; 1 - z]$ and $M[h; z]$ have a common strip of analyticity, say, $a < \operatorname{Re} z < b$, then by inversion we obtain

$$I(x) = \frac{1}{2\pi i} \int_{c - i\infty}^{c + i\infty} x^{-z} M[f; 1 - z]\, M[h; z]\, dz, \tag{1.5}$$

$a < c < b$. This identity (when $x = 1$) is also known as the *Parseval formula*. If $M[f; 1 - z]$ and $M[h; z]$ can be analytically continued to meromorphic functions in a left half-plane and the vertical line of integration can be shifted from $\operatorname{Re} z = c$ to $\operatorname{Re} z = d < c$, then by the Cauchy residue theorem,

$$I(x) = \sum_{d < \operatorname{Re} z < c} \operatorname{Res}\{x^{-z}\, M[f; 1 - z]\, M[h; z]\} + E(x), \tag{1.6}$$

where

$$E(x) = \frac{1}{2\pi i} \int_{d - i\infty}^{d + i\infty} x^{-z}\, M[f; 1 - z]\, M[h; z]\, dz. \tag{1.7}$$

The sum in (1.6) yields the asymptotic expansion of $I(x)$ for small x, the terms of the expansion being the residues. Similarly, if $M[f; 1 - z]$ and $M[h; z]$ can be analytically continued to meromorphic functions in a right half-plane and the vertical line of integration can be shifted to the right, then there results an asymptotic expansion for $I(x)$ with x tending to infinity (see Example 2 in §3).

   The Mellin transform technique outlined above is briefly discussed in Doetsch's book (1955) on Laplace transforms, and is systematically developed by Handelsman and Lew (1969, 1970, 1971). They show, in particular, that this technique can be adapted even to cases in which the functions $M[f; 1 - z]$ and $M[h; z]$ do not have a common strip of analyticity. Furthermore, by using this technique, they have derived asymptotic expansions for a large class of functions expressible in the form (1.3). Generalizations of their results have been given by Handelsman and Bleistein (1973) and Bleistein (1977).

   In §2 we summarize some of the well-known properties of the Mellin transform. Two illustrative examples are given in §3. The work of Handelsman and Lew is presented in §§4 and 5. For a more detailed discussion of their work, we refer to the book by Bleistein and Handelsman (1975a). In §6 we consider methods of constructing explicit expressions for the remainder terms. The final section contains a derivation of an asymptotic expansion of a double integral.

2. Properties of Mellin Transforms

Let $f(t)$ be a locally integrable function on $(0, \infty)$. We first note that the Mellin transform

$$M[f; z] = \int_0^\infty t^{z-1} f(t)\, dt \tag{2.1}$$

can be expressed as a bilateral Laplace transform through the substitution $t = e^{-u}$, which gives

$$M[f; z] = \int_{-\infty}^\infty f(e^{-u}) e^{-zu}\, du. \tag{2.2}$$

Hence many properties of this transform can be obtained from the Laplace transform theory (Widder 1941, Chapter IV). Nevertheless, for our purpose, it is desirable to record them in the Mellin form.

Theorem 1. (i) *Let a and b be given by*

$$a = \sup\{\alpha: f(t) = O(t^\alpha) \text{ as } t \to 0^+\},$$

$$b = \sup\{\beta: f(t) = O(t^{-\beta}) \text{ as } t \to +\infty\}.$$

If $b > -a$ then the integral (2.1) converges absolutely for $-a < \operatorname{Re} z < b$ and defines an analytic function there.

(ii) *For $-a < x < b$, we have*

$$\lim_{y \to \pm\infty} M[f; x + iy] = 0. \tag{2.3}$$

(iii) *If I is a compact interval in $(-a, b)$ and*

$$N(f, I; y) = \sup\{|M[f; x + iy]| : x \in I\}, \tag{2.4}$$

then $N(f, I; y)$ is continuous for $-\infty < y < \infty$ and

$$\lim_{y \to \pm\infty} N(f, I; y) = 0. \tag{2.5}$$

Proof.   Statement (i) is well-known, and can be proved as in Widder (1941, pp. 240–241). Statement (ii) follows from (2.2) and the Riemann–Lebesgue lemma. To prove (iii), we first note that

$$\frac{d}{dz} M[f; z] = \int_0^\infty t^{z-1} (\log t) f(t) \, dt,$$

and hence

$$\left| \frac{d}{dz} M[f; z] \right| \le \int_0^\infty t^{x-1} |f(t) \log t| \, dt.$$

The last integral is continuous, and therefore bounded by some constant $K > 0$, for x in I. A simple estimation then gives

$$|M[f; z_1] - M[f; z_2]| \le K|z_1 - z_2| \tag{2.6}$$

for any z_1 and z_2 with Re z_i in I. From (2.6), it follows that

$$|N(f, I; y_1) - N(f, I; y_2)| \le K|y_1 - y_2|, \tag{2.7}$$

thus proving the continuity of $N(f, I; y)$. The result (2.5) is proved by contradiction. To be definite, we suppose that it is false for the case $y \to +\infty$. Then there exists a positive constant δ and a sequence $\{y_k\}$ such that $y_k \to +\infty$ and $N(f, I; y_k) \ge \delta$. The last inequality implies that there is also a sequence $\{x_k\}$ such that

$$|M[f; x_k + iy_k]| \ge \frac{\delta}{2}. \tag{2.8}$$

Since I is compact, there is a subsequence $\{x_{k_j}\}$ converging to a point x^* in I. Choose J so that $|x_{k_j} - x^*| \le \delta/4K$ for all $k_j \ge J$. From (2.6), we have

$$|M[f; x_{k_j} + iy_{k_j}] - M[f; x^* + iy_{k_j}]| \le \frac{\delta}{4}$$

for all $k_j \ge J$, which coupled with (2.8) yields

$$|M[f; x^* + iy_{k_j}]| \ge \frac{\delta}{4} > 0.$$

This contradicts (2.3) and proves the theorem.   ∎

The next three results require weaker assumptions than Theorem 1.

Theorem 2. *Suppose the integral (2.1) converges absolutely on the line* Re $z = c$. *(a) If $f(t)$ is of bounded variation in a neighbourhood of $t = x$, then*

$$\lim_{T \to \infty} \frac{1}{2\pi i} \int_{c-iT}^{c+iT} x^{-z} M[f; z] \, dz = \frac{1}{2}[f(x^+) + f(x^-)]. \qquad (2.9)$$

(b) *If $M[f; c + iy] \in L^1(-\infty, \infty)$, then*

$$f(t) = \frac{1}{2\pi i} \int_{c-i\infty}^{c+i\infty} t^{-z} M[f; z] \, dz \qquad (2.10)$$

for almost all $t \in [0, \infty)$.

Case (a) follows from (2.2) and the inversion formula for the bilateral Laplace transform (Widder 1941, p. 241 & p. 246), and case (b) may be obtained from the inversion theorem for Fourier transforms (Rudin 1974, p. 199).

Theorem 3. *Let*

$$M[f; 1 - z] = \int_0^\infty t^{-z} f(t) \, dt, \qquad M[h; z] = \int_0^\infty t^{z-1} h(t) \, dt, \quad (2.11)$$

and

$$I(x) = \int_0^\infty f(t)h(xt) \, dt.$$

Then

$$M[I; z] = M[f; 1 - z] \, M[h; z]$$

for all z for which both integrals in (2.11) converge absolutely.

Proof. Let Re $z = \sigma$. By hypothesis

$$\int_0^\infty |f(t)| \int_0^\infty u^{\sigma-1} |h(tu)| \, du \, dt$$

$$= \left(\int_0^\infty t^{-\sigma} |f(t)| \, dt \right) \left(\int_0^\infty t^{\sigma-1} |h(t)| \, dt \right) < \infty.$$

Fubini's theorem then gives

$$M[f; 1 - z] \, M[h; z] = \int_0^\infty t^{-1} f(t) \int_0^\infty \left(\frac{y}{t}\right)^{z-1} h(y) \, dy \, dt$$

$$= \int_0^\infty f(t) \int_0^\infty u^{z-1} h(tu) \, du \, dt = \int_0^\infty u^{z-1} \int_0^\infty f(t) h(ut) \, dt \, du$$

$$= \int_0^\infty u^{z-1} I(u) \, du = M[I; z]. \qquad \blacksquare$$

Theorem 4. *Let $t^{-c} f(t)$ and $t^{c-1} h(t)$ be both absolutely integrable in $(0, \infty)$. If* (i) *$M[h; c + iy] \in L^1(-\infty, \infty)$, or if* (ii) *$M[f; 1 - c - iy] \in L^1(-\infty, \infty)$, then*

$$\int_0^\infty f(t) h(xt) \, dt = \frac{1}{2\pi i} \int_{c-i\infty}^{c+i\infty} x^{-z} M[f; 1 - z] \, M[h; z] \, dz. \qquad (2.12)$$

Proof. We shall prove (2.12) only under the conditions in (i). The argument for the other case is similar. By Fubini's theorem, we have

$$\frac{1}{2\pi i} \int_{c-i\infty}^{c+i\infty} x^{-z} M[f; 1 - z] \, M[h; z] \, dz$$

$$= \frac{1}{2\pi i} \int_{c-i\infty}^{c+i\infty} x^{-z} \int_0^\infty t^{-z} f(t) \, dt \, M[h; z] \, dz$$

$$= \int_0^\infty f(t) \left[\frac{1}{2\pi i} \int_{c-i\infty}^{c+i\infty} (xt)^{-z} M[h; z] \, dz \right] dt$$

$$= \int_0^\infty f(t) h(xt) \, dt,$$

the last step following from the inversion formula (2.10). $\blacksquare$

A well-known property of the Mellin transform, defined by the integral (2.1), is that often it can be continued analytically beyond its original domain of analyticity. Since this fact plays an important role in the work of Handelsman and Lew, we reproduce the following theorem given in Evgrafov (1966, p. 211) as an introduction to some of the later results developed in §4.

Theorem 5. *Let* (i) $f(t)$ *be a locally integrable function on* $(0, \infty)$, (ii) $f(t) = O(t^{-b})$ *as* $t \to \infty$, *and* (iii)

$$f(t) \sim \sum_{s=0}^{\infty} a_s t^{\alpha_s}, \qquad as \ t \to 0^+, \tag{2.13}$$

where $\operatorname{Re} \alpha_s \uparrow +\infty$ *as* $s \to \infty$. *If* $-\operatorname{Re} \alpha_0 < b$ *then the function* $M[f; z]$ *defined in* (2.1) *can be continued analytically to a meromorphic function in the half-plane* $\operatorname{Re} z < b$, *with simple poles at* $z = -\alpha_s$ *of residue* a_s, $s = 0, 1, 2, \ldots$.

Proof. Put

$$f_n(t) = f(t) - \sum_{s=0}^{n-1} a_s t^{\alpha_s}. \tag{2.14}$$

For $-\operatorname{Re} \alpha_0 < \operatorname{Re} z < b$, we have

$$M[f; z] = \int_0^1 t^{z-1} f_n(t) \, dt + \sum_{s=0}^{n-1} \frac{a_s}{z + \alpha_s} + \int_1^{\infty} t^{z-1} f(t) \, dt. \tag{2.15}$$

By Theorem 1, the first integral on the right is analytic for $-\operatorname{Re} \alpha_n < \operatorname{Re} z$, whereas the second integral is analytic for $\operatorname{Re} z < b$. Thus the right-hand side of (2.15) is a meromorphic extension of $M[f; z]$ to the infinite strip $-\operatorname{Re} \alpha_n < \operatorname{Re} z < b$, with poles at $z = -\alpha_s$, $s = 0, 1, \ldots$, $n - 1$. Since $\operatorname{Re} \alpha_n \to +\infty$ as $n \to \infty$, the proof is complete. ∎

3. Examples

In this section, we give two examples. In the first example, we derive an asymptotic expansion for the given integral as the asymptotic variable tends to zero. This problem was proposed by W. T. Yap, a chemist at National Bureau of Standards, who encountered it in an analysis of the chronoamperometry at finite rimless disk electrodes. In the second example, we present an alternative derivation of a known asymptotic expansion first given by Y. L. Luke (1968). The integral in this example arose from a study of free and forced vibration of a circular membrane submerged in a compressible fluid.

Example 1.   Consider the integral

$$I(\lambda) = \int_0^\infty e^{-\lambda t^2} \frac{\sin t}{t^2} J_1(t)\, dt, \tag{3.1}$$

where $J_1(t)$ is the Bessel function of the first kind and of order 1, and λ is a small positive parameter. In the notations of §2, we have $x = \lambda^{1/2}$,

$$h(t) = e^{-t^2} \qquad \text{and} \qquad f(t) = \frac{\sin t}{t^2} J_1(t).$$

The Mellin transform formulas

$$M[h; z] = \frac{1}{2}\Gamma\!\left(\frac{z}{2}\right) \qquad \text{and} \qquad M[f; 1 - z] = \frac{\Gamma\!\left(\frac{3}{2} + z\right)\Gamma\!\left(\frac{1}{2} - \frac{z}{2}\right)}{\Gamma(3 + z)\Gamma\!\left(1 + \frac{z}{2}\right)}$$

can be found in Oberhettinger (1974) or Erdélyi *et al.* (1954). The domain of analyticity of $M[h; z]$ includes the half-plane Re $z > 0$ and that of $M[f; 1 - z]$ includes the strip $-3/2 < $ Re $z < 1$. We may therefore take the common strip of analyticity to be $0 < $ Re $z < 1$. Since $\Gamma(1 + w) = w\Gamma(w)$, it follows from (2.12) that

$$I(\lambda) = \frac{1}{2\pi i} \int_{c-i\infty}^{c+i\infty} \lambda^{-z/2} \frac{\Gamma\!\left(\frac{3}{2} + z\right)\Gamma\!\left(\frac{1}{2} - \frac{z}{2}\right)}{z\Gamma(3 + z)}\, dz \tag{3.2}$$

for any c in $(0, 1)$. In view of the well-known approximation (Lebedev 1965, p. 15),

$$|\Gamma(x + iy)| = \sqrt{2\pi}\, e^{-\pi|y|/2} |y|^{x - 1/2} [1 + r(x, y)], \tag{3.3}$$

where $r(x, y) \to 0$, as $|y| \to \infty$, uniformly for bounded $|x|$, we may use Cauchy's theorem to move the integration contour to the left. Since

$$\text{Res}\{\Gamma(w)\colon w = -s\} = \frac{(-1)^s}{s!}, \qquad s = 0, 1, 2, \ldots, \tag{3.4}$$

we have for any $n \geq 0$

$$I(\lambda) = \frac{\pi}{4} - \sum_{s=0}^{n} \frac{(-1)^s}{s!} \frac{\Gamma\!\left(\frac{s}{2} + \frac{5}{4}\right)}{\left(s + \frac{3}{2}\right)\Gamma\!\left(\frac{3}{2} - s\right)} \lambda^{s/2 + 3/4} + E_n(\lambda), \tag{3.5}$$

where the remainder is given by

$$E_n(\lambda) = \frac{1}{2\pi i} \int_{-n-2-i\infty}^{-n-2+i\infty} \lambda^{-z/2} \frac{\Gamma\left(\frac{3}{2}+z\right)\Gamma\left(\frac{1}{2}-\frac{z}{2}\right)}{z\Gamma(3+z)} \, dz.$$

An explicit estimate for $E_n(\lambda)$ can be obtained as follows. First we recall the well-known identities

$$|\Gamma(iy)|^2 = \frac{\pi}{y \sinh \pi y}, \qquad |\Gamma(\tfrac{1}{2}+iy)|^2 = \frac{\pi}{\cosh \pi y}, \tag{3.6}$$

and

$$\Gamma(z-n) = \frac{\Gamma(z)}{(z-1)(z-2)\cdots(z-n)}. \tag{3.7}$$

From these it follows that

$$\left| \frac{\Gamma(\tfrac{3}{2}-n-2+iy)}{\Gamma(3-n-2+iy)} \right| \le \frac{(y \tanh \pi y)^{1/2}}{\tfrac{1}{4}+y^2}.$$

This, together with the facts that $|\tanh y| \le 1$,

$$|\Gamma(x+iy)| \le |\Gamma(x)|, \tag{3.8}$$

and

$$\int_0^\infty \frac{t^{x-1}}{(1+t)^{x+y}} \, dt = \frac{\Gamma(x)\Gamma(y)}{\Gamma(x+y)}, \tag{3.9}$$

implies

$$|E_n(\lambda)| \le \Gamma\left(\frac{3+n}{2}\right) \lambda^{n/2+1}, \tag{3.10}$$

which shows that (3.5) is an asymptotic expansion.

Example 2. Consider the integral

$$I(x) = \int_0^\infty \frac{J_\nu^2(xt)}{1+t} \, dt, \qquad \nu > -\tfrac{1}{2}, \tag{3.11}$$

where $J_\nu(t)$ is the Bessel function of the first kind and of order ν and x is a large positive parameter. Let $h(t) = J_\nu^2(t)$ and $f(t) = 1/(1+t)$. Then from Oberhettinger (1974, p. 13 & p. 98) we have

$$M[f; 1 - z] = \frac{\pi}{\sin \pi z}$$

for $0 < \text{Re } z < 1$ and

$$M[h; z] = \frac{2^{z-1}\Gamma\left(v + \dfrac{z}{2}\right)}{\Gamma^2\left(1 - \dfrac{z}{2}\right)\Gamma\left(1 + v - \dfrac{z}{2}\right)\Gamma(z)} \frac{\pi}{\sin \pi z}$$

for $-2v < \text{Re } z < 1$. In the last equation, we have used the identity $\Gamma(z)\Gamma(1 - z) = \pi/\sin \pi z$. Thus, in the half-plane $\text{Re } z > \max\{0, -2v\}$, $M[f; 1 - z] M[h; z]$ has a pole of order two at every positive integer. A simple calculation shows that

$$\text{Res}\{x^{-z} M[f; 1 - z] M[h; z]\}|_{z=n} = (a_n \log x + b_n)x^{-n}, \quad (3.12)$$

where

$$a_n = -\frac{2^{n-1}\Gamma\left(v + \dfrac{n}{2}\right)}{\Gamma^2\left(1 - \dfrac{n}{2}\right)\Gamma\left(1 + v - \dfrac{n}{2}\right)\Gamma(n)} \qquad (3.13)$$

and

$$b_n = -a_n\left\{\log 2 + \frac{1}{2}\psi\left(v + \frac{n}{2}\right) + \psi\left(1 - \frac{n}{2}\right) + \frac{1}{2}\psi\left(1 + v - \frac{n}{2}\right) - \psi(n)\right\}.$$

$$(3.14)$$

Note that when n is an even integer, we have $a_n = b_n = 0$.

We now apply the Parseval identity (2.12), i.e.,

$$I(x) = \frac{1}{2\pi i}\int_{c-i\infty}^{c+i\infty} x^{-z} M[f; 1 - z] M[h; z]\, dz.$$

In the present case, c lies between $\max\{0, -2v\}$ and 1, and we shall shift the contour $\text{Re } z = c$ to the right in order to yield the large-x expansion of $I(x)$. This is allowable in view of the asymptotic formula (3.3). Thus, with $d = 2n + 1 - \varepsilon$ $(0 < \varepsilon < 1)$, we have

$$I(x) = - \sum_{c < \text{Re} z < d} \text{Res}\{x^{-z} M[f; 1 - z] M[h; z]\}$$

$$+ \frac{1}{2\pi i}\int_{d-i\infty}^{d+i\infty} x^{-z} M[f; 1 - z] M[h; z]\, dz. \qquad (3.15)$$

The last integral is absolutely convergent again by virtue of (3.3), and
has the order estimate $O(x^{-2n-1+\varepsilon})$. Coupling (3.12) and (3.15) gives

$$I(x) = -\sum_{s=0}^{2n} (a_s \log x + b_s)x^{-s} + O(x^{-2n-1+\varepsilon}). \qquad (3.16)$$

Since $a_{2s} = b_{2s} = 0$, (3.16) reduces to

$$I(x) = \sum_{s=0}^{n-1} (c_s \log x + d_s)x^{-2s-1} + O(x^{-2n-1+\varepsilon}), \qquad (3.17)$$

where

$$c_s = -a_{2s+1} \quad \text{and} \quad d_s = -b_{2s+1}. \qquad (3.18)$$

4. Work of Handelsman and Lew

Let $f(t)$ and $h(t)$ be locally integrable functions on $(0, \infty)$ and satisfy

$$f(t) \sim \sum_{s=0}^{\infty} a_s t^{\alpha_s}, \qquad \text{as } t \to 0^+, \qquad (4.1)$$

where Re $\alpha_s \uparrow +\infty$ as $s \to \infty$, and

$$h(t) \sim \exp\{i\gamma t^p\} \sum_{s=0}^{\infty} b_s t^{-\beta_s}, \qquad \text{as } t \to +\infty, \qquad (4.2)$$

where γ is real, $p > 0$ and Re $\beta_s \uparrow +\infty$ as $s \to \infty$. For the absolute
convergence of the integral

$$I(x) = \int_0^{\infty} f(t)h(xt)\, dt, \qquad (4.3)$$

we also impose the conditions that as $t \to +\infty$,

$$f(t) = O(t^{-b}) \quad \text{with } b + \text{Re } \beta_0 > 1, \qquad (4.4)$$

and that as $t \to 0^+$,

$$h(t) = O(t^c) \quad \text{with } c + \text{Re } \alpha_0 > -1. \qquad (4.5)$$

To apply the Mellin transform technique given in §§1–3, we need the
functions $M[f; 1-z]$ and $M[h; z]$, defined by their corresponding
integrals, to have a common strip of analyticity. This would in turn
require $-b < \text{Re } \alpha_0$, $-c < \text{Re } \beta_0$, and $-c < \text{Re } \alpha_0 + 1$ or $1 - b < \text{Re } \beta_0$.

In what follows, we shall present a method, due to Handelsman and Lew (1971), which requires none of these conditions.

First let us introduce the truncated functions $f_1(t)$ and $f_2(t)$ defined by

$$f_1(t) = \begin{cases} f(t) & 0 < t \leq 1 \\ 0 & 1 < t < \infty, \end{cases} \tag{4.6}$$

$$f_2(t) = f(t) - f_1(t).$$

Similarly we define

$$h_1(t) = \begin{cases} h(t) & 0 < t \leq 1 \\ 0 & 1 < t < \infty, \end{cases} \tag{4.7}$$

$$h_2(t) = h(t) - h_1(t).$$

By Theorem 1, the integrals defining the Mellin transforms of f_i and h_j are absolutely convergent and yield analytic functions in the following half-planes:

$$\begin{array}{ll} M[f_1; z] & \operatorname{Re} z > -\operatorname{Re} \alpha_0 \\[2mm] M[f_2; z] & \operatorname{Re} z < b \\[2mm] M[h_1; z] & \operatorname{Re} z > -c \\[2mm] M[h_2; z] & \operatorname{Re} z < \operatorname{Re} \beta_0. \end{array} \tag{4.8}$$

Lemma 1. *$M[f_1; z]$ can be continued analytically to a meromorphic function on the entire z-plane. The singularities of this function are simple poles at $-\alpha_s$ with principal part*

$$\frac{a_s}{z + \alpha_s}. \tag{4.9}$$

Furthermore, for any compact interval I, we have

$$\lim_{y \to \pm \infty} N(f_1, I; y) = 0, \tag{4.10}$$

where $N(f_1, I; y)$ is the maximum modulus of $M[f_1; z]$ on I; see (2.4).

Proof. This follows from Theorems 1 and 5. ■

By (4.8), $M[f_2; z]$ is an analytic function in the half-plane $\operatorname{Re} z < b$. Hence we can extend the definition of the Mellin transform of f by

$$M[f; z] = M[f_1; z] + M[f_2; z] \tag{4.11}$$

for Re $z < b$. Since

$$\lim_{y \to \pm \infty} N(f_2, I; y) = 0 \qquad (4.12)$$

by (2.5), $M[f; z]$ has the same properties as $M[f_1; z]$ stated in Lemma 1.

Lemma 2. *If $\gamma = 0$ in (4.2) then $M[h_2; z]$ can be continued analytically to a meromorphic function on the entire z-plane. The singularities of this function are simple poles at β_s with principal part*

$$-\frac{b_s}{z - \beta_s}. \qquad (4.13)$$

Furthermore, for any compact interval I, we have

$$\lim_{y \to \pm \infty} N(h_2, I; y) = 0. \qquad (4.14)$$

Proof. Put $\check{h}_2(\tau) = h_2(1/\tau)$ and observe that $M[h_2; z] = M[\check{h}_2; -z]$. As in Lemma 1, the desired result follows from Theorems 1 and 5 with f and z replaced by $\check{h}_2$ and $-z$. ∎

Lemma 3. *If $\gamma \neq 0$ in (4.2) then $M[h_2; z]$ can be continued analytically to an entire function. Furthermore, for any compact interval $I = [x_1, x_2]$ with $x_2 > $ Re β_0, we have*

$$N(h_2, I; y) = O(|y|^{n(I)}), \qquad \text{as } y \to \pm \infty, \qquad (4.15)$$

where $n(I)$ is the smallest integer such that $n(I) \geq (x_2 - $ Re $\beta_0)/p$.

Proof. It suffices to show that for any $k > $ Re β_0, the assertions hold for Re $z < k$. First we set

$$s_n(t) = \begin{cases} 0 & 0 < t \leq 1 \\ \sum_{s=0}^{n-1} b_s \exp\{i\gamma t^p\} t^{-\beta_s} & 1 < t < \infty \end{cases} \qquad (4.16)$$

and

$$h_n^*(t) = h_2(t) - s_n(t). \qquad (4.17)$$

Then

$$M[h_2; z] = M[s_n; z] + M[h_n^*; z]. \qquad (4.18)$$

Choose n so that Re $\beta_n > k$. By Theorem 1, the integral

$$M[h_n^*; z] = \int_0^\infty t^{z-1} h_n^*(t)\, dt$$

converges absolutely and is analytic for Re $z <$ Re β_n. Also, for any interval $I = [x_1, x_2]$ with $x_2 < k$, we have

$$\lim_{y \to \pm\infty} N(h_n^*, I; y) = 0. \tag{4.19}$$

Thus we need consider only the function

$$M[s_n; z] = \sum_{s=0}^{n-1} b_s \int_1^\infty t^{z-\beta_s-1} \exp\{i\gamma t^p\}\, dt.$$

Each integral in the above sum is of the form

$$G(w) = \int_1^\infty t^{w-1} \exp\{i\gamma t^p\}\, dt, \tag{4.20}$$

which is clearly analytic for Re $w < 0$. Integration by parts gives

$$G(w) = \frac{i}{\gamma p} e^{i\gamma} + \frac{i}{\gamma p}(w-p)G(w-p) \tag{4.21}$$

for Re $w < p$, and the right-hand side of (4.21) is analytic there, thus extending $G(w)$ to a larger domain of analyticity. Repeated application of (4.21) leads to

$$G(w) = e^{i\gamma}\left[\frac{i}{\gamma p} + \sum_{s=1}^{N-1}\left(\frac{i}{\gamma p}\right)^s (w-p)(w-2p)\cdots(w-sp)\right]$$
$$+ \left(\frac{i}{\gamma p}\right)^N (w-p)\cdots(w-Np)G(w-Np). \tag{4.22}$$

Since $G(w - NP)$ is analytic for Re $w < Np$ and N is arbitrary, (4.22) provides an analytic continuation of $G(w)$ to an entire function in the w-plane. This, in turn, shows that $M[s_n; z]$ can be continued analytically to an entire function in the whole z-plane, thus proving the first assertion in Lemma 3. Since $G(u + iv) = O(1)$ as $|v| \to \infty$ for $u \le 0$ (Ex. 10), (4.22) also implies that $G(u + iv) = O(|v|^N)$ for $u \le Np$. From this it follows that

$$M[s_n; z] = O(|y|^{n(I)}), \qquad \text{as } y \to \pm\infty, \tag{4.23}$$

where $n(I)$ is as stated in the lemma. Coupling (4.19) and (4.23) gives the required order estimate (4.15). ∎

Since $M[h_1; z]$ is analytic for Re $z > -c$ by (4.8), the above two lemmas allow us to *define* the Mellin transform of h by

$$M[h; z] = M[h_1; z] + M[h_2; z] \qquad (4.24)$$

in the same half-plane. From (4.11) and (4.24) it follows that both $M[f; 1 - z]$ and $M[h; z]$ are defined in the half-plane Re $z > \max\{1 - b, -c\}$.

We are now ready to derive the asymptotic expansion of the integral $I(x)$ given in (4.3) as $x \to +\infty$. First we note that

$$I(x) = \sum_{i,j = 1}^{2} I_{ij}(x), \qquad (4.25)$$

where

$$I_{ij}(x) = \int_{0}^{\infty} f_i(t) h_j(xt) \, dt, \qquad (4.26)$$

and that by direct computation we have

$$I_{21}(x) = 0 \qquad \text{for } x \geq 1. \qquad (4.27)$$

Next we observe that the integrals for the transform pair $M[f_i; 1 - z]$ and $M[h_j; z]$ are absolutely convergent by (4.8) in the domain D_{ij} indicated below, and these domains are nonempty by (4.4) and (4.5).

Transform Pair		Domain D_{ij}
$M[f_1; 1 - z]$ and	$M[h_1; z]$	$-c < \text{Re } z < 1 + \text{Re } \alpha_0$
$M[f_1; 1 - z]$ and	$M[h_2; z]$	$\text{Re } z < \min\{1 + \text{Re } \alpha_0, \text{Re } \beta_0\}$
$M[f_2; 1 - z]$ and	$M[h_1; z]$	$\max\{-c, 1 - b\} < \text{Re } z$
$M[f_2; 1 - z]$ and	$M[h_2; z]$	$1 - b < \text{Re } z < \text{Re } \beta_0$

$$(4.28)$$

For simplicity of presentation, let us introduce the functions

$$G(z) \equiv M[f; 1 - z] \, M[h; z], \qquad G_{ij}(z) \equiv M[f_i; 1 - z] \, M[h_j; z]. \qquad (4.29)$$

By (4.28), each $G_{ij}(z)$ is analytic in its domain D_{ij}. Furthermore, by Lemmas 1, 2 and 3, each $G_{ij}(z)$ has an analytic or meromorphic extension to a half-plane containing D_{ij}. Thus, from (4.11) and (4.24), it follows that

$$G(z) = \sum_{i,j=1}^{2} G_{ij}(z) \qquad (4.30)$$

at least for Re $z > \max\{1 - b, -c\}$. The asymptotic expansion of $I(x)$ is then obtained from (4.25) and the following theorem.

Theorem 6.   *For any $i, j = 1, 2$, suppose that there is a real number p in D_{ij} such that either $M[f_i; 1 - p - iy]$ or $M[h_j; p + iy]$ belongs to $L^1(-\infty, \infty)$. Then*

$$I_{ij}(x) = \frac{1}{2\pi i} \int_{p - i\infty}^{p + i\infty} x^{-z} G_{ij}(z)\, dz. \qquad (4.31)$$

If, in addition, there exists a number $q \geq p$ such that $G_{ij}(q + iy) \in L^1(-\infty, \infty)$ and

$$\sup\{|G_{ij}(x + iy)|: p \leq x \leq q\} \to 0 \qquad as\ y \to \pm\infty, \qquad (4.32)$$

then

$$I_{ij}(x) = \sum_{p < \mathrm{Re}\, z < q} \mathrm{Res}\{-x^{-z} G_{ij}(z)\} + E_{ij}(x), \qquad (4.33)$$

where

$$E_{ij}(x) = \frac{1}{2\pi i} \int_{q - i\infty}^{q + i\infty} x^{-z} G_{ij}(z)\, dz = o(x^{-q}) \qquad (4.34)$$

as $x \to +\infty$.

Proof.   The first assertion follows from Theorem 4, and the second assertion from Cauchy's residue theorem and the Riemann-Lebesgue lemma.   ∎

Here we point out that the result (4.33) is consistent with (4.27). To see this, we first observe that $G_{21}(z)$ is analytic in the half-plane Re $z \geq p$ by (4.28). Hence the residue terms in (4.33) are all zero. Next we observe that $G_{21}(z)$ satisfies (4.32) for any $q \geq p$ by (2.5). Thus, if $G_{21}(q + iy) \in L^1(-\infty, \infty)$, then we have

$$I_{21}(x) = O(x^{-q}) \qquad as\ x \to +\infty \qquad (4.35)$$

for arbitrarily large q.

Theorem 7. *Let f and h satisfy (4.1), (4.4), (4.5) and (4.2) with $\gamma \neq 0$. Suppose that for each i, j = 1, 2, there exists a real number p_{ij} in the domain D_{ij} such that either $M[f_i; 1 - p_{ij} - iy]$ or $M[h_j; p_{ij} + iy]$ belongs to $L^1(-\infty, \infty)$. Suppose further that there are real numbers $q > \max\{p_{ij}: i, j = 1, 2\}$ and $r \geq q$ such that $G_{ij}(q + iy)$ and $G(r + iy)$ belong to $L^1(-\infty, \infty)$. If*

$$\sup\{|G(x + iy)|: q \leq x \leq r\} \to 0 \qquad \text{as } y \to \pm\infty,$$

$$\sup\{|G_{i2}(x + iy)|: p_{i2} \leq x \leq q\} \to 0 \qquad \text{as } y \to \pm\infty, \tag{4.36}$$

for i = 1, 2, then

$$I(x) = \sum_{p_{11} < \operatorname{Re} z \leq q} \operatorname{Res}\{-x^{-z}G_{11}(z)\} + \sum_{p_{12} < \operatorname{Re} z \leq q} \operatorname{Res}\{-x^{-z}G_{12}(z)\}$$

$$+ \sum_{q < \operatorname{Re} z < r} \operatorname{Res}\{-x^{-z}G(z)\} + E(x), \tag{4.37}$$

where

$$E(x) = \frac{1}{2\pi i} \int_{r - i\infty}^{r + i\infty} x^{-z}G(z) \, dz = o(x^{-r}) \tag{4.38}$$

as $x \to +\infty$. All of the residues in (4.37) are taken at the poles of $M[f_1; 1 - z]$. If the above conditions hold for arbitrarily large r, then (4.37) gives an infinite asymptotic expansion for I(x).

Proof. From (4.25) and (4.31), we have

$$I(x) = \sum_{i,j=1}^{2} I_{ij}(x) \qquad \text{and} \qquad I_{ij}(x) = \frac{1}{2\pi i} \int_{p_{ij} - i\infty}^{p_{ij} + i\infty} x^{-z}G_{ij}(z) \, dz,$$

where

$$G_{ij}(z) = M[f_i; 1 - z] \, M[h_j; z].$$

Here we have used the hypothesis that for each i, j = 1, 2, either $M[f_i; 1 - p_{ij} - iy]$ or $M[h_j; p_{ij} + iy]$ belongs to $L^1(-\infty, \infty)$. By Lemmas 1 and 3, $G_{11}(z)$ and $G_{12}(z)$ are meromorphic functions for Re $z > \max\{1 - b, -c\}$. Furthermore, by (2.5) with h_1 replacing f and by (4.10),

$$\sup\{|G_{11}(x + iy)|: p_{11} \leq x \leq q\} \to 0 \qquad \text{as } y \to \pm\infty.$$

Similarly, by (4.28) and Lemma 3, $G_{21}(z)$ and $G_{22}(z)$ are analytic in the half-plane Re $z > \max\{1 - b, -c\}$ and hence the residue terms in (4.33) for $I_{21}(x)$ and $I_{22}(x)$ are all zero. Moreover, by (2.5) with h_1 replacing f and by (4.12), we have

$$\sup\{|G_{21}(x+iy)|: p_{21} \le x \le q\} \to 0 \qquad \text{as } y \to \pm\infty.$$

Thus the second condition in (4.36) ensures that Theorem 6 applies for any $i, j = 1, 2$, and we have

$$I(x) = \sum_{p_{11} < \mathrm{Re}\, z \le q} \mathrm{Res}\{-x^{-z}G_{11}(z)\} + \sum_{p_{12} < \mathrm{Re}\, z \le q} \mathrm{Res}\{-x^{-z}G_{12}(z)\}$$

$$+ \sum_{i,j=1}^{2} \frac{1}{2\pi i} \int_{q-i\infty}^{q+i\infty} x^{-z} G_{ij}(z)\, dz.$$

Since $G(z) = \sum G_{ij}(z)$ by (4.30), the desired expansion (4.37) now follows from the first condition in (4.36) and the Riemann–Lebesgue lemma. ■

By Lemma 1, the third sum in (4.37) can be written as

$$\sum_{q < \mathrm{Re}(1+\alpha_s) < r} a_s\, M[h; 1 + \alpha_s] x^{-\alpha_s - 1}, \tag{4.39}$$

and similar expressions exist for the first two sums. Thus (4.37) gives an asymptotic power series expansion for the integral $I(x)$.

Theorem 8.   *Let f and h satisfy (4.1), (4.4), (4.5), and (4.2) with $\gamma = 0$. If for each $i, j = 1, 2$ there exists a real number p_{ij} in the domain D_{ij} such that either $M[f_i; 1 - p_{ij} - iy]$ or $M[h_j; p_{ij} + iy]$ belongs to $L^1(-\infty, \infty)$, and if in addition there exist real numbers $q \ge \max\{p_{ij}: i, j = 1, 2\}$ and $r \ge q$ such that $G_{ij}(q + iy)$ and $G(r + iy)$ belong to $L^1(-\infty, \infty)$, then*

$$I(x) = \sum{}^* \sum_{p_{ij} < \mathrm{Re}\, z \le q} \mathrm{Res}\{-x^{-z}G_{ij}(z)\}$$

$$+ \sum_{q < \mathrm{Re}\, z < r} \mathrm{Res}\{-x^{-z}G(z)\} + \frac{1}{2\pi i} \int_{r-i\infty}^{r+i\infty} x^{-z} G(z)\, dz, \tag{4.40}$$

where the sum $\sum^*$ is taken over $i = 1, j = 1; i = 1, j = 2; i = 2, j = 2$. All of the residues in (4.40) are taken at the poles of both $M[f_1; 1 - z]$ and $M[h_2; z]$, and the remainder integral is again $o(x^{-r})$.

If, in the above theorem, we have $1 + \alpha_i \ne \beta_j$ for all i and j, then the poles of $M[f_1; 1 - z]$ and $M[h_2; z]$ do not coincide. Hence the last sum in

(4.40) can be written as

$$\sum_{q<1+\operatorname{Re}\alpha_s<r} a_s\, M[h; 1+\alpha_s]x^{-\alpha_s-1} + \sum_{q<\operatorname{Re}\beta_s<r} b_s\, M[f; 1-\beta_s]x^{-\beta_s}. \quad (4.41)$$

Similar expressions can be written for the double sum in (4.40). If $1+\alpha_i=\beta_j$ for some i and j, then $G(z)$ will have a double pole, in which case the expansion for $I(x)$ will include a logarithmic term; see, e.g., (3.12).

Proof of Theorem 8. The argument here is similar to that given in Theorem 7. By (4.35), $I_{21}(x)$ is asymptotically zero and hence does not contribute to the expansion (4.40). Since $M[f_1; 1-z]$ and $M[h_2; z]$ are meromorphic in their respective half-planes, they both contribute to the residue terms in (4.40). Also, we observe that in view of (4.10), (4.12) and (4.14), conditions like those in (4.36) are automatically satisfied. ∎

From (4.15) it is readily seen that $M[h; z]$ has an algebraic growth along vertical lines when $\gamma\neq 0$, and that the rate of growth increases with Re z. Thus, in order to apply Theorem 7, we must establish conditions on f so that $M[f; 1-z]$ decays sufficiently fast for conditions in (4.36) to hold. For this purpose, we give the following three results. The first one is known as the Phragmén-Lindelöf theorem for a strip (Titchmarsh 1939, p. 180).

Lemma 4. *Let $F(z)$ be an analytic function in the strip $x_1\le x\le x_2$, $z=x+iy$, satisfying $F(z)=O(e^{\varepsilon|y|})$ for every $\varepsilon>0$. If $F(x_1+iy)=O(|y|^{k_1})$ and $F(x_2+iy)=O(|y|^{k_2})$, then $F(x+iy)=O(|y|^{k(x)})$ uniformly for $x_1\le x\le x_2$, where*

$$k(x)=\frac{k_2}{x_2-x_1}(x-x_1)+\frac{k_1}{x_2-x_1}(x_2-x).$$

Lemma 5. *Let $f(t)$ be an n-times continuously differentiable function in $(0,\infty)$, and let a and b be defined as in Theorem 1, i.e.,*

$$a=\sup\{\alpha: f(t)=O(t^\alpha)\text{ as }t\to 0^+\}$$
$$b=\sup\{\beta: f(t)=O(t^{-\beta})\text{ as }t\to+\infty\}.$$

Suppose that $-a < b$, *and that* $(t\,d/dt)^j[t^x f(t)] \to 0$ *at both limits* $t \to 0^+$ *and* $t \to +\infty$ *for* $j = 0, 1, \ldots, n-1$ *and* $-a < x < b$. *If* $t^{-1}(t\,d/dt)^n[t^x f(t)] \in L^1(0, \infty)$ *for* $-a < x < b$ *then the integral* (2.1) *satisfies*

$$M[f; z] = O(|y|^{-n}) \qquad (4.42)$$

uniformly for x *in any compact subinterval of* $(-a, b)$.

Proof.   Integration by parts gives

$$M[f; z] = \int_0^\infty t^x f(t) t^{iy-1}\, dt = \left(\frac{i}{y}\right)^n \int_0^\infty t^{iy-1} \left(t\frac{d}{dt}\right)^n [t^x f(t)]\, dt, \quad (4.43)$$

integrated terms all vanishing by hypothesis. Thus (4.42) holds with the O-symbol depending on x. Since $M[f; z]$ is analytic in $-a <$ Re $z < b$ by Theorem 1, the uniformity of (4.42) will follow from Lemma 4. ∎

Lemma 6.   *Let* $f(t)$ *be an n-times continuously differentiable function in* $(0, \infty)$, *and let* $\left(t\dfrac{d}{dt}\right)^j[t^x f(t)] \to 0$, *as* $t \to +\infty$, *for* $j = 0, 1, \ldots, n-1$ *and for all* $x < b$. *Suppose that*

$$f(t) \sim \sum_{s=0}^\infty a_s t^{\alpha_s}, \qquad as \ t \to 0^+ \qquad (4.44)$$

where Re $\alpha_s \uparrow +\infty$ *as* $s \to \infty$, *and that the asymptotic expansion of* $f^{(j)}(t)$, $j = 0, 1, \ldots, n$, *is obtained from* (4.44) *by termwise differentiation. Then*

$$M[f; x + iy] = O(|y|^{-n}) \qquad (4.45)$$

uniformly for x *in any compact interval contained in* $(-\infty, b)$. *Here* $M[f; z]$ *denotes the integral* (2.1) *or its analytic continuation.*

Proof.   For any p, choose $k \equiv k(p)$ such that $k + $ Re $\alpha_0 > $ Re α_p. Set

$$f_k^*(t) = \exp(-t^k) \sum_{s=0}^{p-1} a_s t^{\alpha_s} \qquad (4.46)$$

$$r_k(t) = f(t) - f_k^*(t). \qquad (4.47)$$

It is easily seen that $r_k(t) = O(t^{\mathrm{Re}\,\alpha_p})$ as $t \to 0^+$, and that $r_k(t)$ satisfies all the conditions imposed on $f(t)$ in Lemma 5 with $a = $ Re α_p. Thus, if I is a compact interval in $(-\text{Re }\alpha_p, b)$, then

$$M[r_k; z] = O(|y|^{-n}) \tag{4.48}$$

uniformly for x in I. Furthermore, direct computation gives

$$M[f_k^*; z] = \frac{1}{k}\sum_{s=0}^{p-1} a_s \Gamma\left(\frac{z+\alpha_s}{k}\right) \tag{4.49}$$

for Re $z > -\text{Re }\alpha_0$, and by uniqueness of analytic continuation one readily sees that (4.49) actually holds in the whole z-plane. From (3.3), (4.47), and (4.48), it follows that $M[f; z] = O(|y|^{-n})$ uniformly for x in any compact subinterval of $(-\text{Re }\alpha_p, b)$. Since Re $\alpha_p \to +\infty$ as $p \to \infty$, the lemma is proved. ■

5. Remarks and Examples

The results of the previous section not only give the asymptotic expansion of the integral $I(x)$ in (4.3) for large values of x, but can also be used directly to derive small-parameter expansions for similar integrals (without developing a parallel theory with the vertical line of integration shifted to the left; cf. Section 1). For instance, for any positive parameter s, we can always put $x = 1/s$ and write

$$\int_0^\infty f(st)h(t)\,dt = x\int_0^\infty f(t)h(xt)\,dt. \tag{5.1}$$

The integral on the right is of the form (4.3), and hence we can apply Theorem 7 or Theorem 8 depending whether $\gamma \neq 0$ or $\gamma = 0$.

As an illustration, we consider the Laplace transform

$$L_h(s) = \int_0^\infty h(t)e^{-st}\,dt \tag{5.2}$$

near $s = 0$, where $h(t)$ satisfies (4.2) and (4.5) with $c > -1$. Let $x = 1/s$, and break $h(t)$ at $t = 1$ as in (4.7). Then (5.2) can be written as

$$sL_h(s) = I_1(x) + I_2(x), \tag{5.3}$$

where

$$I_j(x) = \int_0^\infty e^{-t}h_j(xt)\,dt, \qquad j = 1, 2. \tag{5.4}$$

Since $M[e^{-t}; z] = \Gamma(z)$, by Theorem 4, there exist real numbers p_1 and p_2 such that $-c < p_1 < 1$, $p_2 < \min\{1, \operatorname{Re} \beta_0\}$, and

$$I_j(x) = \frac{1}{2\pi i} \int_{p_j - i\infty}^{p_j + i\infty} x^{-z}\Gamma(1 - z)\, M[h_j; z]\, dz, \qquad (5.5)$$

$j = 1, 2$. Since $M[h_1; z]$ is analytic for $\operatorname{Re} z > -c$, by (3.3), we have for any ρ satisfying $1 < \rho < 2$,

$$I_1(x) = M[h_1; 1]x^{-1} + \frac{1}{2\pi i} \int_{\rho - i\infty}^{\rho + i\infty} x^{-z}\Gamma(1 - z)\, M[h_1; z]\, dz.$$

Similarly, by Lemmas 2 and 3, we can choose ρ so that $M[h_2; z]$ has no pole in $1 < \operatorname{Re} z \leq \rho < 2$. Thus

$$I_2(x) = \sum_{\operatorname{Re}\beta_0 \leq \operatorname{Re} z \leq 1} \operatorname{Res}\{-x^{-z}\Gamma(1 - z)\, M[h_2; z]\}$$

$$+ \frac{1}{2\pi i} \int_{\rho - i\infty}^{\rho + i\infty} x^{-z}\Gamma(1 - z)\, M[h_2; z]\, dz.$$

Adding the last two equations together, we have from (5.3)

$$L_h(s) = M[h_1; 1] + \sum_{\operatorname{Re}\beta_0 \leq \operatorname{Re} z \leq 1} \operatorname{Res}\{-s^{z-1}\Gamma(1 - z)\, M[h_2; z]\}$$

$$+ \frac{1}{2\pi i} \int_{\rho - i\infty}^{\rho + i\infty} s^{z-1}\Gamma(1 - z)\, M[h; z]\, dz. \qquad (5.6)$$

Now for any $l > 1$ we can choose δ arbitrarily small and such that $\Gamma(1 - z)\, M[h; z]$ has no poles in the strip $l - \delta \leq \operatorname{Re} z < l$. By translating the contour in (5.6) further to the right, we obtain

$$L_h(s) = M[h_1; 1] + \sum_{\operatorname{Re}\beta_0 \leq \operatorname{Re} z \leq 1} \operatorname{Res}\{-s^{z-1}\Gamma(1 - z)\, M[h_2; z]\}$$

$$+ \sum_{1 < \operatorname{Re} z < l} \operatorname{Res}\{-s^{z-1}\Gamma(1 - z)\, M[h; z]\}$$

$$+ \frac{1}{2\pi i} \int_{l - \delta - i\infty}^{l - \delta + i\infty} s^{z-1}\Gamma(1 - z)\, M[h; z]\, dz. \qquad (5.7)$$

The last term is clearly $O(s^{l-\delta-1})$. Summarizing the above results, we have the following theorem.

Theorem 9.   *Let $h(t)$ be a finite combination of functions each satisfying (4.5) with some $c > -1$ and having an asymptotic expansion of the form (4.2), and let $M[h; z]$ be defined by (4.24) (whether or not the integral* *for $M[h; z]$ converges anywhere). If $l > 1$, $\delta > 0$, and $s > 0$, then (5.7) holds.*

We have derived this result by directly appealing to Theorem 4, instead of deducing it from the general cases stated in Theorems 7 and 8 with $f(t) = e^{-t}$. This is to avoid the absolute integrability conditions imposed on $M[h_i; z]$, $i = 1, 2$. A useful consequence of Theorem 9 is the following corollary.

Corollary 1.   *Let $h(t)$ satisfy (4.2) with $\gamma = 0$ and (4.5) with $c > -1$, and let $M[h; z]$ be defined as in (4.24). If none of the exponents β_n in (4.2) are positive integers, then*

$$L_h(s) \sim \sum_{n=0}^{\infty} b_n \Gamma(1 - \beta_n)s^{\beta_n - 1} + \sum_{n=0}^{\infty} \frac{(-s)^n}{n!} M[h; n + 1], \qquad as \ s \to 0^+.$$

$$(5.8)$$

Proof.   By Lemma 2, the singularities of $M[h_2; z]$ are simple poles at β_n with residue $-b_n$. In view of (4.24), the same is true of $M[h; z]$. Thus

$$\operatorname*{Res}_{z = \beta_n} \{-s^{z-1}\Gamma(1 - z) M[h; z]\} = b_n \Gamma(1 - \beta_n)s^{\beta_n - 1},$$

and from (3.4) we also have

$$\operatorname*{Res}_{z = n+1} \{-s^{z-1}\Gamma(1 - z) M[h; z]\} = \frac{(-s)^n}{n!} M[h; n + 1].$$

Both results of course hold with h replaced by h_2. The conclusion now follows from (5.7).   ∎

If some of the exponents β_n in (4.2) are positive integers, then the expansion of $L_h(s)$ involves logarithmic terms; see, e.g., Ex. 2(iii).

Example 3.   Let v be an arbitrary complex number, and put

$$h(t) = (1 + t^2)^{v - 1/2}. \qquad (5.9)$$

From Gradshteyn and Ryzhik (1980, p. 295, Eq. (2)), we have

$$M[h; z] = \int_0^\infty t^{z-1}h(t) \, dt = \frac{1}{2} \frac{\Gamma(\tfrac{1}{2}z)\Gamma(\tfrac{1}{2} - v - \tfrac{1}{2}z)}{\Gamma(\tfrac{1}{2} - v)} \qquad (5.10)$$

for $0 < \text{Re } z < \text{Re}(1 - 2v)$. Note that if $\text{Re } v > \frac{1}{2}$ then the integral in (5.10) does not converge for any values of z. However, if we split h into h_1 and h_2 as in (4.7), then it is easily seen that $M[h_1; z]$ is analytic in z for $\text{Re } z > 0$ and analytic in v for all v. Furthermore, since

$$h(t) = t^{2v-1} \sum_{n=0}^{\infty} \binom{v - \frac{1}{2}}{n} t^{-2n} \qquad \text{for } t > 1, \qquad (5.11)$$

by Lemma 2, $M[h_2; z]$ is meromorphic in z except for the poles at $z = 2n + 1 - 2v$, $n = 0, 1, 2, \ldots$. By a similar argument (see Theorem 5), it can be shown that $M[h_2; z]$ is also meromorphic in v except at the poles $v = n + \frac{1}{2}(1 - z)$. Thus, by analytic continuation, it follows from (5.10) that the Mellin transform $M[h; z] = M[h_1; z] + M[h_2; z]$ is also given explicitly by

$$M[h; z] = \frac{1}{2} \frac{\Gamma(\frac{1}{2}z)\Gamma(\frac{1}{2} - v - \frac{1}{2}z)}{\Gamma(\frac{1}{2} - v)}$$

for all v except at its poles. If $2v \neq$ integer, then by Corollary 1

$$L_h(s) \sim \sum_{n=0}^{\infty} \binom{v - \frac{1}{2}}{n} \Gamma(2v - 2n) s^{2n - 2v}$$

$$+ \frac{1}{2} \sum_{n=0}^{\infty} \frac{\Gamma(\frac{1}{2}n + \frac{1}{2})\Gamma(-v - \frac{1}{2}n)}{\Gamma(\frac{1}{2} - v)} \frac{(-s)^n}{n!} \qquad (5.12)$$

as $s \to 0^+$. If $2v$ is an integer, then $\Gamma(1 - z) M[h; z]$ has double poles and hence the expansion of $L_h(s)$ involves logarithmic terms; see, for instance, the case $v = 1$ given in Ex. 2(v).

Example 4.   For $\text{Re } v > -\frac{3}{2}$ and $\text{Re}(\mu + \frac{1}{2}v) > -1$, let

$$h(t) = t^\mu J_v(t^{1/2}). \qquad (5.13)$$

From Oberhettinger (1974, p. 93, Eq. (10.1)), we have upon a change of variable

$$M[h; z] = \frac{2^{2z + 2\mu} \Gamma(z + \mu + \frac{1}{2}v)}{\Gamma(1 + \frac{1}{2}v - z - \mu)} \qquad (5.14)$$

for $-\frac{1}{2} \text{Re } v < \text{Re}(z + \mu) < \frac{3}{4}$. Since $M[h; z]$ is analytic in $\text{Re}(z + \mu) > -\frac{1}{2} \text{Re } v$ by Lemma 3 and (4.24), (5.14) continues to hold in this larger domain. From Theorem 9 and (3.4), it then follows that

$$L_h(s) \sim \sum_{n=0}^{\infty} \frac{2^{2n+2\mu+2}\Gamma(n+1+\mu+\frac{1}{2}\nu)\,(-s)^n}{\Gamma(\frac{1}{2}\nu-n-\mu)}\frac{(-s)^n}{n!} \qquad \text{as } s \to 0^+.$$

A useful consequence of Theorem 7 is the derivation of asymptotic expansions of Bessel transforms (Oberhettinger 1972). These are integral transforms whose kernels are Bessel functions.

Example 5.   Consider the integral transform

$$H_1(x) = \int_0^{\infty} f(t)H_{\nu}^{(1)}(xt)\,dt, \tag{5.15}$$

where $H_{\nu}^{(1)}(t)$ is the Hankel function of the first kind and $f(t)$ is infinitely differentiable in $(0, \infty)$. Put $h(t) = H_{\nu}^{(1)}(t)$ and recall the identity $H_{\nu}^{(1)}(t) = J_{\nu}(t) + iY_{\nu}(t)$. From Oberhettinger (1974, p. 93; Eqs. (10.1) and (10.2)), we then have

$$M[h; z] = \frac{2^{z-1}}{\pi} e^{-i(\nu-z+1)\pi/2}\Gamma(\tfrac{1}{2}z+\tfrac{1}{2}\nu)\Gamma(\tfrac{1}{2}z-\tfrac{1}{2}\nu) \tag{5.16}$$

for $\pm\mathrm{Re}\,\nu < \mathrm{Re}\,z < \frac{3}{2}$. (Note that the formulas for $M[h; z]$ as given in Erdélyi *et al.* (1954, p. 330) and Bleistein and Handelsman (1975a, p. 414) are incorrect.) The right-hand side of (5.16) is analytic for $\mathrm{Re}\,z > \pm\mathrm{Re}\,\nu$, and by (3.3),

$$M[h; z] = O(|y|^{x-1}) \qquad \text{as } |y| \to \infty; \tag{5.17}$$

cf. Lemma 3.

Now suppose that $f(t)$ has the asymptotic expansion

$$f(t) \sim \sum_{s=0}^{\infty} a_s t^{\alpha_s} \qquad \text{as } t \to 0^+ \tag{5.18}$$

with $\mathrm{Re}(\alpha_0 \pm \nu) > -1$ and $\mathrm{Re}\,\alpha_s \uparrow +\infty$ as $s \to \infty$, and that the asymptotic expansion of $f^{(j)}(t)$, $j = 0, 1, 2, \ldots$, is obtainable from (5.18) by termwise differentiation. Furthermore, suppose that there exists an $\varepsilon > 0$ such that $f^{(j)}(t) = O(t^{-j-1-\varepsilon})$, as $t \to +\infty$, for $j = 0, 1, 2, \ldots$. By Lemma 6, we have for all n

$$M[f; 1-x-iy] = O(|y|^{-n}) \tag{5.19}$$

uniformly for x in any compact interval contained in $(-\varepsilon, \infty)$.

To avoid the need for absolute integrability of some of the truncated Mellin transforms imposed in Theorem 7, we will break $h(t)$ at $t = 1$ but not $f(t)$. Define h_1 and h_2 as in (4.7). Then the integral (5.5) becomes

$$H_1(x) = I_1(x) + I_2(x), \qquad \text{where} \quad I_j(x) = \int_0^\infty f(t) h_j(xt)\, dt, \quad j = 1, 2.$$

By Theorem 4 and (5.19), for real numbers p_1 and p_2 such that $|\text{Re } v| < p_1 < \text{Re}(\alpha_0 + 1)$ and $0 < p_2 < \frac{1}{2}$, we have

$$I_j(x) = \frac{1}{2\pi i} \int_{p_j - i\infty}^{p_j + i\infty} x^{-z}\, M[f; 1 - z]\, M[h_j; z]\, dz.$$

Since $M[h_1; z]$ is analytic in $\text{Re } z > |\text{Re } v|$ and $M[h_2; z]$ is an entire function by Lemma 3, the residues all come from the poles of $M[f; 1 - z]$. By Theorem 5 these are located at $z = \alpha_s + 1$, $s = 0, 1, \ldots$. Now choose $p_1 = p_2 < \text{Re}(\alpha_0 + 1)$, and let p denote their common value. Then

$$H_1(x) = \frac{1}{2\pi i} \int_{p - i\infty}^{p + i\infty} x^{-z}\, M[f; 1 - z]\, M[h; z]\, dz. \qquad (5.20)$$

(It should be noted that this result does not follow directly from Theorem 4). By (5.17) and (5.19), the integration contour $\text{Re } z = p$ can be moved to the right. This leads to

$$H_1(x) \sim \frac{1}{\pi} \sum_{s=0}^\infty 2^{\alpha_s} e^{i(\alpha_s - v)\pi/2} \Gamma\left(\frac{\alpha_s + v + 1}{2}\right) \Gamma\left(\frac{\alpha_s - v + 1}{2}\right) \frac{a_s}{x^{\alpha_s + 1}} \qquad (5.21)$$

as $x \to +\infty$. There are various other ways of deriving this expansion; the simplest is probably that given in Wong (1979). (Due to an error in Erdélyi *et al.* (1954), the coefficients in expansion (5.4) in Wong (1979) are incorrect.)

Example 6.   The results given thus far in this chapter can be extended easily to the more general integral

$$I(x) = \int_0^\infty f(t) h(x\varphi(t))\, dt, \qquad (5.22)$$

provided that $\varphi(t)$ is a sufficiently smooth real-valued function, $\varphi'(t) > 0$ in $(0, \infty)$ and

$$\varphi(t) \sim \sum_{s=0}^\infty c_s t^{s + \mu} \qquad \text{as } t \to 0^+,$$

where $c_0 \neq 0$ and $\mu > 0$. To see this, we note that the equation $v = \varphi(t)$ determines a one-to-one relation between t and v, and that in terms of variable v, the integral (5.22) transforms to the canonical form (1.3) with t replaced by v, and f replaced by

$$f^*(v) \equiv f(t)\frac{dt}{dv} = \frac{f(t)}{\varphi'(t)}.$$

If $f(t)$ in (5.22) has an expansion of the form

$$f(t) \sim \sum_{s=0}^{\infty} b_s t^{s+\alpha-1} \qquad \text{as } t \to 0^+,$$

then by (1.22) of Chapter II

$$f^*(v) \sim \sum_{s=0}^{\infty} a_s v^{(s+\alpha-\mu)/\mu} \qquad \text{as } v \to 0^+$$

for some constants a_s. Theorems 7 and 8 are now immediately applicable.

6. Explicit Error Terms

The error terms in (1.6), (4.37), and (4.40) are all given in terms of a contour integral involving $M[f; 1 - z]$ and $M[h; z]$. If these transforms are known explicitly, then it may be possible to deduce a bound for the error directly from these representations. However, in general, it is preferable to have the errors expressed explicitly in terms of f and h. There are now several ways of constructing such explicit expressions, but in this section we shall be briefly concerned with only two of these. Some other methods of construction are given in Chapters IV and VI. Before proceeding, we need the following result:

Lemma 7. *Let $f(t)$ satisfy the conditions in Theorem 5, and let $f_n(t)$ denote the n-th remainder term in its asymptotic expansion; see (2.14). Then $M[f; 1 - z]$ is meromorphic in the half-plane $\mathrm{Re}\, z > 1 - b$; moreover, in the strip $1 + \mathrm{Re}\, \alpha_{n-1} < \mathrm{Re}\, z < 1 + \mathrm{Re}\, \alpha_n$, we have*

$$M[f; 1 - z] = \int_0^{\infty} t^{-z} f_n(t)\, dt. \qquad (6.1)$$

Proof.   The first statement is a direct consequence of Theorem 5. Thus we need prove only (6.1). From (2.15), we have

$$M[f; 1 - z] = \int_0^1 t^{-z} f_n(t)\, dt - \sum_{s=0}^{n-1} \frac{a_s}{z - \alpha_s - 1} + \int_1^\infty t^{-z} f(t)\, dt \quad (6.2)$$

for $1 - b < \operatorname{Re} z < 1 + \operatorname{Re} \alpha_n$. From (2.14), we also have

$$\int_1^\infty t^{-z} f(t)\, dt = \int_1^\infty t^{-z} f_n(t)\, dt + \sum_{s=0}^{n-1} \frac{a_s}{z - \alpha_s - 1}$$

for $1 + \operatorname{Re}\alpha_{n-1} < \operatorname{Re} z$. This, together with (6.2), gives the desired result (6.1). ∎

To construct explicit error terms, we divide the discussion into two cases. In one case the kernel $h(t)$ is oscillatory; in the other it is monotonic. Let us first consider the case of oscillatory h, i.e., (4.2) holds with $\gamma \neq 0$. In this case, by Theorem 7, the error term is given by

$$E(x) = \frac{1}{2\pi i} \int_{r - i\infty}^{r + i\infty} x^{-z} M[f; 1 - z] M[h; z]\, dz. \quad (6.3)$$

Without loss of generality, we shall choose n so that $1 + \operatorname{Re}\alpha_{n-1} < r < 1 + \operatorname{Re}\alpha_n$. Then by Lemma 7

$$M[f; 1 - z] = \int_0^\infty t^{-z} f_n(t)\, dt$$

for $\operatorname{Re} z = r$. Assume that $f(t)$ satisfies the conditions of Lemma 6 with n replaced by N and $-\operatorname{Re}\alpha_0 < b$, and in addition that $t^{x+N} f^{(N)}(t) \in L^1[1, \infty)$ for all $x < b - 1$. Then integration by parts yields

$$\begin{aligned} M[f; 1 - z] &= \frac{\Gamma(z - N)}{\Gamma(z)} \int_0^\infty t^{-z+N} f_n^{(N)}(t)\, dt \\ &= \frac{\Gamma(z - N)}{\Gamma(z)} M[f_n^{(N)}; 1 - z + N], \end{aligned} \quad (6.4)$$

all integrated terms vanishing at both limits. Inserting (6.4) into (6.3) and making the change of variable $w = z - N$ gives

$$E(x) = \frac{x^{-N}}{2\pi i} \int_{r - N - i\infty}^{r - N + i\infty} x^{-w} M[h; w + N] \frac{\Gamma(w)}{\Gamma(w + N)} M[f_n^{(N)}; 1 - w]\, dw.$$

$$(6.5)$$

Now suppose that in (4.2) we have $p + \text{Re } \beta_0 > 1$ and in (4.5) $c > -1$. Define $h^{(0)}(t) = h(t)$ and

$$h^{(-j)}(t) = -\int_t^\infty h^{(-j+1)}(\tau)\, d\tau, \qquad j = 1, 2, \dots . \tag{6.6}$$

For $0 < \text{Re } z < p - 1 + \text{Re } \beta_0$, integration by parts gives

$$M[h; z + 1] = -z\, M[h^{(-1)}; z]. \tag{6.7}$$

By analytic continuation, this holds for all z in $\text{Re } z > 0$. Applying (6.7) N times, we obtain

$$M[h^{(-N)}; z] = \frac{(-1)^N}{z(z+1)\cdots(z+N-1)}\, M[h; z + N]. \tag{6.8}$$

Coupling (6.5) and (6.8) yields

$$E(x) = \frac{(-1)^N}{x^N}\frac{1}{2\pi i}\int_{r-N-i\infty}^{r-N+i\infty} x^{-w}\, M[h^{(-N)}; w]\, M[f_n^{(N)}; 1 - w]\, dw. \tag{6.9}$$

A formal application of the Parseval formula (2.12) leads to

$$E(x) = \frac{(-1)^N}{x^N}\int_0^\infty f_n^{(N)}(t) h^{(-N)}(xt)\, dt. \tag{6.10}$$

To justify (6.10), we first note that $t^{-r+N}f_n^{(N)}(t)$ is absolutely integrable on $(0, \infty)$. Thus by interchanging the order of integration,

$$\frac{1}{2\pi i}\int_{r-N-iT}^{r-N+iT} x^{-w}\, M[h^{(-N)}; w]\, M[f_n^{(N)}; 1 - w]\, dw$$
$$= \int_0^\infty f_n^{(N)}(t)\left\{\frac{1}{2\pi i}\int_{r-N-iT}^{r-N+iT}(xt)^{-w}\, M[h^{(-N)}; w]\, dw\right\} dt. \tag{6.11}$$

We note that (6.8) and Exercise 11 give

$$M[h^{(-N)}; r - N + iy] = O(|y|^{-N+(r-\text{Re }\beta_0)/p - 1/2}).$$

Choose r so that $(r - \text{Re } \beta_0)/p + 1/2 < N$. Then the inner integral on the right-hand side of (6.11) is dominated by $(xt)^{N-r}$. Since $h^{(-N)}(t) = O(t^{-N(p-1)-\beta_0})$ as $t \to \infty$ by partial integration (see Ex. 19, Chapter IV), the inner integral also tends to $h^{(-N)}(xt)$ as $T \to \infty$ by Theorem 2. As a consequence of the Lebesgue dominated convergence theorem, the right-hand side of (6.11) approaches

$$\int_0^\infty f_n^{(N)}(t) h^{(-N)}(xt)\, dt$$

as $T \to \infty$. The required result (6.10) now follows from (6.9).

Part of the above analysis is given in Soni (1980). For an alternative and simpler derivation of (6.10), we refer to Section 5 of Chapter IV, where examples can also be found for the construction of bounds for $E(x)$.

We now turn to the case of non-oscillatory kernels, i.e., (4.2) holds with $\gamma = 0$. In this case, Theorem 8 applies and the error term is again of the form

$$E(x) = \frac{1}{2\pi i} \int_{r-i\infty}^{r+i\infty} x^{-z}\, M[f; 1-z]\, M[h; z]\, dz. \tag{6.12}$$

Choose n and m so that $1 + \operatorname{Re} \alpha_{n-1} < r < 1 + \operatorname{Re} \alpha_n$ and $\operatorname{Re} \beta_{m-1} < r < \operatorname{Re} \beta_m$. Then by Lemma 7, for $\operatorname{Re} z = r$, $M[f; 1-z] = M[f_n; 1-z]$ and $M[h; z] = M[h_m; z]$, where h_m denotes the m-th remainder in the asymptotic expansion of $h(t)$; see (4.2). Thus from (6.12) we have

$$E(x) = \frac{1}{2\pi i} \int_{r-i\infty}^{r+i\infty} x^{-z}\, M[f_n; 1-z]\, M[h_m; z]\, dz. \tag{6.13}$$

The defining integrals of $M[f_n; 1-z]$ and $M[h_m; z]$ converge absolutely on the path of integration $\operatorname{Re} z = r$. Therefore by the Parseval identity (2.12)

$$E(x) = \int_0^\infty f_n(t) h_m(xt)\, dt. \tag{6.14}$$

This argument is given in Carlson and Gustafson (1985). For an alternative derivation of (6.14), see Section 7 of Chapter VI, where bounds are also constructed for $E(x)$.

7.   A Double Integral

The Mellin transform technique has so far been applied to integrals of only a single variable. In this section, we shall illustrate its application to integrals of higher dimensions.

Let g be an absolutely integrable function on $[0, \infty)$, and assume that it decays exponentially at infinity. Let $f \in C^\infty(\mathbb{R}^n)$ and $K_n = [0, 1]^n$. Consider the multidimensional integral

$$J(s) = \int_{K_n} g\left(\frac{x^\alpha}{s}\right) x^\beta \log^\gamma x \, f(x) \, dx, \qquad (7.1)$$

where $x \in \mathbb{R}^n$, α and $\beta \in \mathbb{R}^n_+$, $\gamma \in Z^n_+$, $x^\alpha = x_1^{\alpha_1} \cdots x_n^{\alpha_n}$, and

$$\log^\gamma x = \log^{\gamma_1} x_1 \cdots \log^{\gamma_n} x_n.$$

Here, $\mathbb{R}_+ = [0, \infty)$. The following result is given by Brüning (1984).

Theorem 10. *As $s \to 0^+$, we have*

$$J(s) \sim \sum I_{jkl}(f) s^{(\beta_l + j + 1)/\alpha_l} \log^k s \qquad (7.2)$$

where the summation is over all $j \geq 0$, $0 \leq k \leq |\gamma| + n - 1$, and $1 \leq l \leq n$. The I_{jkl} are continuous linear functionals on $C^\infty(\mathbb{R}^n)$ vanishing outside the set $\{x \in K_n : x^\alpha = 0\}$.

Brüning's proof is based on an inductive argument and elementary real-variable techniques. However, it does not lead to an explicit expression for the coefficient $I_{jkl}(f)$. In what follows, we shall reproduce the alternative proof given by McClure and Wong (1987a), which provides exact formulas for the coefficients in (7.2).

For simplicity of presentation, we shall consider only the case $n = 2$, and assume that $\gamma = 0$ in (7.1). The logarithmic factors can always be included in the integral by differentiating both sides of the resulting expansion for $J(s)$ with respect to the exponents of x; see the last paragraph of this section. Thus we are concerned with only the integral

$$J(s) = \int_0^1 \int_0^1 g\left(\frac{x^a y^b}{s}\right) x^\alpha y^\beta f(x, y) \, dx \, dy, \qquad (7.3)$$

where a and b are positive, α and β are nonnegative, $f \in C^\infty(\mathbb{R}^2)$, and g is absolutely integrable on $[0, \infty)$ and exponentially decaying at infinity.

Since g is rapidly decreasing, it is straightforward to show that the Mellin transform $M[J; z]$ exists and is analytic in the strip $-d_{11} < \operatorname{Re} z < 0$, where

$$d_{11} = \min\left\{\frac{1+\alpha}{a}, \frac{1+\beta}{b}\right\}. \qquad (7.4)$$

Furthermore, in this strip,

$$M[J; z] = \int_0^\infty s^{z-1} J(s)\, ds$$

$$= \int_0^1 \int_0^1 x^\alpha y^\beta f(x, y) \left[\int_0^\infty s^{z-1} g\left(\frac{x^a y^b}{s}\right) ds \right] dx\, dy$$

$$= M[g; -z] \int_0^1 \int_0^1 x^{\alpha + az} y^{\beta + bz} f(x, y)\, dx\, dy.$$

Put

$$F(z) = \int_0^1 \int_0^1 x^{\alpha + az} y^{\beta + bz} f(x, y)\, dx\, dy. \qquad (7.5)$$

Then, by the inversion formula (2.9), we have

$$J(s) = \frac{1}{2\pi i} \int_{c-i\infty}^{c+i\infty} s^{-z} M[g; -z] F(z)\, dz, \qquad (7.6)$$

where $-d_{11} < c < 0$. Note that the Mellin transform $M[g; -z]$ is analytic in the half-plane Re $z < 0$ and is bounded in Re $z \le c$ for any $c < 0$. Also, the double integral in (7.5) is analytic in the half-plane $-d_{11} < $ Re z. In the following, we shall show that $F(z)$ can be continued analytically to a meromorphic function in the entire z-plane.

For $-d_{11} <$ Re $z < 0$, integration by parts gives

$$F(z) = \frac{1}{\alpha + az + 1} \left[\int_0^1 y^{\beta + bz} f(1, y)\, dy \right.$$

$$\left. - \int_0^1 \int_0^1 x^{\alpha + az + 1} y^{\beta + bz} f_{1,0}(x, y)\, dx\, dy \right]$$

$$= \frac{1}{(\alpha + az + 1)(\beta + bz + 1)} \left[f(1, 1) - \int_0^1 y^{\beta + bz + 1} f_{0,1}(1, y)\, dy \right. \qquad (7.7)$$

$$- \int_0^1 x^{\alpha + az + 1} f_{1,0}(x, 1)\, dx$$

$$\left. + \int_0^1 \int_0^1 x^{\alpha + az + 1} y^{\beta + bz + 1} f_{1,1}(x, y)\, dx\, dy \right].$$

The last double integral is analytic for Re $z > -d_{22}$, where

$$d_{22} = \min\left\{ \frac{\alpha + 2}{a}, \frac{\beta + 2}{b} \right\}.$$

Thus, through equation (7.7), $F(z)$ is extended to a meromorphic function in the half-plane $\text{Re } z > -d_{22}$ with at least one pole at $z = -d_{11}$ and possibly two poles, one at $z = -(\alpha + 1)/a$ and one at $z = -(\beta + 1)/b$, depending on whether $d_{22} > \max\{(\alpha + 1)/a, (\beta + 1)/b\}$. These poles may or may not be simple; see the discussion following (7.12) below. Equation (7.7) also shows that $F(z) = O((\text{Im } z)^{-2})$ as $z \to \infty$ along vertical lines in $\text{Re } z > -d_{22}$. Now observe that the last double integral in (7.7) is in exactly the same form as $F(z)$. Hence the procedure in (7.7) can be repeated, and n applications of this gives

$$
\begin{aligned}
F(z) = \sum_{k=1}^{n} & \left[\prod_{j=1}^{k} (\alpha + az + j)(\beta + bz + j) \right]^{-1} \bigg\{ f_{k-1,k-1}(1, 1) \\
& - \int_0^1 y^{\beta + bz + k} f_{k-1,k}(1, y)\, dy - \int_0^1 x^{\alpha + az + k} f_{k,k-1}(x, 1)\, dx \bigg\} \\
& + \left[\prod_{j=1}^{n} (\alpha + az + j)(\beta + bz + j) \right]^{-1} \\
& \times \int_0^1 \int_0^1 x^{\alpha + az + n} y^{\beta + bz + n} f_{n,n}(x, y)\, dx\, dy.
\end{aligned}
$$

Further integration by parts shows that for any positive integer n,

$$
\begin{aligned}
F(z) = \sum_{k,l=1}^{n} & (-1)^{k+l} f_{k-1,l-1}(1, 1) \left[\prod_{j=1}^{k} (\alpha + az + j) \prod_{i=1}^{l} (\beta + bz + i) \right]^{-1} \\
& + \sum_{k=1}^{n} (-1)^{k+n-1} \left[\prod_{j=1}^{k} (\alpha + az + j) \prod_{i=1}^{n} (\beta + bz + i) \right]^{-1} \\
& \times \int_0^1 y^{\beta + bz + n} f_{k-1,n}(1, y)\, dy \\
& + \sum_{l=1}^{n} (-1)^{l+n-1} \left[\prod_{j=1}^{n} (\alpha + az + j) \prod_{i=1}^{l} (\beta + bz + i) \right]^{-1} \qquad (7.8) \\
& \times \int_0^1 x^{\alpha + az + n} f_{n,l-1}(x, 1)\, dx \\
& + \left[\prod_{j=1}^{n} (\alpha + az + j) \prod_{i=1}^{n} (\beta + bz + i) \right]^{-1} \\
& \times \int_0^1 \int_0^1 x^{\alpha + az + n} y^{\beta + bz + n} f_{n,n}(x, y)\, dx\, dy.
\end{aligned}
$$

From (7.8) it is evident that $F(z)$ is meromorphic for Re $z > -d_{n+1,n+1}$, where

$$d_{nn} = \min\left\{\frac{\alpha + n}{a}, \frac{\beta + n}{b}\right\}.$$

Since $d_{nn} \to \infty$ as $n \to \infty$, we have proved the following result.

Lemma 8.   *The function $F(z)$, defined by (7.5), can be continued analytically to a meromorphic function in the entire z-plane, with poles at $-(\alpha + n)/a$, $-(\beta + n)/b$ $(n = 1, 2, \ldots)$. Furthermore, $F(z) = O((\text{Im } z)^{-2})$, as $|z| \to \infty$, uniformly in any strip $-\infty < d \le \text{Re } z \le c < 0$.*

To facilitate the calculation of the residues of $F(z)$, we state the following variant of equation (7.8), which is obtained by varying the number of integrations by parts. For any positive integers n and m, we have

$$
\begin{aligned}
F(z) = \;& \sum_{k=1}^{n} \sum_{l=1}^{m} (-1)^{k+l} f_{k-1,l-1}(1, 1) \\
& \times \left[\prod_{j=1}^{k} (az + \alpha + j) \prod_{i=1}^{l} (bz + \beta + i) \right]^{-1} \\
& + \sum_{k=1}^{n} (-1)^{k+m-1} \left[\prod_{j=1}^{k} (az + \alpha + j) \prod_{i=1}^{m} (bz + \beta + i) \right]^{-1} \\
& \times \int_{0}^{1} y^{\beta + bz + m} f_{k-1,m}(1, y)\, dy \\
& + \sum_{l=1}^{m} (-1)^{l+n-1} \left[\prod_{j=1}^{n} (az + \alpha + j) \prod_{i=1}^{l} (bz + \beta + i) \right]^{-1} \\
& \times \int_{0}^{1} x^{\alpha + az + n} f_{n,l-1}(x, 1)\, dx \\
& + (-1)^{n+m} \left[\prod_{j=1}^{n} (az + \alpha + j) \prod_{i=1}^{m} (bz + \beta + i) \right]^{-1} \\
& \times \int_{0}^{1} \int_{0}^{1} x^{\alpha + az + n} y^{\beta + bz + m} f_{n,m}(x, y)\, dx\, dy.
\end{aligned}
\tag{7.9}
$$

Note that each integral on the right-hand side of (7.9) is analytic in Re $z > -d_{n+1,m+1}$, where

$$d_{nm} = \min\left\{\frac{\alpha + n}{a}, \frac{\beta + m}{b}\right\},$$

and that the sequence d_{nm} is monotonically increasing with respect to each of n and m. Moreover, $d_{nm} \to \infty$ as $n, m \to \infty$.

Now let w be a pole of $F(z)$, and let n and m be the smallest positive integers such that $w > -d_{n+1,m+1}$. To be more specific, we suppose that $w = -(\alpha + n)/a$, and that $-(\alpha + n)/a \neq -(\beta + m)/b$ for all positive integers m. Thus w is a simple pole. From (7.9) we have

$$
\begin{aligned}
\operatorname{Res}\left[F; -\frac{\alpha + n}{a} \right] = \frac{1}{a(n-1)!} & \left\{ \sum_{l=1}^{m} (-1)^{l-1} f_{n-1,l-1}(1, 1) \right. \\
& \times \prod_{i=1}^{l} \left(\beta + i - b\,\frac{\alpha + n}{a} \right)^{-1} \\
& + (-1)^m \prod_{i=1}^{m} \left(\beta + i - b\,\frac{\alpha + n}{a} \right)^{-1} \\
& \times \int_0^1 y^{\beta + m - b(n+\alpha)/a} f_{n-1,m}(1, y)\, dy \\
& + \sum_{l=1}^{m} (-1)^l \prod_{i=1}^{l} \left(\beta + i - b\,\frac{\alpha + n}{a} \right)^{-1} \int_0^1 f_{n,l-1}(x, 1)\, dx \\
& + (-1)^{m-1} \prod_{i=1}^{m} \left(\beta + i - b\,\frac{\alpha + n}{a} \right)^{-1} \\
& \left. \times \int_0^1 \int_0^1 y^{\beta + m - b(n+\alpha)/a} f_{n,m}(x, y)\, dx\, dy \right\}.
\end{aligned}
\tag{7.10}
$$

Performing the obvious integration with respect to x in the second sum and the last term on the right simplifies (7.10) to

$$
\begin{aligned}
\operatorname{Res}\left[F; -\frac{\alpha + n}{a} \right] = \frac{1}{a(n-1)!} & \left\{ \sum_{l=1}^{m} (-1)^{l+1} \right. \\
& \times \prod_{i=1}^{l} \left(\beta + i - b\,\frac{\alpha + n}{a} \right)^{-1} f_{n-1,l-1}(0, 1) \\
& + (-1)^m \prod_{i=1}^{m} \left(\beta + i - b\,\frac{\alpha + n}{a} \right)^{-1} \\
& \left. \times \int_0^1 y^{\beta + m - b(n+\alpha)/a} f_{n-1,m}(0, y)\, dy \right\}.
\end{aligned}
\tag{7.11}
$$

Similarly, if $w = -(\beta + m)/b$ is a simple pole, and if n, m are chosen as before, then we have

$$\text{Res}\left[F; -\frac{\beta + m}{b}\right] = \frac{1}{b(m-1)!}\left\{\sum_{k=1}^{n}(-1)^{k+1}\right.$$

$$\times \prod_{j=1}^{k}\left(\alpha + j - a\frac{\beta+m}{b}\right)^{-1}f_{k-1,m-1}(1,0)$$

$$+ (-1)^n\prod_{j=1}^{n}\left(\alpha + j - a\frac{\beta+m}{b}\right)^{-1} \tag{7.12}$$

$$\left. \times \int_0^1 x^{\alpha+n-a(\beta+m)/b}f_{n,m-1}(x,0)\,dx\right\}.$$

Finally, suppose that for some integers n and m, we have $(\alpha + n)/a = (\beta + m)/b$. Then $F(z)$ has a double pole at $-(\alpha + n)/a$. The principal part of $F(z)$ can be calculated from (7.9), and is given by

$$\frac{1}{ab(n-1)!\,(m-1)!}\left\{f_{n-1,m-1}(1,1) - \int_0^1 f_{n-1,m}(1,y)\,dy\right.$$

$$- \int_0^1 f_{n,m-1}(x,1)\,dx$$

$$\left.+ \int_0^1\int_0^1 f_{n,m}(x,y)\,dx\,dy\right\}\left(z + \frac{\alpha+n}{a}\right)^{-2}$$

$$+ \frac{1}{ab(n-1)!\,(m-1)!}\left\{\left[f_{n-1,m-1}(1,1) - \int_0^1 f_{n-1,m}(1,y)\,dy\right.\right.$$

$$\left. - \int_0^1 f_{n,m-1}(x,1)\,dx + \int_0^1\int_0^1 f_{n,m}(x,y)\,dx\,dy\right](aH_{n-1} + bH_{m-1})$$

$$- b\int_0^1 \log y\, f_{n-1,m}(1,y)\,dy - a\int_0^1 \log x\, f_{n,m-1}(x,1)\,dx$$

$$+ \int_0^1\int_0^1 (a\log x + b\log y)f_{n,m}(x,y)\,dx\,dy$$

$$- a\sum_{k=1}^{n-1}(n-k-1)!\,f_{k-1,m-1}(1,1) - b\sum_{l=1}^{m-1}(m-l-1)!\,f_{n-1,l-1}(1,1)$$

$$+ a\sum_{k=1}^{n-1}(n-k-1)!\int_0^1 f_{k-1,m}(1,y)\,dy$$

$$\left.+ b\sum_{l=1}^{m-1}(m-l-1)!\int_0^1 f_{n,l-1}(x,1)\,dx\right\}\left(z + \frac{\alpha+n}{a}\right)^{-1},$$

where $H_n = \sum_{k=1}^{n} 1/k$. Upon simplification, the above expression reduces to

$$
\mathrm{Prin}\left[F; -\frac{\alpha+n}{a}\right] = \frac{1}{ab(n-1)!\,(m-1)!}\, f_{n-1,m-1}(0,0)\left(z+\frac{\alpha+n}{a}\right)^{-2}
$$
$$
+\frac{1}{ab(n-1)!\,(m-1)!}\left\{f_{n-1,m-1}(0,0)(aH_{n-1}+bH_{m-1})\right.
$$
$$
-b\int_0^1 \log y\, f_{n-1,m}(0,y)\,dy \tag{7.13}
$$
$$
-a\int_0^1 \log x\, f_{n,m-1}(x,0)\,dx - a\sum_{k=1}^{n-1}(n-k-1)!\,f_{k-1,m-1}(1,0)
$$
$$
\left.-b\sum_{l=1}^{m-1}(m-l-1)!\,f_{n-1,l-1}(0,1)\right\}\left(z+\frac{\alpha+n}{a}\right)^{-1}.
$$

Observe that the right-hand side of equation (7.11) can be regarded as the result of a continuous linear functional acting on the C^∞-function $f(x,y)$, which vanishes off the coordinate axes. The same remark applies to equations (7.12) and (7.13).

We are now ready to derive the asymptotic expansion of $J(s)$. First we return to the contour integral representation in (7.6). For any positive number $d > -c$ we can choose an arbitrarily small ε such that $F(z)$ has no pole in the strip $-d < \mathrm{Re}\, z \le -d + \varepsilon$. Since $M[g; -z]$ is analytic and bounded for $\mathrm{Re}\, z \le c < 0$, by Lemma 8 we can shift the contour in (7.6) to the left and obtain

$$
J(s) = \sum_{-d+\varepsilon < \mathrm{Re}\, z < c} \mathrm{Res}\{s^{-z}F(z)\,M[g; -z]\}
$$
$$
+\frac{1}{2\pi i}\int_{-d+\varepsilon-i\infty}^{-d+\varepsilon+i\infty} s^{-z}F(z)\,M[g; -z]\,dz. \tag{7.14}
$$

The last integral is clearly bounded by a constant multiple of $s^{d-\varepsilon}$. Thus we have

$$
J(s) = \sum_{-d < \mathrm{Re}\, z < c} \mathrm{Res}\{s^{-z}F(z)\,M[g; -z]\} + O(s^{d-\varepsilon}), \tag{7.15}
$$

as $s \to 0^+$. Each simple pole $w = -(\alpha+n)/a$ of $F(z)$ contributes a term of the form

$$
s^{(\alpha+n)/a}\, M\left[g; \frac{\alpha+n}{a}\right]\mathrm{Res}\left[F; -\frac{\alpha+n}{a}\right] \tag{7.16}
$$

to the asymptotic expansion of $J(s)$. Similarly, each simple pole $w = -(\beta + m)/b$ contributes a term of the form

$$s^{(\beta + m)/b} M\left[g; \frac{\beta + m}{b}\right] \mathrm{Res}\left[F; -\frac{\beta + m}{b}\right]. \tag{7.17}$$

If $w = -(\alpha + n)/a = -(\beta + m)/b$ is a double pole of $F(z)$, then the residue of

$$s^{-z}F(z)\, M[g; -z] = \exp(-z \log s) F(z)\, M[g; -z]$$

at w can be calculated from (7.13), and to the expansion of $J(s)$ this point contributes the term

$$a_n(f)s^{(\alpha + n)/a} \log s + b_n(f)s^{(\alpha + n)/a}, \tag{7.18}$$

where

$$a_n(f) = -\frac{f_{n-1,m-1}(0, 0)}{ab(n-1)!\,(m-1)!} M\left[g; \frac{\alpha + n}{a}\right] \tag{7.19}$$

and

$$\begin{aligned}
b_n(f) = & -\frac{f_{n-1,m-1}(0, 0)}{ab(n-1)!\,(m-1)!} M\left[g; \frac{\alpha + n}{a}\right] + \frac{1}{ab(n-1)!\,(m-1)!} \\
& \times \left\{ f_{n-1,m-1}(0, 0)(aH_{n-1} + bH_{m-1}) - b \int_0^1 \log y\, f_{n-1,m}(0, y)\, dy \right. \\
& - a \int_0^1 \log x\, f_{n,m-1}(x, 0)\, dx - a \sum_{k=1}^{n-1} (n-k-1)!\, f_{k-1,m-1}(1, 0) \\
& \left. - b \sum_{l=1}^{m-1} (m-l-1)!\, f_{n-1,l-1}(0, 1) \right\} M\left[g; \frac{\alpha + n}{a}\right].
\end{aligned} \tag{7.20}$$

Inserting (7.16), (7.17), and (7.18) into (7.15) yields an asymptotic expansion for $J(s)$ in terms of powers and logarithms of s. In view of the statement following (7.13), Theorem 10 is proved for the case $n = 2$ and $\gamma = 0$.

The above method can easily be extended to integrals of higher dimensions. The formulas for the coefficients in the expansion, however, will become overwhelmingly complicated, and hence will not be given. The case $\gamma \neq 0$ in (7.1) can be included by employing a frequently used device in asymptotics, namely, we first replace each term in the sum in (7.14) by its corresponding value given in (7.16), (7.17) or (7.18), and then differentiate both sides of (7.14) with respect to α and β an appropriate number of times. It should be emphasized that this device applies only when we have an equation, and not an asymptotic expansion, i.e., when there is an exact formula for the remainder term. A more detailed discussion of logarithmic singularities is given in Section 2 of Chapter II.

Exercises

1. Let $f(z)$ be analytic in the open sector $S_\theta = \{z: |\arg z| < \theta\}$, and continuous in the closed sector $\bar{S}_\theta = \{z: |\arg z| \leq \theta\}$. Suppose that $z^{-\alpha} f(z) = O(1)$ as $z \to 0$ in $\bar{S}_\theta$ and $z^b f(z) = O(1)$ as $z \to \infty$ in $\bar{S}_\theta$, where $-\alpha < b$. Show that

$$M[f(te^{i\theta}); z] = e^{-i\theta z} M[f; z].$$

   Use this identity to calculate the Mellin transforms of $h(t) = U(a, e^{i\pi/4}t)$ and $h(t) = e^{-it^2/4}U(a, e^{i\pi/4}t)$, where $U(a, t)$ is the parabolic cylinder function and $a > -\frac{1}{2}$; see formulas 49 and 50 in the "Table" given at the end of this chapter. (In Whittaker's notation, $U(a, t) = D_{-a-1/2}(t)$.)

2. Derive the asymptotic expansion of the Laplace integral

$$L(s) = \int_0^\infty f(t)e^{-st}\, dt$$

   as $s \to 0^+$, where

   (i) $f(t) = \mathrm{Ai}(-t)$,   (ii) $f(t) = e^{-it^2/4}U(v - \frac{1}{2}; e^{i\pi/4}t)$,

   (iii) $f(t) = \dfrac{1}{1+t}$,   (iv) $f(t) = \dfrac{1}{\sqrt{1+t}}$,   (v) $f(t) = \sqrt{1+t^2}$.

3. Let $f(t)$ satisfy (4.1) with Re $\alpha_0 > -1$ and (4.4) with $b > 1$. Consider the generalized Stieltjes transform

$$I(x) = \int_0^\infty \frac{f(t)}{(t+x)^\rho}\, dt, \qquad \rho > 0.$$

Show that if $1 + \alpha_s \neq \rho + n$ for all nonnegative integers s and n then

$$I(x) \sim \sum_{s=0}^{\infty} a_s \frac{\Gamma(1+\alpha_s)\Gamma(\rho - 1 - \alpha_s)}{\Gamma(\rho)} x^{1+\alpha_s - \rho}$$

$$+ \sum_{s=0}^{\infty} (-1)^s \frac{\Gamma(s+\rho)\, M[f; 1-\rho - s]}{s!\, \Gamma(\rho)} x^s$$

as $x \to 0^+$.

4. In Ex. 3, let $f(t) = |1 - t|^\nu$, $0 < \mathrm{Re}\, \nu < \rho - 1$. Show that if ρ is not a positive integer, then

$$\frac{\Gamma(\rho)}{\Gamma(\nu + 1)} \int_0^\infty \frac{|1 - t|^\nu}{(x + t)^\rho}\, dt \sim \sum_{s=0}^{\infty} (-1)^s \frac{\Gamma(\rho - s - 1)}{\Gamma(\nu - s + 1)} x^{s+1-\rho}$$

$$+ \sum_{s=0}^{\infty} \frac{\Gamma(s + \rho - \nu - 1)}{\Gamma(s + 1)} \left[1 + \frac{\sin \pi(\nu - \rho)}{\sin \pi \rho} \right] (-x)^s$$

as $x \to 0^+$. Also show that if ρ is a positive integer and ν is not a positive integer, then as $x \to 0^+$ we have

$$\frac{\Gamma(\rho)}{\Gamma(\nu + 1)} \int_0^\infty \frac{|1 - t|^\nu}{(x + t)^\rho}\, dt \sim \sum_{s=0}^{\rho - 2} (-1)^s \frac{\Gamma(\rho - s - 1)}{\Gamma(\nu - s + 1)} x^{s+1-\rho}$$

$$+ \sum_{s=0}^{\infty} \frac{\pi \csc \pi \nu + \psi(\nu + 2 - \rho - s) - \psi(s + 1) + \log s}{\Gamma(\nu + 2 - \rho - s)\Gamma(s + 1)} (-x)^s,$$

where $\psi(z) = \Gamma'(z)/\Gamma(z)$. (Hint: Use Theorem 8.)

5. Show that the integral

$$I(x) = \int_0^{\pi/2} J_\nu^2(x \cos \theta)\, d\theta, \qquad \nu > -\tfrac{1}{2},$$

can be put in the form of a Mellin convolution

$$I(x) = \int_0^\infty f(t) h(xt)\, dt$$

with

$$f(t) = \begin{cases} \dfrac{1}{\sqrt{1 - t^2}} & 0 < t < 1 \\ 0 & t \geq 1 \end{cases}$$

and
$$h(t) = J_v^2(t).$$

Prove that the integration contour in the Parseval identity

$$I(x) = \frac{1}{2\pi i} \int_{c-i\infty}^{c+i\infty} x^{-z}\, M[h; z]\, M[f; 1-z]\, dz,$$

$-2v < c < 1$, however, cannot be shifted to the right beyond the vertical line Re $z = 2$. (A derivation of the asymptotic expansion of $I(x)$ is outlined in Ex. 17, Chapter VI.)

6. Consider the one-sided Hilbert transform

$$I(x) = \int_0^\infty \frac{J_0^2(t)}{t - x}\, dt,$$

where the bar indicates that the integral is a Cauchy principal value. Show that this integral can be written as

$$I(x) = -\int_0^\infty f(t) J_0^2(xt)\, dt \qquad \text{with} \quad f(t) = \frac{1}{1 - t}.$$

Show also that the Parseval identity

$$I(x) = \frac{1}{2\pi i} \int_{c-i\infty}^{c+i\infty} x^{-z}\, M[J_0^2; z]\pi \cot \pi z\, dz$$

holds with $0 < c < 1$. Prove that the integration contour Re $z = c$ cannot be shifted beyond Re $z = \frac{3}{2}$; cf. Example 2. (For a construction of the asymptotic expansion of $I(x)$, see Example 2 in Chapter VI.)

7. Show that for small positive values of p, the integral

$$L(p) = \int_0^\infty \frac{e^{-p^2 x^2}}{x^2} \left\{ J_v(x) J_{1-v}(x) - \frac{x \sin v\pi}{2\pi v(1 - v)} \right\} dx$$

has the asymptotic expansion

$$L(p) \sim -\frac{\sin v\pi}{4\pi v(1 - v)} \{ \psi(2) - \psi(v + 1) - \psi(2 - v)$$

$$- 2 \log 2p\}\{1 + 2v(1 - v)p^2\} - \frac{(1 - 2v)\sin v\pi}{2\pi v(1 - v)} p^2$$

$$+ \sum_{n=3}^\infty \frac{(-1)^n (2p)^n \Gamma(\tfrac{1}{2}n)}{2n(n - 2)\Gamma(v + 1 - \tfrac{1}{2}n)\Gamma(2 - v - \tfrac{1}{2}n)(n - 2)!}.$$

Deduce from this the expansion

$$\lim_{\delta \to 0^+} \left[\frac{\gamma + \log(\frac{1}{2}\delta)}{2} + \int_\delta^\infty \frac{e^{-p^2 x^2}}{x^2} J_0(x) J_1(x) \, dx \right]$$

$$\sim -\frac{1}{2} p^2 + \sum_{m=0}^\infty \frac{[\Gamma(m + \frac{3}{2})]^3 (2p)^{2m+3}}{\pi^2 (2m+1)^2 (2m+3) \cdot (2m+1)!}$$

(Hint: Use Lemma 7.)

8. Let $h(t)$ satisfy (4.5) and (4.2) with $\gamma \neq 0$. Show that the analytic continuation of $M[h; z]$ in the half-plane Re $z > -c$ is given by

$$M[h; z] = \lim_{\varepsilon \to 0^+} \int_0^\infty t^{z-1} h(t) e^{-\varepsilon t} \, dt.$$

(This result is used in §5 of Chapter IV.)

9. Let $f(t)$ be a locally integrable function on $(0, \infty)$ and $f(t) = O(t^b)$ as $t \to 0^+$. Assume that for each $n \geq 1$

$$f(t) = \sum_{s=0}^{n-1} a_s t^{-s-1} + f_n(t),$$

where $f_n(t) = O(t^{-n-1})$ as $t \to +\infty$. Show that

$$\int_0^1 t^n f_n(t) \, dt + \int_1^\infty t^n f_{n+1}(t) \, dt = \lim_{z \to n+1} \left\{ M[f; z] + \frac{a_n}{z - n - 1} \right\}.$$

10. Consider the function

$$G(z) = \int_1^\infty t^{z-1} \exp\{i\gamma t^p\} \, dt$$

where $\gamma \neq 0$ is real, $z = x + iy$ and $x \leq 0$. Show that

$$G(x + iy) = O(|y|^{-1/2}), \qquad \text{as } |y| \to \infty,$$

uniformly for x in any compact interval contained in $(-\infty, 0]$. (Hint: Use (3.3).)

11. Let $h(t)$ satisfy (4.2) with $\gamma \neq 0$. Use Ex. 10 and Lemma 4 to show that

$$M[h; x + iy] = O(|y|^{(x - \operatorname{Re}\beta_0)/p - 1/2})$$

as $|y| \to \infty$. (Hint: Modify the proof of Lemma 3.)

12. Let $f(t) = O(t^\alpha)$ as $t \to 0^+$ with $\alpha > -1$, and let $f(t) \in L^1(0, \infty)$. Assume that the integral defining $M[h; z]$ converges for $0 < \text{Re } z < \sigma$ and that there exist nonnegative constants C_1, C_2, and ρ such that $\rho < \alpha + 1$,

$$\left| \int_{c-iT}^{c+iT} t^{-z} M[h; z] \, dz \right| \le C_1 t^{-\rho} + C_2, \qquad 0 < c < \sigma, \quad T \ge 0,$$

and

$$\lim_{T \to \infty} \int_{c-iT}^{c+iT} t^{-z} M[h; z] \, dz = h(t), \qquad 0 < c < \sigma, \quad 0 < t < \infty.$$

Show that the Parseval identity (2.12) holds under these conditions. Furthermore, show that the functions $h(t) = e^{it}$ and $h(t) = H_0^{(1)}(t)$ satisfy these conditions but not those in Theorem 4.

13. Let f and h satisfy (4.1), (4.2), (4.4), and (4.5), and let D_{ij} be the domains defined in (4.28). Let X be a set of real numbers, whose intersection with each of the D_{ij}'s is nonempty and which contains arbitrarily large positive numbers. Suppose that for each $x \in X$, either $M[f_i; 1 - x - iy]$ for $i = 1, 2$, or $M[h_j; x + iy]$ for $j = 1$, 2, is $o(|y|^{-n})$ as $|y| \to \infty$ for arbitarily large n. Show that the assumptions, and hence the conclusions, of Theorems 7 and 8 hold.

14. Let f and h satisfy (4.1), (4.2), (4.4), and (4.5), and let X be the set defined in Ex. 13. Suppose that for each $x \in X$, either $M[f_i; 1 - x - iy]$ for $i = 1, 2$, or $M[h_j; x + iy]$ for $j = 1, 2$, is $O(e^{-\alpha|y|})$ as $|y| \to \infty$ for some $\alpha > 0$. Show that the results of Theorems 7 and 8 hold for complex x in the sector $S_\alpha = \{x : |\arg x| < \alpha\}$.

15. Show that (i) if $h(t) = K_\nu(\sqrt{t})$ and α is any constant in $(0, \pi)$, then $M[h; x + iy] = O(e^{-\alpha|y|})$; (ii) if $h(t) = 1/\sqrt{1 + t^2}$ then $M[h; x + iy] = O(e^{-\pi|y|/2})$; and (iii) if $h(t) = e^{-t} I_\nu(t)$ and $0 < \beta < \pi/2$, then $M[h; x + iy] = O(e^{-\beta|y|})$.

16. Let $f(t)$ satisfy the conditions of Lemma 6 with $\text{Re } \alpha_0 > -1$ and $b \ge 1$. Consider the integral

$$I(x) = \int_0^\infty f(t) J_\nu(xt) Y_\nu(xt) \, dt.$$

Show that for any given $n \geq 0$,

$$I(x) = -\frac{1}{2\sqrt{\pi}} \sum_{s=0}^{N} \frac{\Gamma[\frac{1}{2}(\alpha_s + 1)]\Gamma[\frac{1}{2}(\alpha_s + 1) + v]}{\Gamma[\frac{1}{2}(\alpha_s + 2)]\Gamma(\frac{1}{2} + v - \frac{1}{2}\alpha_s)} \frac{a_s}{x^{\alpha_s + 1}}$$

$$+ O(x^{-\alpha_N - 1 - \varepsilon}),$$

as $x \to +\infty$, where N is the largest integer such that Re $\alpha_N < n + \frac{1}{2}$ and ε is an arbitrary positive number.

17. Let $f(t)$ be a locally integrable function in $(0, \infty)$ satisfying

$$f(t) \sim \sum_{s=0}^{\infty} a_s t^{s + \alpha} \qquad \text{as } t \to 0^+,$$

where $-1 < \alpha < 0$. Suppose that $f(t) = O(t^{-b})$ as $t \to +\infty$, where $0 < b < -\alpha$. Consider the integral

$$I(x) = \int_0^{\infty} f(t) I_v(xt) K_v(xt)\, dt, \qquad v \geq 0.$$

Show that

$$I(x) \sim \sum_{s=0}^{\infty} a_s \frac{\Gamma[\frac{1}{2}(\alpha + s + 1) + v]\Gamma[\frac{1}{2}(\alpha + s + 1)]\Gamma[-\frac{1}{2}(\alpha + s)]}{4\sqrt{\pi}\Gamma[\frac{1}{2} + v - \frac{1}{2}(\alpha + s)]}$$

$$\times x^{-\alpha - s - 1}$$

$$+ \sum_{s=0}^{\infty} \frac{(-1)^s}{s!} M[f; -2s] \frac{\Gamma(s + v + \frac{1}{2})\Gamma(s + \frac{1}{2})}{2\sqrt{\pi}\Gamma(\frac{1}{2} + v - s)} x^{-2s - 1},$$

as $x \to \infty$ in the sector $|\arg x| \leq \pi/2 - \delta < \pi/2$.

18. Let $g \in C^{\infty}(\mathbb{R})$ and $f \in C^{\infty}(\mathbb{R}^2)$. Let $a, b > 0$, and $\alpha, \beta \in \mathbb{R}$. Suppose that $g = 0$ in a neighborhood of 0 and $g = 1$ in a neighborhood of ∞. Show that there exist a constant C and sequences $\{a_j\}$, $\{b_j\}$, $\{c_j\}$, and $\{d_j\}$ such that

$$\int_0^1 \int_0^1 g\left(\frac{x^a y^b}{s}\right) x^{\alpha} y^{\beta} f(x, y)\, dx\, dy$$

$$\sim C + \sum_{j=1}^{\infty} [a_j \log s + b_j] s^{(\alpha + j)/a} + \sum_{j=1}^{\infty} [c_j \log s + d_j] s^{(\beta + j)/b}$$

as $s \to 0^+$. (Hint: Write $g = 1 + (g - 1)$ and apply Theorem 10 to $g - 1$ with $n = 2$ and $\gamma = 0$.)

19. Let $g \in C^\infty(\mathbb{R})$ have compact support, and let $f \in C^\infty(\mathbb{R}^2)$ have the support $[0,1] \times [0,1]$. Suppose that $g = 1$ in a neighborhood of 0. Show that for a, $b > 0$ and α, $\beta \in \mathbb{R}$, there exist a constant C and sequences $\{a_j\}$, $\{b_j\}$, $\{c_j\}$, and $\{d_j\}$ such that

$$\int_0^1 \int_0^1 g\left(\frac{s}{xy}\right) x^{-\alpha} y^{-\beta} f(x^a, y^b) \, dx \, dy$$

$$\sim C + \sum_{j=1}^{\infty} [a_j \log s + b_j] s^{a(j-1)-\alpha+1}$$

$$+ \sum_{j=1}^{\infty} [c_j \log s + d_j] s^{b(j-1)-\beta+1}$$

as $s \to 0^+$. (Hint: Make the substitutions $u = x^a$, $v = y^b$, and apply the preceding exercise with $g(t)$ replaced by $g(1/t)$ for $t \neq 0$.)

Supplementary Notes

§2. Statement (iii) in Theorem 1 is taken from Handelsman and Lew (1970). Theorem 4 is given in Titchmarsh (1959, p. 60, Theorem 42). Titchmarsh's hypotheses, however, seem to be insufficient for the application of the inversion formula used in the proof. We have therefore imposed stronger conditions, namely, that both $t^{-c}f(t)$ and $t^{c-1}h(t)$ be absolutely integrable in $(0, \infty)$.

§3. Example 1 is given in Wong (1983). There is some resemblance between this example and Exercise 7.

§4. Most of the results in this section can be found in Handelsman and Lew (1971). Lemmas 5 and 6 are given in Handelsman and Bleistein (1973).

§5. Examples 3 and 4 are taken from Handelsman and Lew (1970). An alternative derivation of (5.21) can be found in Soni and Soni (1985).

§7. The proof of Theorem 10 given by Brüning (1984) is difficult to follow. The argument of §§5, 6, and 7 in his paper, which involves passing back and forth between one- and two-variable expansions, is more subtle than is indicated.

Exercises.   For Ex. 4, see Handelsman and Lew (1971). The integral in Ex. 5 occurs in studies of crystallography and diffraction theory. The asymptotic expansion of this integral has been obtained by

Stoyanov and Farrell (1987) and Wong (1988). The integral in Ex. 6 arose in some work on water waves, and its asymptotic behavior has been discussed by Ursell (1983). For Ex. 7, see Watson (1944, p. 437). A result similar to Ex. 11 was given in Handelsman and Bleistein (1973). Exercise 12 is based on the material in Soni (1980). The results in Exs. 13 and 14 are proved in Handelsman and Lew (1971). For Exs. 18 and 19, see Brüning (1984).

Short Table.   The first eleven equations are properties of the Mellin transform. The table begins with equation (12). Most of these equations are taken from Oberhettinger (1974). The rest of the equations are taken from Erdélyi *et al.* (1954) and Titchmarsh (1959). An extensive tabulation can also be found in Marichev (1983).

SHORT TABLE OF MELLIN TRANSFORMS

$f(t)$		$M[f; z] = \displaystyle\int_0^\infty t^{z-1} f(t)\, dt$
1. $f(at)$	$a > 0$	$a^{-z} M[f; z]$
2. $t^\alpha f(t)$		$M[f; z + \alpha]$
3. $f(t^p)$	$p > 0$	$p^{-1} M[f; z/p]$
4. $f(t^{-p})$	$p > 0$	$p^{-1} M[f; -z/p]$
5. $t^\alpha f(at^p)$	$a, p > 0$	$p^{-1} a^{-(z+\alpha)/p} M[f; (z+\alpha)/p]$
6. $t^\alpha f(at^{-p})$	$a, p > 0$	$p^{-1} a^{(z+\alpha)/p} M[f; -(z+\alpha)/p]$
7. $f(t)(\log t)^n$		$\left(\dfrac{d}{dz}\right)^n M[f; z]$
8. $f^{(n)}(t)$, provided that $$\lim_{t\to 0} t^{z-k-1} f^{(k)}(t) = 0,$$ $$k = 0, 1, \ldots, n-1.$$		$\dfrac{\Gamma(n+1-z)}{\Gamma(1-z)} M[f; z-n]$
9. $\left(t\dfrac{d}{dt}\right)^n f(t)$		$(-z)^n M[f; z]$
10. $\displaystyle\int_0^t f(\tau)\, d\tau$		$-z^{-1} M[f; z+1]$
11. $\displaystyle\int_t^\infty f(\tau)\, d\tau$		$z^{-1} M[f; z+1]$
12. t^ν $t < 1$ 0 $t > 1$		$(\nu + z)^{-1}$ $\qquad$ Re $z > -$Re ν
13. 0 $t < 1$ t^ν $t > 1$		$-(\nu + z)^{-1}$ $\qquad$ Re $z < -$Re ν
14. $(1 + t)^{-1}$ $t < 1$ $\phantom{(1+t)^{-1}}$ 0 $t > 1$		$\frac{1}{2}[\psi(\frac{1}{2} + \frac{1}{2}z) - \psi(\frac{1}{2}z)]$ $\qquad$ Re $z > 0$

	$f(t)$		$M[f; z] = \displaystyle\int_0^\infty t^{z-1} f(t)\, dt$	
15.	$(1 + t)^{-\nu}$		$B(z, \nu - z)$	$0 < \operatorname{Re} z < \operatorname{Re} \nu$
16.	$(1 - t)^{-1}$		$\pi \cot \pi z$	$0 < \operatorname{Re} z < 1$
	Cauchy principal value			
17.	$(1 - t)^\nu$	$t < 1$	$B(\nu + 1, z)$	$\operatorname{Re} z > 0$
	0	$t > 1$		
	$\operatorname{Re} \nu > -1$			
18.	0	$t < 1$	$B(-\nu - z, \nu + 1)$	$\operatorname{Re} z < -\operatorname{Re} \nu$
	$(t - 1)^\nu$	$t > 1$		
	$\operatorname{Re} \nu > -1$			
19.	$(1 + t^2)^{-1}$		$\tfrac{1}{2}\pi \csc(\tfrac{1}{2}\pi z)$	$0 < \operatorname{Re} z < 2$
20.	$[t + (1 + t^2)^{1/2}]^{-\nu}$		$\nu(\nu + z)^{-1} 2^{-z} B(z, \tfrac{1}{2}\nu - \tfrac{1}{2}z)$	
				$0 < \operatorname{Re} z < \operatorname{Re} \nu$
21.	$(1 + t^2)^{-1/2}$		$2^{-z} B(z, \tfrac{1}{2} - \tfrac{1}{2}z)$	$0 < \operatorname{Re} z < 1$
22.	$(t^2 + 2t \cos\theta + 1)^{-1}$		$\pi \csc\theta \, \csc(\pi z) \sin[(1 - z)\theta]$	
	$-\pi < \theta < \pi$			$0 < \operatorname{Re} z < 2$
23.	$(t^2 + 2t \cos\theta + 1)^{-1/2}$		$\pi \csc(\pi z) P_{z-1}(\cos\theta)$	
	$-\pi < \theta < \pi$			$0 < \operatorname{Re} z < 1$
24.	e^{-t}		$\Gamma(z)$	$\operatorname{Re} z > 0$
25.	e^{-bt}	$t < 1$	$b^{-z}\gamma(z, b)$	$\operatorname{Re} z > 0$
	0	$t > 1$		
26.	0	$t < 1$	$b^{-z}\Gamma(z, b)$	
	e^{-bt}	$t > 1$		
27.	$\log t$	$t < 1$	$-z^{-2}$	$\operatorname{Re} z > 0$
	0	$t > 1$		
28.	0	$t < 1$	z^{-2}	$\operatorname{Re} z < 0$
	$\log t$	$t > 1$		
29.	$\log(1 + t)$		$\pi z^{-1} \csc(\pi z)$	$-1 < \operatorname{Re} z < 0$
30.	$\sin at$	$a > 0$	$a^{-z}\Gamma(z) \sin(\tfrac{1}{2}\pi z)$	$-1 < \operatorname{Re} z < 1$
31.	$\cos at$	$a > 0$	$a^{-z}\Gamma(z) \cos(\tfrac{1}{2}\pi z)$	$0 < \operatorname{Re} z < 1$
32.	e^{it}		$e^{iz\pi/2}\Gamma(z)$	$0 < \operatorname{Re} z < 1$
33.	$J_\nu(t)$		$2^{z-1}\Gamma(\tfrac{1}{2}\nu + \tfrac{1}{2}z)[\Gamma(1 + \tfrac{1}{2}\nu - \tfrac{1}{2}z)]^{-1}$	
				$-\operatorname{Re} \nu < \operatorname{Re} z < \tfrac{3}{2}$
34.	$Y_\nu(t)$		$-\pi^{-1} 2^{z-1} \cos[\tfrac{1}{2}\pi(z - \nu)]$	
			$\times \Gamma(\tfrac{1}{2}z - \tfrac{1}{2}\nu)\Gamma(\tfrac{1}{2}\nu + \tfrac{1}{2}z)$	
				$\pm \operatorname{Re} \nu < \operatorname{Re} z < \tfrac{3}{2}$
35.	$H_\nu^{(1)}(t)$		$\pi^{-1} 2^{z-1} e^{-i(\nu - z + 1)\pi/2}$	
			$\times \Gamma(\tfrac{1}{2}z - \tfrac{1}{2}\nu)\Gamma(\tfrac{1}{2}\nu + \tfrac{1}{2}z)$	
				$\pm \operatorname{Re} \nu < \operatorname{Re} z < \tfrac{3}{2}$
36.	$\sin t\, J_\nu(t)$		$\dfrac{2^{\nu-1}\Gamma(\tfrac{1}{2} - z)\Gamma(\tfrac{1}{2} + \tfrac{1}{2}\nu + \tfrac{1}{2}z)}{\Gamma(1 + \nu - z)\Gamma(1 - \tfrac{1}{2}\nu - \tfrac{1}{2}z)}$	
				$-1 - \operatorname{Re} \nu < \operatorname{Re} z < \tfrac{1}{2}$
37.	$\cos t\, J_\nu(t)$		$\dfrac{2^{\nu-1}\Gamma(\tfrac{1}{2} - z)\Gamma(\tfrac{1}{2}\nu + \tfrac{1}{2}z)}{\Gamma(\tfrac{1}{2} - \tfrac{1}{2}\nu - \tfrac{1}{2}z)\Gamma(1 + \nu - z)}$	
				$-\operatorname{Re} \nu < \operatorname{Re} z < \tfrac{1}{2}$

$f(t)$	$M[f; z] = \int_0^\infty t^{z-1} f(t)\, dt$
38. $J_\nu^2(t)$	$2^{z-1} \Gamma(1-z) \Gamma(\nu + \tfrac{1}{2}z)$ $\times [\Gamma(1 - \tfrac{1}{2}z)]^{-2} [\Gamma(1 + \nu - \tfrac{1}{2}z)]^{-1}$ $-2 \operatorname{Re} \nu < \operatorname{Re} z < 1$
39. $J_\nu(t) Y_\nu(t)$	$-\dfrac{1}{2\sqrt{\pi}} \dfrac{\Gamma(\tfrac{1}{2}z) \Gamma(\tfrac{1}{2}z + \nu)}{\Gamma(\tfrac{1}{2}z + \tfrac{1}{2}) \Gamma(1 + \nu - \tfrac{1}{2}z)}$ $\max\{0, -2\operatorname{Re}\nu\} < \operatorname{Re} z < 1$
40. $J_\nu^2(t) + Y_\nu^2(t)$	$\pi^{-5/2} \cos(\pi\nu) \Gamma(\tfrac{1}{2}z) \Gamma(\tfrac{1}{2} - \tfrac{1}{2}z)$ $\times \Gamma(\nu + \tfrac{1}{2}z) \Gamma(\tfrac{1}{2}z - \nu)$ $2|\operatorname{Re}\nu| < \operatorname{Re} z < 1$
41. $J_\nu(t) J_{-\nu}(t)$	$\tfrac{1}{2}\pi^{-1/2} \Gamma(\tfrac{1}{2}z) \Gamma(\tfrac{1}{2} - \tfrac{1}{2}z)$ $\times [\Gamma(1 + \nu - \tfrac{1}{2}z) \Gamma(1 - \nu - \tfrac{1}{2}z)]^{-1}$ $0 < \operatorname{Re} z < 1$
42. $K_\nu(t)$	$2^{z-2} \Gamma(\tfrac{1}{2}z + \tfrac{1}{2}\nu) \Gamma(\tfrac{1}{2}z - \tfrac{1}{2}\nu)$ $\operatorname{Re} z > \pm \operatorname{Re} \nu$
43. $e^t K_\nu(t)$	$\pi^{-1/2} \cos(\pi\nu) 2^{-z} \Gamma(\tfrac{1}{2} - z)$ $\times \Gamma(z + \nu) \Gamma(z - \nu)$ $\pm \operatorname{Re} \nu < \operatorname{Re} z < \tfrac{1}{2}$
44. $e^{-t} I_\nu(t)$	$\dfrac{\Gamma(\tfrac{1}{2} - z) \Gamma(z + \nu)}{2^z \pi^{1/2} \Gamma(1 + \nu - z)}$ $-\operatorname{Re}\nu < \operatorname{Re} z < \tfrac{1}{2}$
45. $I_\nu(t) K_\nu(t)$	$\tfrac{1}{4}\pi^{-1/2} \Gamma(\tfrac{1}{2}z) \Gamma(\tfrac{1}{2} - \tfrac{1}{2}z) \Gamma(\nu + \tfrac{1}{2}z)$ $\times [\Gamma(1 - \tfrac{1}{2}z + \nu)]^{-1}$ $\max\{0, -2\operatorname{Re}\nu\} < \operatorname{Re} z < 1$
46. $K_\nu^2(t)$	$\tfrac{1}{4}\pi^{1/2} \Gamma(\tfrac{1}{2}z + \nu) \Gamma(\tfrac{1}{2}z - \nu) \Gamma(\tfrac{1}{2}z)$ $\times [\Gamma(\tfrac{1}{2} + \tfrac{1}{2}z)]^{-1}$ $\operatorname{Re} z > 2|\operatorname{Re}\nu|$
47. $\operatorname{Ai}(t)$	$(2\pi)^{-1} 3^{2z/3 - 7/6} \Gamma\!\left(\dfrac{z}{3}\right) \Gamma\!\left(\dfrac{z+1}{3}\right)$ $\operatorname{Re} z > 0$
48. $\operatorname{Ai}(-t)$	$\pi^{-1} 3^{2z/3 - 7/6} \Gamma\!\left(\dfrac{z}{3}\right) \Gamma\!\left(\dfrac{z+1}{3}\right)$ $\times \sin\!\left(\dfrac{\pi z}{3} + \dfrac{\pi}{6}\right)$ $0 < \operatorname{Re} z < \tfrac{7}{4}$
49. $U(a, t)$	$\dfrac{\sqrt{\pi}\, \Gamma(z) 2^{(1-2a)/4}}{\Gamma[(2z + 2a + 3)/4]}$ $\times {}_2F_1\!\left(\dfrac{z+1}{2}, \dfrac{a}{2} + \dfrac{3}{4}; \dfrac{z+a}{2} + \dfrac{3}{4}; -1\right)$ $\operatorname{Re} z > 0$
50. $e^{-t^2/4} U(a, t)$	$\dfrac{\sqrt{\pi}\, 2^{-(z+a+1/2)/2} \Gamma(z)}{\Gamma[\tfrac{1}{2}(z + a + \tfrac{3}{2})]}$ $\operatorname{Re} z > 0$

IV

The Summability Method

1. Introduction

In his book *The Theory of Functions*, Titchmarsh considered the asymptotic behavior of the Fourier integral

$$F(x) = \int_0^\infty e^{ixt - t^{1/4}} \sin t^{1/4} \, dt. \tag{1.1}$$

Clearly the function $e^{-t^{1/4}} \sin t^{1/4}$ cannot be put in the form $t^{\alpha-1} f(t)$ for some $\alpha > 0$ and for some sufficiently smooth function $f(t)$ in $[0, \infty)$, and hence the method of repeated integration by parts, as given in Chapter I, Section 4, does not apply. By using a combination of integration by parts and contour integrations, Titchmarsh (1939, p. 109) was able to show that as $x \to \infty$,

$$F(x) \sim \frac{i}{4} \Gamma\left(\frac{1}{4}\right) e^{i\pi/8} x^{-5/4}. \tag{1.2}$$

He also remarked that contour integration alone, without partial integration, will lead only to the result $F(x) = o(x^{-1})$ as $x \to \infty$.

Let us now proceed formally to derive this result in an entirely different manner. For all $t > 0$, we can write

$$e^{-t^{1/4}} \sin t^{1/4} = \sum_{s=0}^{\infty} a_s t^{(s+1)/4}, \tag{1.3}$$

where

$$a_s = \frac{(-1)^s}{(s+1)!} \sum_{0 \le l \le s/2} (-1)^l \binom{s+1}{2l+1}. \tag{1.4}$$

If we ignore all consideration of convergence and proceed to integrate term by term, we arrive at the divergent integrals

$$\int_0^{\infty} t^{s/4} e^{ixt}\, dt, \qquad s = 1, 2, 3, \ldots.$$

If we replace z by $-ix$ in the well-known formula

$$\int_0^{\infty} t^{\alpha-1} e^{-zt}\, dt = \frac{\Gamma(\alpha)}{z^\alpha}, \qquad \alpha > 0, \quad \operatorname{Re} z > 0, \tag{1.5}$$

then we obtain the formal identity

$$\int_0^{\infty} t^{\alpha-1} e^{ixt}\, dt = \exp\left(\frac{\alpha\pi i}{2}\right) \frac{\Gamma(\alpha)}{x^\alpha}, \qquad x > 0. \tag{1.6}$$

Inserting (1.3) in (1.1) and integrating term by term, we obtain from (1.6)

$$F(x) \sim i \sum_{s=1}^{\infty} \exp\left(\frac{s\pi i}{8}\right) \Gamma\left(\frac{s}{4}+1\right) \frac{a_{s-1}}{x^{s/4+1}}, \qquad \text{as } x \to \infty. \tag{1.7}$$

Taking the first few terms in the expansion, we have

$$\begin{aligned}
F(x) \sim {}& \frac{i}{4}\Gamma\left(\frac{1}{4}\right) e^{i\pi/8} x^{-5/4} - \frac{i}{2}\Gamma\left(\frac{1}{2}\right) e^{i\pi/4} x^{-3/2} \\
& + \frac{i}{4}\Gamma\left(\frac{3}{4}\right) e^{3\pi i/8} x^{-7/4} + \cdots,
\end{aligned} \tag{1.8}$$

which is in agreement with the approximation given in (1.2).

The above procedure was first made rigorous by Olver (1974b), and has subsequently generated a great deal of research in this direction.

This method not only has the advantage of giving an easy derivation of the desired asymptotic expansions but also leads to the construction of error bounds associated with these expansions. In the following section, we shall give a rigorous proof of the result (1.8). Further applications of this method to integrals with oscillatory kernels are given in §§3-7.

To conclude this section, we present the following well-known result in summability, (see, for example, Titchmarsh (1939, p. 26) and Hardy (1949, p. 151)), which is crucial to our arguments throughout this chapter.

Lemma 1.   *If the integral $\int_0^\infty f(t)\, dt$ exists as an improper Riemann integral, then*

$$\lim_{\varepsilon \to 0^+} \int_0^\infty e^{-\varepsilon t} f(t)\, dt = \int_0^\infty f(t)\, dt.$$

Proof.   Let $F(t) = \int_t^\infty f(\tau)\, d\tau$. For any $\eta > 0$, choose $t_0 > 0$ such that $|F(t)| < \eta/5$ for all $t \geq t_0$. By integration by parts, we have for $t \geq t_0$,

$$\left| \int_{t_0}^t f(\tau) e^{-\varepsilon\tau}\, d\tau \right| = \left| F(t_0) e^{-\varepsilon t_0} - F(t) e^{-\varepsilon t} - \varepsilon \int_{t_0}^t F(\tau) e^{-\varepsilon\tau}\, d\tau \right|$$

$$< \frac{\eta}{5} + \frac{\eta}{5} + \frac{\eta}{5} = \frac{3\eta}{5}.$$

Letting $t \to \infty$ gives

$$\left| \int_{t_0}^\infty f(\tau) e^{-\varepsilon\tau}\, d\tau \right| \leq \frac{3\eta}{5}.$$

Now choose ε sufficiently small so that

$$\left| \int_0^{t_0} f(\tau) e^{-\varepsilon\tau}\, d\tau - \int_0^{t_0} f(\tau)\, d\tau \right| < \frac{\eta}{5}.$$

Since $|F(t_0)| < \eta/5$, it is clear that

$$\left| \int_0^\infty f(t) e^{-\varepsilon t}\, dt - \int_0^\infty f(t)\, dt \right| \leq \eta.$$

Since η is arbitrary, the lemma is proved.   ∎

2.   A Fourier Integral

We now return to the Fourier integral

$$F(x) = \int_0^\infty f(t)e^{ixt}\, dt \tag{2.1}$$

considered in Chapter I, Section 4, and assume that $f(t)$ has an asymptotic expansion of the form

$$f(t) \sim \sum_{s=0}^\infty a_s t^{\lambda_s - 1} \tag{2.2}$$

as $t \to 0^+$, where $\operatorname{Re} \lambda_0 > 0$ and $\operatorname{Re} \lambda_{s+1} > \operatorname{Re} \lambda_s$ for $s = 0, 1, 2, \ldots$.
    The *Abel limit* of a function $f(t)$ is defined to be

$$\lim_{\varepsilon \to 0^+} \int_0^\infty f(t)e^{-\varepsilon t}\, dt,$$

if this limit exists, in which case we say that f is *Abel summable*. If f is absolutely integrable in $[0, \infty)$, then it is easily seen that the Abel limit is simply the integral of f. The lemma in §1 in fact shows that this is true as long as $f(t)$ is improper Riemann integrable on $(0, \infty)$. The following result demonstrates that the Abel limit of a function may exist even when the improper Riemann integral of this function does not converge.

Lemma 2.   *For $x > 0$ and* $\operatorname{Re} \lambda > 0$,

$$\lim_{\varepsilon \to 0^+} \int_0^\infty t^{\lambda - 1} e^{-(\varepsilon - ix)t}\, dt = \frac{e^{\lambda \pi i/2}\Gamma(\lambda)}{x^\lambda}. \tag{2.3}$$

Proof.   We recall the identity in (1.5)

$$\int_0^\infty t^{\lambda - 1} e^{-zt}\, dt = \frac{\Gamma(\lambda)}{z^\lambda}, \qquad \operatorname{Re} \lambda > 0, \quad \operatorname{Re} z > 0.$$

Replacing z by $\varepsilon - ix$ with $\varepsilon > 0$ gives

$$\int_0^\infty t^{\lambda - 1} e^{-(\varepsilon - ix)t}\, dt = \frac{\Gamma(\lambda)}{(\varepsilon - ix)^\lambda}.$$

The desired result now follows, when we let ε tend to zero.   ∎

We shall further impose the following conditions on $f(t)$.

(F_1) $f(t)$ is m times continuously differentiable in $(0, \infty)$, m being a nonnegative integer.

(F_2) $f(t)$ has the asymptotic expansion (2.2), and this expansion can be differentiated term by term m times.

(F_3)   Each of the integrals

$$\int_1^\infty f^{(s)}(t)e^{ixt}\, dt, \qquad s = 0, 1, \ldots, m,$$

converges uniformly for all sufficiently large x.

Theorem 1. *Let n be the smallest nonnegative integer such that*

$$\operatorname{Re} \lambda_n > m, \tag{2.4}$$

and define $f_n(t)$ by

$$f(t) = \sum_{s=0}^{n-1} a_s t^{\lambda_s - 1} + f_n(t). \tag{2.5}$$

Then the Fourier integral (2.1) satisfies

$$F(x) = \sum_{s=0}^{n-1} a_s \exp\left(i\frac{\pi}{2}\lambda_s\right)\frac{\Gamma(\lambda_s)}{x^{\lambda_s}} + F_n(x), \tag{2.6}$$

where the remainder is given by

$$F_n(x) = \left(\frac{i}{x}\right)^m \int_0^\infty f_n^{(m)}(t)e^{ixt}\, dt. \tag{2.7}$$

Furthermore, as $x \to \infty$, $F_n(x) = o(x^{-m})$.

Proof.   We first observe that for $0 \le j \le m - 1$,

$$\int_1^c f^{(j)}(t)e^{ixt}\, dt = \frac{1}{ix}\left[e^{ixc}f^{(j)}(c) - e^{ix}f^{(j)}(1)\right] - \frac{1}{ix}\int_1^c f^{(j+1)}(t)e^{ixt}\, dt. \tag{2.8}$$

By condition (F_3), both integrals converge as $c \to +\infty$. Therefore, it follows from (2.8) that

$$f^{(j)}(c) \to 0, \qquad \text{as } c \to \infty, \qquad \text{for } 0 \le j \le m - 1. \tag{2.9}$$

We next observe that condition (F_2) gives, as $t \to 0^+$,

$$f_n^{(j)}(t) = O(t^{\lambda_n - j - 1}), \qquad j = 0, 1, \ldots, m. \tag{2.10}$$

Since Re $\lambda_n > m$, (2.10) implies

$$f_n^{(j)}(0) = 0, \qquad j = 0, 1, \ldots, m - 1, \tag{2.11}$$

and also that $f_n^{(m)}(t)$ is absolutely integrable at $t = 0$.
   We now return to Eq. (2.5) and write

$$\int_0^\infty f(t)e^{-(\varepsilon - ix)t} \, dt = \sum_{s=0}^{n-1} a_s \int_0^\infty t^{\lambda_s - 1} e^{-(\varepsilon - ix)t} \, dt + E_n(\varepsilon, x), \tag{2.12}$$

where

$$E_n(\varepsilon, x) = \int_0^\infty f_n(t)e^{-(\varepsilon - ix)t} \, dt. \tag{2.13}$$

Integration by parts gives

$$E_n(\varepsilon, x) = \frac{1}{(\varepsilon - ix)} \int_0^\infty f_n'(t)e^{-(\varepsilon - ix)t} \, dt, \tag{2.14}$$

the integrated term vanishing at ∞ and 0 on account of (2.9) and (2.11),
respectively. Repeating this process $m - 1$ times, we obtain

$$E_n(\varepsilon, x) = \frac{1}{(\varepsilon - ix)^m} \int_0^\infty f_n^{(m)}(t)e^{-(\varepsilon - ix)t} \, dt. \tag{2.15}$$

Note that equation (2.5) gives

$$f_n^{(m)}(t) = f^{(m)}(t) - \sum_{s=0}^{n-1} \frac{\Gamma(\lambda_s)}{\Gamma(\lambda_s - m)} a_s t^{\lambda_s - m - 1}. \tag{2.16}$$

Since n is the smallest positive integer satisfying (2.4), it follows that
all powers of t in the above sum have negative real parts. This together
with condition (F_3) implies that the integral $\int_0^\infty f_n^{(m)}(t)e^{ixt} \, dt$ converges
uniformly for all large values of x. The desired result (2.6)–(2.7) now
follows from Lemmas 1 and 2, by letting $\varepsilon \to 0$ in (2.12) and (2.15). The
fact that $F_n(x)$ in (2.7) satisfies $o(x^{-m})$, as $x \to +\infty$, is by virtue of the
following generalization of the Riemann-Lebesgue lemma (Olver 1974a,
p. 73). This completes the proof of the theorem. ■

Lemma 3.   *Let $q(t)$ be continuous in $(0, \infty)$. Then*

$$\int_0^\infty q(t)e^{ixt} \, dt = o(1), \qquad \text{as } x \to \infty, \tag{2.17}$$

provided that the integral converges uniformly at 0 and ∞ for all sufficiently large x.

Proof. Let ε be an arbitrary positive number. Then, by hypothesis, there are positive numbers a and b such that

$$\left| \int_0^a q(t)e^{ixt}\,dt \right| < \frac{\varepsilon}{3} \quad \text{and} \quad \left| \int_b^\infty q(t)e^{ixt}\,dt \right| < \frac{\varepsilon}{3}. \tag{2.18}$$

Since q is continuous on $(0, \infty)$, it is bounded and uniformly continuous on $[a, b]$. Let M denote the maximum of $q(t)$ on $[a, b]$, and pick a partition $a = t_0 < t_1 < \cdots < t_n = b$ such that for $j = 0, 1, \ldots, n-1$,

$$|q(t) - q(t_j)| < \frac{\varepsilon}{6(b-a)}, \qquad t \in [t_j, t_{j+1}]. \tag{2.19}$$

Now we write

$$\int_a^b q(t)e^{ixt}\,dt = \sum_{j=0}^{n-1} \int_{t_j}^{t_{j+1}} [q(t) - q(t_j)]e^{ixt}\,dt + \sum_{j=0}^{n-1} q(t_j) \int_{t_j}^{t_{j+1}} e^{ixt}\,dt. \tag{2.20}$$

By (2.19), it is easy to see that the first sum in (2.20) is dominated by $\varepsilon/6$. It is also easy to show that the second sum is bounded by $2nM/x$. Thus, by choosing $x > 12nM/\varepsilon$, we have

$$\left| \int_a^b q(t)e^{ixt}\,dt \right| < \frac{\varepsilon}{3}. \tag{2.21}$$

Since ε is arbitrary, the desired result (2.17) now follows from (2.18) and (2.21). ∎

An obvious bound for the error term in (2.7) is provided by

$$|F_n(x)| \le \frac{1}{x^m} \int_0^\infty |f_n^{(m)}(t)|\,dt. \tag{2.22}$$

However, the integral on the right may not converge under the conditions imposed above. To ensure the finiteness of this integral, we must in addition require

$$\int_1^\infty |f^{(m)}(t)|\,dt < \infty. \tag{2.23}$$

Even with this additional assumption, the integral in (2.22) may still be infinite, since the terms under the summation in (2.16) may not decrease sufficiently fast. If Re λ_{n-1} happens to be *not* equal to m, in which case we have Re $\lambda_{n-1} < m <$ Re λ_n, then the exponents of t in (2.16) all have real parts less than -1 and hence

$$\int_0^\infty |f_n^{(m)}(t)|\,dt < \infty. \tag{2.24}$$

Another possibility for (2.24) to hold is when the last term in the series in (2.16) vanishes. This occurs either when $a_{n-1} = 0$ or when $\lambda_{n-1} = m$.

Let us now illustrate the above result with the integral (1.1) and construct an error bound for the approximation (1.8). Here we have

$$f(t) = e^{-t^{1/4}} \sin t^{1/4},$$

and $\lambda_s = \frac{1}{4}(s + 5)$. Since f is infinitely differentiable on $(0, \infty)$, we may take $m = 2$ and $n = 4$ so that the inequality (2.4) is satisfied. From (1.4) and (2.7), it now follows that

$$F(x) = \frac{i}{4}\,\Gamma\!\left(\frac{1}{4}\right)e^{i\pi/8}x^{-5/4} - \frac{i}{2}\,\Gamma\!\left(\frac{1}{2}\right)e^{i\pi/4}x^{-3/2}$$
$$+ \frac{i}{4}\,\Gamma\!\left(\frac{3}{4}\right)e^{i3\pi/8}x^{-7/4} + F_4(x), \tag{2.25}$$

where

$$F_4(x) = \left(\frac{i}{x}\right)^2 \int_0^\infty f_4''(t)e^{ixt}\,dt. \tag{2.26}$$

(Note that equation (1.4) gives $a_3 = 0$.) Simple calculation gives the quantity $f_4''(t)$, and shows that

$$|f_4''(t)| \le \frac{3}{8}\,t^{-3/2} + \frac{9}{16}\,t^{-7/4} + \frac{1}{16}\,t^{-5/4} \tag{2.27}$$

for $1 < t < \infty$. By using Taylor's theorem with remainder, we also have

$$|f_4''(t)| \le \frac{7}{48}\,t^{-3/4} \tag{2.28}$$

for $0 < t < 1$. A combination of (2.26), (2.27), and (2.28) yields

$$|F_4(x)| \le \frac{7}{3}\,x^{-2}. \tag{2.29}$$

3.   Hankel Transform

The above summability method can easily be extended to the Hankel transform

$$H(x) = \int_0^\infty f(t)J_\nu(xt)\,dt, \tag{3.1}$$

where $J_\nu(t)$ is the Bessel function of the first kind and ν is a fixed real or complex number. As in §2, we shall again assume that $f(t)$ has an asymptotic expansion of the form

$$f(t) \sim \sum_{s=0}^\infty a_s t^{\lambda_s - 1}, \qquad \text{as } t \to 0^+, \tag{3.2}$$

where $\mathrm{Re}(\lambda_0 + \nu) > 0$ and $\mathrm{Re}\ \lambda_{s+1} > \mathrm{Re}\ \lambda_s$ for $s = 0,\ 1,\ 2,\dots$. The following lemma is an analogue of (2.3).

Lemma 4.   *For $x > 0$, ε real and $\mathrm{Re}(\mu + \alpha) > 0$,*

$$\lim_{\varepsilon \to 0^+} \int_0^\infty t^{\mu - 1} J_\alpha(xt) e^{-\varepsilon t}\,dt = \frac{\Gamma(\tfrac{1}{2}\alpha + \tfrac{1}{2}\mu)2^{\mu - 1}}{\Gamma(\tfrac{1}{2}\alpha - \tfrac{1}{2}\mu + 1)x^\mu}. \tag{3.3}$$

Proof.   The integral in (3.3) can be evaluated by means of the hypergeometric function; see Watson (1944, p. 385). The result is

$$\int_0^\infty t^{\mu - 1} J_\alpha(xt) e^{-\varepsilon t}\,dt$$

$$= \frac{(\tfrac{1}{2}x)^\alpha \Gamma(\mu + \alpha)}{(x^2 + \varepsilon^2)^{(\mu + \alpha)/2}\Gamma(\alpha + 1)}\,{}_2F_1\!\left(\frac{\mu + \alpha}{2}, \frac{1 - \mu + \alpha}{2}; \alpha + 1; \frac{x^2}{x^2 + \varepsilon^2}\right).$$

In view of the Gauss sum (Olver 1974a, p. 161)

$${}_2F_1(a, b; c; 1) = \frac{\Gamma(c)\Gamma(c - a - b)}{\Gamma(c - a)\Gamma(c - b)}$$

and the duplication formula for the Γ-function, we immediately obtain (3.3).   ■

We now impose conditions which correspond to those given in §2.

(H$_1$) $f(t)$ is m times continuously differentiable in $(0, \infty)$, m being a nonnegative integer.

(H$_2$) $f(t)$ has the asymptotic expansion (3.2), and this expansion can be differentiated term by term m times.

(H$_3$)   Each of the integrals

$$\int_1^\infty f^{(j)}(t)t^{-1/2}e^{ixt}\,dt, \qquad j = 0, 1, \ldots, m,$$

converges uniformly for all large values of x.

Theorem 2.   *Let n be chosen as in (2.4), and let $f_n(t)$ be defined as in (2.5). Then the Hankel transform (3.1) satisfies*

$$H(x) = \sum_{s=0}^{n-1} a_s \frac{\Gamma(\frac{1}{2}v + \frac{1}{2}\lambda_s)2^{\lambda_s - 1}}{\Gamma(\frac{1}{2}v - \frac{1}{2}\lambda_s + 1)x^{\lambda_s}} + E_n(x), \tag{3.4}$$

where

$$E_n(x) = \left(\frac{-1}{x}\right)^m \int_0^\infty f_{n,m}(t)J_{v+m}(xt)\,dt \tag{3.5}$$

and the functions $f_{n,j+1}(t)$, $j = 0, 1, \ldots, m$, are defined recursively by $f_{n,0}(t) = f_n(t)$ and

$$f_{n,j+1}(t) = f'_{n,j}(t) - (v + j + 1)f_{n,j}(t)t^{-1}. \tag{3.6}$$

Proof.   For any $\varepsilon > 0$, we have from (2.5)

$$\int_0^\infty f(t)J_v(xt)e^{-\varepsilon t}\,dt = \sum_{s=0}^{n-1} a_s \int_0^\infty t^{\lambda_s - 1}J_v(xt)e^{-\varepsilon t}\,dt + E_n(\varepsilon, x), \tag{3.7}$$

where

$$E_n(\varepsilon, x) = \int_0^\infty f_n(t)J_v(xt)e^{-\varepsilon t}\,dt. \tag{3.8}$$

Applying Lemmas 1 and 4, we obtain, by passing to the limit as $\varepsilon \to 0$,

$$H(x) = \sum_{s=0}^{n-1} a_s \frac{\Gamma(\frac{1}{2}v + \frac{1}{2}\lambda_s)2^{\lambda_s - 1}}{\Gamma(\frac{1}{2}v - \frac{1}{2}\lambda_s + 1)x^{\lambda_s}} + E_n(x), \tag{3.9}$$

where

$$E_n(x) = \lim_{\varepsilon \to 0^+} E_n(\varepsilon, x). \tag{3.10}$$

From the well-known identity

$$\frac{d}{dt}[t^{\nu+1}J_{\nu+1}(t)] = t^{\nu+1}J_{\nu}(t), \tag{3.11}$$

it follows by integration by parts that

$$\int f_{n,j}(t)J_{\nu+j}(xt)e^{-\varepsilon t}\,dt = \frac{1}{x}f_{n,j}(t)J_{\nu+j+1}(xt)e^{-\varepsilon t}$$

$$-\frac{1}{x}\int f_{n,j+1}(t)J_{\nu+j+1}(xt)e^{-\varepsilon t}\,dt + \frac{\varepsilon}{x}\int f_{n,j}(t)J_{\nu+j+1}(xt)e^{-\varepsilon t}\,dt. \tag{3.12}$$

From (3.6), it is easy to see that there are constants $c_1, \ldots, c_m$ such that

$$f_{n,j}(t) = f_n^{(j)}(t) + \sum_{i=1}^{j} c_i f_n^{(j-i)}(t)t^{-i}, \qquad j = 0, 1, \ldots, m. \tag{3.13}$$

(As usual, empty sums are understood to be zero.) Equation (2.10) then gives

$$f_{n,j}(t) = O(t^{\lambda_n - j - 1}), \qquad \text{as } t \to 0^+, \tag{3.14}$$

for $j = 0, 1, \ldots, m$. This, together with the asymptotic formula

$$J_\alpha(t) \sim \frac{t^\alpha}{2^\alpha \Gamma(\alpha+1)}, \qquad \text{as } t \to 0^+, \tag{3.15}$$

implies

$$f_{n,j}(t)J_{\nu+j+1}(xt)e^{-\varepsilon t} \to 0, \qquad \text{as } t \to 0^+, \tag{3.16}$$

for $j = 0, 1, \ldots, m$. From (2.16) we also have constants $d_0, d_1, \ldots, d_{n-1}$ such that

$$f_{n,j}(t) = f^{(j)}(t) + \sum_{i=1}^{j} c_i f^{(j-i)}(t)t^{-i} + \sum_{s=0}^{n-1} d_s t^{\lambda_s - j - 1}, \tag{3.17}$$

$j = 0, 1, \ldots, m$. By integration by parts,

$$\int_1^c f^{(j)}(t)t^{-1/2}e^{ixt}\,dt = \frac{1}{ix}[f^{(j)}(c)c^{-1/2}e^{ixc} - f^{(j)}(1)e^{ix}]$$

$$+ \frac{i}{x}\int_1^c \left[f^{(j+1)}(t)t^{-1/2} - \frac{1}{2}f^{(j)}(t)t^{-3/2}\right]e^{ixt}\,dt$$

for any $c > 1$. Furthermore, as a consequence of condition (H_3), the integrals

$$\int_1^\infty f^{(j)}(t) t^{-3/2} e^{ixt} \, dt, \qquad j = 0, 1, \ldots, m, \tag{3.18}$$

are all uniformly convergent for sufficiently large x; see Ex. 1. These two statements together imply

$$f^{(j)}(c) c^{-1/2} \to 0, \qquad \text{as } c \to \infty, \tag{3.19}$$

for $j = 0, 1, \ldots, m - 1$. From this and (3.17), it follows that

$$f_{n,j}(t) J_{\nu+j+1}(xt) e^{-\varepsilon t} \to 0, \qquad \text{as } t \to \infty, \tag{3.20}$$

$j = 0, 1, \ldots, m - 1$. Here we have also used the asymptotic formula

$$J_\alpha(t) = \sqrt{\frac{2}{\pi t}} \cos\left(t - \frac{1}{2}\alpha\pi - \frac{1}{4}\pi\right) + O(t^{-3/2}), \qquad \text{as } t \to \infty. \tag{3.21}$$

A combination of (3.12), (3.16), and (3.20) gives

$$\begin{aligned}
\int_0^\infty f_{n,j}(t) J_{\nu+j}(xt) e^{-\varepsilon t} \, dt &= -\frac{1}{x} \int_0^\infty f_{n,j+1}(t) J_{\nu+j+1}(xt) e^{-\varepsilon t} \, dt \\
&\quad + \frac{\varepsilon}{x} \int_0^\infty f_{n,j}(t) J_{\nu+j+1}(xt) e^{-\varepsilon t} \, dt
\end{aligned} \tag{3.22}$$

for $j = 0, 1, \ldots, m - 1$. The last term in (3.22) tends to zero, as $\varepsilon \to 0$, by using (3.3), (3.17), (3.19), and (3.21); see Ex. 8. Thus we obtain

$$\lim_{\varepsilon \to 0^+} \int_0^\infty f_{n,j}(t) J_{\nu+j}(xt) e^{-\varepsilon t} \, dt = \left(\frac{-1}{x}\right) \lim_{\varepsilon \to 0^+} \int_0^\infty f_{n,j+1}(t) J_{\nu+j+1}(xt) e^{-\varepsilon t} \, dt.$$

Applying this identity m times gives

$$E_n(x) = \left(\frac{-1}{x}\right)^m \lim_{\varepsilon \to 0^+} \int_0^\infty f_{n,m}(t) J_{\nu+m}(xt) e^{-\varepsilon t} \, dt \tag{3.23}$$

on account of (3.8) and (3.10). The desired result (3.5) will follow from (3.23) and Lemma 1, if we can show that the integral in (3.5) exists as an improper integral. By (3.14) and (3.15), this integral converges absolutely at $t = 0^+$. Since the powers of t in the last sum in (3.17) all have real parts less than or equal to -1 in view of (2.4), condition (H_3) and (3.21) imply that this integral converges also at $t = \infty$ uniformly for all sufficiently large values of x. This completes the proof. ∎

To show that expansion (3.4) is asymptotic, it is enough to prove that

$$E_n(x) = o(x^{-m}), \qquad \text{as } x \to +\infty, \tag{3.24}$$

or equivalently

$$\lim_{x \to \infty} \int_0^\infty f_{n,m}(t) J_{v+m}(xt) \, dt = 0. \tag{3.25}$$

In the proof of Theorem 2 we have shown that the integral in (3.25) converges uniformly for all sufficiently large values of x. Hence for any $\varepsilon > 0$, there is a constant c independent of large x such that

$$\left| \int_c^\infty f_{n,m}(t) J_{v+m}(xt) \, dt \right| < \frac{\varepsilon}{2}. \tag{3.26}$$

In view of (2.4), we have $\text{Re } v + m \geq \text{Re}(v + \lambda_0) > 0$. Thus, for $t \geq 0$, $J_{v+m}(t)$ is bounded. From (3.21), we also have $J_{v+m}(xt) \to 0$ as $x \to \infty$ for every fixed $t > 0$. The Lebesgue dominated convergence theorem (Rudin 1974, p. 27) then gives

$$\left| \int_0^c f_{n,m}(t) J_{v+m}(xt) \, dt \right| < \frac{\varepsilon}{2} \tag{3.27}$$

for sufficiently large x. The result in (3.25) now follows from (3.26) and (3.27).

In order to construct an explicit bound for the error term $E_n(x)$, we define

$$A_\alpha = \sup_{0 \leq t < \infty} |J_\alpha(t)| \tag{3.28}$$

and

$$B_\alpha = \sup_{0 \leq t < \infty} |t^{1/2} J_\alpha(t)|. \tag{3.29}$$

Since $J_\alpha(t)$ is continuous on $[0, \infty)$ for $\text{Re } \alpha \geq 0$, relations (3.15) and (3.21) show that A_α and B_α are finite. If α is real and $\alpha \geq 0$, then it is known (Watson 1944, p. 406) that

$$A_\alpha \leq 1 \qquad \text{and} \qquad A_{\alpha+1} \leq \frac{1}{\sqrt{2}}. \tag{3.30}$$

From Szegö (1967, p. 167), we also have

$$B_\alpha = \begin{cases} \left(\dfrac{2}{\pi} \right)^{1/2} & \text{if } -\tfrac{1}{2} \leq \alpha \leq \tfrac{1}{2} \\ m_{\alpha 1}^{1/2} J_\alpha(m_{\alpha 1}) & \text{if } \alpha > \tfrac{1}{2}, \end{cases} \tag{3.31}$$

where $m_{\alpha 1}$ is the first positive maximum of $x^{1/2}J_{\alpha}(x)$; see Lorch (1966). Numerical computations yield

$$
\begin{aligned}
A_0 &= 1.00000, & B_0 &= 0.79788, \\
A_1 &= 0.58187, & B_1 &= 0.82503, \\
A_2 &= 0.48650, & B_2 &= 0.86842, \\
A_3 &= 0.43439, & B_3 &= 0.90238.
\end{aligned}
\tag{3.32}
$$

Theorem 3.   *Assume that conditions* (H_1) *and* (H_2) *hold, and replace condition* (H_3) *by*

(H_3') *Each of the integrals*

$$
\int_1^\infty f^{(j)}(t)t^{-1/2}e^{ixt}\, dt
$$

$j = 0, 1, \ldots, m$, *converges uniformly for all sufficiently large* x, *and for each* $j = 0, 1, \ldots, m$,

$$
\int_1^\infty |f^{(j)}(t)|t^{j-m}\, dt < \infty.
$$

Choose n *again as in* (2.4). *If* $\operatorname{Re} \lambda_{n-1} < m$, *then*

$$
|E_n(x)| \le \frac{A_{v+m}}{x^m}\int_0^\infty |f_{n,m}(t)|\, dt.
\tag{3.33}
$$

If $\operatorname{Re} \lambda_{n-1} = m$ *and* $\operatorname{Re} \lambda_n > m + \tfrac{1}{2}$, *then*

$$
|E_n(x)| \le \frac{B_{v+m}}{x^{m+1/2}}\int_0^\infty t^{-1/2}|f_{n,m}(t)|\, dt.
\tag{3.34}
$$

Proof.   From (3.14), we have $f_{n,m}(t) = O(t^{\lambda_n - m - 1})$ as $t \to 0^+$. Since $\operatorname{Re} \lambda_n > m$, the integral $\int |f_{n,m}(t)|\, dt$ converges at $t = 0$. Furthermore, (3.17) gives

$$
|f_{n,m}(t)| \le |f^{(m)}(t)| + \sum_{i=1}^m |c_i||f^{(m-i)}(t)|t^{-i} + \sum_{s=0}^{n-1} |d_s|t^{\operatorname{Re}\lambda_s - m - 1}.
\tag{3.35}
$$

If $\operatorname{Re}\lambda_{n-1} < m$, then the exponents in the last sum are all less than -1. Therefore, by condition (H_3'), the integral $\int |f_{n,m}(t)|\, dt$ is also convergent at $t = \infty$. That is, $\int_0^\infty |f_{n,m}(t)|\, dt < \infty$. The error bound (3.33) now follows from (3.5) and (3.28).

The proof of (3.34) is similar. From (3.14), $f_{n,m}(t) = O(t^{\lambda_n - m - 1})$ as $t \to 0^+$, and hence $\int t^{-1/2} |f_{n,m}(t)| \, dt$ is convergent at $t = 0$. Since the exponents in the last sum of (3.35) are less than or equal to -1, the integral $\int t^{-1/2} |f_{n,m}(t)| \, dt$ also converges at $t = \infty$. Therefore, (3.34) holds in view of (3.5) and (3.29). ∎

In some cases a combination of estimates like (3.33) and (3.34) is needed. For instance, if Re $\lambda_{n-1} = m$ and $m < $ Re $\lambda_n \leq m + \frac{1}{2}$, then we have

$$|E_n(x)| \leq \frac{1}{x^m} \left[A_{v+m} \int_0^1 |f_{n,m}(t)| \, dt + \frac{B_{v+m}}{\sqrt{x}} \int_1^\infty t^{-1/2} |f_{n,m}(t)| \, dt \right]. \qquad (3.36)$$

Example 1. Consider the Hankel transform

$$H(x) = \int_0^\infty \sin \sqrt{t} \, J_{-1}(xt) \, dt. \qquad (3.37)$$

Here we have $v = -1$ and

$$f(t) = \sin \sqrt{t} = \sum_{s=0}^\infty \frac{(-1)^s}{(2s+1)!} t^{s+1/2}$$

for all values of t. The conditions of Theorem 2 are clearly satisfied with $\lambda_s = s + \frac{3}{2}$, m being an arbitrary positive integer and $n = m - 1$. Thus (3.4) yields

$$H(x) = \sum_{s=0}^{m-2} \frac{(-1)^s}{(2s+1)!} \frac{\Gamma(\frac{1}{2}s + \frac{1}{4}) 2^{s+1/2}}{\Gamma(-\frac{1}{2}s - \frac{1}{4}) x^{s+(3/2)}} + \delta_m(x), \qquad (3.38)$$

where

$$\delta_m(x) = \left(\frac{-1}{x} \right)^m \int_0^\infty f_{m-1,m}(t) J_{m-1}(xt) \, dt. \qquad (3.39)$$

To illustrate the calculation of error bounds, we suppose, for example, $m = 3$. Equation (2.5) and (3.6) give

$$f_2(t) = \sin \sqrt{t} - \sqrt{t} + \frac{1}{3!} t^{3/2},$$

$$f_{2,3}(t) = f_2'''(t) - 3f_2''(t)t^{-1} + 3f_2'(t)t^{-2}.$$

Elementary estimations show that

$$|f_{2,3}(t)| \le \frac{1}{6} t^{-1/2}, \qquad 0 < t \le 1,$$

$$|f_{2,3}(t)| \le \frac{15}{4} t^{-3/2}, \qquad t \ge 1,$$

and hence

$$\int_0^\infty |f_{2,3}(t)| \, dt \le \frac{47}{6};$$

see Ex. 10. From (3.28) and (3.39), we derive

$$|\delta_3(x)| \le 3.8109 x^{-3}. \tag{3.40}$$

4.   Hankel Transform (continued)

Some problems of high energy nuclear physics (see Glauber (1959) and Gabutti and Minetti (1981)) involved integrals of the form

$$I_g(x) = \int_0^\infty e^{-t^2} J_0(xt) g(t^2) t \, dt, \tag{4.1}$$

where $J_0(t)$ is the Bessel function of order zero and $g(t)$ is a continuous function in $(0, \infty)$. Asymptotic expansions were required for large positive x. Clearly these integrals are special cases of the Hankel transform discussed in §3. However, the result of Theorem 2 applied to the odd function $e^{-t^2} g(t^2) t$ only gives an asymptotic expansion whose coefficients are all zero. This phenomenon reminds one of the Fourier integrals

$$F_1(x) = \int_{-\infty}^\infty \frac{t \sin xt}{t^2 + \alpha^2} h(t) \, dt \tag{4.2}$$

and

$$F_2(x) = \int_0^\infty \frac{t \exp(-\rho^2 t)}{t^2 + \alpha^2} \sin xt \, dt, \tag{4.3}$$

where α and ρ are positive constants and x is a large positive parameter. The function $h(t)$ in (4.2) is real when t is real and holomorphic in a domain containing the strip $|\operatorname{Im} t| \le \beta$, where $\beta > \alpha$. These integrals have been considered by Olver (1974a, p. 78–79). Since $\sin t$ is the imaginary part of e^{it}, the results of §2 can be applied to both of these integrals. The coefficients of their asymptotic expansions, however, turn out also to be all zero. Olver showed that

$$F_1(x) = \pi e^{-\alpha x} \operatorname{Re}\{h(i\alpha)\} + O(e^{-\beta x}) \tag{4.4}$$

and

$$F_2(x) = \frac{\pi}{2} \exp(\alpha^2 \rho^2 - \alpha x) + O\{x^{-1} e^{-x^2/4\rho^2}\}, \tag{4.5}$$

both as $x \to \infty$. Thus the integrals $F_1(x)$ and $F_2(x)$ are exponentially decaying.

In this section we shall derive similar results for the integral $I_g(x)$ in (4.1), when $g(t)$ is a meromorphic function satisfying

$$|g(t)| \le M e^{\xi \operatorname{Re} t + \eta |\operatorname{Im} t|} \tag{4.6}$$

for all sufficiently large $|t|$, where $\eta \ge 0$, $\xi < 1$, and $M > 0$. First we prove the following theorem.

Theorem 4. *Let $g(t)$ be an entire function satisfying the growth condition (4.6), and define*

$$\alpha = 1 - \xi + \frac{\eta^2}{1 - \xi}. \tag{4.7}$$

Then we have

$$I_g(x) = O(e^{-x^2/4\alpha}), \qquad as \ x \to +\infty. \tag{4.8}$$

Proof. It is well known that

$$H_\nu^{(1)}(z e^{m\pi i}) = e^{-m\nu\pi i} H_\nu^{(1)}(z) - 2e^{-\nu\pi i} \frac{\sin m\nu\pi}{\sin \nu\pi} J_\nu(z),$$

where $H_\nu^{(1)}(z)$ is a Hankel function of order ν, see Watson (1944, p. 75, (5)). Letting $m = 1$ and ν tend to zero, we obtain

$$H_0^{(1)}(z e^{i\pi}) = H_0^{(1)}(z) - 2J_0(z).$$

Note that the Hankel function $H_0^{(1)}(z)$ has a logarithmic branch point at $z = 0$. By taking a branch cut along the negative imaginary axis, it can be shown that

$$I_g(x) = \frac{1}{2} \int_{-\infty}^{\infty} e^{-t^2} H_0^{(1)}(xt) g(t^2) t \, dt. \tag{4.9}$$

Now consider the contour integral

$$C_R(x) = \frac{1}{2} \int_\Gamma e^{-t^2} H_0^{(1)}(xt) g(t^2) t \, dt, \tag{4.10}$$

where the path Γ, traversed in the positive direction, is the boundary of the rectangle whose vertices are at $(-R, 0)$, $(R, 0)$, $(R, x/2\alpha)$ and $(-R, x/2\alpha)$; here $R > 0$ and α is given in (4.7). With the cut along the negative imaginary axis, by Cauchy's theorem, we have $C_R(x) = 0$ for all $R > 0$. Letting $R \to +\infty$, the contribution from the portion of Γ parallel to the imaginary axis tends to zero by condition (4.6). The remaining horizontal pieces of (4.10) reduce to

$$I_g(x) = \frac{1}{2} \int_{-\infty + ix/2\alpha}^{\infty + ix/2\alpha} e^{-t^2} H_0^{(1)}(xt) g(t^2) t \, dt. \tag{4.11}$$

In view of the well-known result (Watson 1944, p. 219)

$$|H_0^{(1)}(t)| \le \left| \left(\frac{2}{\pi t} \right)^{1/2} e^{i(t - \pi/4)} \right|, \tag{4.12}$$

the right-hand side of (4.11) is easily seen to be dominated by

$$\frac{M}{\sqrt{2\pi x}} \exp\left\{ x^2 \left(\frac{1 - \xi}{4\alpha^2} - \frac{1}{2\alpha} \right) \right\} \int_{-\infty}^{\infty} e^{-(1-\xi)t^2 + \eta(x/\alpha)|t|} \left(t^2 + \frac{x^2}{4\alpha^2} \right)^{1/4} dt. \tag{4.13}$$

Since

$$\left(t^2 + \frac{x^2}{4\alpha^2} \right)^{1/4} \le \left(\frac{2\alpha}{x} \right)^{3/2} \left(t^2 + \frac{x^2}{4\alpha^2} \right),$$

upon splitting the last integral at $t = 0$, completing the square in the exponents of each of the two integrals and recombining them, we can show that the integral in (4.13) is dominated by

$$2\left(\frac{2\alpha}{x} \right)^{3/2} \exp\left\{ \frac{\eta^2(x/\alpha)^2}{4(1 - \xi)} \right\} \int_0^{\infty} \left(t^2 + \frac{x^2}{4\alpha^2} \right) \exp\left\{ -\left(\sqrt{1 - \xi} t - \frac{\eta x/\alpha}{2\sqrt{1 - \xi}} \right)^2 \right\} dt.$$

Making the change of variable

$$u = \sqrt{1 - \xi} t - \frac{\eta x/\alpha}{2\sqrt{1 - \xi}},$$

the above integral becomes, by using (4.7),

$$\frac{1}{\sqrt{1 - \xi}} \int_{-\eta(x/\alpha)/2\sqrt{1-\xi}}^{\infty} e^{-u^2} \left[\frac{1}{1 - \xi} u^2 + \frac{\eta x/\alpha}{(1 - \xi)^{3/2}} u + \frac{x^2}{4\alpha(1 - \xi)} \right] du,$$

which is clearly dominated by

$$\frac{1}{\sqrt{1-\xi}}\left\{\frac{1}{1-\xi}\int_{-\infty}^{\infty} u^2 e^{-u^2}\, du + \frac{\eta x/\alpha}{(1-\xi)^{3/2}}\int_{-\infty}^{\infty} |u| e^{-u^2}\, du\right.$$

$$\left. + \frac{x^2}{4\alpha(1-\xi)}\int_{-\infty}^{\infty} e^{-u^2}\, du\right\}.$$

A combination of the above results shows

$$|I_g(x)| \le M' \exp\!\left(-\frac{x^2}{4\alpha}\right),$$

where

$$M' = \frac{M\alpha^{1/2}}{(1-\xi)^{3/2}}\left[1 + \frac{4\eta}{\sqrt{\pi(1-\xi)}}\frac{1}{x} + \frac{2\alpha}{x^2}\right].$$

This completes the proof of the theorem. ∎

   As an example of Theorem 4, we consider the integral

$$I^{(1)}(x) = \int_0^{\infty} e^{-t^2} J_0(xt)(\sin t^2)t\, dt. \tag{4.14}$$

Since $f(t) = \sin t$ is entire and bounded by $e^{|\mathrm{Im}\, t|}$, we have $\xi = 0$ and $\eta = 1$. Equation (4.7) then gives $\alpha = 2$. From Theorem 4, it follows that

$$I^{(1)}(x) = O(e^{-x^2/8}). \tag{4.15}$$

The integral $I^{(1)}(x)$ can actually be evaluated in closed form. To see this, we recall the well-known identity (Watson 1944, p. 393)

$$\int_0^{\infty} t e^{-\lambda t^2} J_0(xt)\, dt = \frac{1}{2\lambda} e^{-x^2/4\lambda}, \tag{4.16}$$

Re $\lambda > 0$. Put $\lambda = 1 - i$ and equate imaginary parts on both sides of (4.16). This yields

$$I^{(1)}(x) = \frac{1}{4} e^{-x^2/8}\left(\cos\frac{x^2}{8} - \sin\frac{x^2}{8}\right), \tag{4.17}$$

which is of course in agreement with (4.15).

   Now we turn to the case of meromorphic functions. First we note that if $g(t)$ is a meromorphic function with a finite number m, say, of poles located away from the nonnegative real axis, $g(t^2)$ is a meromorphic function with $2m$ poles located away from the real axis, with m

poles in the upper half plane and m poles in the lower half plane. For convenience, we shall state the following result in terms of $g(t^2)$ and its poles in the upper half plane.

Theorem 5.   *Let $g(t)$ be a meromorphic function such that $g(t^2)$ has a finite number of poles located in the upper half plane, say, at $a_1, \ldots, a_m$, and let $g(t)$ satisfy the growth condition (4.6). Then, as $x \to \infty$,*

$$I_g(x) = \pi i \sum_{j=1}^{m} \mathrm{Res}\{e^{-t^2} H_0^{(1)}(xt) g(t^2) t; a_j\} + O(e^{-x^2/4\alpha}), \qquad (4.18)$$

α being the same as in Theorem 4.

Proof.   The proof is identical to that of Theorem 4, except that as we deform the contour of $C_R(x)$ in (4.10), we pick up the residues from the poles of $g(t^2)$.   ∎

An immediate consequence of (4.18) is that as $x \to +\infty$,

$$I_g(x) = O(e^{-\delta x}), \qquad (4.19)$$

where $\delta = \min\{\mathrm{Im}\; a_i\colon i = 1, \ldots, m\}$. This follows from the fact that $H_0^{(1)}(xt)$ decays faster than $e^{-x\,\mathrm{Im}\,t}$ in the upper half plane; see Eq. (4.12).

As an example of Theorem 5, we consider the integral

$$I^{(2)}(x) = \int_0^\infty e^{-t^2} J_0(xt) \frac{t}{1+t^2}\, dt. \qquad (4.20)$$

Since $g(t^2) = (1 + t^2)^{-1}$ has a simple pole in the upper half plane at $z = i$, Theorem 5 gives

$$I^{(2)}(x) = e K_0(x) + O(e^{-x^2/4}), \qquad (4.21)$$

where $K_0(x)$ is the modified Bessel function. The well-known asymptotic expansion (Olver 1974a, p. 250) of $K_0(x)$ then yields

$$I^{(2)}(x) \sim \sqrt{\frac{\pi}{2}} e^{-x+1} \left[x^{-1/2} - \frac{1}{8} x^{-3/2} + \frac{9}{128} x^{-5/2} + \cdots \right]; \qquad (4.22)$$

cf. Ex. 18.

The above method based on Cauchy's residue theorem cannot be repeated when $g(t)$ has a branch point singularity. In such a case, a completely different line of attack may be required. We illustrate this remark with the following specific example.

Example 2. Consider

$$I^{(3)}(x) = \int_0^\infty e^{-t^2} J_0(xt) \frac{t}{\sqrt{1+t^2}} dt. \qquad (4.23)$$

Here the function $g(t^2) = 1/\sqrt{1+t^2}$ has the integral representation

$$\frac{1}{\sqrt{1+t^2}} = \frac{1}{\sqrt{\pi}} \int_0^\infty \tau^{-1/2} e^{-(1+t^2)\tau} d\tau. \qquad (4.24)$$

Inserting (4.24) in (4.23) and reversing the order of integration, we have from (4.16)

$$I^{(3)}(x) = \frac{1}{2\sqrt{\pi}} \int_0^\infty \frac{\tau^{-1/2}}{1+\tau} \exp\left\{ -\tau - \frac{x^2}{4(1+\tau)} \right\} d\tau.$$

The substitution $1 + \tau = xu/2$ then gives

$$I^{(3)}(x) = \frac{e}{\sqrt{2\pi x}} \int_{2/x}^\infty \left(u - \frac{2}{x} \right)^{-1/2} u^{-1} \exp\left\{ -\frac{x}{2}\left(u + \frac{1}{u} \right) \right\} du.$$

The last integral is of the form considered in Example 2 of Chapter II, and its asymptotic expansion can be obtained directly from Exercise 4 in that chapter. Upon simplification, the result is

$$I^{(3)}(x) \sim e^{-x+1} \left[\frac{1}{x} + \frac{1}{x^2} + \frac{5}{2x^3} + \cdots \right]. \qquad (4.25)$$

This approach can be extended to give asymptotic expansions of the integrals

$$I^{(4)}(x) = \int_0^\infty e^{-t^2} J_0(xt) \frac{t^{2n+1}}{(a^2+t^2)^\nu} dt \qquad (4.26)$$

and

$$I^{(5)}(x) = \int_0^\infty e^{-t^2} J_0(xt) \log(a^2+t^2) t\, dt, \qquad (4.27)$$

where ν is real, a is positive and $n = 1, 2, \ldots$. For details of these results, see Frenzen and Wong (1985).

5.   Oscillatory Kernels: General Case

The summability method used in §§2 and 3 can actually be applied to the more general integral

$$I(x) = \int_0^\infty f(t)h(xt)\,dt, \tag{5.1}$$

where $h(t)$ is an oscillatory function. Here we again assume that $f(t)$ has an expansion of the form

$$f(t) \sim \sum_{s=0}^\infty a_s t^{\lambda_s - 1}, \qquad \text{as } t \to 0^+, \tag{5.2}$$

where $\operatorname{Re} \lambda_{s+1} > \operatorname{Re} \lambda_s$ for $s = 0, 1, \dots$. Regarding the function h, we also assume that as $t \to 0^+$,

$$h(t) = O(t^b), \qquad b + \operatorname{Re} \lambda_0 > 0, \tag{5.3}$$

and that as $t \to \infty$,

$$h(t) \sim \exp\{ict^p\} \sum_{s=0}^\infty b_s t^{-s-\beta}, \tag{5.4}$$

where $c \neq 0$ is real, $p \geq 1$, and $\beta > 0$. If $M[h; z]$ denotes, as before, the generalized Mellin transform of h defined by (4.24) in Chapter III, then Ex. 8 in that chapter gives

$$M[h; z] = \lim_{\varepsilon \to 0} \int_0^\infty t^{z-1} h(t) \exp\{-\varepsilon t^p\}\,dt. \tag{5.5}$$

This, together with (a slight modification of) Lemma 1 in the present chapter, gives

$$I(x) = \sum_{s=0}^{n-1} a_s M[h; \lambda_s] x^{-\lambda_s} + \delta_n(x), \tag{5.6}$$

where

$$\delta_n(x) = \lim_{\varepsilon \to 0^+} \int_0^\infty f_n(t) h(xt) \exp\{-\varepsilon t^p\}\,dt, \tag{5.7}$$

$f_n(t)$ being defined as in (2.5). We now define recursively $h^{(0)}(t) = h(t)$ and

$$h^{(-j)}(t) = -\int_t^\infty h^{(-j+1)}(\tau)\,d\tau, \qquad j = 1, 2, \dots . \tag{5.8}$$

Repeated integration by parts shows that

$$h^{(-j)}(t) \sim \exp\{ict^p\} \sum_{s=0}^{\infty} b_s^{(j)} t^{-\mu_{s,j}}, \qquad \text{as } t \to \infty, \tag{5.9}$$

where $b_s^{(j)}$ are some constants and for each j, and $\{\mu_{s,j}\}$ is a monotonically increasing sequence of positive numbers depending on p and β.

Theorem 6. *Assume that (i) $f^{(m)}(t)$ is continuous on $(0, \infty)$, where m is a nonnegative integer; (ii) $f(t)$ has an expansion of the form (5.2), and the expansion is m times differentiable; (iii) $h(t)$ satisfies (5.3) and (5.4), and (iv) as $t \to \infty$, $t^{-\beta} f^{(j)}(t) = O(t^{-1-\varepsilon})$ for $j = 0, 1, \ldots, m$ and for some $\varepsilon > 0$. Under these conditions, the result (5.6) holds with*

$$\delta_n(x) = \frac{(-1)^m}{x^m} \int_0^\infty f_n^{(m)}(t) h^{(-m)}(xt)\, dt, \tag{5.10}$$

where n is the smallest positive integer such that $\mathrm{Re}\ \lambda_n > m$.

Proof. Integration by parts gives

$$\int_0^\infty f_n(t) h(xt) e^{-\varepsilon t^p}\, dt = -\frac{1}{x} \int_0^\infty f_n'(t) h^{(-1)}(xt) e^{-\varepsilon t^p}\, dt$$

$$+ \frac{\varepsilon p}{x} \int_0^\infty f_n(t) h^{(-1)}(xt) t^{p-1} e^{-\varepsilon t^p}\, dt, \tag{5.11}$$

the integrated term vanishing because of $\mathrm{Re}\ \lambda_n + b > 0$ and condition (iv) and the asymptotic behavior in (5.9). The same reasoning, together with Lemma 1 and a modification of Lemma 2, ensures that the second term on the right-hand side of (5.11) tends to zero as $\varepsilon \to 0^+$; see Ex. 8. Thus

$$\delta_n(x) = \left(-\frac{1}{x} \right) \lim_{\varepsilon \to 0^+} \int_0^\infty f_n'(t) h^{(-1)}(xt) e^{-\varepsilon t^p}\, dt. \tag{5.12}$$

Repeated application of this technique shows that

$$\delta_n(x) = \frac{(-1)^m}{x^m} \lim_{\varepsilon \to 0^+} \int_0^\infty f_n^{(m)}(t) h^{(-m)}(xt) e^{-\varepsilon t^p}\, dt$$

$$= \frac{(-1)^m}{x^m} \int_0^\infty f_n^{(m)}(t) h^{(-m)}(xt)\, dt. \tag{5.13}$$

The last equality again follows from Lemma 1. ■

Although a quite general theorem has been established by means of which an explicit expression is obtained for the remainder $\delta_n(x)$ in (5.6), the conditions in this theorem appear to be more restrictive than the corresponding ones for the Fourier and Hankel transform. Owing to the fact that $h^{(-m)}(t)$ is the m-th iterated integral of h, defined in the particular manner given in (5.8), it is also difficult to estimate the error $\delta_n(x)$ in (5.10), and modifications are often needed in some specific instances. Here we wish to emphasize the fact that it is the pattern, not the detail, of proof which is important. The pattern of proof need not be abandoned in a specific example just because one or more of the conditions of the theorem happen to be violated.

Example 3.   In the calculation of flux integrals associated with double-scattering waves, Servadio (1982) has investigated the asymptotic behavior of the integral

$$I(x) = \int_0^\infty f(t)F(xt)\, dt, \tag{5.14}$$

where $F(t)$ is the Fresnel integral defined by

$$F(t) = \int_t^\infty e^{-i\tau^2}\, d\tau. \tag{5.15}$$

This integral can be expressed in terms of the more familiar complementary error function via the identity

$$F(t) = \frac{\sqrt{\pi}}{2} e^{-i\pi/4} \operatorname{erfc}(e^{i\pi/4}t). \tag{5.16}$$

Exercise 1 in Chapter I then gives

$$|F(t)| \le \frac{1}{2|t| \sin \delta} |\exp(-it^2)| \tag{5.17}$$

for $-3\pi/4 + \delta \le \arg t \le \pi/4 - \delta,\ \delta > 0$, and

$$F(t) \sim e^{-it^2}\left[-\frac{i}{2t} + \frac{1}{2^2 t^3} + i\frac{1\cdot 3}{2^3 t^5} + \cdots \right] \tag{5.18}$$

as $t \to \infty$ in the same sector. (5.18) of course implies that $F(t)$ is an oscillatory function of the type (5.4) with $c = -1$, $p = 2$, and $\beta = 1$, when t is real and positive. Now define $F^{(0)}(t) = F(t)$ and

$$F^{(-j)}(t) = -\int_{t}^{t+\infty e^{-i\pi/4}} F^{(-j+1)}(\tau)\, d\tau, \qquad j = 1, 2, \ldots, \qquad (5.19)$$

where the path of integration is the ray $\tau = t + \rho e^{-i\pi/4}$; compare with (5.8). By induction, it can be shown that

$$F^{(-m)}(t) = \frac{(-1)^m}{(m-1)!} \int_{t}^{t+\infty e^{-i\pi/4}} (\tau - t)^{m-1} F(\tau)\, d\tau. \qquad (5.20)$$

Rotating the path of integration, one also has from Erdélyi *et al.* (1954, p. 325)

$$M[F; z] = \frac{1}{2z} e^{-i\pi(z+1)/4} \Gamma\left(\frac{z+1}{2}\right). \qquad (5.21)$$

The asymptotic expansion of the integral in (5.14) is obtained by replacing h by F in (5.6). The remainder $\delta_n(x)$ is given by

$$\delta_n(x) = \frac{(-1)^m}{x^m} \int_{0}^{\infty} f_n^{(m)}(t) F^{(-m)}(xt)\, dt. \qquad (5.22)$$

Taking $\delta = \pi/4$ in (5.17), we get from (5.20)

$$|F^{(-m)}(t)| \le \frac{\Gamma(m/2)}{\Gamma(m)} \frac{\sqrt{2}}{4t}. \qquad (5.23)$$

Therefore

$$|\delta_n(x)| \le \frac{\sqrt{2}\,\Gamma(m/2)}{4\Gamma(m)x^{m+1}} \int_{0}^{\infty} |f_n^{(m)}(t)| t^{-1}\, dt, \qquad (5.24)$$

provided the integral on the right exists. As a further specialization, we take

$$f(t) = \frac{1}{\sqrt{t(1+t)}} = \sum_{s=0}^{n-1} (-1)^s t^{s+1/2-1} + f_n(t),$$

where

$$f_n(t) = \frac{(-1)^n t^{n-1/2}}{1+t}.$$

In the notation of (5.2), $\lambda_s = s + \frac{1}{2}$. Therefore, according to the statement of Theorem 6, $m = n$ in (5.10). Using Leibniz's rule, we obtain

$$|f_n^{(n)}(t)| \le \Gamma(n + \tfrac{1}{2}) \frac{2^2 \cdot n!}{t^{1/2}(1+t)}.$$

If $f_n^{(n)}(t)$ in (5.24) is replaced by its upper bound, then the integral in (5.24) does not converge. In this case, what we should have done is to stop the repeated integration-by-parts procedure leading to (5.13) one step earlier; that is, to take

$$\delta_n(x) = \frac{(-1)^{n-1}}{x^{n-1}} \int_0^\infty f_n^{(n-1)}(t) F^{(-n+1)}(xt)\, dt.$$

Since

$$|f_n^{(n-1)}(t)| \leq 2\Gamma(n)\Gamma(n+\tfrac{1}{2}) \frac{t^{1/2}}{1+t},$$

it follows that

$$|\delta_n(x)| \leq \frac{\pi}{\sqrt{2}} \Gamma(n+\tfrac{1}{2})\Gamma\left(\frac{n-1}{2}\right) \frac{(n-1)}{x^n}.$$

Example 4.   In Olver (1974a, p. 342), an asymptotic expansion was derived for the integral

$$I(x) = \int_0^\infty \mathrm{Ai}(-xv) f(v)\, dv, \qquad\qquad (5.25)$$

where $f(v)$ is infinitely differentiable in $[0, \infty)$. Olver's method is based on repeated integration by parts. In view of the oscillatory behavior

$$\mathrm{Ai}(-v) \sim \cos(\tfrac{2}{3}v^{3/2} - \tfrac{1}{4}\pi) \sum_{s=0}^\infty (-1)^s \frac{\alpha_{2s}}{v^{3s+1/4}}$$

$$+ \sin(\tfrac{2}{3}v^{3/2} - \tfrac{1}{4}\pi) \sum_{s=0}^\infty (-1)^s \frac{\alpha_{2s+1}}{v^{3s+7/4}},$$

α_s being constants, it is easily seen that Olver's expansion can also be obtained by using Theorem 6 above. However, it does not seem easy to construct a numerical bound for the error term associated with the expansion by using either one of the two methods. In this example, we shall indicate how the results of §3 can be used to derive such an error bound. First we recall the identity (Olver 1974a, p. 61)

$$\mathrm{Ai}(-v) = \frac{1}{3} v^{1/2}\{J_{-1/3}(\xi) + J_{1/3}(\xi)\}$$

where $\zeta = \frac{2}{3}v^{3/2}$. Put $\rho = \frac{2}{3}x^{3/2}$ and

$$I^{\pm}(\rho) = \int_0^{\infty} f(v)v^{1/2}J_{\pm 1/3}(\rho v^{3/2})\, dv.$$

Then clearly

$$I(x) = \frac{\sqrt{x}}{3}[I^+(\rho) + I^-(\rho)].$$

Since the change of variable $v^{3/2} = t$ gives

$$I^{\pm}(\rho) = \frac{2}{3}\int_0^{\infty} f(t^{2/3})J_{\pm 1/3}(\rho t)\, dt,$$

Theorem 3 can be used to yield a similar result for the integral in (5.25).

6. Some Quadrature Formulas

If $w(t)$ is a positive function on $(0, \infty)$ which is rapidly decreasing at infinity, and if $f(t)$ is sufficiently smooth in $(0, \infty)$, then it is well known that integrals of the form

$$\int_0^{\infty} f(t)w(t)\, dt \tag{6.1}$$

can be evaluated numerically by Gaussian quadrature rules; see, for example, Hildebrand (1974). However, if $w(t)$ is an oscillatory function, such as e^{it} or the Bessel function $J_\nu(t)$, and if $f(t)$ decreases slowly at infinity, then the problem of numerical computation of these integrals becomes considerably more difficult. In this section we shall present some quadrature formulas for the Fourier and the Bessel transforms

$$F(x) = \int_0^{\infty} t^{\mu}f(t)e^{ixt}\, dt \tag{6.2}$$

$$H_i(x) = \int_0^{\infty} t^{\mu}f(t)H_\nu^{(i)}(xt)\, dt, \qquad i = 1, 2, \tag{6.3}$$

where x is a real parameter and $H_\nu^{(i)}(t)$, $i = 1$, 2, are the Hankel functions. In (6.2) we require $\mu > -1$ and in (6.3) we require $\mu \pm \nu > -1$. It turns out that these quadrature formulas also provide asymptotic approximations for the integrals in (6.2) and (6.3), as $x \to \infty$, complete with error bounds.

We first recall the Gauss–Laguerre formula (Davis and Rabinowitz 1975, p. 174)

$$\int_0^\infty t^\mu f(t) e^{-t}\, dt = \sum_{k=1}^n w_k f(t_k) + E_n(f), \tag{6.4}$$

where

$$E_n(f) = \frac{n!\,\Gamma(\mu + n + 1)}{(2n)!}\, f^{(2n)}(\xi), \qquad 0 < \xi < \infty. \tag{6.5}$$

The abscissas t_k are the zeros of the Laguerre polynomial

$$L_n^{(\mu)}(t) = e^t t^{-\mu}\, \frac{d^n}{dt^n}\, (e^{-t} t^{\mu + n}),$$

and the weights w_k are given by

$$w_k = \frac{n!\,\Gamma(n + \mu + 1) t_k}{[L_{n+1}^{(\mu)}(t_k)]^2}.$$

Some tables of t_k and w_k can be found in Rabinowitz and Weiss (1959) and Abramowitz and Stegun (1964, p. 923). If z is real and positive, then it is easy to see that (6.4) can be written in the more general form

$$\int_0^\infty t^\mu f(t) e^{-zt}\, dt = z^{-\mu - 1} \sum_{k=1}^n w_k f\!\left(\frac{t_k}{z}\right) + E_n(f; z), \tag{6.6}$$

where

$$E_n(f; z) = \frac{n!\,\Gamma(n + \mu + 1)}{(2n)!\, z^{2n + \mu + 1}}\, f^{(2n)}\!\left(\frac{\xi}{z}\right), \qquad 0 < \xi < \infty. \tag{6.7}$$

The following result shows that (6.6)–(6.7) in fact holds when z is purely imaginary, provided that $f(t)$ is an analytic function in the half-plane $\mathrm{Re}\, t > 0$.

Theorem 7.   Let $f(t)$ be an analytic function in $\mathrm{Re}\, t > 0$, and suppose that $f^{(2n)}(t)$ is continuous in $\mathrm{Re}\, t \geq 0$. If the Fourier transform $F(x)$ in (6.2) converges as an improper Riemann integral, then we have

$$F(x) = \frac{e^{(\mu + 1)\pi i/2}}{x^{\mu + 1}} \sum_{k=1}^n w_k f\!\left(\frac{it_k}{x}\right) + \varepsilon_n(f; x), \tag{6.8}$$

where

$$\varepsilon_n(f; x) = \frac{n! \, \Gamma(n + \mu + 1)}{(2n)! \, x^{2n + \mu + 1}} \, e^{(2n + \mu + 1)\pi i/2} f^{(2n)}\left(\frac{i\xi}{x}\right), \qquad 0 < \xi < \infty. \quad (6.9)$$

Proof. The integral on the left-hand side of (6.6) can be considered as the Laplace transform of $t^\mu f(t)$. Thus, by a well-known result from Laplace transform theory, this integral converges for Re $z > 0$ and defines an analytic function there; see Widder (1941, p. 37 and p. 57). Since $f(t)$ is analytic for Re $t > 0$, the terms on the right-hand side of (6.6) are all analytic for Re $z > 0$. By analytic continuation, the identity in (6.6) holds for all z in Re $z > 0$. Now, write $z = \varepsilon - ix$ and let $\varepsilon \to 0$ in (6.6). The right-hand side of (6.6) clearly tends to the right-hand side of (6.8), as desired. The fact that the Laplace integral in (6.6) tends to the Fourier integral in (6.8) follows from Lemma 1. This proves (6.8). ∎

It is sometimes advantageous to express the truncation error in (6.4) in the form

$$E_n(f) = n! \, \Gamma(n + \mu + 1) f[t_1, t_1, t_2, t_2, \ldots, t_n, t_n, \xi_1] \quad (6.10)$$

where $0 < \xi_1 < \infty$ and $f[t_1, t_1, \ldots, t_n, t_n, \xi_1]$ is the $2n$-th divided difference of $f(t)$, relative to the abscissas $t_1, t_1, \ldots, t_n, t_n$, and ξ_1; see Hildebrand (1974, p. 397, Eq. (8.7.12)). Divided differences of orders $0, 1, \ldots, k$ are defined recursively by the relations

$$f[x_0] = f(x_0), \quad f[x_0, x_1] = \frac{f[x_1] - f[x_0]}{x_1 - x_0}, \ldots,$$

$$f[x_0, \ldots, x_k] = \frac{f[x_1, \ldots, x_k] - f[x_0, \ldots, x_{k-1}]}{x_k - x_0}. \quad (6.11)$$

If two or more arguments in a divided difference become the same, then appropriate limiting processes must be taken. For example, if $x_1 = x + \varepsilon$, then

$$f[x_1, x] = f[x + \varepsilon, x] = \frac{f(x + \varepsilon) - f(x)}{\varepsilon}.$$

Thus, as $\varepsilon \to 0$, we have $f[x, x] = f'(x)$ if $f'(x)$ exists. Similarly it can be shown that

$$\frac{d}{dx} f[x_0, \ldots, x_k, x] = f[x_0, \ldots, x_k, x, x] \quad (6.12)$$

if $x_0, \ldots, x_k$ are constants. In terms of divided differences, the error term in (6.8) can be written as

$$\varepsilon_n(f; x) = \frac{n! \, \Gamma(n + \mu + 1)}{x^{2n + \mu + 1}} \, e^{(2n + \mu + 1)\pi i/2} f\left[i\frac{t_1}{x}, i\frac{t_1}{x}, \ldots, i\frac{t_n}{x}, i\frac{t_n}{x}, i\frac{\xi_1}{x} \right],$$

(6.13)

where $0 < \xi_1 < \infty$.

If $f^{(2n)}(t)$ is bounded on the imaginary axis, say by M_{2n}, then from (6.9) we have

$$|\varepsilon_n(f; x)| \le \frac{n! \, \Gamma(n + \mu + 1)}{(2n)! \, x^{2n + \mu + 1}} \, M_{2n}.$$

(6.14)

A similar estimate holds if the divided difference in (6.13) is bounded. In either case, (6.8) provides an attractive asymptotic approximation for the Fourier integral $F(x)$.

Example 5.   Consider the integral

$$S(x) = \int_0^\infty t^{1/2} \frac{\sin xt}{1 + t} \, dt.$$

(6.15)

With $\mu = \frac{1}{2}$ and $f(t) = 1/(1 + t)$, the quadrature formula (6.8) gives

$$S(x) = \frac{1}{\sqrt{2x}} \sum_{k=1}^n w_k \frac{x + t_k}{x^2 + t_k^2} + \varepsilon_n(x),$$

(6.16)

where

$$|\varepsilon_n(x)| \le \frac{n! \, \Gamma(n + \frac{1}{2} + 1)}{x^{3/2}(x^2 + t_1^2) \cdots (x^2 + t_n^2)};$$

(6.17)

see Ex. 22. The weights w_k and t_k are given in Shao, Chen, and Frank (1964). Let $s_n(x)$ denote the sum on the right-hand side of (6.16). In Table 4.1 we tabulate the values of $s_n(x)$ for $n = 4, 8, 16$ and $x = 2, 3, 4, 5$.

x n	2	3	4	5
4	0.231422	0.13392163	0.088647098	0.0636329754
8	0.232113	0.13411233	0.088672001	0.0636330084
16	0.232087	0.13410937	0.088672219	0.0636330809

Table 4.1

The estimate (6.17) shows that $|\varepsilon_{16}(5)| \leq 2 \times 10^{-10}$. Thus, we have $S(5) = 0.0636330809$ accurate to at least eight decimal places. For comparison, we also refer to Stenger (1981, p. 202) for a different way of computing $S(x)$.

We now turn to the consideration of the Bessel transforms (6.3). Let $K_\nu(t)$ denote the modified Bessel function of the third kind, and let

$$w(t) = t^\mu K_\nu(t), \qquad \mu \pm \nu > -1, \tag{6.18}$$

be the weight function in (6.1). The moments

$$\rho_n = \int_0^\infty t^{n+\mu} K_\nu(t)\,dt, \qquad n = 0, 1, 2, \dots, \tag{6.19}$$

can easily be found (Olver 1974a, p. 254) to be

$$\rho_n = 2^{n+\mu-1}\Gamma(\tfrac{1}{2}n + \tfrac{1}{2}\mu + \tfrac{1}{2} + \tfrac{1}{2}\nu)\Gamma(\tfrac{1}{2}n + \tfrac{1}{2}\mu + \tfrac{1}{2} - \tfrac{1}{2}\nu). \tag{6.20}$$

A sequence of polynomials, orthonormal with respect to the weight function $w(t)$ in $(0, \infty)$, can be constructed as follows; see Szegö (1967, §§2.1 and 2.2). Put $D_{-1} = 1$, $D_0 = \rho_0$, and

$$D_n = \begin{vmatrix} \rho_0 & \rho_1 & \cdots & \rho_n \\ \rho_1 & \rho_2 & \cdots & \rho_{n+1} \\ \vdots & \ddots & & \vdots \\ \rho_{n-1} & \rho_n & \cdots & \rho_{2n-1} \\ \rho_n & \rho_{n+1} & \cdots & \rho_{2n} \end{vmatrix}, \qquad n \geq 1; \tag{6.21}$$

and define the polynomials $p_n(t)$ by $p_0(t) = D_0^{-1/2} = \rho_0^{-1/2}$ and

$$p_n(t) = (D_{n-1}D_n)^{-1/2} \begin{vmatrix} \rho_0 & \rho_1 & \rho_2 & \cdots & \rho_n \\ \rho_1 & \rho_2 & \rho_3 & \cdots & \rho_{n+1} \\ \vdots & \ddots & & & \vdots \\ \rho_{n-1} & \rho_n & \rho_{n+1} & \cdots & \rho_{2n-1} \\ 1 & t & t^2 & \cdots & t^n \end{vmatrix}, \qquad n \geq 1. \tag{6.22}$$

From the general theory of orthogonal polynomials, it follows that these polynomials are orthonormal with respect to the weight function $w(t)$ given in (6.18), and that the zeros of these polynomials are positive and distinct. Furthermore, between two consecutive zeros of $p_n(t)$, there is exactly one zero of $p_{n+1}(t)$; see Szegö (1967, §3.3).

For fixed n, let $t_1, \ldots, t_n$ denote the zeros of $p_n(t)$, and let A_n be the coefficient of t^n in $p_n(t)$. The general quadrature formula (Hildebrand 1974, Eqs. (8.4.6), (8.4.9), and (8.4.17)) then gives

$$\int_0^\infty t^\mu f(t) K_\nu(zt)\, dt = z^{-\mu-1} \sum_{k=1}^n w_k f\left(\frac{t_k}{z}\right) + E_n(f; z), \qquad (6.23)$$

where z is real and positive,

$$w_k = \frac{A_n}{A_{n-1} p_n'(t_k) p_{n-1}(t_k)} \qquad (6.24)$$

and

$$E_n(f; z) = \frac{f^{(2n)}(\xi/z)}{A_n^2 (2n)!\, z^{2n+\mu+1}}, \qquad 0 < \xi < \infty. \qquad (6.25)$$

If $f(t)$ is an analytic function in $\operatorname{Re} t > 0$ then, by analytic continuation, (6.23) also holds for complex z as long as $\operatorname{Re} z > 0$. The following result is a generalization of (6.8)–(6.9).

Theorem 8. *Let $f(t)$ be an analytic function in the half plane $\operatorname{Re} t > 0$, and suppose that $f^{(2n)}(t)$ is continuous in $\operatorname{Re} t \geq 0$. If the Bessel transforms in (6.3) converge as improper Riemann integrals, then we have*

$$H_1(x) = \frac{2}{\pi} \frac{e^{i(\mu-\nu)\pi/2}}{x^{\mu+1}} \sum_{k=1}^n w_k f\left(i\frac{t_k}{x}\right) + \delta_n^{(1)}(f; x), \qquad (6.26)$$

where

$$\delta_n^{(1)}(f; x) = \frac{2}{\pi} \frac{e^{i(\mu+2n-\nu)\pi/2}}{x^{2n+\mu+1}} \frac{f^{(2n)}(i\xi_1/x)}{A_n^2 (2n)!}, \qquad 0 < \xi_1 < \infty. \qquad (6.27)$$

The corresponding formula for $H_2(x)$ is obtained by replacing i by $-i$ in (6.26) and (6.27).

Proof. In (6.23), we first put $z = \varepsilon - ix$ and then let $\varepsilon \to 0$. The right-hand side of (6.23) clearly tends to the right-hand side of (6.26), except for the factor $(2/\pi i)e^{-i\pi\nu/2}$. In view of the connecting formula

$$H_\nu^{(1)}(t) = \frac{2}{\pi i} e^{-i\pi\nu/2} K_\nu(-it), \qquad 0 < t < \infty, \qquad (6.28)$$

the left-hand side of (6.23) also tends to the left-hand side of (6.26) provided that the limit (as $\varepsilon \to 0$) can be taken inside the integral sign.

The fact that the limit and the integral can indeed be interchanged is justified by the asymptotic expansion (Olver 1974a, p. 250)

$$K_\nu(z) \sim \left(\frac{\pi}{2z}\right)^{1/2} e^{-z} \sum_{s=0}^{\infty} \frac{A_s(\nu)}{z^s}, \tag{6.29}$$

as $z \to \infty$ in $|\arg z| \le 3\pi/2 - \delta$, and Lemma 1. The corresponding formula for $H_2(x)$ is obtained by using, instead of (6.28), the connecting formula

$$H_\nu^{(2)}(t) = -\frac{2}{\pi i} e^{i\pi\nu/2} K_\nu(it), \qquad 0 < t < \infty, \tag{6.30}$$

and putting $z = \varepsilon + ix$ in (6.23). This completes the proof of Theorem 8. ■

It is well-known that the Bessel function of the first kind, $J_\nu(t)$, can be written as

$$J_\nu(t) = \tfrac{1}{2}\{H_\nu^{(1)}(t) + H_\nu^{(2)}(t)\}. \tag{6.31}$$

Hence, the quadrature formulas for $H_1(x)$ and $H_2(x)$ can also be used to compute numerically the Hankel transform

$$\int_0^\infty t^\mu f(t) J_\nu(xt)\, dt. \tag{6.32}$$

In this case, the condition $\mu \pm \nu > -1$ in (6.3) can be weakened to $\mu + \nu > -1$. The more restrictive assumption is needed only to ensure the convergence of the integrals in (6.3) and (6.19).

The derivative form of the error term in (6.26) can also be expressed in terms of the divided difference of $f(t)$. More explicitly, we have

$$\delta_n^{(1)}(f; x) = \frac{2}{\pi} \frac{e^{i(\mu + 2n - \nu)\pi/2}}{A_n^2 x^{2n + \mu + 1}} f\left[i\frac{t_1}{x}, i\frac{t_1}{x}, \ldots, i\frac{t_n}{x}, i\frac{t_n}{x}, \frac{\xi}{x}\right] \tag{6.33}$$

for some $\xi \in (0, \infty)$; cf. Hildebrand (1974, p. 397, (8.7.12)). The use of (6.33) is usually preferable to that of (6.27).

Example 6.   Consider the integral

$$I(x) = \int_0^\infty \frac{J_0(xt)}{\sqrt{t}(1 + t)}\, dt. \tag{6.34}$$

n	Coefficients	Abscissas	Weights
1	1.0429856349	0.2284732905	4.6474760094
2	0.6958420717	0.1272660741	4.4186595245
		2.1828789401	0.2288164849
3	0.2776778397	0.08975811418	4.1825838264
		1.4038905409	0.4551726046
		4.8758067266	0.009719578404
4	0.0791554398	0.06990008100	3.9858407490
		1.04587495139	0.6239374973
		3.4232461358	0.03733775108
		7.8792758079	0.0003600119525
5	0.0175560146	0.05750827921	3.8221033421
		0.8366348329	0.7471552593
		2.6726988390	0.07574750466
		5.8349740709	0.002457777509
		11.0589492082	0.00001212583461

Table 4.2

In the notation of (6.32), we have $\mu = -\frac{1}{2}$, $\nu = 0$, and $f(t) = 1/(1 + t)$. From (6.31) and Theorem 8 it follows that

$$I(x) = \frac{\sqrt{2x}}{\pi} \sum_{k=1}^{n} w_k \frac{x - t_k}{x^2 + t_k^2} + \delta_n(x) \tag{6.35}$$

with

$$|\delta_n(x)| \le \frac{2}{\pi A_n^2 x^{2n + 1/2}}. \tag{6.36}$$

The values of the coefficients A_n, the abscissas t_k and the weights w_k are listed in Table 4.2. From (6.35), with $n = 2$, we have $I(50) = 0.2944872$ accurate to seven decimal places.

There are various ways of constructing quadrature formulas. We have constructed Table 4.2 by using the method based on determinants; see (6.19)–(6.22). For a more effective procedure, we refer to Gautschi (1982, Example 4.10).

7.   Mellin–Barnes Type Integrals

In applied mathematics, one often solves a linear boundary value problem by using integral transform methods. Upon applying an inversion formula, the final solution is then expressed in terms of an integral of the form

$$f(t) = \int_C F(z)K(z, t)\, dz, \tag{7.1}$$

where t is real, $K(z, t)$ is some given kernel, C is a finite or infinite contour in the complex z-plane, and $F(z)$ is a known function. Usually $F(z)$ is the integral transform of $f(t)$, and equation (7.1) defines its inverse.

In this context, the most common transforms are the Fourier and Laplace transforms. In the case of the Fourier transform, the inverse transform is given by

$$f(t) = \frac{1}{\sqrt{2\pi}} \int_{-\infty}^{\infty} F(x)e^{-itx}\, dx, \tag{7.2}$$

where t is real. To obtain the large-t behavior of $f(t)$, we can use either the integration by parts technique described in Chapter I or the summability method discussed in Section 2 of the present chapter. In the case of Laplace transform, the inverse is

$$f(t) = \frac{1}{2\pi i} \int_{c-i\infty}^{c+i\infty} F(z)e^{tz}\, dz, \tag{7.3}$$

where t is real, $c > 0$ and $F(z)$ is analytic in the half-plane $\operatorname{Re} z > c$. The contour in (7.3) is the infinite vertical line $\operatorname{Re} z = c$. If the function $F(z)$ happens to be also analytic in the other half of the complex z-plane except for a branch point singularity at $z = 0$, and if the contour can be deformed into a loop as indicated in Figure 4.1, then the asymptotic behavior of $f(t)$ as $t \to \infty$ can be obtained from Barnes' lemma given in Chapter I (Ex. 14).

However, in some more delicate problems, the function $F(z)$ is not easily extendable to the half-plane $\operatorname{Re} z < c$, and hence a loop-integral representation for $f(t)$ is not always possible. In such cases, we would have to rely completely on the properties of $F(z)$ in the half plane $\operatorname{Re} z \geq c$.

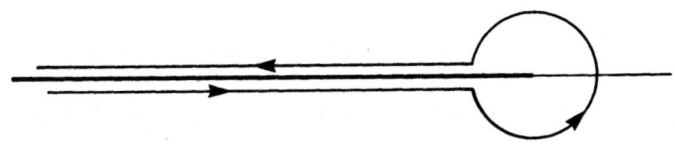

Fig. 4.1

In a recent investigation of the asymptotic behavior of the Schrödinger spectral kernel (Osborn and Wong 1983), there arose a contour
integral of the form

$$I(t; \mu) = \frac{1}{2\pi i} \int_{c-i\infty}^{c+i\infty} \exp\left(tz - \frac{\mu}{z}\right) F(z)\, dz, \qquad (7.4)$$

where c is positive, μ is real, and $F(z)$ is an analytic function in Re $z > 0$
and continuous for Re $z \geq 0$ except possibly at the origin. At the origin,
$F(z)$ has an expansion of the form

$$F(z) = \sum_{s=0}^{n-1} a_s z^{s+\alpha-1} + F_n(z), \qquad (7.5)$$

where $F_n(z) = O(|z|^{n+\alpha-1})$ as $|z| \to 0$. Observe that if $\mu \neq 0$ in (7.4), then
the origin is not only a branch-point singularity of $F(z)$ but also an
essential singularity of the integrand in (7.4).

Theorem 9.   *Assume*   (i)   *for   each $j = 0, 1, \ldots, m$, $F_n^{(j)}(z) =$*
$O(|z|^{n+\alpha-j-1})$ as $|z| \to 0$, where m and n are related via the inequalities

$$2m - \alpha + 1 \leq n < 2m - \alpha + 2; \qquad (7.6)$$

(ii) *in the strip $0 \leq x \leq c$, $F(x + iy) \to 0$ as $|y| \to \infty$ uniformly in x and
$F^{(j)}(iy) = O(1)$ as $|y| \to \infty$ for $j = 1, \ldots, m-1$;* (iii) *each of the integrals*

$$\int_1^\infty \exp\left(ity \pm i\frac{\mu}{y}\right) F^{(j)}(\pm iy)\, dy, \qquad j = 0, 1, \ldots, m, \qquad (7.7)$$

*converges uniformly for all large t and for all bounded μ. Then there
exists a function $I_m(t; \mu)$ such that*

$$I(t; \mu) = \sum_{s=0}^{n-1} a_s J_{-s-\alpha}(2\sqrt{t\mu})\left(\frac{\mu}{t}\right)^{(s+\alpha)/2} + \frac{1}{t^m} I_m(t; \mu) \qquad (7.8)$$

and

$$\lim_{t \to \infty} I_m(t; \mu) = 0, \qquad (7.9)$$

the limit being uniformly valid for μ in any bounded subset of $(-\infty, \infty)$.

The proof of this theorem is given in a sequence of exercises at the
end of the chapter. Here we present only the following corollary, which
is similar to the Tauberian theorem for the Laplace transform due to
von Stachò; see Doetsch (1944, p. 277).

Corollary. *Assume that conditions* (i) *and* (ii) *in Theorem 9 hold, and that condition* (iii) *is replaced by* (iii′) *each of the integrals*

$$\int_1^\infty e^{ity} F^{(j)}(\pm iy)\, dy, \qquad j = 0, 1, \ldots, m, \tag{7.10}$$

converges uniformly for all large t. Then the integral

$$f(t) = \frac{1}{2\pi i} \int_{c-i\infty}^{c+i\infty} e^{tz} F(z)\, dz \tag{7.11}$$

has the asymptotic expansion

$$f(t) \sim \frac{\sin \alpha\pi}{\pi} \sum_{s=0}^\infty (-1)^s a_s \frac{\Gamma(s+\alpha)}{t^{s+\alpha}}, \qquad \text{as } t \to \infty. \tag{7.12}$$

Proof. We first note that if $g(y)$ is improper Riemann integrable on $[1, \infty)$, then so is $e^{i\mu/y} g(y)$ for all bounded real values of μ. Thus condition (iii′) is actually equivalent to condition (iii). We also note that from the series representation

$$J_\nu(z) = \left(\frac{1}{2} z\right)^\nu \sum_{k=0}^\infty \frac{(-1)^k (\frac{1}{4} z^2)^k}{k!\, \Gamma(\nu + k + 1)},$$

it follows that

$$\lim_{\mu \to 0} \left(\frac{\mu}{t}\right)^{(s+\alpha)/2} J_{-s-\alpha}(2\sqrt{t\mu}) = (-1)^s \frac{\sin \alpha\pi}{\pi} \frac{\Gamma(s+\alpha)}{t^{s+\alpha}}.$$

Thus, by Lemma 1 and Theorem 9, we have for any $m = 1, 2, \ldots$,

$$f(t) = \frac{\sin \alpha\pi}{\pi} \sum_{s=0}^{m-1} (-1)^s a_s \frac{\Gamma(s+\alpha)}{t^{s+\alpha}} + o(t^{-m}), \qquad \text{as } t \to \infty. \quad \blacksquare$$

Exercises

1. Suppose that (i) $f(t)$ has a continuous derivative, (ii) $f(t)$ decreases monotonically to zero as $t \to \infty$, and (iii) $G(t, x) = \int_a^t g(\tau, x)\, d\tau$ is uniformly bounded for $x \geq X$ and $t \geq a$. Show that $\int_a^\infty f(t) g(t, x)\, dt$ is uniformly convergent for $x \geq X$.

2. Give an example to show that the convergence of $\int_0^\infty f(t) \cos xt\, dt$ does not imply that $f(t) \to 0$ as $t \to \infty$. (Hint: Try $f(t) = e^t \cos e^{2t}$.)

3.   Show that

$$\int_0^\infty \frac{\sin t}{t}\, dt = \int_0^\infty \left(\frac{\sin t}{t}\right)^2 dt,$$

and note that the integral on the right is absolutely convergent whereas the integral on the left is not.

4.   (a)   Let $f(t)$ be m times continuously differentiable in $0 < t \le b$, where m is a nonnegative integer and b is finite, and let

$$f(t) \sim \sum_{s=0}^\infty a_s t^{\lambda_s - 1}, \qquad \text{as } t \to 0^+,$$

where $\lambda_0 > 0$ and $\lambda_{s+1} > \lambda_s$ for $s = 0, 1, 2, \dots$. Suppose that this expansion can be differentiated term by term m times, and that n is a nonnegative integer such that $m < \lambda_n$. Show that

$$\int_0^b f(t)e^{ixt}\, dt = \sum_{s=0}^{n-1} a_s \Gamma(\lambda_s)e^{i\lambda_s \pi/2} x^{-\lambda_s}$$

$$- e^{ibx} \sum_{s=0}^{m-1} f^{(s)}(b)\left(\frac{i}{x}\right)^{s+1} + \delta_{m,n}(x) - \varepsilon_{m,n}(x).$$

The error terms are given by

$$\delta_{m,n}(x) = \left(\frac{i}{x}\right)^m \int_0^b f_n^{(m)}(t)e^{ixt}\, dt$$

and

$$\varepsilon_{m,n}(x) = \sum_{s=0}^{n-1} a_s e^{i\lambda_s \pi/2} \frac{\Gamma(\lambda_s)}{\Gamma(\lambda_s - m)} \Gamma(\lambda_s - m, -ixb)x^{-\lambda_s},$$

where the function $f_n(t)$ is defined by

$$f(t) = \sum_{s=0}^{n-1} a_s t^{\lambda_s - 1} + f_n(t)$$

and $\Gamma(\alpha, z)$ is the complementary incomplete gamma function (Chapter I, Exs. 3 and 4) $\Gamma(\alpha, z) = \int_z^\infty t^{\alpha - 1}e^{-t}\, dt$.

(b)   Show that for any nonnegative integer m,

$$\Gamma(\alpha, z) = e^{-z}z^{\alpha - 1}\sum_{j=0}^{m-1} \frac{\Gamma(\alpha)}{\Gamma(\alpha - j)}\frac{1}{z^j} + \frac{\Gamma(\alpha)}{\Gamma(\alpha - m)}\Gamma(\alpha - m, z),$$

and that by deforming the path of integration and then integrating by parts $|\Gamma(\alpha, \pm iy)| \le 2y^{\alpha - 1}$, $\alpha < 1$, $y > 0$.

(c)   From (a) and (b) construct the error bounds

$$|\delta_{m,n}(x)| \le \frac{1}{x^m} \int_0^b |f_n^{(m)}(t)| \, dt,$$

$$|\varepsilon_{m,n}(x)| \le \frac{2}{x^{m+1}} \sum_{s=0}^{n-1} \frac{\Gamma(\lambda_s)}{\Gamma(\lambda_s - m)} |a_s| b^{\lambda_s - m - 1}.$$

If $\lambda_n \ge m + 1$, show that $\delta_{m,n}(x)$ also satisfies

$$|\delta_{m,n}(x)| \le \frac{1}{x^{m+1}} \left[|f_n^{(m)}(0)| + |f_n^{(m)}(b)| + \int_0^b |f_n^{(m+1)}(t)| \, dt \right].$$

5.   Put

$$f(t) = \frac{1}{t \log t} \sum_{n=1}^{\infty} \frac{1}{n^2} \sin nt,$$

and note that $f(t) = o(t^{-1})$ as $t \to \infty$. (i) Show that $F_s(x) = \int_2^\infty f(t) \sin xt \, dt$ does not converge for $x = 1, 2, \dots$. (ii) Show that $F_c(x) = \int_2^\infty f(t) \cos xt \, dt$ converges for every $x > 0$ but not uniformly.

6.   Construct a function $f(t)$ such that $f^{(k)}(t) = o(t^{-1})$, as $t \to \infty$, for all $k \ge 0$, but the Fourier integral $\int_0^\infty f(t) e^{ixt} \, dt$ does not converge for all large values of x.

7.   Let $f(t)$ be continuous in $[1, \infty)$, and suppose that $\int_1^\infty f(t) \, dt$ converges. Show that $\int_1^\infty f(t) e^{ix/t} \, dt$ converges uniformly for all x.

8.   Let $f(t)$ be absolutely integrable on $[0, \delta]$ for some $\delta > 0$, and be bounded for $t \ge \delta$. Show that for any $p \ge 1$, we have

$$\lim_{\varepsilon \to 0^+} \varepsilon \int_0^\infty f(t) e^{-\varepsilon t^p} \, dt = 0.$$

9.   Show that the integral

$$\int_1^\infty \frac{\sin \sqrt{t}}{\sqrt{t}} e^{ixt} \, dt$$

converges uniformly for large values of $|x|$.

10.   Let $f_2(t) = \sin \sqrt{t} - \sqrt{t} + t^{3/2}/3!$. Calculate the derivatives f_2', f_2'', and f_2'''. Show that for $t \ge 0$, $|f_2(t)| \le t^{5/2}/5!$ and $|f_2'(t)| \le t^{3/2}/2 \cdot 4!$.

Furthermore, show that for $t \geq 1$, $|f_2''(t)| \leq t^{-1/2}/2$ and $|f_2'''(t)| \leq 3t^{-3/2}/4$. Prove that there exist ξ and η in $(0, \sqrt{t})$ such that

$$f_2''(t) = \frac{1}{4 \cdot 4!} t^{1/2} (\cos \xi - \xi \sin \xi),$$

$$f_2'''(t) = \frac{3}{8 \cdot 4!} t^{-1/2} \left[\left(1 - \frac{1}{3} \eta^2 \right) \cos \eta - \frac{5}{4} \eta \sin \eta \right].$$

Deduce from the last two equations that for $0 < t < 1$, $|f_2''(t)| \leq t^{1/2}/2 \cdot 4!$ and $|f_2'''(t)| \leq t^{-1/2}/4!$.

11. Let $f(t)$ be an m times continuously differentiable function in $(0, \infty)$, m being a positive integer, and let $F(x) = \int_0^\infty f(t) e^{ixt} \, dt$. Suppose that

$$f(t) = \sum_{s=0}^{n-1} a_s t^{\lambda_s - 1} (\log t) + f_n(t),$$

where n is the smallest nonnegative integer such that Re $\lambda_n > m$, and $f_n^{(j)}(t) = O(t^{\lambda_n - j - 1} \log t)$, as $t \to 0^+$, for $j = 0, 1, \ldots, m$. Also suppose that each of the integrals $\int_1^\infty f^{(j)}(t) e^{ixt} \, dt$, $j = 0, 1, \ldots, m$, converge uniformly for all sufficiently large x. Show that

$$F(x) = \sum_{s=0}^{n-1} a_s \Gamma(\lambda_s) e^{i\lambda_s \pi/2} \left\{ \psi(\lambda_s) + i\frac{\pi}{2} - \log x \right\} x^{-\lambda_s} + F_n(x),$$

where

$$F_n(x) = \left(\frac{i}{x} \right)^n \int_0^\infty f_n^{(m)}(t) e^{ixt} \, dt = o(x^{-m}), \qquad \text{as } x \to \infty.$$

12. Derive the two-term expansion

$$\int_0^\infty \frac{J_0(xt)}{1+t} \, dt = \frac{1}{x} - \frac{1}{x^3} + E_3(x),$$

and show that

$$|E_3(x)| \leq \frac{15}{8} \pi B_3 x^{-7/2},$$

where B_3 is given in (3.32).

13. Prove that

$$\int_0^\infty \frac{J_0(xt)}{\sqrt{t}(1+t)} \, dt = \frac{\Gamma^2(\frac{1}{4})}{2\pi\sqrt{x}} - \frac{2\pi}{\Gamma^2(\frac{1}{4}) x^{3/2}} + E_2(x),$$

where $|E_2(x)| \leq 1.00138 x^{-2}$. Compare it with (6.35)–(6.36).

14. Show that if the growth condition (4.6) is weakened to

$$|g(t)| \le M|t|^n e^{\xi \operatorname{Re} t + \eta |\operatorname{Im} t|}$$

for all large $|t|$, where $\eta \ge 0$, $\xi < 1$ and n is a positive integer, then the order estimate (4.8) is given by $I_g(x) = O(x^{2n} e^{-x^2/4\alpha})$.

15. Explain why the method given in §4 cannot be extended to the more general integral

$$\int_0^\infty e^{-t^2} J_\nu(xt) g(t^2) t \, dt,$$

unless ν is an even integer.

16. Restate Theorem 5 in terms of the poles of $g(t^2)$ in the lower half plane, and prove it.

17. Evaluate the integrals

$$\int_0^\infty e^{-t^2} J_0(xt) t^{2n+1} \begin{Bmatrix} \sin at^2 \\ \cos at^2 \end{Bmatrix} dt, \qquad n = 0, 1, 2, \ldots.$$

18. Show that the function

$$y(\varepsilon) = \int_0^\infty e^{-\varepsilon x^2} J_0(wx) \frac{x}{1+x^2} \, dx$$

satisfies a first order linear nonhomogeneous ordinary differential equation. Solve this equation and recover the result in (4.21) by setting $\varepsilon = 1$.

19. Let $p \ge 1$, $\alpha > 0$, and $c \ne 0$ be real. Define

$$E_1^*(t) = \int_t^\infty \exp\{ic\tau^p\} \tau^{-\alpha} \, d\tau.$$

Show that as $t \to \infty$,

$$E_1^*(t) \sim \exp\{ict^p\} \sum_{k=1}^\infty c_k t^{1-\alpha-kp},$$

where $c_1 = 1$ and $c_k = (i/cp)^k (1-\alpha-p) \cdots (1-\alpha-[k-1]p)$, $k > 1$. Also show that the iterated integral

$$E_j^*(t) = \int_t^\infty \int_{t_j}^\infty \cdots \int_{t_2}^\infty \exp\{ict_1^p\} t_1^{-\alpha} \, dt_1 \ldots dt_j$$

has an asymptotic expansion of the form

$$E_j^*(t) \sim \exp\{ict^p\} \sum_{k=j}^{\infty} c_{k,j} t^{j-\alpha-kp},$$

as $t \to \infty$, for some constants $c_{k,j}$. Use this result to prove (5.9), and show that the sequence $\{\mu_{s,j}\}$ of exponents in (5.9) can be arranged so that it increases monotonically with s and tends to infinity as $s \to +\infty$.

20. The Weber parabolic cylinder function $U(a, x)$ has the integral representation

$$U(a, t) = \frac{\exp(-\frac{1}{4}t^2)}{\Gamma(a + \frac{1}{2})} \int_0^{\infty} \exp(-tx - \frac{1}{2}x^2) x^{a-1/2} \, dx,$$

$a > -\frac{1}{2}$; see Olver (1974a, p. 208). Put $h(t) = U(a, e^{\pi i/4}t)$, and show that

$$|h(t)| \le \frac{\sqrt{\pi} 2^{-(2a+1)/4}}{\Gamma[(2a+3)/4]} e^{\mathrm{Im}(t^2)/4}$$

for $-\pi/2 \le \arg t \le 0$. Let $h^{(-n)}(t)$ denote the n-th iterated integral of $h(t)$; cf. (5.8). Show that $h^{(-n)}(t)$ can be written as

$$h^{(-n)}(t) = \frac{(-1)^n}{(n-1)!} \int_t^{t+\infty e^{-i\pi/4}} (\tau - t)^{n-1} U(a, e^{\pi i/4}\tau) \, d\tau,$$

for $t > 0$, where the path of integration is taken to be the ray $\tau = t + \rho e^{-i\pi/4}$. Impose appropriate conditions on $f(t)$, similar to those given in Theorem 6, so that an asymptotic expansion of the integral

$$I(x) = \int_0^{\infty} f(t) U(a, xe^{\pi i/4}t) \, dt$$

can be constructed, complete with error bounds.

21. Derive an asymptotic expansion for the integral $I(x) = \int_0^{\infty} f(t) Y_\nu(xt) \, dt$, where $Y_\nu(t)$ is the Bessel function of the second kind and $f(t)$ satisfies (5.2). Give precise conditions for which the resulting expansion is valid.

22. Let $f(t) = 1/(t + a)$. Show that

$$f[t_1, t_1, \ldots, t_n, t_n, \xi_1] = \frac{1}{[(a + t_1) \cdots (a + t_n)]^2 (a + \xi_1)}.$$

Use this result to prove (6.17).

23. Consider

$$I(x) = \int_0^\infty \frac{J_0(xt)}{1+t}\,dt.$$

Derive the quadrature formula

$$I(x) = \frac{1}{\pi}\sum_{k=1}^{n} w_k \frac{2x}{x^2+t_k^2} + \delta_n(x),$$

where the abscissas t_k and weights w_k associate with the weight function $K_0(t)$; cf. (6.23) and (6.24). Show that the remainder satisfies the estimate

$$|\delta_n(x)| \le \frac{2}{\pi A_n^2}\frac{1}{x(x^2+t_1^2)\cdots(x^2+t_n^2)},$$

where A_n is the leading coefficient of the n-th polynomial in the orthonormal sequence associated with the weight function $K_0(t)$. Use Table 4.3 to calculate $I(10)$ accurate to seven decimal places.

24. Derive the quadrature formula in (6.8) by rotating the path of integration in (6.2) through an angle of $\pi/2$, and state precise conditions for the validity of the result. Show that this result does not apply to the integral in Example 5.

n	Coefficients	Abscissas	Weights
1	1.0346310752	0.6366197724	1.5707963268
2	0.5839852962	0.3672186882	1.3999512373
		2.8441656971	0.1708450895
3	0.2108566158	0.2609612883	1.2294421665
		1.8802425952	0.3313894656
		5.6269259843	0.00996494674
4	0.05596246029	0.2034678616	1.0948309833
		1.4202585051	0.4377391923
		4.0139876796	0.0377769612
		8.6778718603	0.0004491900176
5	0.01174097062	0.1672481118	0.9888810597
		1.1456294188	0.5037500094
		3.1628739845	0.07510923965
		6.4939278147	0.003038578952
		11.8874874319	0.00001743909733

Table 4.3

25. Use the integral representation (Erdélyi *et al.* 1954, (4.5.29))

$$\int_0^\infty y^{\nu-1} \exp\left(-py - \frac{a}{4y}\right) dy = 2\left(\frac{a}{4p}\right)^{\nu/2} K_\nu(\sqrt{ap}),$$

where Re $a > 0$ and Re $p > 0$, to show that

$$\lim_{\eta \to 0^+} \frac{1}{2\pi} \int_{-\infty}^\infty (iy)^{\nu-1} \exp\left\{-\eta|y| + ity + i\frac{\mu}{y}\right\} dy = \left(\frac{\mu}{t}\right)^{\nu/2} J_{-\nu}(2\sqrt{t\mu})$$

for any positive number ν.

26. Let $q(x, \mu)$ be continuous in any closed rectangle $[a, b] \times [c, d]$, where $0 < a < b < \infty$ and $-\infty < c < d < \infty$, and suppose that the integral $F(t, \mu) = \int_0^\infty q(x, \mu)e^{itx} dx$ converges uniformly at $x = 0$ and $x = \infty$ for μ in $[c, d]$ and for all sufficiently large t. Show that $F(t, \mu) = o(1)$, as $t \to \infty$, uniformly for μ in $[c, d]$.

27. Let m be a positive integer, G be an m times continuously differentiable function on $(0, \infty)$, and

$$G(y) = \sum_{s=0}^{n-1} a_s y^{s+\alpha-1} + G_n(y),$$

where α is a positive number and m, n satisfy $2m + 1 - \alpha \le n < 2m + 2 - \alpha$. Assume that $G^{(j)}(y)$ is bounded at $y = \infty$ for $j = 0, 1, \ldots, m - 1$, and that each of the integrals

$$\int_1^\infty \exp\left(ity + i\frac{\mu}{y}\right) G^{(j)}(y) \, dy, \qquad j = 0, 1, \ldots, m,$$

converges uniformly for $t \ge T$ and for $-\infty < c \le \mu \le d < \infty$. Furthermore, assume that for $j = 0, 1, \ldots, m$, $G_n^{(j)}(y) = O(y^{n+\alpha-j-1})$, as $y \to 0^+$.

(a) Show that

$$\int_0^\infty G_n(y)e^{-(\eta-it)y + i\mu/y} \, dy = \frac{1}{(\eta - it)^m} \int_0^\infty [G_n(y)e^{i\mu/y}]^{(m)} e^{-(\eta-it)y} \, dy.$$

(b) Use Lemma 1 to prove that each of the limits

$$\lim_{\eta \to 0^+} \int_0^\infty G_n^{(m-k)}(y) y^{-k-l} e^{-(\eta-it)y + i\mu/y} \, dy, \qquad k = 0, 1, \ldots, m,$$

exists, where $l = 0$ when $k = 0$ and $0 \le l \le k$ when $k \ge 1$. (Hint: First consider the two extremal cases $k = 0$ and $l = k = m$.)

(c)   Use (b) and the identity

$$\frac{d^k}{dy^k} e^{a/y} = \frac{(-1)^k}{y^k} e^{a/y} \sum_{j=0}^{k-1} \frac{\Gamma(k)}{\Gamma(k-j)} \binom{k}{j}\left(\frac{a}{y}\right)^{k-j},$$

where k is a positive integer and a is any complex number, to show that there exists a function $B_{mn}(t, \mu)$, which is continuous for $t \ge T$ and $c \le \mu \le d$, such that

$$\lim_{\eta \to 0^+} \int_0^\infty G_n(y) e^{-(\eta - it)y + i\mu/y}\, dy = \left(\frac{i}{t}\right)^m B_{mn}(t, \mu).$$

(d)   By using Ex. 26, prove that $B_{mn}(t, \mu) = o(1)$, as $t \to \infty$, uniformly for $\mu \in [c, d]$.

28.   (Proof of Theorem 9) Split the sum in (7.5) into two and write

$$F(z) = \sum_{s=0}^{<1-\alpha} a_s z^{s+\alpha-1} + F_\alpha(z)$$

with

$$F_\alpha(z) = \sum_{s \ge 1-\alpha}^{n-1} a_s z^{s+\alpha-1} + F_n(z).$$

Show that (7.4) can be expressed as

$$I(t; \mu) = \sum_{s=0}^{<1-\alpha} a_s J_{-s-\alpha}(2\sqrt{t\mu})\left(\frac{\mu}{t}\right)^{(s+\alpha)/2} + I_\alpha(t; \mu),$$

where

$$I_\alpha(t; \mu) = \frac{1}{2\pi i} \int_{c-i\infty}^{c+i\infty} F_\alpha(z) \exp\left(tz - \frac{\mu}{z}\right) dz.$$

Furthermore, show that the vertical line of integration $\operatorname{Re} z = c$ can be shifted to the imaginary axis, i.e.,

$$I_\alpha(t; \mu) = \frac{1}{2\pi} \int_{-\infty}^{\infty} F_\alpha(iy) \exp\left\{ity + i\frac{\mu}{y}\right\} dy.$$

Finally, use Ex. 27 to prove the result stated in Theorem 9.

Supplementary Notes

§1. Prior to Olver's paper in 1974, the idea described in this section
     had already been used by Lighthill (1958). However, his discussion
     is in the context of generalized functions; see, in particular,
     Theorem 19 on p. 52 of his book and the comments at the end of
     that page.

§2. The summability method in Olver's paper (1974b) was actually
     used to construct error bounds for the stationary phase approxi-
     mation. Since the integral considered in the stationary phase
     principle can be transformed into a Fourier integral (see Chapter
     II, §3), the essence of his method is all contained here.

§3. The results in this section are taken from Wong (1976). Condition
     (Q_3) of that paper is insufficient for the validity of Theorem 2, and
     should be replaced by condition (H_3) given in the present section.
     For a discussion of the two conditions (Q_3) and (H_3), see Soni and
     Soni (1985). Alternative derivations of the expansion (3.4) with
     remainder (3.5) have been given by Soni (1982), and Soni and Soni
     (1981, 1985).

§4. Numerical evaluation of the integral in (4.1) has been carried out
     by Gabutti (1979), and Gabutti and Minetti (1981). Theorems 4 and
     5 are taken from Frenzen and Wong (1985). An entirely different
     approach to the problem considered in this section has been given
     by Gabutti (1985). Extensions of the idea used in Example 2 can be
     found in Gabutti and Lepora (1987).

§6. The material in this section is taken from Wong (1982).

Exercises.   Exercises 5 and 6 are related to Example 2 in Soni and
Soni (1985). The results in Exs. 12 and 13 are given in Wong (1976). For
Ex. 20, see Wong (1980b). The problem in Ex. 23 is taken from Wong
(1982). Solutions to Exs. 25–28 can be found in Osborn and Wong (1983).

V

Elementary Theory of Distributions

1. Introduction

The theory of distributions has its origins in problems of mathematical physics, and for many years it has been known that the development of this theory is closely connected with the theory of partial differential equations. But, as frequently happens, it has gone far beyond the immediate applications and has become an independent discipline. In recent years, this theory has also been introduced into the field of asymptotic expansion of integrals. One of the advantages of this approach is that it enables one to interpret and assign values to some divergent integrals which cannot be made meaningful by the more classical methods.

The purpose of this chapter is to give a condensed exposition of those definitions and results of distribution theory which will be required in the later chapters. Most results are stated without proof, but the proofs of many of them are implicit in the description of the theory. For the complete proofs, as well as for complementary information, we refer the reader to the book of Gelfand and Shilov (1964). Our definitions of convolutions of distributions and surface distributions, however, differ from those given in that book. They are taken from Jones (1966) and de Jager (1970), respectively.

Before going into details, let us first indicate the basic idea behind the theory of distributions. Let f be a fixed function on the real line $\mathbb{R}$, integrable on every finite interval, and let φ be any continuous function on $\mathbb{R}$ which vanishes outside some finite interval. To each φ let us assign the number

$$\langle f, \varphi \rangle = \int_{-\infty}^{\infty} f(x)\varphi(x)\, dx. \tag{1.1}$$

Thus, f can be regarded as a linear functional defined on some space $\mathscr{C}_c$ of continuous functions vanishing outside some finite interval. However, there are many other linear functionals on $\mathscr{C}_c$ besides functionals of the form (1.1). For example, by assigning each function φ its value at the point $x = 0$, we get a linear functional which cannot be represented in the form (1.1); see Example 3 in §2. In this sense, the function f can be regarded as part of a much larger set, namely the set of all possible linear functionals on $\mathscr{C}_c$. The space $\mathscr{C}_c$ of so-called test functions φ can be chosen in various ways. However, as we shall see in the following sections, it makes sense to require the test functions to satisfy a more stringent smoothness condition than just being continuous.

2.   Test Functions and Distributions

Turning now to details, let I be an open interval (finite or infinite) in the real line $\mathbb{R}$, and let φ be a function defined on I. By the support of φ we mean the closure of the set $\{x : \varphi(x) \neq 0\}$. The support is thus the smallest relatively closed subset of I outside which φ vanishes. Let $\mathscr{D}(I)$ be the set of all C^∞ (i.e., infinitely differentiable) functions φ with compact support in I. Clearly, $\mathscr{D}(I)$ is a linear space under the usual operations of addition of functions and multiplication of functions by constants.

Definition 1.   A sequence $\{\varphi_n\}$ of functions in $\mathscr{D}(I)$ is said to *converge* to a function φ *in* $\mathscr{D}(I)$ if (i) the support of every φ_n is contained in a fixed compact set $K \subset I$, and (ii) the sequence $\{\varphi_n^{(k)}\}$ converges uniformly on K to $\varphi^{(k)}$ for every $k = 0, 1, 2, \ldots$.

Example 1. A classical example of a function in $\mathscr{D}(\mathbb{R})$ is given by

$$\varphi_\varepsilon(x) = \begin{cases} \exp\left(-\dfrac{\varepsilon^2}{\varepsilon^2 - x^2}\right) & \text{for } |x| < \varepsilon, \\ 0 & \text{for } |x| \geq \varepsilon. \end{cases} \tag{2.1}$$

The sequence $\{(1/n)\varphi_\varepsilon(x): n = 1, 2, \ldots\}$ converges to zero in $\mathscr{D}(\mathbb{R})$, but the sequence $\{(1/n)\varphi_\varepsilon(x/n): n = 1, 2, \ldots\}$ does not converge to zero in $\mathscr{D}(\mathbb{R})$, since the support of $\varphi_\varepsilon(x/n)$ is the interval $[-n\varepsilon, n\varepsilon]$ and hence there is no common support outside which all these functions vanish.

Definition 2. The linear space $\mathscr{D}(I)$ equipped with the above notion of convergence is called the *test function space*, and the functions φ in $\mathscr{D}(I)$ are called the *test functions*.

Definition 3. A *distribution* Λ *in* I is a linear functional on $\mathscr{D}(I)$ which is continuous in the sense that if $\varphi_n \to \varphi$ in $\mathscr{D}(I)$ then

$$\lim_{n \to \infty} \langle \Lambda, \varphi_n \rangle = \langle \Lambda, \varphi \rangle \tag{2.2}$$

where $\langle \Lambda, \varphi \rangle$ denotes the action of the functional Λ on a test function φ. The space of all distributions in I is denoted by $\mathscr{D}'(I)$.

Example 2. Let $f(x)$ be a locally integrable function on I, i.e., a function which is absolutely integrable on every compact $K \subset I$. Then $f(x)$ generates a distribution Λ_f in I via the expression

$$\langle \Lambda_f, \varphi \rangle = \langle f, \varphi \rangle = \int_I f(x)\varphi(x) \, dx. \tag{2.3}$$

Distributions of this type are called *regular*, and all other distributions are called *singular*.

Example 3. The functional δ defined by

$$\langle \delta, \varphi \rangle = \varphi(0) \tag{2.4}$$

is clearly a distribution in $\mathbb{R}$ in the sense of Definition 3. To see that the distribution δ is singular, let us assume that there exists some locally integrable function $f(x)$ such that for every $\varphi \in \mathscr{D}(\mathbb{R})$ we have

$$\int_{-\infty}^{\infty} f(x)\varphi(x) \, dx = \varphi(0). \tag{2.5}$$

Then, in particular, we have

$$\int_{-\infty}^{\infty} f(x)\varphi_\varepsilon(x)\, dx = \varphi_\varepsilon(0) = e^{-1}, \tag{2.6}$$

where $\varphi_\varepsilon(x)$ is the function given in (2.1). As $\varepsilon \to 0$, the integral on the left tends to zero by the Lebesgue dominated convergence theorem, which contradicts (2.6).

Example 4. Generalizing (2.4), we define the functional δ_a by

$$\langle \delta_a, \varphi \rangle = \varphi(a). \tag{2.7}$$

Clearly δ_a is a distribution in I for each given $a \in I$.

Remark. It is customary to write $\delta(x)$ for δ and $\delta(x - a)$ for δ_a and to call $\delta(x)$ the Dirac delta function and $\delta(x - a)$ the translated delta function. However, this should not be interpreted as implying that $\delta(x - a)$ is an ordinary function. It is also frequent, especially in physics books, to write (2.7) in the form

$$\int_{-\infty}^{\infty} \delta(x - a)\varphi(x)\, dx = \varphi(a), \tag{2.8}$$

if $I = \mathbb{R}$ and $a \in I$. But it must be remembered that the integral on the left is only a symbolic representation of the value of the functional δ_a. There is another way of interpreting (2.8). That is to regard it as a representation for

$$\lim_{n \to \infty} \sqrt{\frac{n}{\pi}} \int_{-\infty}^{\infty} e^{-n(x - a)^2}\varphi(x)\, dx = \varphi(a). \tag{2.9}$$

This identity can be shown to hold not only for $\varphi \in \mathscr{D}(\mathbb{R})$ but also for a bounded continuous function φ on $\mathbb{R}$.

3. Support of Distributions

A distribution Λ in I is said to be *zero* in an open set $\Omega \subseteq I$ if $\langle \Lambda, \varphi \rangle = 0$ for all $\varphi \in \mathscr{D}(I)$ whose support is contained in Ω. Thus, for example, the regular distribution Λ_f associated with an ordinary function $f(x)$ is zero in an open set Ω if $f(x)$ itself is zero (almost everywhere) in Ω.

The singular distribution δ is zero in $(-\infty, 0)$ and $(0, \infty)$. In fact, it is for this reason and equation (2.8) or (2.9), with $\varphi(x) \equiv 1$, that physicists continue to use the definition

$$\begin{cases} \delta(x) = 0 & \text{for } x \neq 0 \\ \displaystyle\int_{-\infty}^{\infty} \delta(x)\, dx = 1 \end{cases}$$

for the delta function, although the two conditions are inconsistent in the classical sense of a function and an integral.

Definition 4. If $\Lambda \in \mathscr{D}'(I)$, then the *support* of Λ is the complement of the union of all open subsets of I in which Λ is zero.

Let Λ be a distribution in I, and let Ω be the union of all open sets $\Omega_\alpha \subset I$ such that Λ is zero in Ω_α. We shall show, in the following, that Λ is also zero in Ω. Since Ω contains all open subsets of I in which Λ is zero, the support of Λ is hence the complement of the largest open subset of I in which Λ is zero.

Remark. The support of the regular distribution associated with a continuous (or piecewise continuous) function $f(x)$ is the closure of the set on which $f(x) \neq 0$, i.e., the support of $f(x)$. The support of the distribution δ_a is the single point a.

Two distributions Λ_1 and Λ_2 in I are said to be *equal in an open set* $\Omega \subset I$ if $\langle \Lambda_1, \varphi \rangle = \langle \Lambda_2, \varphi \rangle$ for all $\varphi \in \mathscr{D}(I)$ whose support is contained in Ω.

Lemma 1. *For any $a < b$, there exists a C^∞-function $\rho(x)$ in $[a, b]$ such that* (i) $\rho(a) = 1$ *and* $\rho(b) = 0$, *and* (ii) $\rho^{(k)}(a) = \rho^{(k)}(b) = 0$ *for* $k = 1$, $2, \ldots$.

Proof. Put

$$v(x) = \exp\left\{ -\frac{1}{x-a} - \frac{1}{b-x} \right\}, \qquad a \leq x \leq b,$$

and define

$$\rho(x) = \frac{\int_x^b v(t)\, dt}{\int_a^b v(t)\, dt}.$$

Since the function

$$f(x) = \begin{cases} e^{-1/x} & \text{if } x > 0 \\ 0 & \text{if } x \leq 0 \end{cases}$$

is C^∞ and satisfies $f^{(k)}(0) = 0$ for all $k \geq 1$, the function $\rho(x)$ clearly meets all the required conditions. ∎

The function $\rho(x)$ was called a *neutralizer* in asymptotics by van der Corput (1948).

An *open covering* of a set E is a collection of open sets $\{\Omega_i\}$, where i ranges in an index set A, such that each point $x \in E$ is contained in at least one Ω_i.

Theorem 1.   *Let* $\{\Omega_1, \ldots, \Omega_s\}$ *be a finite open covering of a compact set* K. *Then there exist* C^∞*-functions* $\psi_1, \ldots, \psi_s$ *such that* (i) $0 \leq \psi_i \leq 1$ *for* $1 \leq i \leq s$, (ii) *each* ψ_i *has its support in* Ω_i, *and* (iii) $\sum_{i=1}^s \psi_i(x) = 1$ *for all* x *in* K.

Proof.   To each $x \in K$ we assign an index $i(x)$ so that $x \in \Omega_{i(x)}$. Then there are open intervals $I(x)$ and $J(x)$, centered at x, with

$$\overline{I(x)} \subset J(x) \subset \overline{J(x)} \subset \Omega_{i(x)}.$$

Since K is compact, there exist $x_1, \ldots, x_s$ in K such that

$$K \subset I(x_1) \cup \cdots \cup I(x_s).$$

By Lemma 1, there are C^∞-functions $\varphi_1, \ldots, \varphi_s$ such that $\varphi_i(x) = 1$ on $I(x_i)$, $\varphi_i(x) = 0$ outside $J(x_i)$, and $0 \leq \varphi_i(x) \leq 1$ on $\mathbb{R}$. Set

$$\psi_1 = \varphi_1; \qquad \psi_i = \varphi_i(1 - \varphi_1) \cdots (1 - \varphi_{i-1}), \qquad i = 2, \ldots, s.$$

Then we have

$$\sum_{i=1}^s \psi_i = 1 - (1 - \varphi_1) \cdots (1 - \varphi_s),$$

so that all the statements in the theorem are obviously valid. ∎

The set $\{\psi_i\}$ is called a *partition of unity*, or more accurately a *partition of unity subordinate to the covering* $\{\Omega_i\}$. The above result is used not only in the following theorem, but also in isolating critical points and singularities in the asymptotic evaluation of integrals to be discussed in later chapters.

Theorem 2. *Let $\Lambda \in \mathscr{D}'(I)$, Ω be an open subset of I, and $\{\Omega_i\}$ be an open covering of Ω. If Λ vanishes in each Ω_i, then Λ vanishes in Ω.*

Proof. Let $\varphi \in \mathscr{D}(I)$, K be the support of φ, and let $K \subset \Omega$. Since $\varphi \in \mathscr{D}(I)$, K is compact. Also $\{\Omega_i\}$ is an open covering of Ω, hence of K. Since K is compact, there is a finite subcovering $\{\Omega_{i_1}, \ldots, \Omega_{i_s}\}$ of K. By Theorem 1, there exists a partition of unity $\psi_1, \ldots, \psi_s$ subordinate to $\{\Omega_{i_1}, \ldots, \Omega_{i_s}\}$. Thus

$$\langle \Lambda, \varphi \rangle = \left\langle \Lambda, \sum_{k=1}^{s} \psi_k \varphi \right\rangle = \sum_{k=1}^{s} \langle \Lambda, \psi_k \varphi \rangle,$$

where $\psi_k \varphi$ has support in Ω_{i_k}. Since $\langle \Lambda, \psi_k \varphi \rangle = 0$ for $k = 1, \ldots, s$, we have $\langle \Lambda, \varphi \rangle = 0$. This proves the theorem. ∎

Corollary. *If $\Lambda \in \mathscr{D}'(I)$ vanishes in each open subset $\Omega \subset I$, then $\Lambda = 0$.*

4. Operations on Distributions

Addition of distributions and multiplication of a distribution by a number c are defined in the same way as for linear functionals in general. Thus, if Λ_1 and Λ_2 are two distributions in $\mathscr{D}'(I)$ and c is a number then $\Lambda_1 + \Lambda_2$ and $c\Lambda_1$ are defined by

$$\langle \Lambda_1 + \Lambda_2, \varphi \rangle = \langle \Lambda_1, \varphi \rangle + \langle \Lambda_2, \varphi \rangle, \tag{4.1}$$

$$\langle c\Lambda_1, \varphi \rangle = c\langle \Lambda_1, \varphi \rangle = \langle \Lambda_1, c\varphi \rangle, \tag{4.2}$$

for all $\varphi \in \mathscr{D}(I)$. Clearly these functionals are continuous and linear.

Definition 5. A sequence of distributions $\{\Lambda_n\}$ in $\mathscr{D}'(I)$ is said to *converge* to a distribution $\Lambda \in \mathscr{D}'(I)$ if

$$\lim_{n \to \infty} \langle \Lambda_n, \varphi \rangle = \langle \Lambda, \varphi \rangle \tag{4.3}$$

for every test function $\varphi \in \mathscr{D}(I)$. We call Λ the *distributional* or the *weak limit* of the sequence $\{\Lambda_n\}$.

It is easy to show that the operation of passing to a limit is linear, i.e., if $\Lambda_1 = \lim_{n \to \infty} \Lambda_{n,1}$ and $\Lambda_2 = \lim_{n \to \infty} \Lambda_{n,2}$ and c is a constant, then $\lim_{n \to \infty}(\Lambda_{n,1} + \Lambda_{n,2}) = \Lambda_1 + \Lambda_2$ and $\lim_{n \to \infty} c\Lambda_{n,1} = c\Lambda_1$. It is also easy to verify that if a sequence of locally integrable functions $\{f_n\}$ in I converges uniformly to a locally integrable function f in every compact subset of I, then the corresponding functionals Λf_n converge to the regular functional Λ_f.

However, a sequence of regular functionals may converge to a singular functional. For instance, the regular functionals corresponding to the ordinary functions

$$f_n(x) = \sqrt{\frac{n}{\pi}}\, e^{-nx^2}$$

converge to the singular functional δ; see Eq. (2.9). In fact, it can be shown in general that every singular functional is the limit of a sequence of regular functionals (Rudin 1973, p. 158).

An important property of the space $\mathscr{D}'(I)$ is its completeness with respect to the convergence defined above. That is, if a sequence of distributions $\Lambda_n \in \mathscr{D}'(I)$ is such that for every $\varphi \in \mathscr{D}(I)$ the sequence of numbers $\langle \Lambda_n, \varphi \rangle$ has a limit, then $\{\Lambda_n\}$ converges to some $\Lambda \in \mathscr{D}'(I)$. An elementary proof of this result can be found in Gelfand and Shilov (1964, p. 368).

Let $\Lambda \in \mathscr{D}'(I)$ and $\alpha(x)$ be a C^∞-function on I. The *product* $\alpha\Lambda$ is defined by

$$\langle \alpha\Lambda, \varphi \rangle = \langle \Lambda, \alpha\varphi \rangle, \qquad \varphi \in \mathscr{D}(I). \tag{4.4}$$

The right side of this equation makes sense because $\alpha\varphi \in \mathscr{D}(I)$ when $\varphi \in \mathscr{D}(I)$. Clearly, $\alpha\Lambda$ is linear. It is also continuous, since if $\varphi_n \to \varphi$ then $\alpha\varphi_n \to \alpha\varphi$ and

$$\langle \alpha\Lambda, \varphi_n \rangle = \langle \Lambda, \alpha\varphi_n \rangle \to \langle \Lambda, \alpha\varphi \rangle.$$

Therefore, $\alpha\Lambda$ is a distribution in I.

If Λ_f is a regular distribution corresponding to a locally integrable function f in I, then (4.4) implies

$$\langle \alpha\Lambda_f, \varphi \rangle = \langle \Lambda_f, \alpha\varphi \rangle = \int_I f(x)\alpha(x)\varphi(x)\, dx$$

$$= \int_I \alpha(x)f(x)\varphi(x)\, dx = \langle \Lambda_{\alpha f}, \varphi \rangle,$$

thus showing $\alpha\Lambda_f = \Lambda_{\alpha f}$.

We shall often denote a distribution by the symbol f, as if a representation of the form

$$\langle f, \varphi \rangle = \int_I f(x)\varphi(x)\, dx \tag{4.5}$$

exists, even in the case where the distribution is singular; cf. Eq. (2.8). This motivates us to define the translation of a distribution f in $\mathbb{R}$ through a number h by writing

$$\langle f(x - h), \varphi \rangle = \langle f, \varphi(x + h) \rangle, \qquad \varphi \in \mathscr{D}(\mathbb{R}). \tag{4.6}$$

5. Differentiation of Distributions

Let $f(x)$ be a continuously differentiable function on I, and let $\varphi \in \mathscr{D}(I)$. An integration by parts gives

$$\int_I f'(x)\varphi(x)\, dx = -\int_I f(x)\varphi'(x)\, dx, \tag{5.1}$$

since φ vanishes outside a compact subset of I. Observe that the integral on the right side of (5.1) makes sense whether f is differentiable or not and that it defines a linear functional on $\mathscr{D}(I)$. This suggests that we should define the derivative of distributions as follows.

Definition 6.   Let $\Lambda \in \mathscr{D}'(I)$. The *derivative* Λ' of Λ is the functional defined by the formula

$$\langle \Lambda', \varphi \rangle = -\langle \Lambda, \varphi' \rangle, \qquad \varphi \in \mathscr{D}(I). \tag{5.2}$$

Clearly the functional (5.2) is linear and continuous, and hence is itself a distribution. Higher order derivatives are defined in the same way. Thus

$$\langle \Lambda^{(k)}, \varphi \rangle = (-1)^k \langle \Lambda, \varphi^{(k)} \rangle, \qquad k = 1, 2, \ldots . \tag{5.3}$$

From Definition 6, it follows immediately that every distribution has derivatives of all orders. Furthermore, if a sequence of distributions $\{\Lambda_n\}$ converges to a distribution Λ, then the sequence of derivatives $\{\Lambda_n'\}$ converges to the derivative Λ' of the limit.

Equation (5.1) shows that when $f(x)$ is a continuously differentiable function in I, i.e., $f \in C^1(I)$, then the derivative Λ_f' of the distribution Λ_f

corresponding to the function $f(x)$ coincides with the distribution $\Lambda_{f'}$ associated with the ordinary derivative $f'(x)$, i.e.,

$$\langle \Lambda'_f, \varphi \rangle = \langle \Lambda_{f'}, \varphi \rangle. \tag{5.4}$$

In fact, this is true even under the weaker conditions that $f(x)$ is continuous and $f'(x)$ is merely piecewise continuous, since Eq. (5.1) remains valid in this case.

Now, let $L_{\text{loc}}(I)$ denote the collection of all locally integrable functions on I, and we are concerned with the question: for which $f \in L_{\text{loc}}(I)$ does (5.4) hold? If Λ is a distribution whose domain includes I, then we say $\Lambda \in L_{\text{loc}}(I)$ if there is a function $f \in L_{\text{loc}}(I)$ such that $\langle \Lambda, \varphi \rangle = \langle f, \varphi \rangle$ for all $\varphi \in \mathscr{D}(I)$. For convenience here, the distribution Λ_f generated by a function $f \in L_{\text{loc}}(I)$ has been simply denoted by f, and we shall continue to adopt this simplification hereafter; see Eq. (2.3). If $f \in L_{\text{loc}}(I)$, we shall write Df for the distributional derivative of f to distinguish it from the ordinary derivative f'. With these notations, (5.4) now becomes

$$\langle Df, \varphi \rangle = \langle f', \varphi \rangle, \qquad \varphi \in \mathscr{D}(I). \tag{5.5}$$

This equation, however, does not hold for $f \in L_{\text{loc}}(I)$ in general, even if the obvious necessary conditions $f'(x)$ exists a.e. on I, and $f' \in L_{\text{loc}}(I)$ are satisfied. For instance, the function

$$f(x) = \begin{cases} 0 & \text{if } x \le 0 \\ 1 & \text{if } x > 0 \end{cases} \tag{5.6}$$

clearly belongs to $L_{\text{loc}}(\mathbb{R})$, and $f'(x)$ exists everywhere except at $x = 0$ and also belongs to $L_{\text{loc}}(\mathbb{R})$, but

$$\langle Df, \varphi \rangle = - \int_{-\infty}^{\infty} f(x)\varphi'(x)\, dx = \varphi(0)$$

and

$$\langle f', \varphi \rangle = \int_{-\infty}^{\infty} f'(x)\varphi(x)\, dx = 0.$$

However, we have the following lemma.

A function f is said to be *locally absolutely continuous* in I if it is absolutely continuous in every compact subinterval of I.

Lemma 2. *Suppose $f \in L_{\mathrm{loc}}(I)$. Then $Df \in L_{\mathrm{loc}}(I)$ if and only if f is (equal almost everywhere to a) locally absolutely continuous (function) in I, and in that case $Df = f'$.*

Proof. Let (a, b) be any finite interval contained in I, and consider all test functions φ whose supports lie inside (a, b). Suppose $Df \in L_{\mathrm{loc}}(I)$. Then there exists a $g \in L_{\mathrm{loc}}(I)$ such that

$$\langle Df, \varphi \rangle = \langle g, \varphi \rangle = \int_a^b g(x)\varphi(x)\, dx. \qquad (5.7)$$

But by definition

$$\langle Df, \varphi \rangle = -\langle f, \varphi' \rangle = -\int_a^b f(x)\varphi'(x)\, dx. \qquad (5.8)$$

Therefore

$$\int_a^b f(x)\varphi'(x)\, dx = -\int_a^b g(x)\varphi(x)\, dx.$$

The member on the right side of this equation can be written as

$$\int_a^b g(x) \int_x^b \varphi'(t)\, dt\, dx,$$

which, by Fubini's theorem, is in turn equal to

$$\int_a^b \varphi'(t) \int_a^t g(x)\, dx\, dt = \int_a^b \varphi'(t)\left[g(a) + \int_a^t g(x)\, dx \right] dt.$$

It now follows that for almost all x in (a, b)

$$f(x) = g(a) + \int_a^x g(t)\, dt. \qquad (5.9)$$

The integral on the right is obviously an absolutely continuous function. Thus $f'(x) = g(x)$ almost everywhere in (a, b), and

$$\int_a^b f'(x)\varphi(x)\, dx = \int_a^b g(x)\varphi(x)\, dx. \qquad (5.10)$$

Coupling (5.7) and (5.10) gives $Df = f'$ in (a, b). Since the interval (a, b) is arbitrary, by Theorem 2, we conclude that $Df = f'$ in I.

To prove the converse statement, we insert (5.9) in (5.8) and apply Fubini's theorem. The result is $Df = g = f'$ in (a, b) and hence also in I. ■

Remark.   It should be mentioned that an everywhere differentiable function with a locally integrable derivative is automatically locally absolutely continuous; see Rudin (1974, Theorem 8.21). The problem with the function (5.6) is that it is not even continuous in any interval around 0.

Example 5.   Consider the delta function $\langle \delta, \varphi \rangle = \varphi(0)$. It follows from (5.3) that the derivatives $\delta^{(s)}$ of the distribution are defined by

$$\langle \delta^{(s)}, \varphi \rangle = (-1)^s \varphi^{(s)}(0), \qquad s = 1, 2, \ldots . \tag{5.11}$$

Similarly, for $s = 0, 1, 2, \ldots,$

$$\langle \delta_a^{(s)}, \varphi \rangle = (-1)^s \varphi^{(s)}(a). \tag{5.12}$$

Example 6.   Consider the Heaviside function

$$H(x) = \begin{cases} 0 & \text{if } x \le 0 \\ 1 & \text{if } x > 0. \end{cases}$$

The distribution associated with $H(x)$ is the functional

$$\langle H, \varphi \rangle = \int_0^\infty \varphi(x)\, dx.$$

The distributional derivative of H is, by Definition 6,

$$\langle DH, \varphi \rangle = -\int_0^\infty \varphi'(x)\, dx = \varphi(0),$$

since φ vanishes at infinity. Therefore, we have

$$\langle DH, \varphi \rangle = \langle \delta, \varphi \rangle \tag{5.13}$$

for all $\varphi \in \mathscr{D}(\mathbb{R})$. Similarly, it is easily shown that

$$DH(x - a) = \delta_a \qquad \text{in } \mathscr{D}(\mathbb{R}). \tag{5.14}$$

Example 7.   Let $f(x)$ be a C^∞-function on $\mathbb{R}$ except at $x = x_0$. Suppose that the right and the left derivatives of all orders of $f(x)$ exist at $x = x_0$. Denote by σ_s the "jump" in the s-th derivative $f^{(s)}(x)$ at $x = x_0$, i.e.,

$$\sigma_s = f^{(s)}(x_0^+) - f^{(s)}(x_0^-). \tag{5.15}$$

We shall obtain the derivative of the regular distribution associated with $f(x)$, or simply Df. By definition,

$$\langle Df, \varphi \rangle = -\langle f, \varphi' \rangle = - \int_{-\infty}^{\infty} f(x)\varphi'(x)\, dx.$$

An integration by parts gives

$$- \int_{x_0}^{\infty} f(x)\varphi'(x)\, dx = f(x_0^+)\varphi(x_0) + \int_{x_0}^{\infty} f'(x)\varphi(x)\, dx$$

and

$$- \int_{-\infty}^{x_0} f(x)\varphi'(x)\, dx = -f(x_0^-)\varphi(x_0) + \int_{-\infty}^{x_0} f'(x)\varphi(x)\, dx.$$

Adding the last two identities, we have

$$\langle Df, \varphi \rangle = \sigma_0 \varphi(x_0) + \int_{-\infty}^{\infty} f'(x)\varphi(x)\, dx,$$

i.e.,

$$Df = f' + \sigma_0 \delta_{x_0}. \tag{5.16}$$

Successive differentiation in the sense of distributions, using (5.16) at each stage, gives

$$D^m f = f^{(m)} + \sigma_0 \delta_{x_0}^{(m-1)} + \sigma_1 \delta_{x_0}^{(m-2)} + \cdots + \sigma_{m-1} \delta_{x_0}, \qquad m = 1, 2, \ldots . \tag{5.17}$$

Example 8.   Consider the function

$$x_+^{-\alpha} = \begin{cases} 0 & \text{if } x \leq 0 \\ x^{-\alpha} & \text{if } x > 0. \end{cases} \tag{5.18}$$

For $0 < \alpha < 1$, this function is locally integrable and defines a regular distribution on $\mathbb{R}$ given by

$$\langle x_+^{-\alpha}, \varphi \rangle = \int_0^{\infty} x^{-\alpha}\varphi(x)\, dx. \tag{5.19}$$

Note that its ordinary derivative $(-\alpha)x_+^{-\alpha-1}$ is not integrable near $x = 0$ and hence does not define a distribution in the same manner.

However, $x^{-\alpha}$ is locally absolutely continuous in $(0, \infty)$, so $Dx_+^{-\alpha} = -\alpha x^{-\alpha-1}$ in $(0, \infty)$; see Lemma 2. We use this equation to *define* $x_+^{-\alpha-1}$ on $\mathbb{R}$:

$$\langle x_+^{-\alpha-1}, \varphi \rangle = -\frac{1}{\alpha} \langle Dx_+^{-\alpha}, \varphi \rangle = \frac{1}{\alpha} \int_0^\infty x^{-\alpha} \varphi'(x) \, dx.$$

More generally, we define

$$\langle x_+^{-\alpha-s}, \varphi \rangle = \frac{1}{(\alpha)_s} \int_0^\infty x^{-\alpha} \varphi^{(s)}(x) \, dx, \tag{5.20}$$

where

$$(\alpha)_s = \alpha(\alpha+1) \cdots (\alpha+s-1), \qquad s = 1, 2, \ldots . \tag{5.21}$$

Similarly, we regard $(-1)^s s! \, x_+^{-1-s}$ as the $(s+1)$-th distributional derivative of the function

$$\log_+ x = \begin{cases} 0 & \text{if } x \le 0 \\ \log x & x > 0. \end{cases} \tag{5.22}$$

That is, for $s = 0, 1, 2, \ldots$, we define

$$\langle x_+^{-1-s}, \varphi \rangle = \frac{(-1)^s}{s!} \langle D^{s+1} \log_+ x, \varphi \rangle = -\frac{1}{s!} \int_0^\infty (\log x) \varphi^{(s+1)}(x) \, dx. \tag{5.23}$$

In Examples 6 and 8, we have illustrated that some singular distributions may be represented as the derivatives of some regular distributions. While this is not true in general, many distributions encountered in practice have this property. For a general result about representation of distributions with compact support as derivatives, see Rudin (1973, p. 153).

6. Convolutions

Let $L_{\text{loc}}^+(\mathbb{R})$ denote the class of functions which are locally integrable on $\mathbb{R}$ and which vanish on $(-\infty, 0]$. Thus the distributions associated with functions in $L_{\text{loc}}^+(\mathbb{R})$ all have support contained in $[0, \infty)$. From Lemma 2, we have immediately the first part of the following corollary.

Corollary. *Suppose $f \in L_{loc}^+(\mathbb{R})$. Then $Df \in L_{loc}^+(\mathbb{R})$ if and only if f is equal a.e. to a locally absolutely continuous function g in $\mathbb{R}$. In particular, $Df \in L_{loc}^+(\mathbb{R})$ implies $g(0) = 0$.*

The second assertion follows from the fact that $Df \in L_{loc}^+(\mathbb{R})$ implies

$$f(x) = g(x) \quad \text{a.e.}$$

$$= \int_a^x g'(t)\, dt, \qquad x \in [a, 0],$$

where $a < 0$ is chosen so that $g(a) = 0$. Since $g'(x) = 0$ almost everywhere on $[a, 0]$, $g(0) = 0$.

If f and g belong to $L_{loc}^+(\mathbb{R})$, then

$$(f * g)(x) = \int_0^x f(x - t)g(t)\, dt \tag{6.1}$$

defines $f * g \in L_{loc}^+(\mathbb{R})$. Clearly, this product, together with the usual addition and scalar multiplication, makes $L_{loc}^+(\mathbb{R})$ a commutative, associative linear algebra over the complex field. Here we also note that $f * g$ is continuous at 0, and at any point of continuity of either f or g.

Equation (6.1) may be regarded as the definition of the convolution of two distributions, each associated with a function in $L_{loc}^+(\mathbb{R})$. We now wish to extend this to a definition of a convolution product in a space of distributions.

Definition 7. Let K^+ denote the space of all distributions of the form $D^n f$, where n is a nonnegative integer, and $f \in L_{loc}^+(\mathbb{R})$. For F and G in K^+, say $F = D^n f$ and $G = D^m g$, we define

$$F * G = D^{n+m}(f * g). \tag{6.2}$$

The obvious consistency question is whether the equation

$$D(f * g) = Df * g \tag{6.3}$$

holds if Df also belongs to $L_{loc}^+(\mathbb{R})$. The answer to the question is given in the following lemma, which also serves as the motivation for Definition 7.

Lemma 3. *If f, g, and Df all belong to $L_{loc}^+(\mathbb{R})$, then $f * g$ is locally absolutely continuous in $\mathbb{R}$, and (6.3) holds.*

Proof.   By the corollary to Lemma 2, there exists a locally absolutely continuous function $h(x)$ in $\mathbb{R}$ such that $f(x) = h(x)$ a.e. in $\mathbb{R}$, and $h(0) = 0$. Therefore

$$f(x) = \int_0^x h'(t)\,dt \quad \text{a.e.} \qquad \text{for } x \geq 0, \tag{6.4}$$

and $f'(x) = h'(x)$ a.e. in $\mathbb{R}$. Equation (6.4) can be written as $f = h' * H$, where H is the Heaviside function defined in Example 6. Thus for $x \geq 0$ we have a.e.

$$f * g = (h' * H) * g = (h' * g) * H \qquad \text{and} \qquad (f * g)(x) = \int_0^x (h' * g)(t)\,dt.$$

This implies that $f * g$ is equal a.e. to a locally absolutely continuous function, and $(f * g)' = h' * g$ a.e. in $[0, \infty)$. By Lemma 2, $D(f * g) = (f * g)'$, and $Df = f' = h'$. This proves the lemma. ∎

The following theorem follows immediately from the definition in (6.2).

Theorem 3. K^+ *is a commutative, associative linear algebra. Thus, for all* F, G, *and* $H \in K^+$, *we have*

$$F * G = G * F \tag{6.5}$$

$$(F * G) * H = F * (G * H) \tag{6.6}$$

$$F * (G + H) = F * G + F * H \tag{6.7}$$

$$(F + G) * H = F * H + G * H. \tag{6.8}$$

The algebra K^+ clearly contains the distributions H, δ, and x_+^λ for any real (or complex) number λ. The last mentioned distribution was discussed in Example 8. Using the fact that $\delta = DH$, we can derive the following equations. For any nonnegative integer n and $F \in K^+$,

$$D^n \delta * F = D^n F. \tag{6.9}$$

In particular,

$$D^n \delta * x_+^\lambda = \lambda(\lambda - 1) \cdots (\lambda - n + 1) x_+^{\lambda - n} \tag{6.10}$$

for all n, if λ is not a positive integer or zero. If λ is a nonnegative integer, then (6.10) remains valid for $n \leq \lambda$ (understood that $x_+^0 = H$), and is replaced by

$$D^n\delta * x_+^\lambda = \lambda! \, D^{n-\lambda}H = \lambda! \, D^{n-\lambda-1}\delta \qquad \text{for } n > \lambda. \qquad (6.11)$$

Using the fact that

$$\int_0^1 (1-x)^\mu x^\lambda \, dx = \frac{\Gamma(\mu+1)\Gamma(\lambda+1)}{\Gamma(\mu+\lambda+2)} \qquad (6.12)$$

whenever $\mu > -1$ and $\lambda > -1$, one can also show that

$$x_+^\mu * x_+^\lambda = \frac{\Gamma(\mu+1)\Gamma(\lambda+1)}{\Gamma(\mu+\lambda+2)} x_+^{\mu+\lambda+1} \qquad (6.13)$$

provided that μ, λ, and $\mu + \lambda$ are not negative integers. If μ and λ are not negative integers, but $\mu + \lambda$ is a negative integer (≤ -2), then (6.13) must be replaced by

$$x_+^\mu * x_+^\lambda = \Gamma(\mu+1)\Gamma(\lambda+1)D^{-\mu-\lambda-1}H. \qquad (6.14)$$

Finally, if $0 < \sigma < 1$, then

$$x_+^{-\sigma} * \log_+ x = \frac{x_+^{1-\sigma}}{1-\sigma}(\log_+ x - \gamma - \psi(2-\sigma)), \qquad (6.15)$$

where $\gamma = 0.57721\ldots$ is the Euler–Mascheroni constant and ψ is the logarithmic derivative of the Gamma function. This leads to the formula

$$x_+^{-n-\sigma} * x_+^{-m} = \frac{(-1)^{m-1}\Gamma(1-\sigma-n)}{(m-1)! \, \Gamma(1-\sigma)}$$
$$\times D^{m+n}\left\{\frac{x_+^{1-\sigma}}{1-\sigma}[\log_+ x - \gamma - \psi(2-\sigma)]\right\}. \qquad (6.16)$$

7. Regularization of Divergent Integrals

Consider a function $f(x)$ that has only one non-integrable singularity at x_0, e.g., $f(x) = 1/x$ on $\mathbb{R}$. In general, the integral

$$\int_{-\infty}^\infty f(x)\varphi(x) \, dx, \qquad (7.1)$$

where $\varphi(x)$ belongs to $\mathscr{D}(\mathbb{R})$, diverges. However, it will converge if $\varphi(x)$ vanishes in a neighborhood of x_0. A distribution Λ which has the property that for all $\varphi \in \mathscr{D}(\mathbb{R})$ vanishing in a neighborhood of x_0, $\langle \Lambda, \varphi \rangle$ has the value given by (7.1) is called a *regularization of the function* $f(x)$, and, when φ does not vanish near x_0, $\langle \Lambda, \varphi \rangle$ is called a *regularization of the integral* (7.1). For example, $\Lambda = D \log |x|$ is a regularization of $1/x$. So is $\Lambda + P(D)\delta$, where P is any polynomial.

In this section we shall construct regularizations of some particularly important functions. Let us begin with the function

$$x_+^\lambda = \begin{cases} 0 & \text{for } x \leq 0 \\ x^\lambda & \text{for } x > 0, \end{cases}$$

where λ is a complex number. For $\operatorname{Re} \lambda > -1$ this function defines the regular distribution

$$\langle x_+^\lambda, \varphi \rangle = \int_0^\infty x^\lambda \varphi(x) \, dx, \qquad \varphi \in \mathscr{D}(\mathbb{R}). \tag{7.2}$$

The right-hand side is obviously an analytic function of λ for $\operatorname{Re} \lambda > -1$. To define a regularization of x_+^λ for $\operatorname{Re} \lambda \leq -1$, we rewrite (7.2) in the form

$$\int_0^1 x^\lambda [\varphi(x) - \varphi(0)] \, dx + \int_1^\infty x^\lambda \varphi(x) \, dx + \frac{\varphi(0)}{\lambda + 1}.$$

The first term here is defined for $\operatorname{Re} \lambda > -2$, the second for all λ, and the third for $\lambda \neq -1$. Thus the functional defined in (7.2) can be analytically continued to $\operatorname{Re} \lambda > -2$, $\lambda \neq -1$. Proceeding similarly, we can continue x_+^λ into the region $\operatorname{Re} \lambda > -n - 1$, $\lambda \neq -1, -2, \ldots, -n$, by using

$$\int_0^\infty x^\lambda \varphi(x) \, dx = \int_0^1 x^\lambda \left[\varphi(x) - \varphi(0) - x\varphi'(0) - \cdots - \frac{x^{n-1}}{(n-1)!} \varphi^{(n-1)}(0) \right] dx$$

$$+ \int_1^\infty x^\lambda \varphi(x) \, dx + \sum_{k=1}^n \frac{\varphi^{(k-1)}(0)}{(k-1)! \, (\lambda + k)}. \tag{7.3}$$

Note that for $\operatorname{Re} \lambda > -n - 1$, $\lambda \neq -1, -2, \ldots, -n$, the right-hand side exists and defines a regularization of the integral on the left. Also, if $-n - 1 < \operatorname{Re} \lambda \leq -n$, $\lambda \neq -n$, and if the test function vanishes in a neighborhood of the origin, what remains on the right is $\int_0^\infty x^\lambda \varphi(x) \, dx$. Furthermore, in any strip of the form $-n - 1 < \operatorname{Re} \lambda < -n$, Eq. (7.3) can be written in the simple form

$$\langle x_+^\lambda, \varphi \rangle = \int_0^\infty x^\lambda \left[\varphi(x) - \varphi(0) - x\varphi'(0) - \cdots - \frac{x^{n-1}}{(n-1)!} \varphi^{(n-1)}(0) \right] dx, \tag{7.4}$$

since for $1 \le k \le n$

$$\int_1^\infty x^{\lambda + k - 1}\, dx = -\frac{1}{\lambda + k}. \tag{7.5}$$

Equation (7.3) shows that when $\langle x_+^\lambda, \varphi \rangle$ is considered as a function of λ, it has simple poles at $\lambda = -1, -2, \ldots$, and that its residue at $\lambda = -k$ is $\varphi^{(k-1)}(0)/(k-1)!$. This may be expressed by saying that the distribution x_+^λ itself has a simple pole at $\lambda = -k$ with residue

$$\frac{(-1)^{k-1}}{(k-1)!}\, \delta^{(k-1)}(x), \qquad k = 1, 2, \ldots .$$

For $\operatorname{Re} \lambda > 0$, we have by integration by parts

$$\langle x_+^\lambda, \varphi' \rangle = -\langle \lambda x_+^{\lambda-1}, \varphi \rangle, \qquad \varphi \in \mathscr{D}(\mathbb{R}).$$

Since both sides of this equation can be analytically continued to the entire λ-plane except for the points $0, -1, -2, \ldots$, by uniqueness of analytic continuation, this equation holds in the whole λ-plane except for these points. Thus $Dx_+^\lambda = \lambda x_+^{\lambda-1}$, in agreement with (5.20) with $s = 1$ and $\alpha = -\lambda$.

Let us now consider the problem of regularizing the function

$$x_{0 \le x \le b}^\lambda = \begin{cases} x^\lambda & \text{for } 0 \le x \le b \\ 0 & \text{otherwise,} \end{cases}$$

or equivalently the finite integral

$$\int_0^b x^\lambda \varphi(x)\, dx, \qquad 0 < b < \infty, \quad \varphi \in \mathscr{D}(\mathbb{R}). \tag{7.6}$$

For $\operatorname{Re} \lambda > -1$, this integral converges and is an analytic function of λ. Furthermore, it can be continued analytically to the entire λ-plane via the identity

$$\int_0^b x^\lambda \varphi(x)\, dx = \int_0^b x^\lambda \left[\varphi(x) - \varphi(0) - \varphi'(0)x - \cdots - \varphi^{(n-1)}(0)\frac{x^{n-1}}{(n-1)!} \right] dx$$
$$+ \varphi(0)\frac{b^{\lambda+1}}{\lambda+1} + \varphi'(0)\frac{b^{\lambda+2}}{\lambda+2} + \cdots + \varphi^{(n-1)}(0)\frac{b^{\lambda+n}}{(n-1)!\,(\lambda+n)}, \tag{7.7}$$

which is valid for all λ with $\operatorname{Re} \lambda > -n - 1$ except for $\lambda = -1$, $-2, \ldots, -n$. Note that n can be chosen arbitrarily large. We shall use

(7.7) to define the distribution which regularizes the integral (7.6) for $\lambda \neq -1, -2, \ldots$. That is, we define

$$\langle x_{0 \leq x \leq b}^{\lambda}, \varphi \rangle = \int_0^b x^{\lambda} \left[\varphi(x) - \varphi(0) - \varphi'(0)x - \cdots - \varphi^{(n-1)}(0) \frac{x^{n-1}}{(n-1)!} \right] dx$$

$$+ \varphi(0) \frac{b^{\lambda+1}}{\lambda+1} + \varphi'(0) \frac{b^{\lambda+2}}{\lambda+2} + \cdots + \varphi^{(n-1)}(0) \frac{b^{\lambda+n}}{(n-1)!(\lambda+n)},$$

$$(7.8)$$

for $\operatorname{Re} \lambda > -n - 1$ and $\lambda \neq -1, -2, \ldots, -n$.

Note that if $\varphi(x)$ vanishes in a neighborhood of $x = 0$, then (7.8) shows that the value $\langle x_{0 \leq x \leq b}^{\lambda}, \varphi \rangle$ is indeed the same as that given by the integral (7.6). If we take $\varphi^*(x)$ to be a test function in $\mathscr{D}(\mathbb{R})$ which equals "1" on the interval $[0, b]$, then we have

$$\langle x_{0 \leq x \leq b}^{\lambda}, \varphi^* \rangle = \frac{b^{\lambda+1}}{\lambda+1}. \tag{7.9}$$

Interpreting the integral $\int_0^b x^{\lambda} \, dx$ as the result of applying the distribution $x_{0 \leq x \leq b}^{\lambda}$ to the test function $\varphi^*(x)$ gives

$$\int_0^b x^{\lambda} \, dx = \frac{b^{\lambda+1}}{\lambda+1}, \tag{7.10}$$

valid for all $\lambda \neq -1, -2, -3, \ldots$. Since the right-hand side has only one singularity at $\lambda = -1$, we use (7.10) to define this integral also for $\lambda = -2, -3, \ldots$.

Having given a meaning to the integral $\int_0^b x^{\lambda} \, dx$, one now naturally wishes to give a meaning to the integral

$$\int_b^{\infty} x^{\lambda} \, dx, \qquad 0 < b < \infty, \tag{7.11}$$

in a similar way. However, there is unfortunately no test function in $\mathscr{D}(\mathbb{R})$ which is equal to "1" on the infinite interval $[b, \infty)$. Therefore, we must proceed in a different manner. Let $\hat{\mathscr{D}}(b, \infty)$ denote the space of all C^{∞}-functions $\psi(x)$ on $[b, \infty)$ such that for $x \in (0, 1/b)$, we have $\psi(1/x) = \varphi(x)$ for some $\varphi \in \mathscr{D}(\mathbb{R})$. Define a functional $x_{b \leq x < \infty}^{\lambda}$ on $\hat{\mathscr{D}}(b, \infty)$ by

$$\langle x_{b \leq x < \infty}^{\lambda}, \psi \rangle = \int_0^{1/b} y^{-\lambda-2} \varphi(y) \, dy, \tag{7.12}$$

where $\psi \in \hat{\mathscr{D}}(b, \infty)$, $\varphi(y) = \psi(1/y)$ and the integral on the right is of course interpreted in the sense of regularization. Equation (7.12) makes sense for all $\lambda \neq -1, 0, 1, \ldots$. We now interpret the integral $\int_b^\infty x^\lambda \, dx$ as the result of the action of $x_{b \leq x < \infty}^\lambda$ on a test function $\psi^*(x)$ such that $\psi^*(1/x) = \varphi^*(x) \equiv 1$ for $0 \leq x \leq 1/b$. Thus

$$\int_b^\infty x^\lambda \, dx = -\frac{b^{\lambda+1}}{\lambda+1} \tag{7.13}$$

for $\lambda \neq -1, 0, 1, \ldots$. As before, we use (7.13) to define $\int_b^\infty x^\lambda \, dx$ for $\lambda = 0$, $1, 2, \ldots$. Therefore, the only singularity is again at $\lambda = -1$.

In view of (7.10) and (7.13), we put

$$\int_0^\infty x^\lambda \, dx = \int_0^b x^\lambda \, dx + \int_b^\infty x^\lambda \, dx = 0. \tag{7.14}$$

The last equality is valid for all λ except for $\lambda = -1$. This result will play an important role in the derivation of the asymptotic expansions of a Mellin-type convolution integral; see Section 7, Chapter VI.

8. Tempered Distributions

There are many possible choices of the test function space other than the space of infinitely differentiable functions with compact supports. A space which very often occurs in applications is $\mathscr{S}$, the space of infinitely differentiable functions that, together with their derivatives, approach zero more rapidly than any power of $1/|x|$ as $|x| \to \infty$. More precisely, a function $\varphi(x)$ belongs to $\mathscr{S}$ or $\mathscr{S}(\mathbb{R})$ if and only if, given any $p, q = 0, 1, 2, \ldots$, there is a constant C_{pq} (depending on p, q, and φ) such that

$$|x^p \varphi^{(q)}(x)| \leq C_{pq}, \qquad -\infty < x < \infty. \tag{8.1}$$

For example, the function $\exp(-x^2)$ belongs to $\mathscr{S}$. The members of $\mathscr{S}$ are called *rapidly decreasing functions*.

A sequence $\{\varphi_n\}$ of functions in $\mathscr{S}$ is said to converge to a function $\varphi \in \mathscr{S}$ if the sequence $\{\varphi_n^{(q)}\}$ converges uniformly to $\varphi^{(q)}$ on every finite interval and if the constants in the inequalities

$$|x^p \varphi_n^{(q)}(x)| \leq C_{pq} \tag{8.2}$$

can be chosen independently of n. For instance, the sequence $\varphi_n(x) = (1/n)\exp(-x^2)$ converges to zero in $\mathscr{S}$ as $n \to \infty$, whereas the sequence $\varphi_n(x) = (1/n)\exp(-x^2/n^2)$ does not. This is due to the fact that the latter sequence does not satisfy (8.2) with C_{pq} independent of n.

Clearly, every function in $\mathscr{D}(\mathbb{R})$ also belongs to $\mathscr{S}$. Furthermore, $\mathscr{D}(\mathbb{R})$ is dense in $\mathscr{S}$. This can be proved as follows. Take an arbitrary C^∞-function $e(x)$ which equals 1 for $|x| \le 1$ and is zero for $|x| \ge 2$. If $\varphi \in \mathscr{S}$, then the functions $\varphi_n(x) = e(x/n)\varphi(x)$, $n = 1, 2, \ldots$, all belong to $\mathscr{D}(\mathbb{R})$, and the sequence $\{\varphi_n\}$ converges to φ in the sense of $\mathscr{S}$; hence $\mathscr{D}(\mathbb{R})$ is dense in $\mathscr{S}$. It is also clear that convergence in $\mathscr{D}(\mathbb{R})$ implies convergence in $\mathscr{S}$, i.e., if $\varphi_n, \varphi \in \mathscr{D}(\mathbb{R})$ and $\varphi_n \to \varphi$ in $\mathscr{D}(\mathbb{R})$ then $\varphi_n \to \varphi$ in $\mathscr{S}$; see Rudin (1973, p. 173, Theorem 7.10).

Definition 8.   A *tempered distribution* is a continuous linear functional Λ on $\mathscr{S}$. That is, $\varphi_n \to \varphi$ in $\mathscr{S}$ implies $\langle \Lambda, \varphi_n \rangle \to \langle \Lambda, \varphi \rangle$. The set of tempered distributions on $\mathscr{S}$ is denoted by $\mathscr{S}'$.

Since $\mathscr{D}$ is continuously embedded in $\mathscr{S}$, every continuous linear functional on $\mathscr{S}$ is also a continuous linear functional on $\mathscr{D}(\mathbb{R})$. However, the converse is not true. For example, the function $f(x) = \exp(x^2)$ gives a continuous linear functional $\langle f, \varphi \rangle$ on $\mathscr{D}(\mathbb{R})$ but not on $\mathscr{S}$. One also readily sees that $\mathscr{S}'$ is a linear space and hence is a linear subspace of $\mathscr{D}'(\mathbb{R})$. Convergence in $\mathscr{S}'$ can be defined in an analogous manner to that in $\mathscr{D}'(\mathbb{R})$.

Definition 9.   A sequence of functionals Λ_n in $\mathscr{S}'$ *converges* to a functional Λ in $\mathscr{S}'$ if for every $\varphi \in \mathscr{S}$ we have

$$\langle \Lambda_n, \varphi \rangle \to \langle \Lambda, \varphi \rangle. \tag{8.3}$$

We call Λ the *distributional* or the *weak limit* of Λ_n.

As before, $\langle \Lambda_n, \varphi \rangle$ converges for all $\varphi \in \mathscr{S}$ implies $\{\Lambda_n\}$ has a weak limit in $\mathscr{S}'$; see the paragraph preceding (4.4).

Since $\mathscr{D}(\mathbb{R}) \subset \mathscr{S}$, it is clear that Λ_n, Λ in $\mathscr{S}'$, and $\Lambda_n \to \Lambda$ in $\mathscr{S}'$ implies $\Lambda_n \to \Lambda$ in $\mathscr{D}'(\mathbb{R})$. Summarizing, we have the following theorem.

Theorem 4. $\mathscr{D}(\mathbb{R}) \subset \mathscr{S}$ *and* $\mathscr{S}' \subset \mathscr{D}'(\mathbb{R})$; *convergence in* $\mathscr{D}(\mathbb{R})$ *implies convergence in* $\mathscr{S}$ *and weak convergence in* $\mathscr{S}'$ *implies weak convergence in* $\mathscr{D}'(\mathbb{R})$.

Example 9. Let $f(x)$ be a locally integrable function with finite algebraic growth at infinity, i.e., $f(x) = O(|x|^\alpha)$, as $|x| \to \infty$, for some $\alpha \geq 0$. In view of the inequality (8.1), we can again associate $f(x)$ with the functional Λ_f or simply f given by

$$\langle \Lambda_f, \varphi \rangle = \langle f, \varphi \rangle = \int_{-\infty}^{\infty} f(x)\varphi(x)\, dx. \tag{8.4}$$

It is easy to show that (8.4) defines a continuous linear functional on $\mathscr{S}$, and hence that Λ_f is a tempered distribution.

Example 10. The delta function $\delta(x)$, defined as

$$\langle \delta(x), \varphi(x) \rangle = \varphi(0), \qquad \varphi \in \mathscr{S}, \tag{8.5}$$

is also a distribution in $\mathscr{S}'$. Since the function $\varphi_\varepsilon(x)$ in (2.1) also belongs to $\mathscr{S}$, Example 3 shows that the δ-function again cannot be written in the form (8.4).

We define derivatives as before.

Definition 10. Let $\Lambda \in \mathscr{S}'$. Then we define

$$\langle \Lambda^{(k)}, \varphi \rangle = (-1)^k \langle \Lambda, \varphi^{(k)} \rangle \qquad \text{for all } \varphi \in \mathscr{S}. \tag{8.6}$$

Theorem 5. *Let* $\Lambda \in \mathscr{S}'$. *Then* $\Lambda^{(k)} \in \mathscr{S}'$ *for any* $k > 0$.

Proof. If $\varphi \in \mathscr{S}$, then $\varphi^{(k)} \in \mathscr{S}$. Also, if $\varphi_n \to \varphi$ in $\mathscr{S}$, then $\varphi_n^{(k)} \to \varphi^{(k)}$ in $\mathscr{S}$. Therefore, $\Lambda^{(k)} \in \mathscr{S}'$. ∎

Example 11. For $0 < \alpha < 1$, the function $x_+^{-\alpha}$ defined in (5.19) clearly generates a tempered distribution via

$$\langle x_+^{-\alpha}, \varphi \rangle = \int_0^\infty x^{-\alpha}\varphi(x)\, dx, \qquad \varphi \in \mathscr{S}. \tag{8.7}$$

In analogy with (5.20), we shall regard $x_+^{-s-\alpha}$ as the s-th distributional derivative of $x_+^{-\alpha}$ in $\mathscr{S}'$ (modulo a constant factor). More exactly, we define for $s = 1, 2, \ldots$,

$$\langle x_+^{-s-\alpha}, \varphi \rangle = \frac{1}{(\alpha)_s} \int_0^\infty x^{-\alpha}\varphi^{(s)}(x)\, dx, \qquad \varphi \in \mathscr{S}, \tag{8.8}$$

where $(\alpha)_s = \alpha(\alpha + 1) \cdots (\alpha + s - 1)$. Similarly, we put

$$\langle x_+^{-s-1}, \varphi \rangle = -\frac{1}{s!} \int_0^\infty (\log x) \varphi^{(s+1)}(x)\, dx, \qquad \varphi \in \mathscr{S}. \qquad (8.9)$$

Note that this defines x_+^λ when λ is a negative integer, a case not considered in §7. The relationship of this definition and the suggestion, made in §7, that x_+^λ be considered meromorphic in λ with poles at the negative integers, will be clarified later in §11.

Example 12.   Let η be a positive number, and let $\varphi_\eta(x)$ denote a test function in $\mathscr{S}$ which satisfies

$$\varphi_\eta(x) = \frac{e^{-\eta x}}{x + \lambda} \qquad \text{for } x \geq 0, \qquad (8.10)$$

where λ is a fixed complex number. Clearly, $\varphi_\eta(x)$ is a rapidly decreasing function for each $\eta > 0$. From (8.8), it follows that

$$\lim_{\eta \to 0} \langle x_+^{-s-\alpha}, \varphi_\eta \rangle = \frac{\pi}{\sin \alpha \pi} \frac{(-1)^s}{\lambda^{s+\alpha}} \qquad (8.11)$$

for $0 < \alpha < 1$ and $s = 0, 1, 2, \ldots$. Here we have made use of the fact that

$$\int_0^\infty \frac{x^{\lambda-1}}{(1+x)^{\lambda+\mu}}\, dx = \frac{\Gamma(\lambda)\Gamma(\mu)}{\Gamma(\lambda+\mu)}$$

for $\mu, \lambda > 0$. From (8.9), we also have

$$\lim_{\eta \to 0} \langle x_+^{-s-1}, \varphi_\eta \rangle = \frac{(-1)^{s+1}}{\lambda^{s+1}} \sum_{k=1}^{s} \frac{1}{k} + \frac{(-1)^s}{\lambda^{s+1}} \log \lambda, \qquad (8.12)$$

$s = 0, 1, 2, \ldots$. To obtain the last identity, we note that

$$\int_0^\infty \frac{(\log x)}{x + \lambda} e^{-\eta x}\, dx = O(\log^2 \eta) \qquad \text{as } \eta \to 0^+,$$

and

$$\int_0^\infty \frac{\log u}{(u+1)^{s+2}}\, du = -\frac{1}{s+1} \sum_{k=1}^{s} \frac{1}{k}, \qquad s = 0, 1, \ldots,$$

empty sum being understood to be zero; see Ex. 18. The results (8.11) and (8.12) are crucial to our derivation of the asymptotic expansion of the Stieltjes transforms.

To conclude this section, we also mention the following characterization of tempered distributions, a proof of which can be found in Bremermann (1965, p. 121).

Theorem 6. *For every distribution Λ in $\mathscr{S}'$, there is a continuous function f of finite algebraic growth at infinity and an integer m such that $f^{(m)} = \Lambda$.*

9.   Distributions of Several Variables

The one-dimensional distribution theory discussed so far can easily be extended to higher dimensions, and in this section we are concerned with this extension. Since many of the results to be presented here are straightforward generalizations of those already given, we shall keep our discussion brief.

A multi-index is an ordered n-tuple

$$p = (p_1, \ldots, p_n) \tag{9.1}$$

of nonnegative integers. With each multi-index p we associate the differential operator

$$D^p = \left(\frac{\partial}{\partial x_1}\right)^{p_1} \cdots \left(\frac{\partial}{\partial x_n}\right)^{p_n}. \tag{9.2}$$

The order of D^p is

$$|p| = p_1 + \cdots + p_n. \tag{9.3}$$

If $|p| = 0$, then $D^p f = f$. A complex function f defined in $\mathbb{R}^n$ is said to be C^∞ if $D^p f$ is continuous in $\mathbb{R}^n$ for every multi-index p.

Let $\mathscr{D}(\mathbb{R}^n)$ or simply $\mathscr{D}_n$ denote the space of all C^∞-functions $\varphi(x_1, \ldots, x_n)$ in $\mathbb{R}^n$ with compact support. A sequence $\{\varphi_k\}$ of functions in $\mathscr{D}_n$ is said to *converge* to a function φ in $\mathscr{D}_n$ if the supports of the functions φ_k all lie within a fixed compact set $K \subset \mathbb{R}^n$, and $D^p \varphi_k \to D^p \varphi$ uniformly on K, as $k \to \infty$, for every multi-index p. Every continuous linear functional on $\mathscr{D}_n$ is called a *distribution* in $\mathbb{R}^n$. The space of all distributions in $\mathbb{R}^n$ is denoted by $\mathscr{D}'_n$. Convergence of distributions is defined by the obvious analogue of Definition 5.

Any locally integrable function $f(x)$ on $\mathbb{R}^n$ generates a distribution Λ_f or simply f in $\mathbb{R}^n$ by means of the definition

$$\langle \Lambda_f, \varphi \rangle = \langle f, \varphi \rangle = \int_{\mathbb{R}^n} f(x)\varphi(x)\, dx, \qquad (9.4)$$

where $x = (x_1, \dots, x_n)$ and $dx = dx_1 \cdots dx_n$. Another example of a distribution in $\mathbb{R}^n$ is the delta "function" given by

$$\langle \delta, \varphi \rangle = \varphi(0) = \varphi(0, \dots, 0). \qquad (9.5)$$

Differentiation of distributions is defined by the formula

$$\langle D^p \Lambda, \varphi \rangle = (-1)^{|p|} \langle \Lambda, D^p \varphi \rangle, \qquad \varphi \in \mathscr{D}_n. \qquad (9.6)$$

From this, it is clear that every distribution in $\mathbb{R}^n$ has partial derivatives of all orders. If f has continuous partial derivatives of all orders up to and including $|p|$, then integration by parts gives

$$(-1)^{|p|} \int_{\mathbb{R}^n} f(x)(D^p \varphi)(x)\, dx = \int_{\mathbb{R}^n} (D^p f)(x)\varphi(x)\, dx \qquad (9.7)$$

for every $\varphi \in \mathscr{D}_n$. Thus we have

$$D^p \Lambda_f = \Lambda_{D^p f} \qquad \text{on } \mathscr{D}_n. \qquad (9.8)$$

The identity (9.7) in fact holds under the weaker conditions that f has continuous partial derivatives of all orders up to $|p| - 1$ and that all partial derivatives of order $|p| - 1$ are locally absolutely continuous in the variables $x_1, \dots, x_n$; see the one-dimensional case given in Lemma 2.

If α is a C^∞-function on $\mathbb{R}^n$ and $\Lambda \in \mathscr{D}_n'$, then the product $\alpha\Lambda$ is defined by

$$\langle \alpha\Lambda, \varphi \rangle = \langle \Lambda, \alpha\varphi \rangle \qquad \text{for all } \varphi \in \mathscr{D}_n. \qquad (9.9)$$

Similarly, we can extend the concept of tempered distributions in $\mathbb{R}^1$ to $\mathbb{R}^n$. Thus we let $\mathscr{S}_n$ denote the space of all *rapidly decreasing functions* $\varphi(x_1, \dots, x_n)$ on $\mathbb{R}^n$, i.e., $\varphi \in C^\infty(\mathbb{R}^n)$ and φ satisfies

$$|x^k D^q \varphi(x)| \le C_{kq}, \qquad x \in \mathbb{R}^n, \qquad (9.10)$$

for some constants C_{kq}, where $k = (k_1, \dots, k_n)$ and $q = (q_1, \dots, q_n)$ are any multi-indices and

$$x^k = x_1^{k_1} \cdots x_n^{k_n}. \qquad (9.11)$$

A sequence $\{\varphi_j\}$, $\varphi_j \in \mathscr{S}_n$, is said to *converge to* $\varphi \in \mathscr{S}_n$ if and only if $D^q\varphi_j \to D^q\varphi$ uniformly on every compact set in $\mathbb{R}^n$ for any multi-index q and the constants C_{kq} in (9.10) can be chosen independent of j, i.e.,

$$|x^k D^q\varphi_j(x)| \le C_{kq} \qquad (9.12)$$

for all j. The space of all continuous linear functionals on $\mathscr{S}_n$ is denoted by $\mathscr{S}'_n$, and members of $\mathscr{S}'_n$ are called *tempered* distributions.

As before, a locally integrable function $f(x)$ on $\mathbb{R}^n$ of finite algebraic growth at infinity, i.e., $f(x) = O(|x|^\alpha)$ as $|x| = \sqrt{x_1^2 + \cdots + x_n^2} \to \infty$ for some $\alpha \ge 0$, generates a tempered distribution Λ_f by means of the expression

$$\langle \Lambda_f, \varphi \rangle = \int_{\mathbb{R}^n} f(x)\varphi(x)\, dx. \qquad (9.13)$$

In fact, every tempered distribution on $\mathbb{R}^1$ is a distributional derivative of such a function; cf. Theorem 6.

If p is a multi-index and $\Lambda \in \mathscr{S}'_n$ then the formula

$$\langle D^p\Lambda, \varphi \rangle = (-1)^{|p|}\langle \Lambda, D^p\varphi \rangle, \qquad \varphi \in \mathscr{S}_n, \qquad (9.14)$$

defines, as usual, a distribution $D^p\Lambda \in \mathscr{S}'_n$. If f and D^pf are both locally integrable in $\mathbb{R}^n$ and of finite algebraic growth at infinity then the remark following (9.8) also holds for the distribution $D^p\Lambda_f$ and Λ_{D^pf} in $\mathscr{S}'_n$.

Let Δ denote the Laplacian operator

$$\Delta = \frac{\partial^2}{\partial x_1^2} + \cdots + \frac{\partial^2}{\partial x_n^2}. \qquad (9.15)$$

A variant of (9.7) is the identity

$$\langle \Delta f, \varphi \rangle = \langle f, \Delta\varphi \rangle \qquad (9.16)$$

or more explicitly

$$\int_{\mathbb{R}^n} (\Delta f)(x)\varphi(x)\, dx = \int_{\mathbb{R}^n} f(x)(\Delta\varphi)(x)\, dx. \qquad (9.17)$$

The last equation also holds under the conditions: (i) f has continuous second partial derivatives in $\mathbb{R}^n$ except possibly at the origin; (ii) $f(x) = O(|x|^\lambda)$ and $(\Delta f)(x) = O(|x|^{\lambda-2})$, as $|x| \to 0$, with $\lambda + n > 2$; and

(iii) f grows at most algebraically at infinity. To prove this, we let S denote the shell domain $\{x \in \mathbb{R}^n : \varepsilon \le |x| \le \rho\}$. Then by Green's theorem

$$\int_S (f\Delta\varphi - \varphi\Delta f)\, dx = \int_{\partial S} \left(f\frac{\partial\varphi}{\partial n} - \varphi\frac{\partial f}{\partial n} \right) d\sigma, \tag{9.18}$$

where ∂S is the boundary of S, $\partial/\partial n$ denotes differentiation in the direction of the outward normal to ∂S, and $d\sigma$ is the element of surface area on ∂S. It can be easily shown that the right-hand side of (9.18) tends to zero as $\varepsilon \to 0$ and $\rho \to +\infty$. Thus, relation (9.17) results as a limit.

10. The Distribution r^λ

Let $r = \sqrt{x_1^2 + \cdots + x_n^2}$. We consider the regular functional defined by

$$\langle r^\lambda, \varphi \rangle = \int_{\mathbb{R}^n} r^\lambda \varphi(x)\, dx, \qquad \varphi \in \mathscr{D}_n, \tag{10.1}$$

where $\operatorname{Re} \lambda > -n$. Since the derivative

$$\frac{d}{d\lambda}\langle r^\lambda, \varphi \rangle = \int_{\mathbb{R}^n} r^\lambda (\log r)\varphi(x)\, dx$$

can be easily shown to exist, the functional r^λ is an analytic function of λ for $\operatorname{Re} \lambda > -n$. We shall show that this function can be continued analytically to the entire λ-plane except for the points $\lambda = -n, -n-2, -n-4, \ldots$.

In (10.1), we make the change of variables $(x_1, \ldots, x_n) = r(\xi_1, \ldots, \xi_n)$. This yields

$$\langle r^\lambda, \varphi \rangle = \int_0^\infty r^\lambda \left\{ \int_{|\xi|=1} \varphi(r\xi)\, d\sigma \right\} r^{n-1}\, dr, \tag{10.2}$$

where $d\sigma$ is the surface element on the unit sphere. In terms of the polar coordinates $\xi = (1, \theta_1, \ldots, \theta_{n-1})$, where

$$\xi_1 = \sin\theta_{n-1}\sin\theta_{n-2}\cdots\sin\theta_2\sin\theta_1,$$

$$\xi_2 = \sin\theta_{n-1}\sin\theta_{n-2}\cdots\sin\theta_2\cos\theta_1,$$

$$\xi_3 = \sin\theta_{n-1}\sin\theta_{n-2}\cdots\cos\theta_2,$$

$$\vdots$$

$$\xi_n = \cos\theta_{n-1}, \tag{10.3}$$

we have

$$d\sigma = \sin^{n-2}\theta_{n-1} \sin^{n-3}\theta_{n-2} \cdots \sin\theta_2 \, d\theta_1 \, d\theta_2 \cdots d\theta_{n-1}. \quad (10.4)$$

As ξ varies on the unit sphere, $\theta_2, \ldots, \theta_{n-1}$ each vary in the interval $(0, \pi)$; but θ_1 varies in the interval $(0, 2\pi)$.

From the mean-value theorem for multiple integrals, it follows that

$$\int_{|\xi|=1} \varphi(r\xi) \, d\sigma = \Omega_n S_\varphi(r), \quad (10.5)$$

where Ω_n is the area of the n-dimensional unit sphere, i.e.,

$$\Omega_n = \frac{2\pi^{n/2}}{\Gamma(n/2)}, \quad (10.6)$$

and $S_\varphi(r)$ is the mean value of $\varphi(x)$ on the sphere of radius r. Coupling (10.2) and (10.5) gives

$$\langle r^\lambda, \varphi \rangle = \Omega_n \int_0^\infty r^{\lambda+n-1} S_\varphi(r) \, dr. \quad (10.7)$$

Clearly, $S_\varphi(r)$ is infinitely differentiable for $r > 0$ and, since $\varphi(x)$ vanishes for sufficiently large r, $S_\varphi(r)$ has compact support.

Expanding $\varphi(x)$ in a Taylor series, we have from (10.5)

$$\Omega_n S_\varphi(r) = \int_{|\xi|=1} \left[\varphi(0) + \sum \frac{\partial\varphi(0)}{\partial x_i} x_i + \frac{1}{2} \sum \frac{\partial^2\varphi(0)}{\partial x_i \, \partial x_j} x_i x_j \right.$$
$$\left. + \frac{1}{3!} \sum \frac{\partial^3\varphi(0)}{\partial x_i \, \partial x_j \, \partial x_k} x_i x_j x_k + \cdots \right] d\sigma. \quad (10.8)$$

Note that, in view of (10.3) and (10.4), the terms involving an odd number of factors of the x_i do not contribute to the integral on the right-hand side, and that the sum of terms involving $2k$ factors of the x_i contributes an expression of the form $a_k r^{2k}$. Thus we obtain

$$\Omega_n S_\varphi(r) = \Omega_n \varphi(0) + a_1 r^2 + a_2 r^4 + \cdots + a_k r^{2k} + o(r^{2k}), \quad (10.9)$$

as $r \to 0^+$. Since k is arbitrary, equation (10.9) in fact shows that $S_\varphi(r)$ has derivatives of all orders at $r = 0$, and that all of its derivatives of odd order vanish. From here, it follows that $S_\varphi(r)$ is an even function of the single variable r, and that $S_\varphi(r)$ belongs to $\mathscr{D}(\mathbb{R})$. The integral in (10.7) hence represents the action of the functional $x_+^{\lambda+n-1}$ on the test function $S_\varphi(x)$. But we have already shown that $x_+^{\lambda+n-1}$ is analytic in

Re $\lambda > -n$ and can be analytically continued to the entire λ-plane except for $\lambda = -n, -n - 1, -n - 2, \ldots$. Furthermore, at $\lambda = -n - m + 1, m = 1, 2, \ldots$, the integral in (10.7) has a simple pole with residue

$$\frac{(-1)^{m-1}}{(m-1)!} \langle \delta^{(m-1)}(x), S_\varphi(x) \rangle = \frac{1}{(m-1)!} S_\varphi^{(m-1)}(0).$$

Since the odd-order derivatives of $S_\varphi(x)$ vanish at $x = 0$, the poles at $\lambda = -n - m + 1$, when m is even, do not actually occur. This implies that $\langle r^\lambda, \varphi \rangle$ has poles only at $\lambda = -n, -n - 2, -n - 4, \ldots$, and further-more that its residue at $\lambda = -n - 2k, k = 0, 1, 2, \ldots$, is given by

$$\frac{\Omega_n}{(2k)!} \langle \delta^{(2k)}(x), S_\varphi(x) \rangle = \frac{\Omega_n}{(2k)!} S_\varphi^{(2k)}(0). \tag{10.10}$$

In particular, the residue of $\langle r^\lambda, \varphi \rangle$ at $\lambda = -n$ is $\Omega_n S_\varphi(0) = \Omega_n \varphi(0)$. Since $\varphi(0) = \langle \delta, \varphi \rangle$, we may say that the distribution r^λ itself has a simple pole at $\lambda = -n$ with residue $\Omega_n \delta(x)$.

11.   Taylor and Laurent Series for r^λ

We now turn to the problem of obtaining the Taylor and Laurent series for r^λ, which, in view of (10.7), is equivalent to that of finding these series for x_+^λ.

In the neighborhood of a regular point λ_0, x_+^λ has the Taylor expansion

$$x_+^\lambda = x_+^{\lambda_0} + (\lambda - \lambda_0) \frac{\partial}{\partial \lambda} x_+^\lambda \bigg|_{\lambda = \lambda_0} + \frac{1}{2}(\lambda - \lambda_0)^2 \frac{\partial^2}{\partial \lambda^2} x_+^\lambda \bigg|_{\lambda = \lambda_0} + \cdots$$

$$= x_+^{\lambda_0} + (\lambda - \lambda_0)x_+^{\lambda_0} \log x_+ + \frac{1}{2}(\lambda - \lambda_0)^2 x_+^{\lambda_0} \log^2 x_+ + \cdots, \tag{11.1}$$

where the distributions $x_+^\lambda \log^m x_+$, $m = 1, 2, \ldots$, are defined by

$$\langle x_+^\lambda \log^m x_+, \varphi \rangle = \int_0^1 x^\lambda \log^m x \left[\varphi(x) - \varphi(0) - x\varphi'(0) - \cdots \right.$$

$$\left. - \frac{x^{n-1}}{(n-1)!} \varphi^{(n-1)}(0) \right] dx \tag{11.2}$$

$$+ \int_1^\infty x^\lambda \log^m x \, \varphi(x) \, dx + \sum_{k=1}^n \frac{(-1)^m m! \, \varphi^{(k-1)}(0)}{(k-1)! \, (\lambda + k)^{m+1}}$$

for Re $\lambda > -n - 1$ and $\lambda \neq -1, -2, \ldots, -n$. Equation (11.2) can either be formally obtained from (7.3) by m-fold differentiation with respect to λ or by regularizing the function $x_+^{\lambda} \log^m x$ using the method in §7. As in (7.4), Eq. (11.2) simplifies to

$$
\langle x_+^{\lambda} \log^m x_+ , \varphi \rangle = \int_0^{\infty} x^{\lambda} \log^m x \left[\varphi(x) - \varphi(0) - x\varphi'(0) - \cdots \right.
$$
$$
\left. - \frac{x^{n-1}}{(n-1)!} \varphi^{(n-1)}(0) \right] dx \tag{11.3}
$$

in the strip $-n - 1 < \mathrm{Re}\ \lambda < -n$.

In the neighbourhood of the pole $\lambda = -n$, x_+^{λ} has a Laurent expansion whose principal part consists of only the term $c_{-1}/(\lambda + n)$, c_{-1} being the residue at $\lambda = -n$; see the remark following (7.5). To derive this expansion explicitly, we first recall Eq. (7.4):

$$
\langle x_+^{\lambda} , \varphi \rangle = \int_0^{\infty} x^{\lambda} \left[\varphi(x) - \varphi(0) - x\varphi'(0) - \cdots - \frac{x^{n-1}}{(n-1)!} \varphi^{(n-1)}(0) \right] dx,
$$

which is valid for $-n - 1 < \mathrm{Re}\ \lambda < -n$. Let us rewrite this equation as

$$
\langle x_+^{\lambda} , \varphi \rangle = \int_0^1 x^{\lambda} \left[\varphi(x) - \cdots - \frac{x^{n-1}}{(n-1)!} \varphi^{(n-1)}(0) \right] dx
$$
$$
+ \int_1^{\infty} x^{\lambda} \left[\varphi(x) - \cdots - \frac{x^{n-2}}{(n-2)!} \varphi^{(n-2)}(0) \right] dx \tag{11.4}
$$
$$
+ \frac{\varphi^{(n-1)}(0)}{(n-1)!\ (\lambda + n)}.
$$

The first integral defines an analytic function in Re $\lambda > -n - 1$, and the second integral gives an analytic function in Re $\lambda < -n + 1$. Thus the sum of the two integrals is an analytic function in the strip $|\mathrm{Re}\ \lambda + n| < 1$ and represents the regular part of the Laurent expansion of x_+^{λ}, in the neighbourhood of $\lambda = -n$. We shall denote this function by $R(x_+^{\lambda} ; -n)$. That is, we shall define

$$
\langle R(x_+^{\lambda} ; -n), \varphi \rangle = \int_0^{\infty} x^{\lambda} \left[\varphi(x) - \varphi(0) - x\varphi'(0) - \cdots - \frac{x^{n-2}}{(n-2)!} \varphi^{(n-2)}(0) \right.
$$
$$
\left. - \frac{x^{n-1}}{(n-1)!} \varphi^{(n-1)}(0) H(1 - x) \right] dx, \tag{11.5}
$$

where $H(x)$ is the Heaviside function given in Example 6.

The value of the functional $R(x_+^\lambda ; -n)$ at $\lambda = -n$ will be denoted by x_+^{-n}. From (11.5), we have

$$\langle x_+^{-n}, \varphi \rangle = \int_0^\infty x^{-n} \left[\varphi(x) - \varphi(0) - \cdots - \frac{x^{n-1}}{(n-1)!} \varphi^{(n-1)}(0) H(1-x) \right] dx.$$

$$(11.6)$$

This is, of course, also the value given in the following formula:

$$x_+^{-n} = \lim_{\lambda \to -n} \frac{\partial}{\partial \lambda} [(\lambda + n) x_+^\lambda].$$

$$(11.7)$$

It can be shown that Eq. (11.6) differs from the definition given in (8.9) by a constant depending on n; see Ex. 19. (Recall that regularization of a function is not unique.)

The derivatives of $R(x_+^\lambda ; -n)$ with respect to λ at $\lambda = -n$ can be obtained from (11.5) by differentiation. If we put

$$\left. \frac{\partial}{\partial \lambda} R(x_+^\lambda ; -n) \right|_{\lambda = -n} = x_+^{-n} \log x_+$$

$$\left. \frac{\partial^2}{\partial \lambda^2} R(x_+^\lambda ; -n) \right|_{\lambda = -n} = x_+^{-n} \log^2 x_+$$

$$\vdots$$

then it follows that

$$\langle x_+^{-n} \log^k x_+ , \varphi \rangle = \int_0^\infty x^{-n} \log^k x \left[\varphi(x) - \varphi(0) - \cdots \right.$$

$$\left. - \frac{x^{n-1}}{(n-1)!} \varphi^{(n-1)}(0) H(1-x) \right] dx, \qquad k = 1, 2, \ldots .$$

$$(11.8)$$

The Taylor series of $R(x_+^\lambda ; -n)$ at $\lambda = -n$ is therefore given by

$$R(x_+^\lambda ; -n) = x_+^{-n} + (\lambda + n) x_+^{-n} \log x_+ + \cdots$$

$$+ \frac{(\lambda + n)^k}{k!} x_+^{-n} \log^k x_+ + \cdots .$$

$$(11.9)$$

A combination of (11.4), (11.5), and (11.9) yields

$$x_+^\lambda = \frac{(-1)^{n-1} \delta^{(n-1)}(x)}{(n-1)! (\lambda + n)} + x_+^{-n} + (\lambda + n) x_+^{-n} \log x_+ + \cdots$$

$$+ \frac{(\lambda + n)^k}{k!} x_+^{-n} \log^k x_+ + \cdots .$$

$$(11.10)$$

On account of (10.7), the Taylor and Laurent series for r^λ can be obtained directly from the corresponding expansions for $\Omega_n x_+^{\lambda + n - 1}$. Thus, for example, it follows from (11.1) that the Taylor series at a regular point λ_0 is

$$r^\lambda = r^{\lambda_0} + (\lambda - \lambda_0) r^{\lambda_0} \log r + \frac{1}{2} (\lambda - \lambda_0)^2 r^{\lambda_0} \log^2 r + \cdots, \quad (11.11)$$

where $r^{\lambda_0} \log^k r$ is the distribution defined by

$$\langle r^{\lambda_0} \log^k r, \varphi \rangle = \Omega_n \int_0^\infty r^{\lambda_0 + n - 1} \log^k r \left[S_\varphi(r) - \varphi(0) - \cdots \right.$$

$$\left. - \frac{r^{m-1}}{(m-1)!} S_\varphi^{(m-1)}(0) \right] dr$$

for $-m - n < \operatorname{Re} \lambda_0 < -m - n + 1$. (Recall: $S_\varphi(0) = \varphi(0)$.) Also, it follows from (11.10) that the Laurent series of r^λ at $\lambda = -n - 2k$ is given by

$$r^\lambda = \frac{\Lambda(r)}{\lambda + n + 2k} + r^{-2k - n} + (\lambda + n + 2k) r^{-2k - n} \log r + \cdots, \quad (11.12)$$

where the distributions $\Lambda(r)$ and $r^{-2k - n} \log^m r$ are defined by

$$\langle \Lambda(r), \varphi \rangle = \frac{\Omega_n}{(2k)!} \langle \delta^{(2k)}(r), S_\varphi(r) \rangle, \quad (11.13)$$

and

$$\langle r^{-2k - n} \log^m r, \varphi \rangle = \Omega_n \langle r_+^{-2k - 1} \log^m r_+, S_\varphi(r) \rangle$$

$$= \Omega_n \int_0^\infty r^{-2k - 1} \log^m r \left[S_\varphi(r) - \varphi(0) - \cdots \right. \quad (11.14)$$

$$\left. - \frac{r^{2k - 2}}{(2k - 2)!} S_\varphi^{(2k - 2)}(0) - \frac{r^{2k}}{(2k)!} S_\varphi^{(2k)}(0) H(1 - r) \right] dr.$$

Note that the distributions $\Lambda(r)$ and $r^{-2k - n} \log^m r$ on the left-hand side of (11.13) and (11.14) act on $\varphi \in \mathscr{D}(\mathbb{R}^n)$, whereas the distributions $\delta^{(2k)}(r)$ and $r_+^{-2k - 1} \log^m r_+$ on the right-hand side of (11.13) and (11.14) act on $S_\varphi(r) \in \mathscr{D}(\mathbb{R}^1)$. Furthermore, we again remark that $r^{-2k - n}$ is not the value of the distribution r^λ at $\lambda = -2k - n$, but the value at this point of the regular part of r^λ; cf. Eq. (11.6).

Here it should also be pointed out that there is a slight error on page 99 of Gelfand and Shilov (1964). The distributions $r^{-2k - n} \log^m r$ in Eqs.

(2) and (3) there should be replaced by $r_+^{-2k-1} \log^m r_+$, since they are acting on $S_\varphi(r)$. Moreover, the integral in Eq. (3), as it stands, does not exist, but will exist if n is replaced by 1.

12.   Fourier Transforms

If $f \in L^1(\mathbb{R}^n)$, then the Fourier transform of f is defined by

$$\hat{f}(t) = \frac{1}{(2\pi)^{n/2}} \int_{\mathbb{R}^n} f(x) e^{it \cdot x} \, dx, \tag{12.1}$$

where $t = (t_1, \ldots, t_n)$ and $t \cdot x = t_1 x_1 + \cdots + t_n x_n$. If $\varphi \in \mathscr{S}_n$ then $\hat{\varphi}$ exists, since $\mathscr{S}_n \subset L^1(\mathbb{R}^n)$. We shall show that, in fact, $\varphi \in \mathscr{S}_n$ implies $\hat{\varphi} \in \mathscr{S}_n$.

Let $p = (p_1, \ldots, p_n)$ be a multi-index and put

$$D_p = (i)^{-|p|} D^p = \left(\frac{1}{i} \frac{\partial}{\partial x_1} \right)^{p_1} \cdots \left(\frac{1}{i} \frac{\partial}{\partial x_n} \right)^{p_n}.$$

If P is a polynomial in n variables, say

$$P(x) = \sum c_p x^p = \sum c_p x_1^{p_1} \cdots x_n^{p_n},$$

then we define the differential operator

$$P(D) = \sum c_p D_p.$$

Lemma 4.   *If $\varphi \in \mathscr{S}_n$ and P is a polynomial, then*

$$(P(D)\varphi)\hat{\ } = P(-t)\hat{\varphi} \qquad and \qquad (P\varphi)\hat{\ } = P(D)\hat{\varphi}. \tag{12.2}$$

Proof.   Without loss of generality, we may assume that $P(x)$ is the monomial $P(x) = x^p$. The first part of (12.2) then follows immediately by repeated integration by parts on each of the variables $x_1, \ldots, x_n$. The second part of (12.2) is a consequence of repeated differentiation under the integral sign, which is legitimate since φ is a rapidly decreasing function (Fleming 1965, p. 199).   ∎

Theorem 7.   *If $\varphi \in \mathscr{S}_n$, then $\hat{\varphi} \in \mathscr{S}_n$. Furthermore, if $\varphi_j \to \varphi$ in $\mathscr{S}_n$ as $j \to \infty$, then $\hat{\varphi}_j \to \hat{\varphi}$ in $\mathscr{S}_n$.*

Proof. If $\varphi \in \mathscr{S}_n$, then $D_k \varphi \in \mathscr{S}_n$ for any multi-index k. By Lemma 4, $(D_k \varphi)^{\wedge} = (-1)^{|k|} t^k \hat{\varphi}$. Since $(D_k \varphi)^{\wedge}$ is uniformly bounded, we have

$$|t^k \hat{\varphi}(t)| \le C_k \tag{12.3}$$

for some constant C_k and for all $t \in \mathbb{R}^n$. Also, by Lemma 4, $D_q \hat{\varphi} = (x^q \varphi)^{\wedge}$ for any multi-index q. Since $\varphi \in \mathscr{S}_n$ implies $x^q \varphi \in \mathscr{S}_n$, we have, from (12.3) with φ replaced by $x^q \varphi$,

$$|t^k D_q \hat{\varphi}(t)| \le C_{kq}$$

for some constant C_{kq}. Hence $\hat{\varphi} \in \mathscr{S}_n$.

Now consider a sequence $\{\varphi_j\}$, which converges to φ in $\mathscr{S}_n$. Since the Fourier transform defined in (12.1) is clearly linear, we may assume without loss of generality that φ is identically zero. Since $\varphi_j \to 0$ in $\mathscr{S}_n$ implies $\varphi_j \to 0$ in $L^1(\mathbb{R}^n)$ and also $x^p \varphi_j \to 0$ in $\mathscr{S}_n$ for any multi-index p, we have $x^p \varphi_j \to 0$ in $L^1(\mathbb{R}^n)$ and $(x^p \varphi_j)^{\wedge} \to 0$ uniformly in $t \in \mathbb{R}^n$. By Lemma 4, $D_p \hat{\varphi}_j = (x^p \varphi_j)^{\wedge}$. Consequently, $D^p \hat{\varphi}_j \to 0$ uniformly in $t \in \mathbb{R}^n$. We must also show that for any multi-indices k and q, there exists a constant A_{kq} independent of j such that

$$|t^k D^q \hat{\varphi}_j| \le A_{kq} \qquad \text{for all } j. \tag{12.4}$$

By Lemma 4,

$$(-t)^k D_q \hat{\varphi}_j = (-t)^k (x^q \varphi_j)^{\wedge} = [D_k(x^q \varphi_j)]^{\wedge}$$

and hence

$$(-it)^k D_q \hat{\varphi}_j = \frac{1}{(2\pi)^{n/2}} \int_{\mathbb{R}^n} D^k [x^q \varphi_j] e^{it \cdot x} \, dx,$$

where $(-it)^k$ denotes $(-it_1)^{k_1} \cdots (-it_n)^{k_n}$. The Leibniz rule then gives

$$(-it)^k D_q \hat{\varphi}_j = \frac{1}{(2\pi)^{n/2}} \sum_l c_{kl} \int D^l x^q D^{k-l} \varphi_j(x) e^{it \cdot x} \, dx.$$

Here, c_{kl} are some positive numbers and $k - l = (k_1 - l_1, \ldots, k_n - l_n)$. A simple estimation yields

$$|t^k D^q \hat{\varphi}_j| \le \sum_l b_{kl} \int |x^{q-l} D^{k-l} \varphi_j(x)| \, dx$$

$$\le \sum_{l,p} b_{kl} \int \frac{|x^{q-l+p} D^{k-l} \varphi_j(x)|}{\prod_{i=1}^n (x_i^2 + 1)} \, dx,$$

where the last inequality is obtained by expanding $\prod_{i=1}^{n}(x_i^2 + 1)$ in the numerator, and where the multi-index p satisfies $0 \le |p| \le 2n$. Since $\varphi_j \to 0$ in $\mathscr{S}_n$, we have from (9.12)

$$|x^{q-l+p}D^{k-l}\varphi_j| \le C_{kqlp}$$

for some constant C_{kqlp} independent of j. The required result (12.4) now follows from a combination of the last two inequalities. This completes the proof of the theorem.   ∎

Definition 11.   If $u \in \mathscr{S}_n'$, define $\hat{u}$ by the relation

$$\langle \hat{u}, \varphi \rangle = \langle u, \hat{\varphi} \rangle, \qquad \varphi \in \mathscr{S}_n. \tag{12.5}$$

Theorem 8.   *If $u \in \mathscr{S}_n'$ then $\hat{u} \in \mathscr{S}_n'$. Furthermore, if P is a polynomial then*

$$(P(D)u)^{\wedge} = P(-t)\hat{u} \qquad and \qquad (Pu)^{\wedge} = P(D)\hat{u}. \tag{12.6}$$

Proof.   Clearly, $\hat{u}$ is a linear functional on $\mathscr{S}_n$. If $\varphi_j \to \varphi$ in $\mathscr{S}_n$ then, by Theorem 7, $\hat{\varphi}_j \to \hat{\varphi}$ in $\mathscr{S}_n$. Hence $\hat{u}$ is continuous in $\mathscr{S}_n$, i.e., $\hat{u} \in \mathscr{S}_n'$.

Let $P(x) = \sum c_p x^p$. The following computations are obvious. For any $\varphi \in \mathscr{S}_n$,

$$\langle (P(D)u)^{\wedge}, \varphi \rangle = \langle P(D)u, \hat{\varphi} \rangle = \sum c_p \langle D_p u, \hat{\varphi} \rangle = \sum (i)^{-|p|} c_p \langle D^p u, \hat{\varphi} \rangle$$

$$= \sum \left(-\frac{1}{i} \right)^{|p|} c_p \langle u, D^p \hat{\varphi} \rangle = \sum (-1)^{|p|} c_p \langle u, D_p \hat{\varphi} \rangle$$

$$= \sum (-1)^{|p|} c_p \langle u, (x^p \varphi)^{\wedge} \rangle$$

$$= \langle u, (P(-x)\varphi)^{\wedge} \rangle = \langle \hat{u}, P(-x)\varphi \rangle = \langle P(-x)\hat{u}, \varphi \rangle.$$

The fourth last equality follows from (12.2). This proves the first identity in (12.6). The second identity is proved similarly.   ∎

We now turn to the question of consistency. Let f be a locally integrable function on $\mathbb{R}^n$, which increases no more rapidly than an algebraic function. The ordinary Fourier transform of f can be defined by

$$\hat{f}(t) = \lim_{\nu \to \infty} \frac{1}{(2\pi)^{n/2}} \int_{|x| < \nu} f(x)e^{it \cdot x} \, dx, \tag{12.7}$$

provided that the limit exists. If $f \in L^1(\mathbb{R}^n)$ then, of course, this limit agrees with the integral in (12.1) by the Lebesgue dominated convergence theorem. Now, since f defines a tempered distribution u_f, f also has the distributional Fourier transform $\hat{u}_f$. A natural question which arises is whether the two Fourier transforms coincide. The answer is yes. To show this, we assume that the limit in (12.7) exists uniformly in $t \in \mathbb{R}^n$, and put

$$f_v(x) = \begin{cases} f(x) & \text{if } |x| \le v \\ 0 & \text{if } |x| > v. \end{cases}$$

From (12.7), it follows that $\hat{f}_v(t)$ converges to $\hat{f}(t)$ uniformly in $t \in \mathbb{R}^n$ and consequently

$$\lim_{v \to \infty} \langle \hat{f}_v, \varphi \rangle = \langle \hat{f}, \varphi \rangle, \qquad \varphi \in \mathscr{S}_n.$$

Therefore, for all $\varphi \in \mathscr{S}_n$, we have

$$\langle \hat{u}_f, \varphi \rangle = \langle u_f, \hat{\varphi} \rangle = \lim_{v \to \infty} \langle u_{f_v}, \hat{\varphi} \rangle = \lim_{v \to \infty} \langle \hat{f}_v, \varphi \rangle = \langle \hat{f}, \varphi \rangle.$$

In the third equality, we have used Fubini's theorem to interchange the order of integrations.

Example 13. Let us calculate the Fourier transform of x_+^λ. For $-1 < \operatorname{Re} \lambda < 0$, it is well-known that

$$\int_0^\infty x^\lambda e^{itx}\, dx = e^{i(\lambda + 1)\pi/2} \Gamma(\lambda + 1) t^{-\lambda - 1} \tag{12.8}$$

for all $t > 0$; see Erdélyi *et al.* (1953, p. 1). By the consistency of Fourier transforms, we have

$$(x_+^\lambda)^\wedge = \frac{\Gamma(\lambda + 1)}{\sqrt{2\pi}} e^{i(\lambda + 1)\pi/2} t_+^{-\lambda - 1} \qquad \text{in } \mathscr{S}'(\mathbb{R}) \tag{12.9}$$

for $-1 < \operatorname{Re} \lambda < 0$. Written in full, (12.9) means that

$$\langle (x_+^\lambda)^\wedge, \varphi \rangle = \frac{\Gamma(\lambda + 1)}{\sqrt{2\pi}} e^{i(\lambda + 1)\pi/2} \langle t_+^{-\lambda - 1}, \varphi \rangle \tag{12.10}$$

for $-1 < \operatorname{Re} \lambda < 0$ and for all $\varphi \in \mathscr{S}(\mathbb{R})$. The left-hand side of (12.10) is, by definition, equal to

$$\langle x_+^\lambda, \hat{\varphi} \rangle = \int_0^\infty x^\lambda \hat{\varphi}(x)\, dx, \qquad \hat{\varphi} \in \mathscr{S}(\mathbb{R}),$$

which is obviously analytic in Re $\lambda > -1$. Since $-1 < \text{Re } \lambda < 0$ implies $-1 < -\text{Re } \lambda - 1 < 0$, the function

$$\langle t_+^{-\lambda-1}, \varphi \rangle = \int_0^\infty t^{-\lambda-1}\varphi(t)\, dt$$

is analytic in the strip $-1 < \text{Re } \lambda < 0$. If we restrict φ to those test functions which vanish in a neighbourhood of the origin, then $\langle t_+^{-\lambda-1}, \varphi \rangle$ is an entire function of λ. Thus, for such $\varphi \in \mathscr{S}(\mathbb{R})$, Eq. (12.10) holds for all λ in Re $\lambda > -1$. Let $\mathscr{S}_0(\mathbb{R})$ denote the subspace of $\mathscr{S}(\mathbb{R})$ which consists of all rapidly decreasing functions vanishing in a neighbourhood of the origin. Then we have, as functionals acting on the space $\mathscr{S}_0(\mathbb{R})$,

$$(x_+^\lambda)^\wedge = \frac{\Gamma(\lambda+1)}{\sqrt{2\pi}} e^{i(\lambda+1)\pi/2} t_+^{-\lambda-1} \qquad (12.11)$$

for Re $\lambda > -1$. This result can be used to derive the asymptotic expansions of the one-sided Fourier transform; cf. Chapter IV, Section 2.

Example 14.   To compute the Fourier transform of r^λ, we first need the following result: If $f(x)$ is spherically symmetric, i.e., $f(x)$ is a function of r only, say $f(x) = g(r)$, then its Fourier transform $\hat{f}(t)$ is also spherically symmetric; more specifically, we have

$$\hat{f}(t) = \rho^{(2-n)/2} \int_0^\infty r^{n/2} g(r) J_{(n-2)/2}(\rho r)\, dr,$$

where $\rho = |t|$ and $J_\nu(r)$ denotes the Bessel function of the first kind. For a proof of this formula, we refer to Schwartz (1966b, pp. 201–203). Since

$$\int_0^\infty r^\mu J_\nu(\rho r)\, dr = 2^\mu \frac{\Gamma(\frac{1}{2}\mu + \frac{1}{2}\nu + \frac{1}{2})}{\Gamma(\frac{1}{2}\nu - \frac{1}{2}\mu + \frac{1}{2})} \rho^{-\mu-1},$$

for $-\text{Re } \nu - 1 < \text{Re } \mu < 0$, we have

$$(r^\lambda)^\wedge = 2^{\lambda+n/2} \frac{\Gamma(\frac{1}{2}\lambda + \frac{1}{2}n)}{\Gamma(-\frac{1}{2}\lambda)} \rho^{-\lambda-n} \qquad (12.12)$$

for $-n < \text{Re } \lambda < -n/2$. In view of the consistency of Fourier transforms, this equation also holds in the sense of tempered distributions. As in Example 13, the range of validity of (12.12) can be extended by analytic continuation to the whole λ-plane except $\lambda = -n, -n-2, \ldots,$ at which points r^λ has a simple pole.

To calculate the Fourier transform of r^λ at $\lambda = -n - 2m$, we observe that in the neighbourhood of this point

$$r^\lambda = \frac{a_{-1}}{\lambda + n + m} + a_0 + a_1(\lambda + n + m) + \cdots, \qquad (12.13)$$

where $a_{-1}, a_0, a_1, \ldots,$ are given by

$$a_{-1} = \Lambda(r), \quad a_0 = r^{-2m-n}, \quad a_1 = r^{-2m-n} \log r, \ldots;$$

see Eq. (11.12). Taking Fourier transforms term by term, we obtain

$$(r^\lambda)^\wedge = \frac{[\Lambda(r)]^\wedge}{\lambda + n + 2m} + (r^{-2m-n})^\wedge + (\lambda + n + 2m)(r^{-2m-n} \log r)^\wedge + \cdots.$$
$$(12.14)$$

On the other hand, the distribution $\rho^{-\lambda-n}$ on the right-hand side of (12.12) is analytic at $\lambda = -n - 2m$ and has the Taylor expansion

$$\rho^{-\lambda-n} = \rho^{2m} - (\lambda + n + 2m)\rho^{2m} \log \rho + \tfrac{1}{2}(\lambda + n + m)^2 \rho^{2m} \log^2 \rho + \cdots;$$
$$(12.15)$$

see Eq. (11.11). The coefficient $C_\lambda = 2^{\lambda + n/2}\Gamma(\tfrac{1}{2}\lambda + \tfrac{1}{2}n)/\Gamma(-\tfrac{1}{2}\lambda)$ in (12.12), however, has a simple pole at $\lambda = -n - 2m$. Its Laurent expansion there is given by

$$C_\lambda = \frac{c_{-1}}{\lambda + n + 2m} + c_0 + c_1(\lambda + n + 2m) + \cdots, \qquad (12.16)$$

where

$$c_{-1} = \frac{(-1)^m 2^{1-n/2-2m}}{m!\,\Gamma(\tfrac{1}{2}n + m)}, \qquad (12.17)$$

c_0 and c_1 are some fixed constants whose exact value can be calculated but are irrelevant to our needs. Multiplying (12.15) and (12.16) together gives

$$C_\lambda \rho^{-\lambda-n} = \frac{c_{-1}\rho^{2m}}{\lambda + n + 2m} - (c_{-1}\rho^{2m} \log \rho - c_0\rho^{2m}) + \cdots. \qquad (12.18)$$

Comparing coefficients of (12.14) and (12.18), we obtain

$$(r^{-n-2m})^\wedge = -c_{-1}\rho^{2m} \log \rho + c_0\rho^{2m}. \qquad (12.19)$$

This equation holds, of course, only in the sense of tempered distributions. Both results (12.12) and (12.19) will be used in our derivation of the asymptotic expansion of multiple Fourier transforms.

13. Surface Distributions

Undoubtedly, among all distributions the so-called singular distributions are the most interesting ones, since they are genuinely different from the ordinary functions. An example of such a distribution is provided by the Dirac δ-function, which is zero everywhere except at a single point. Distributions such as these are said to be *concentrated* at a single point. In this section we are concerned with the analogue of these distributions in $\mathbb{R}^n$. That is, we shall study nonzero distributions which are zero outside an $(n-1)$-dimensional surface S given by

$$P(x_1, \ldots, x_n) = 0.$$

Distributions of this kind are said to be *concentrated* on S and are called *surface distributions*.

Before proceeding further, let us first digress to recall the following result concerning resolution of multiple integrals, a discussion of which can be found in Courant and John (1974, p. 445–455).

Theorem 9.   *Let Ω be a bounded domain in $\mathbb{R}^n$ and let $\varphi(x_1, \ldots, x_n)$ be a C^2-function in Ω. Denote by M and m the supremum and infimum of φ in Ω, respectively. If* (i) *Ω can be covered by the family of surfaces determined by*

$$\varphi(x_1, \ldots, x_n) = t, \qquad m < t < M,$$

in such a way that through each point $(x_1, \ldots, x_n)$ of Ω there passes one, and only one, surface, and (ii) *the gradient*

$$\nabla \varphi = (\varphi_{x_1}, \ldots, \varphi_{x_n}) \tag{13.1}$$

is nowhere zero on $\varphi(x_1, \ldots, x_n) = t$ for $t \in (m, M)$, then for any continuous function f in Ω the multiple integral

$$I = \int \cdots \int_{\Omega} f(x_1, \ldots, x_n) \, dx_1 \ldots dx_n \tag{13.2}$$

can be reduced to the single integral

$$I = \int_m^M h(t) \, dt,$$ (13.3)

where h(t) is the surface integral given by

$$h(t) = \int_{\varphi=t} \frac{f(x_1, \ldots, x_n)}{|\nabla\varphi|} \, d\sigma,$$ (13.4)

$d\sigma$ being the surface element on $\varphi = t$ and $|\nabla\varphi| = \sqrt{\varphi_{x_1}^2 + \cdots + \varphi_{x_n}^2}$.

The difficulty in using the above reduction procedure lies in the verification of the condition that the domain Ω can be covered by a family of surfaces as described above. To facilitate the application of this method, we also mention the following *nesting property* for positive definite functions. A C^1-function $V(x) = V(x_1, \ldots, x_n)$ in a neighbour-hood N of the origin is said to be *positive definite* there if it is positive at all points of N except the origin, and $V(0) = 0$. Clearly, if V is positive definite in N, then V has a minimum at the origin, and the origin is a *critical point* of V, that is, a point at which all the partial derivatives

$$\frac{\partial V}{\partial x_1}, \ldots, \frac{\partial V}{\partial x_n}$$

vanish.

Theorem 10. *Let Ω be a bounded domain containing the origin 0 in $\mathbb{R}^n$, and let its boundary Γ be a closed $(n-1)$-dimensional C^1-surface in $\mathbb{R}^n$. Let $V(x) = V(x_1, \ldots, x_n)$ be a positive definite C^1-function in $\Omega \cup \Gamma$, and put*

$$\delta = \inf\{V(x): x \in \Gamma\} > 0, \quad V_\delta = \{x \in \bar\Omega: V(x) = \delta\}.$$

If Ω_δ denotes the domain bounded by V_δ, then $\Omega_\delta - \{0\}$ is the union of level sets

$$V_t = \{x \in \Omega: V(x) = t\}, \quad 0 < t < \delta,$$

and each V_t is an $(n-1)$-dimensional closed C^1-surface which separates $\mathbb{R}^n - V_t$ into two components one of which $B(V_t)$ contains 0. Furthermore, if $0 < t_1 < t_2 < \delta$ then $V_{t_1} \subset B(V_{t_2})$, and if $x \in B(V_t)$ then $V(x) < t$.

The above nesting property is well-known in the study of stability theory in ordinary differential equations; see, for example, LaSalle and Lefschetz (1961) and Leighton (1976, p. 198). For a more precise statement and a rigorous proof of this result, we refer to Lam (1976).

We are now in the position to introduce surface distributions. Consider a hypersurface S given by $P(x_1, \ldots, x_n) = 0$, where $P(x)$ is a C^∞-function with $\nabla P = (P_{x_1}, \ldots, P_{x_n})$ nowhere zero on S, i.e., S has no singularities. Let $H(P)$ denote the characteristic function for the region $P(x) \geq 0$:

$$H(P) = \begin{cases} 0 & \text{for } P < 0 \\ 1 & \text{for } P \geq 0, \end{cases}$$

and define the distribution associated with this function by

$$\langle H(P), \varphi \rangle = \int_{P \geq 0} \varphi(x)\, dx, \qquad \varphi \in \mathscr{D}(\mathbb{R}^n). \tag{13.5}$$

Since the one-dimensional δ-function is the (distributional) derivative of the Heaviside step function, it is natural to define the "Dirac distribution" $\delta(P)$ by

$$\langle \delta(P), \varphi \rangle = \lim_{c \to 0} \frac{1}{c} \langle H(P+c) - H(P), \varphi \rangle = \lim_{c \to 0} \frac{1}{c} \int_{-c \leq P < 0} \varphi(x)\, dx. \tag{13.6}$$

By Theorem 9, the last integral may be written as $\int_{-c}^0 h(t)\, dt$, where

$$h(t) = \int_{P=t} \frac{\varphi(x)}{|\nabla P|}\, d\sigma.$$

Since it can be verified that $h(t)$ is continuous at $t = 0$, (see Ex. 26), we have

$$\lim_{c \to 0} \frac{1}{c} \int_{-c}^0 h(t)\, dt = h(0).$$

Therefore, Eq. (13.6) becomes

$$\langle \delta(P), \varphi \rangle = \int_{P=0} \frac{\varphi(x)}{|\nabla P|}\, d\sigma. \tag{13.7}$$

This is clearly a continuous linear functional on $\mathscr{D}(\mathbb{R}^n)$.

The derivatives of $\delta(P)$ can be defined successively by

$$\delta^{(k)}(P) = \lim_{c \to 0} \frac{1}{c}[\delta^{(k-1)}(P+c) - \delta^{(k-1)}(P)], \qquad (13.8)$$

$k = 1, 2, \ldots$. From (13.7)

$$\langle P(x)\delta(P), \varphi(x)\rangle = \langle \delta(P), P(x)\varphi(x)\rangle = 0,$$

and hence

$$P\delta(P) = 0. \qquad (13.9)$$

Differentiation with respect to P in the sense of (13.6) gives

$$P\delta'(P) = -\delta(P).$$

Repeating the process, we obtain

$$P\delta^{(k)}(P) = -k\delta^{(k-1)}(P), \qquad k = 1, 2, \ldots. \qquad (13.10)$$

Example 15. Let $P(x, y) = t - x$. Then $\nabla P = (-1, 0)$ and $|\nabla P| = 1$. Hence, for any $\varphi \in \mathscr{D}(\mathbb{R}^2)$,

$$\int_{P=0} \frac{\varphi(x, y)}{|\nabla P|}\, d\sigma = \int_{-\infty}^{\infty} \varphi(t, y)\, dy.$$

That is,

$$\langle \delta(t - x), \varphi\rangle = \int_{-\infty}^{\infty} \varphi(t, y)\, dy. \qquad (13.11)$$

Coupling (13.8) and (13.11), we have

$$\langle \delta'(t - x), \varphi\rangle = \int_{-\infty}^{\infty} \varphi_t(t, y)\, dy$$

and more generally

$$\langle \delta^{(k)}(t - x), \varphi\rangle = \int_{-\infty}^{\infty} \frac{d^k}{dt^k} \varphi(t, y)\, dy, \qquad k = 1, 2, \ldots. \qquad (13.12)$$

Exercises

1. If φ is a test function on (a, b), then $\varphi = \psi'$ for some test function ψ if and only if $\int_a^b \varphi(x)\, dx = 0$. Here a and/or b may be infinite.

2.   Fix $\varepsilon > 0$; take $\varphi_\varepsilon(x)$ as in Example 1, and put $\psi_n(x) = \varphi_\varepsilon(x/n)/n$. Show that for each compact set K in $\mathbb{R}$ and each nonnegative integer k, the sequence $\psi_n^{(k)}$ converges to zero uniformly on K as $n \to \infty$. Thus, convergence in $\mathscr{D}(\mathbb{R})$ is not equivalent to convergence in $C^\infty(\mathbb{R})$.

3.   Show that each of the following sequences

$$\delta_n(x) = \frac{n}{\sqrt{\pi}} e^{-n^2 x^2}, \qquad \delta_n(x) = \frac{n}{\pi} \frac{1}{1 + n^2 x^2},$$

$$\delta_n(x) = \frac{\sin nx}{\pi x} = \frac{1}{2\pi} \int_{-n}^{n} e^{ixt}\, dt,$$

satisfies

$$\varphi(a) = \lim_{n \to \infty} \int_{-\infty}^{\infty} \delta_n(x - a)\varphi(x)\, dx \qquad \text{for any } \varphi \in \mathscr{D}(\mathbb{R}).$$

4.   If $\varphi \in \mathscr{D}(I)$ and $\Lambda \in \mathscr{D}'(I)$, does either of the statements

$$\varphi\Lambda = 0, \qquad \langle \Lambda, \varphi \rangle = 0$$

imply the other?

5.   Find a distribution $\Lambda \in \mathscr{D}'(\mathbb{R})$ such that $\Lambda = 1/x$ on $(0, \infty)$ and $\Lambda = 0$ on $(-\infty, 0)$.

6.   Verify the first statement in the *Remark* in §3 and give an example to show that it is not true when f is only locally integrable.

7.   Use the function ρ in Lemma 1 to find a function of (x, y) with derivatives of all orders which is one in a circle, zero outside a larger concentric circle, and between zero and one in the ring between the circles.

8.   Let f_n and f be locally integrable functions such that

$$\int_{-N}^{N} |f_n - f|\, dx \to 0,$$

as $n \to \infty$, for every N. Show that $\Lambda_{f_n} \to \Lambda_f$ in the sense of Definition 5, as $n \to \infty$.

9.   Let $I = (0, \infty)$, and define

$$\langle \Lambda, \varphi \rangle = \sum_{k=0}^{\infty} \varphi^{(k)}\!\left(\frac{1}{k}\right), \qquad \varphi \in \mathscr{D}(I).$$

Show that Λ is a distribution in $(0, \infty)$, but that there is no $\Lambda_1 \in \mathscr{D}'(\mathbb{R})$ such that $\Lambda_1 = \Lambda$ in $(0, \infty)$.

10. Show that with δ defined by (2.4) and translation by (4.6), we indeed have $\langle \delta(x - a), \varphi \rangle = \varphi(a)$.

11. Prove that $x^m \delta^{(n)} = 0$ for $m \geq n + 1$.

12. Prove that the support of Λ' is contained in the support of Λ.

13. (a) Use Ex. 1 to show that if $D\Lambda = 0$, then $\Lambda = \Lambda_c$ for some constant function c. (Hint: if two linear functionals have the same null space, then one is a multiple of the other.)

    (b) If $D\Lambda$ is a regular distribution, then so is Λ.

14. Verify the identities in (6.9), (6.10), (6.11), (6.13), (6.14), (6.15), and (6.16).

15. Using the motivations for (7.10) and (7.13), show that the identity

$$\int_0^b t^{-1}\varphi(t)\, dt = \varphi(b) \log b - \int_0^b \varphi'(t) \log t\, dt,$$

valid for all test functions φ whose supports do not contain 0, suggests that

$$\int_0^b t^{-1}\, dt = \log b, \qquad \int_b^\infty t^{-1}\, dt = -\log b,$$

and hence,

$$\int_0^\infty t^{-1}\, dt = 0;$$

i.e., (7.14) holds also for $\lambda = -1$.

16. Using arguments similar to those for (7.10) and (7.13), derive the formulas

$$\int_0^b t^\lambda (\log t)^r\, dt = \sum_{k=0}^r \frac{(-1)^k r!}{(r-k)!} \frac{b^{\lambda+1}}{(\lambda+1)^{k+1}} (\log b)^{r-k}$$

$$\int_b^\infty t^\lambda (\log t)^r\, dt = - \sum_{k=0}^r \frac{(-1)^k r!}{(r-k)!} \frac{b^{\lambda+1}}{(\lambda+1)^{k+1}} (\log b)^{r-k}$$

for all complex $\lambda \neq -1$ and all nonnegative integers r. Furthermore, show that

$$\int_0^b t^{-1}(\log t)^r \, dt = \frac{1}{r+1}(\log b)^{r+1}$$

$$\int_b^\infty t^{-1}(\log t)^r \, dt = -\frac{1}{r+1}(\log b)^{r+1}$$

for any nonnegative integer r. Thus,

$$\int_0^\infty t^\lambda (\log t)^r \, dt = \left(\int_0^b + \int_b^\infty\right) t^\lambda (\log t)^r \, dt = 0$$

for all complex λ and all nonnegative integers r. Note that this result generalizes (7.14).

17. Show that $f(x) = e^x \cos(e^x)$ is a tempered distribution, but $f(x) = e^x$ is not.

18. Prove that

$$\int_0^\infty \frac{\log x}{x + \lambda} e^{-\eta x} \, dx = O(\log^2 \eta), \qquad \text{as } \eta \to 0^+,$$

and

$$\int_0^\infty \frac{\log u}{(u+1)^{s+2}} \, du = -\frac{1}{s+1} \sum_{k=1}^s \frac{1}{k}, \qquad s = 0, 1, 2, \ldots,$$

where λ is a fixed complex number with $|\arg \lambda| < \pi$, and an empty sum is understood to be zero.

19. Show that Eq. (11.6) differs from the definition given in (8.9) by

$$\left(1 + \frac{1}{2} + \cdots + \frac{1}{n-1}\right) \frac{1}{(n-1)!} \varphi^{(n-1)}(0).$$

20. (a) Let f be a scalar function on $\mathbb{R}^n$, and define the translation operator τ_x by

$$(\tau_x f)(y) = f(y - x).$$

(b) Let $u \in \mathscr{D}'_n$, and define the translate $\tau_x u$ by

$$\langle \tau_x u, \varphi \rangle = \langle u, \tau_{-x}\varphi \rangle, \qquad \varphi \in \mathscr{D}_n, \quad x \in \mathbb{R}^n;$$

cf. (4.6). Show that $D[\Lambda(x - a)] = D\Lambda(x - a)$ for any distribution Λ and any a. (Thus, differentiation and translation are commuting operators on $\mathscr{D}'_n$.)

21. Show that if $u \in \mathscr{S}'_n$, then

$$(\tau_x u)^\wedge = \hat{u} e^{it \cdot x} \qquad \text{and} \qquad (e^{ix \cdot t} u)^\wedge = \tau_{-x} \hat{u}$$

for every $x \in \mathbb{R}^n$.

22. If φ is defined by

$$\varphi(x) = \exp\left(-\frac{1}{2}|x|^2\right), \qquad x \in \mathbb{R}^n,$$

show that $\hat{\varphi} = \varphi$.

23. Show that if f and g are absolutely integrable, then

$$\int_{\mathbb{R}^n} \hat{f}(t) g(t)\, dt = \int_{\mathbb{R}^n} f(x) \hat{g}(x)\, dx.$$

This is known as the *Parseval identity* for Fourier transforms.

24. Let $\varphi \in \mathscr{S}_n$. Use Exercises 22 and 23 to show that

$$\int_{\mathbb{R}^n} \hat{\varphi}(t) \exp\left(-\frac{1}{2} \varepsilon^2 |t|^2\right) dt = \int_{\mathbb{R}^n} \varphi(\varepsilon x) \exp\left(-\frac{1}{2}|x|^2\right) dx.$$

Deduce that

$$\varphi(0) = \frac{1}{(2\pi)^{n/2}} \int_{\mathbb{R}^n} \hat{\varphi}(t)\, dt.$$

Use Ex. 20(a) to conclude that

$$\varphi(x) = \frac{1}{(2\pi)^{n/2}} \int_{\mathbb{R}^n} \hat{\varphi}(t) e^{-ix \cdot t}\, dt.$$

This is known as the *Fourier inversion formula,* and $\varphi(x)$ is referred to as the *inverse Fourier transform* of $\hat{\varphi}$.

25. Show that $\hat{\delta} = 1$ and $\hat{1} = \delta$ in $\mathscr{S}'_n$.

26. (a) Let $f \in C^2(\mathbb{R}^2)$ and $g \in \mathscr{D}(\mathbb{R}^2)$. Suppose that $\nabla f \neq 0$ on $f(x, y) = t$ for all t in $[t_0 - \varepsilon, t_0 + \varepsilon]$, and put

$$h(t) = \int_{f(x,y) = t} g(x, y)\, d\sigma,$$

where $d\sigma$ is an element of the arc length of the curve $f(x, y) = t$. Show that $h(t)$ is continuous at t_0. (Hint: show that $d\sigma = \pm(f_y\, dy - f_x\, dx)/|\nabla f|$, and apply Green's theorem.)

(b) Extend this result to $\mathbb{R}^n$. (Hint: The surface element on $f(x_1, \ldots, x_n) = t$ is given by

$$d\sigma = \sum (-1)^{i-1}\eta_i(x)\, dx_1 \wedge \cdots \wedge dx_{i-1} \wedge dx_{i+1} \wedge \cdots \wedge dx_n,$$

where $(\eta_1, \ldots, \eta_n) = \pm\nabla f/|\nabla f|$; see Edwards (1973, p. 380). Apply Stokes' theorem.)

27. Let f and g be C^∞-functions, and let n be a nonnegative integer. Show that the sum

$$[f(x)g^{n+1}(x)]^{(n+1)} + \sum_{p=0}^{n} \binom{n+1}{p+1}[f(x)g^{n-p}(x)]^{(n+1)}(-g(x))^{p+1}$$

is equal to

$$\frac{d^{n+1}}{dt^{n+1}}[f(t)(g(t) - g(x))^{n+1}]\Big|_{t=x}.$$

Moreover, by using the identity

$$g(t) - g(x) = (t - x)[g'(x) + \varepsilon(t, x)],$$

where $\lim_{t \to x} \varepsilon(t, x) = 0$, show that

$$[f(x)g^{n+1}(x)]^{(n+1)} = -\sum_{p=0}^{n} \binom{n+1}{p+1}[f(x)g^{n-p}(x)]^{(n+1)}(-g(x))^{p+1}$$

$$+ (n+1)!\, f(x)(g'(x))^{n+1}.$$

28. Let $\Phi(\xi, \eta)$ be a C^∞-function with compact support, and let $F(\xi, \eta)$ be a C^∞-function with $\partial F/\partial\xi > -1$ on the support of Φ. Show that

$$\langle \delta(t - \xi - F), \Phi \rangle = \int_{-\infty}^{\infty} \frac{\Phi(\xi_t, \eta)}{1 + F_\xi(\xi_t, \eta)}\, d\eta,$$

where ξ_t is the solution to the equation $t = \xi + F(\xi, \eta)$. Furthermore, derive the Taylor-type series

$$\langle \delta(t - \xi - F), \Phi \rangle = \sum_{r=0}^{n} \frac{(-1)^r}{r!} \langle F^r(\xi, \eta)\delta^{(r)}(t - \xi), \Phi \rangle + R_{n+1}(t),$$

where

$$R_{n+1}(t) = \int_{-\infty}^{\infty} r_{n+1}(t, \eta) \, d\eta$$

and

$$r_{n+1}(t, \eta) = \frac{\Phi(\xi_t, \eta)}{1 + F_\xi(\xi_t, \eta)} - \sum_{r=0}^{n} \frac{(-1)^r}{r!} \frac{\partial^r}{\partial t^r} [\Phi(t, \eta)F^r(t, \eta)].$$

Finally, use induction and Ex. 27 to show that

$$r_{n+1}(t, \eta) = \frac{\Phi(\xi_t, \eta)[-F_\xi(\xi_t, \eta)]^{n+1}}{1 + F_\xi(\xi_t, \eta)}$$

$$+ \sum_{r=0}^{n} \frac{1}{r!\,(n-r)!} \int_t^{\xi_t} (t-\mu)^r$$

$$\times \frac{\partial^{n+1}}{\partial \mu^{n+1}} \{\Phi(\mu, \eta)[-F(\mu, \eta)]^{n-r}\} \, d\mu.$$

(Hint: First insert the last equation (with $n = k$) in

$$r_{k+2}(t, \eta) = r_{k+1}(t, \eta) - \frac{(-1)^{k+1}}{(k+1)!} \frac{\partial^{k+1}}{\partial t^{k+1}} [\Phi(t, \eta)F^{k+1}(t, \eta)].$$

Next, integrate the terms in the sum by parts. Finally, add and subtract the terms

$$\Phi(\xi_t, \eta)[-F_\xi(\xi_t, \eta)]^{k+1}$$

and

$$\frac{(-1)^{k+1}}{(k+1)!} \left[\frac{\partial^{k+1}}{\partial \xi^{k+1}} \{\Phi(\xi, \eta)F^{k+1}(\xi, \eta)\} \right]_{\xi = \xi_t} .)$$

29. Let f be a real-valued and C^∞-function in $\mathbb{R}^2$, and let g be a C^∞-function with compact support. Consider the double integral

$$I(\lambda) = \iint_{\mathbb{R}^2} g(x, y)e^{i\lambda f(x,y)} \, dx \, dy.$$

Suppose that the minimum of f is zero and is attained at the origin.

(a) Show that

$$I(\lambda) = \int_0^\infty h(t)e^{i\lambda t} \, dt, \qquad \text{where } h(t) = \langle \delta(t - f), g \rangle.$$

(b)   Suppose that the Maclaurin expansion of $f(x, y)$ takes the form

$$f(x, y) = f_{20}x^2 + f_{02}y^2 + \text{cubic terms} + \cdots,$$

where f_{20} and f_{02} are both positive, and let

$$x = \xi^{1/2} \frac{\cos \eta}{f_{20}^{1/2}}, \qquad y = \xi^{1/2} \frac{\sin \eta}{f_{02}^{1/2}}$$

so that $f(x, y) = \xi + F(\xi, \eta)$. Show that the function $h(t)$ in (a) can be written as $h(t) = \langle \delta(t - \xi - F), \Phi \rangle$, where $\Phi(\xi, \eta) = g(x, y) \times \partial(x, y)/\partial(\xi, \eta)$.

(c)   Use Ex. 28 to show that

$$h(t) = \sum_{r=0}^{n} \frac{\partial^r}{\partial t^r} \int_0^{2\pi} \frac{(-1)^r}{r!} \, [\Phi(t, \eta) F^r(t, \eta)] \, d\eta + R_{n+1}(t),$$

where

$$R_{n+1}(t) = O(t^{(n+1)/2}), \qquad \text{as } t \to 0^+.$$

(d)   Deduce from (c) the asymptotic expansion

$$h(t) \sim \sum_{v=0}^{\infty} a_v t^v, \qquad \text{as } t \to 0^+,$$

where the coefficients are constants. Finally, show that

$$I(\lambda) \sim \sum_{v=0}^{\infty} a_v e^{i(v+1)\pi/2} \frac{\Gamma(v+1)}{\lambda^{v+1}}, \qquad \text{as } \lambda \to +\infty.$$

Supplementary Notes

Standard references for distribution theory are Gelfand and Shilov (1964) and Schwartz (1966a), and a large part of this chapter is taken from the first reference. However, use has also been made of the books by Bremermann (1965) and Rudin (1973).

§6.   The material in this section is based on McClure and Wong (1979). The results obtained here will be used in Chapter VI, §6.

§7.   The contents of this section are taken from Gelfand and Shilov (1964, Chapter I, §3). For further discussion of regularization of divergent integrals, see Wong and McClure (1984).

§§8 & 9.   The definition of convergence in $\mathscr{S}$ given in these sections differs from the usual one (see Bremermann (1965, p. 86)). Our definition is taken from Gelfand and Shilov (1964, pp. 16–17).

§§10 & 11.   The contents of these sections are essentially an adaptation of Chapter I, §§3.9, 4.2, and 4.6, of Gelfand and Shilov (1964).

§12.   For an application of the results in this section, see Chapter IX, §6.

§13.   The results in this section will be used in the derivation of asymptotic expansions of double integrals (Chapter VIII).

Exercises.   Exercises 4 and 9 are taken from Rudin (1973, p. 163). For a solution to Ex. 16, see McClure and Wong (1984). If $f \in C^\infty(\mathbb{R}^2)$, then, in fact, it is known that the function $h(t)$ defined by a line integral in Ex. 26(a) is infinitely differentiable at t_0; see Jeanquartier (1970). Exercise 27 is based on the short note of Meinardus (1985). The results in Exs. 28 and 29 can be found in Wong and McClure (1981); for more general and abstract results, see also Malgrange (1974).

VI

The Distributional Approach

1. Introduction

The construction of an asymptotic expansion usually consists of three steps. The first is to derive a formal expansion, the second is to put the result on a rigorous basis, and the third is to construct error bounds. The two most frequently used methods for constructing formal asymptotic expansions for integrals are undoubtedly "integration by parts" and "termwise integration". However, as evidenced in Example 3, Chapter I, a purely formal application of partial integration may lead to an incorrect result. The integral considered there is

$$S(x) = \int_0^\infty \frac{1}{\sqrt[3]{1 + t(x + t)}}\, dt, \tag{1.1}$$

and the incorrect result given is

$$S(x) \sim - \sum_{s=1}^\infty \frac{3^s(s-1)!}{2 \cdot 5 \cdots (3s-1)} \frac{1}{x^s}. \tag{1.2}$$

We shall now illustrate, by using the same example, that a formal application of termwise integration may also yield an incorrect result. For $t > 1$, we have

$$(1 + t)^{-1/3} = \sum_{s=0}^{\infty} \binom{-\frac{1}{3}}{s} t^{-s-1/3}. \tag{1.3}$$

Substituting (1.3) in (1.1), and integrating term by term, we obtain divergent integrals of the form

$$\int_0^{\infty} \frac{t^{-s-1/3}}{x+t} \, dt, \qquad s = 0, 1, 2, \dots. \tag{1.4}$$

Motivated by the use of Abel limits for the divergent integrals in Chapter IV, it is tempting to interpret the above integrals also in some generalized sense. A natural analogue of (2.3) in that chapter is to consider the integrals in (1.4) as limits of the distributions $t_+^{-s-1/3}$ acting on the test function $\varphi_\eta(t) = e^{-\eta t}/(x + t)$ as $\eta \to 0^+$, i.e., we set

$$\int_0^{\infty} \frac{t^{-s-1/3}}{x+t} \, dt = \frac{2\pi}{\sqrt{3}} \frac{(-1)^s}{x^{s+1/3}}, \qquad s = 0, 1, 2, \dots; \tag{1.5}$$

see (8.11), Chapter V. With this interpretation, termwise integration gives

$$S(x) \sim \frac{2\pi}{\sqrt{3}} \sum_{s=0}^{\infty} (-1)^s \binom{-\frac{1}{3}}{s} x^{-s-1/3}, \qquad \text{as } x \to \infty, \tag{1.6}$$

a result which is again incorrect, in view of the correct result

$$S(x) \sim \frac{2\pi}{\sqrt{3}} \sum_{s=0}^{\infty} (-1)^s \binom{-\frac{1}{3}}{s} x^{-s-1/3} - \sum_{s=1}^{\infty} \frac{3^s(s-1)!}{2 \cdot 5 \cdots (3s-1)} x^{-s} \tag{1.7}$$

given in §2 below.

The fact that expansion (1.6) misses all the terms in the second series in (1.7) raises the question: what went wrong in the derivation of (1.6), knowing that the result (8.11) in Chapter V is correct? In the present chapter, we shall study in great detail the proper use of divergent integrals. In particular, the question just raised will be answered in §2. Infinite asymptotic expansions are derived for the Stieltjes transform in §§2 and 3. Similar results are obtained in §5 for the Laplace and Fourier transforms near the origin. In §§4 and 6, we apply the results in §2 to construct asymptotic expansions for the Hilbert transform and the Riemann–Liouville fractional integrals, respectively. The final section is devoted to a further discussion of the regularization technique introduced in §7, Chapter V. Asymptotic expansions are then obtained, by using this technique, for integrals that can be put in the form of a Mellin convolution.

2. The Stieltjes Transform

The Stieltjes transform of a locally integrable function $f(t)$ on $[0, \infty)$ is defined by

$$S_f(z) = \int_0^\infty \frac{f(t)}{t + z}\, dt, \tag{2.1}$$

where z is a complex variable in the cut plane $|\arg z| < \pi$. If $f(t)$ has an exponential rate of decay at ∞, then all the moments of f,

$$\mu_s = \int_0^\infty t^s f(t)\, dt, \qquad s = 0, 1, 2, \ldots, \tag{2.2}$$

exist. Furthermore, if $z = x$ is real and positive, then it is easily shown that

$$S_f(x) = \sum_{s=0}^{n-1} (-1)^s \mu_s x^{-s-1} + \varepsilon_n(x), \tag{2.3}$$

where

$$|\varepsilon_n(x)| \le x^{-n-1} \sup_{(0,\infty)} \left| \int_0^t \tau^n f(\tau)\, d\tau \right|; \tag{2.4}$$

see Ex. 1. However, if $f(t)$ decays only algebraically, then the above result is no longer meaningful.

In what follows, we shall assume that $f(t)$ possesses an asymptotic expansion of the form

$$f(t) \sim \sum_{s=0}^\infty a_s t^{-s-\alpha}, \qquad \text{as } t \to \infty, \tag{2.5}$$

where $0 < \alpha \le 1$. For each $n \ge 1$, we set

$$f(t) = \sum_{s=0}^{n-1} a_s t^{-s-\alpha} + f_n(t). \tag{2.6}$$

We shall now assign to each function in (2.6) a distribution on $\mathscr{S}$. Since $f(t)$ is locally integrable on $[0, \infty)$, it defines a distribution by

$$\langle f, \varphi \rangle = \int_0^\infty f(t)\varphi(t)\, dt. \tag{2.7}$$

The distributions associated with $t^{-s-\alpha}$, $s = 0, 1, \ldots, n-1$, are given by

$$\langle t_+^{-s-\alpha}, \varphi \rangle = \frac{1}{(\alpha)_s} \int_0^\infty t^{-\alpha}\varphi^{(s)}(t)\, dt, \tag{2.8}$$

when $0 < \alpha < 1$, where $(\alpha)_s = \alpha(\alpha + 1)\cdots(\alpha + s - 1)$, and similarly

$$\langle t_+^{-s-1}, \varphi \rangle = -\frac{1}{s!} \int_0^\infty (\log t)\varphi^{(s+1)}(t)\, dt; \tag{2.9}$$

see Example 8, Chapter V. To assign a distribution to the function $f_n(t)$, we first define inductively $f_{n,0}(t) = f_n(t)$ and

$$f_{n,j+1}(t) = -\int_t^\infty f_{n,j}(\tau)\, d\tau = \frac{(-1)^{j+1}}{j!} \int_t^\infty (\tau - t)^j f_n(\tau)\, d\tau, \tag{2.10}$$

$j = 0, 1, \ldots, n-1$. For $0 < \alpha < 1$, it is easily seen that $f_{n,n}(t)$ is bounded on $[0, R]$ for any positive R and $O(t^{-\alpha})$ as $t \to \infty$. For $\alpha = 1$, we have $f_{n,n}(t) = O(t^{-1})$ as $t \to \infty$ and $f_{n,n}(t) = O(\log t)$ as $t \to 0^+$. In either case, we can define the distribution associated with $f_n(t)$ by

$$\langle f_n, \varphi \rangle = (-1)^n \langle f_{n,n}, \varphi^{(n)} \rangle = (-1)^n \int_0^\infty f_{n,n}(t)\varphi^{(n)}(t)\, dt, \qquad \varphi \in \mathscr{S}. \tag{2.11}$$

We have now assigned a distribution to each function in (2.6). A natural question that arises is: what is the exact relation between these distributions? The answer to this question is provided by the following lemmas.

Lemma 1. *For $0 < \alpha < 1$ and $n \geq 1$, the identity*

$$\langle f, \varphi \rangle = \sum_{s=0}^{n-1} a_s \langle t_+^{-s-\alpha}, \varphi \rangle - \sum_{s=1}^{n} c_s \langle \delta^{(s-1)}, \varphi \rangle + \langle f_n, \varphi \rangle \tag{2.12}$$

holds for any rapidly decreasing function $\varphi \in \mathscr{S}$, where

$$c_s = \frac{(-1)^s}{(s-1)!} M[f; s], \tag{2.13}$$

$M[f; z]$ being the Mellin transform of $f(t)$, or its analytic continuation.

Proof. Let $f_0(t) = f(t)$. Then

$$f_{n+1}(t) = f_n(t) - a_n t^{-n-\alpha} \qquad (2.14)$$

and

$$f_{n+1,n}(t) = f_{n,n}(t) - \frac{(-1)^n a_n}{(\alpha)_n} t^{-\alpha}, \qquad (2.15)$$

$n = 0, 1, 2, \ldots$. From this it follows, by integration by parts, that

$$\langle f_n, \varphi \rangle = -f_{n+1,n+1}(0)\langle \delta^{(n)}, \varphi \rangle + \langle f_{n+1}, \varphi \rangle + a_n \langle t_+^{-n-\alpha}, \varphi \rangle$$

($f_{n+1,n+1}(0)$ exists, by the remark following (2.10)). Therefore

$$\langle f_{n+1}, \varphi \rangle + \sum_{s=0}^{n} a_s \langle t_+^{-s-\alpha}, \varphi \rangle = \langle f_n, \varphi \rangle + \sum_{s=0}^{n-1} a_s \langle t_+^{-s-\alpha}, \varphi \rangle$$
$$+ f_{n+1,n+1}(0)\langle \delta^{(n)}, \varphi \rangle.$$

Repeated application of this identity gives

$$\langle f_{n+1}, \varphi \rangle + \sum_{s=0}^{n} a_s \langle t_+^{-s-\alpha}, \varphi \rangle = \langle f, \varphi \rangle + \sum_{s=0}^{n} f_{s+1,s+1}(0)\langle \delta^{(s)}, \varphi \rangle. \quad (2.16)$$

Upon rearranging the terms in (2.16), we obtain the desired result (2.12) with $c_s = f_{s,s}(0)$. Equation (2.13) now follows from (2.10) and Lemma 7, Chapter III. ∎

Lemma 2. *If $\alpha = 1$ in (2.6) then for each integer $n \geq 1$ and for any function $\varphi \in \mathscr{S}$ we have*

$$\langle f, \varphi \rangle = \sum_{s=0}^{n-1} a_s \langle t_+^{-s-1}, \varphi \rangle - \sum_{s=1}^{n} d_s \langle \delta^{(s-1)}, \varphi \rangle + \langle f_n, \varphi \rangle, \quad (2.17)$$

where

$$d_s = \lim_{t \to 0} \left[f_{s,s}(t) + \frac{(-1)^{s-1}}{(s-1)!} a_{s-1} \log t \right]. \qquad (2.18)$$

(For a more tractable expression for d_s, see Eqs. (2.32) and (2.34) below.)

Proof. Let $f_{0,0}(t) = f_0(t) = f(t)$. Then for any $n \geq 0$

$$f_{n+1}(t) + a_n t^{-n-1} = f_n(t) \qquad (2.19)$$

and hence

$$f_{n+1,n}(t) + \frac{(-1)^n}{n!} a_n t^{-1} = f_{n,n}(t). \qquad (2.20)$$

Note that from (2.20) it follows that

$$f_{n+1,n+1}(t) + \frac{(-1)^n}{n!} a_n \log t$$

is an antiderivative of $f_{n,n}(t)$ on $(0, \infty)$, but so is the integral $\int_0^t f_{n,n}(\tau)\, d\tau$. Thus we have

$$\int_0^t f_{n,n}(\tau)\, d\tau = f_{n+1,n+1}(t) + \frac{(-1)^n}{n!} a_n \log t + \text{constant}, \qquad (2.21)$$

and the limits

$$d_{n+1} = \lim_{t \to 0} \left[f_{n+1,n+1}(t) + \frac{(-1)^n}{n!} a_n \log t \right]$$

exist, $n = 0, 1, 2, \ldots$. From (2.20) we also have, by integration by parts,

$$\langle f_n, \varphi \rangle = -d_{n+1} \langle \delta^{(n)}, \varphi \rangle + \langle f_{n+1}, \varphi \rangle + a_n \langle t_+^{-n-1}, \varphi \rangle.$$

Thus

$$\langle f_{n+1}, \varphi \rangle + \sum_{s=0}^{n} a_s \langle t_+^{-s-1}, \varphi \rangle = \langle f_n, \varphi \rangle + \sum_{s=0}^{n-1} a_s \langle t_+^{-s-1}, \varphi \rangle$$

$$+ d_{n+1} \langle \delta^{(n)}, \varphi \rangle.$$

Repeated application of these two identities gives

$$\langle f_{n+1}, \varphi \rangle + \sum_{s=0}^{n} a_s \langle t_+^{-s-1}, \varphi \rangle = \langle f, \varphi \rangle + \sum_{s=1}^{n+1} d_s \langle \delta^{(s-1)}, \varphi \rangle, \qquad (2.22)$$

which is of course equivalent to the statement of the lemma. ∎

To apply the above results to the Stieltjes transform, we shall take a specific function $\varphi \in \mathscr{S}$ in (2.12) and (2.17). Thus, let η be a positive number, and let $\varphi_\eta(t)$ denote a function in $\mathscr{S}$ which satisfies

$$\varphi_\eta(t) = \frac{e^{-\eta t}}{t + z} \qquad \text{for } t \in (0, \infty).$$

From Example 12 in Chapter V, we recall the identities

$$\lim_{\eta \to 0} \langle t_+^{-s-\alpha}, \varphi_\eta \rangle = \frac{\pi}{\sin \alpha \pi} \frac{(-1)^s}{z^{s+\alpha}}, \tag{2.23}$$

$$\lim_{\eta \to 0} \langle t_+^{-s-1}, \varphi_\eta \rangle = \frac{(-1)^{s+1}}{z^{s+1}} \sum_{k=1}^{s} \frac{1}{k} + \frac{(-1)^s}{z^{s+1}} \log z, \tag{2.24}$$

$s = 0, 1, 2, \ldots$, empty sum being understood to be zero. According to (2.7) and (2.11), we also have

$$\lim_{\eta \to 0} \langle f, \varphi_\eta \rangle = S_f(z) \tag{2.25}$$

$$\lim_{\eta \to 0} \langle f_n, \varphi_\eta \rangle = n! \int_0^\infty \frac{f_{n,n}(t)}{(t+z)^{n+1}} \, dt. \tag{2.26}$$

We are now ready to state and prove the following results.

Theorem 1. *Let f be a locally integrable function on $[0, \infty)$ and let f satisfy (2.5) with $0 < \alpha < 1$. Then for any $n \geq 1$*

$$S_f(z) = \frac{\pi}{\sin \pi\alpha} \sum_{s=0}^{n-1} (-1)^s \frac{a_s}{z^{s+\alpha}} - \sum_{s=1}^{n} (s-1)! \frac{c_s}{z^s} + R_n(z), \tag{2.27}$$

where the coefficients c_s are given in (2.13). The error term satisfies

$$R_n(z) = n! \int_0^\infty \frac{f_{n,n}(t)}{(t+z)^{n+1}} \, dt \tag{2.28}$$

with $f_{n,n}(t)$ being defined by (2.10).

Proof. By Lemma 1,

$$\langle f, \varphi_\eta \rangle = \sum_{s=0}^{n-1} a_s \langle t_+^{-s-\alpha}, \varphi_\eta \rangle - \sum_{s=1}^{n} c_s \langle \delta^{(s-1)}, \varphi_\eta \rangle + \langle f_n, \varphi_\eta \rangle. \tag{2.29}$$

Since

$$\lim_{\eta \to 0} \langle \delta^{(s-1)}, \varphi_\eta \rangle = \frac{(s-1)!}{z^s}, \qquad s = 1, 2, \ldots, \tag{2.30}$$

equation (2.27) follows from (2.23), (2.25), and (2.26) by letting $\eta \to 0$ in (2.29). ∎

Theorem 2.   *Let f be a locally integrable function on $[0, \infty)$ and let f satisfy (2.5) with $\alpha = 1$. Then for any $n \geq 1$*

$$S_f(z) = \log z \sum_{s=0}^{n-1} (-1)^s a_s z^{-s-1} + \sum_{s=0}^{n-1} (-1)^s a_s^* z^{-s-1} + R_n(z), \quad (2.31)$$

where

$$a_s^* = \lim_{z \to s+1} \left\{ M[f; z] + \frac{a_s}{z - s - 1} \right\} \quad (2.32)$$

and $M[f; z]$ is (the analytic continuation of) the Mellin transform of $f(t)$. The remainder term satisfies

$$R_n(z) = n! \int_0^\infty \frac{f_{n,n}(t)}{(t + z)^{n+1}} \, dt \quad (2.33)$$

with $f_{n,n}(t)$ being defined by (2.10).

Proof.   Equation (2.31) follows from Lemma 2 by replacing φ by φ_η and letting $\eta \to 0$. The coefficient a_s^* in (2.31) is first expressed in the form

$$(-1)^s a_s^* = (-1)^{s+1} a_s \sum_{k=1}^{s} \frac{1}{k} - s! \, d_{s+1}, \quad (2.34)$$

where d_{s+1} is as given in (2.18). Thus it remains to prove only the identity in (2.32). Observe that equation (2.21) is equivalent to

$$\int_0^t f_{n,n}(\tau) \, d\tau = - \int_t^\infty f_{n+1,n}(\tau) \, d\tau + \frac{(-1)^n}{n!} a_n \log t - d_{n+1}. \quad (2.35)$$

Taking $t = 1$ gives

$$d_{s+1} = - \int_1^\infty f_{s+1,s}(\tau) \, d\tau - \int_0^1 f_{s,s}(\tau) \, d\tau. \quad (2.36)$$

By integration by parts,

$$\int_0^1 t^s f_s(t) \, dt = f_{s,1}(1) + \sum_{k=2}^{s} (-1)^{k-1} s(s-1) \cdots (s-k+2) f_{s,k}(1)$$
$$+ (-1)^s s! \int_0^1 f_{s,s}(t) \, dt \quad (2.37)$$

and

$$\int_1^\infty t^s f_{s+1}(t)\, dt = -f_{s+1,1}(1) + \sum_{k=2}^s (-1)^k s(s-1)\cdots(s-k+2) f_{s+1,k}(1)$$

$$+ (-1)^s s! \int_1^\infty f_{s+1,s}(t)\, dt. \tag{2.38}$$

From (2.19), we also have $f_{s+1,1}(1) = f_{s,1}(1) + a_s/s$ and

$$f_{s+1,k}(1) = f_{s,k}(1) - \frac{(-1)^k a_s}{s(s-1)\cdots(s-k+1)},$$

$k = 2, \ldots, s$. Thus, coupling (2.37) and (2.38), we obtain

$$\int_0^1 t^s f_s(t)\, dt + \int_1^\infty t^s f_{s+1}(t)\, dt$$

$$= -a_s \sum_{k=1}^s \frac{1}{k} + (-1)^s s! \left[\int_0^1 f_{s,s}(t)\, dt + \int_1^\infty f_{s+1,s}(t)\, dt \right]. \tag{2.39}$$

Equations (2.36) and (2.39) together yield

$$s!\, d_{s+1} = (-1)^{s+1} a_s \sum_{k=1}^s \frac{1}{k} + (-1)^{s+1} \left[\int_0^1 t^s f_s(t)\, dt + \int_1^\infty t^s f_{s+1}(t)\, dt \right].$$

$$\tag{2.40}$$

The quantity inside the square bracket in (2.40) is equal to

$$\lim_{z \to s+1} \left\{ M[f; z] + \frac{a_s}{z - s - 1} \right\},$$

by virtue of Ex. 9, Chapter III. Therefore, (2.32) follows immediately from (2.34) and (2.40). ∎

We are now ready to answer the question raised in §1. First observe that the integral (1.1) is a special case of (2.1) with $f(t) = (1 + t)^{-1/3}$, and that the expansion (1.3) can be put in the form of (2.6) with $\alpha = \frac{1}{3}$. However, if each function in (2.6) is to be considered as a distribution, then the equation satisfied by these distributions is not (2.6) but (2.12), i.e.,

$$f = \sum_{s=0}^{n-1} a_s t_+^{-s-\alpha} - \sum_{s=1}^n c_s \delta^{(s-1)} + f_n. \tag{2.41}$$

Thus, if one replaces f by its series expansion (2.6) and takes the Stieltjes transform term by term, one obtains only the contribution from the first series in (2.41) and misses completely the contribution from the second series. This is why we obtained in §1 the incorrect result (1.6), instead of the correct result (1.7).

To prove (1.7), we apply Theorem 1 with $f(t) = (1 + t)^{-1/3}$ and $\alpha = \frac{1}{3}$. If, as before, we let $f_n(t)$ denote the n-th remainder in the asymptotic expansion of $f(t)$ and $f_{n,n}(t)$ denote its n-th iterated integral, then it is evident that

$$|f_n(t)| \le \left|\binom{-\frac{1}{3}}{n}\right| t^{-n-1/3}$$

and

$$|f_{n,n}(t)| \le \frac{1}{(\frac{1}{3})_n} \left|\binom{-\frac{1}{3}}{n}\right| t^{-1/3} \tag{2.42}$$

for $t > 0$. Since

$$M[f; z] = \int_0^\infty \frac{t^{z-1}}{(1+t)^{1/3}}\, dt = \frac{\Gamma(z)\Gamma(\frac{1}{3} - z)}{\Gamma(\frac{1}{3})},$$

the coefficient c_s in (2.13) is given by

$$c_s = \frac{3^s}{2 \cdot 5 \cdots (3s - 1)}.$$

Thus, by Theorem 1, we obtain

$$\int_0^\infty \frac{1}{\sqrt[3]{1 + t}(x + t)}\, dt = \frac{2\pi}{\sqrt{3}} \sum_{s=0}^{n-1} (-1)^s \binom{-\frac{1}{3}}{s} x^{-s-1/3}$$
$$- \sum_{s=1}^{n} \frac{3^s (s - 1)!}{2 \cdot 5 \cdots (3s - 1)} x^{-s} + R_n(x), \tag{2.43}$$

where in view of (2.42) the remainder satisfies

$$|R_n(x)| \le \frac{2\pi}{\sqrt{3}} \left|\binom{-\frac{1}{3}}{n}\right| x^{-n-1/3}, \qquad x > 0. \tag{2.44}$$

Numerical bounds similar to (2.44) can be constructed for the remainder (2.28) in general, provided that

$$M_n = \sup_{(0,\infty)} \{t^{n+\alpha} |f_n(t)|\} < +\infty. \tag{2.45}$$

Note that this condition does not follow from (2.5) and the local integrability of f, but will hold in most applications. From (2.45) it follows immediately that the function $f_{n,n}(t)$ satisfies

$$|f_{n,n}(t)| \leq \frac{M_n}{(\alpha)_n} t^{-\alpha} \qquad (2.46)$$

for $t > 0$ and $0 < \alpha < 1$. If z is real and positive, say $z = x > 0$, then the remainder in (2.28) is bounded by

$$|R_n(x)| \leq \frac{M_n}{x^{n+\alpha}} \frac{\pi}{\sin \alpha\pi}, \qquad 0 < \alpha < 1. \qquad (2.47)$$

If z is complex, then it can be shown that

$$|R_n(z)| \leq M_n \frac{C_\alpha(\theta)}{|z|^{n+\alpha}}, \qquad 0 < \alpha < 1, \qquad (2.48)$$

where the constant $C_\alpha(\theta)$ depends on $\theta = \arg z$, $-\pi < \theta < \pi$, but is independent of $|z|$. The explicit value of $C_\alpha(\theta)$ is given by

$$C_\alpha(\theta) = \pi \operatorname{cosec} \alpha\pi \, F\left(\alpha, 1 - \alpha; 1; \sin^2 \frac{\theta}{2}\right), \qquad (2.49)$$

$F(\alpha, \beta; \gamma; z)$ being the hypergeometric function; see Ex. 4. Inequalities (2.47) and (2.48) provide strict error bounds for the asymptotic expansion (2.27). Thus,

$$S_f(z) \sim \frac{\pi}{\sin \alpha\pi} \sum_{s=0}^{\infty} (-1)^s \frac{a_s}{z^{s+\alpha}} - \sum_{s=1}^{\infty} (s-1)! \frac{c_s}{z^s}, \qquad (2.50)$$

as $|z| \to \infty$ in $|\arg z| \leq \pi - \delta(<\pi)$. Note that the quantity in (2.49) tends to infinity as $\theta \to \pm\pi$. Hence the asymptotic expansion (2.50) is not valid in $|\arg z| \leq \pi$.

In the case when f satisfies (2.5) with $\alpha = 1$, a natural way of extending the error analysis (2.47) and (2.48) is to *assume* that for some $\sigma \in (0, 1)$,

$$M_n(\sigma) = \sup_{(0,\infty)} \{t^{n+\sigma} |f_n(t)|\} < \infty. \qquad (2.51)$$

The remainder in (2.33) then satisfies

$$|R_n(x)| \leq \frac{M_n(\sigma)}{x^{n+\sigma}} \frac{\pi}{\sin \sigma\pi} \qquad (2.52)$$

if $z = x$ is real and positive, and

$$|R_n(z)| \le M_n(\sigma) \frac{C_\sigma(\theta)}{|z|^{n+\sigma}} \qquad (2.53)$$

if $0 < |\arg z| < \pi$. Relations (2.52) and (2.53) confirm the asymptotic nature of the expansion (2.31)

$$S_f(z) \sim \log z \sum_{s=0}^\infty (-1)^s a_s z^{-s-1} + \sum_{s=0}^\infty (-1)^s a_s^* z^{-s-1}, \qquad (2.54)$$

as $|z| \to \infty$ in $|\arg z| \le \pi - \delta (<\pi)$. Note that the actual error $R_n(z)$ in the approximation (2.31) is $O(\log z/z^{n+1})$. Hence the estimates (2.52) and (2.53) fall short of the actual result. For the construction of a strict error bound when $z = x$ is real and positive, see Ex. 5.

3. Stieltjes Transform: An Oscillatory Case

The analysis of the previous section can be extended to handle cases in which the function f is ocillatory rather than algebraic near ∞. In this section, we shall consider a function f with the asymptotic expansion

$$f(t) \sim e^{ict} \sum_{s=0}^\infty a_s t^{-s-1}, \qquad c \ne 0, \quad c \text{ real.} \qquad (3.1)$$

For simplicity let us introduce the notations

$$e_s(t) = e^{ict} t^{-s-1} \qquad (3.2)$$

and

$$E_s(t) = (-1)^{s+1} \int_t^\infty \int_{\tau_s}^\infty \cdots \int_{\tau_1}^\infty e_s(\tau_0) \, d\tau_0 \ldots d\tau_s. \qquad (3.3)$$

Clearly $E_s(t)$ is locally integrable, and $O(t^{-s-1})$ as $t \to \infty$. As before we set, for each $n \ge 1$,

$$f(t) = \sum_{s=0}^{n-1} a_s e_s(t) + f_n(t) \qquad (3.4)$$

and define $f_{n,n}(t)$ as in (2.10). Note that $f_{n,n}(t)$ is locally integrable on $[0, \infty)$, and is $O(t^{-n-1})$ as $t \to \infty$. The distributions associated with $e_s(t)$, $s = 0, 1, 2, \ldots$, and $f_n(t)$ are, respectively, defined by

$$\langle e_s, \varphi \rangle = (-1)^{s+1} \langle E_s, \varphi^{(s+1)} \rangle = (-1)^{s+1} \int_0^\infty E_s(t) \varphi^{(s+1)}(t) \, dt, \quad (3.5)$$

$s = 0, 1, 2, \ldots,$ and

$$\langle f_n, \varphi \rangle = (-1)^n \langle f_{n,n}, \varphi^{(n)} \rangle = (-1)^n \int_0^\infty f_{n,n}(t) \varphi^{(n)}(t)\, dt. \qquad (3.6)$$

Lemma 3. *For each $n \geq 1$ and any $\varphi \in \mathscr{S}$,*

$$\langle f, \varphi \rangle = \sum_{s=0}^{n-1} a_s \langle e_s, \varphi \rangle - \sum_{s=0}^{n-1} b_s \langle \delta^{(s)}, \varphi \rangle + \langle f_n, \varphi \rangle, \qquad (3.7)$$

where

$$b_s = \frac{(-1)^{s+1}}{s!} \left\{ M[f; s+1] - \sum_{k=0}^{s-1} a_k \exp\left[\frac{(s-k)\pi i}{2} \right] \frac{\Gamma(s-k)}{c^{s-k}} \right\}, \qquad (3.8)$$

$M[f; z]$ being the Mellin transform of f.

Proof. We again let $f_{0,0}(t) = f_0(t) = f(t)$. Then for each $n \geq 0$, we have

$$f_{n+1}(t) + a_n e_n(t) = f_n(t)$$

and hence

$$f_{n+1,n}(t) + a_n E'_n(t) = f_{n,n}(t). \qquad (3.9)$$

Since $f_{n,n}(t) = O(\log t)$ as $t \to 0^+$, it follows from (3.9) that the limit

$$b_n = \lim_{t \to 0^+} [f_{n+1,n+1}(t) + a_n E_n(t)] \qquad (3.10)$$

exists; see a similar argument in the proof of Lemma 2. An integration by parts yields

$$\langle f_n, \varphi \rangle = -b_n \langle \delta^{(n)}, \varphi \rangle + \langle f_{n+1}, \varphi \rangle + a_n \langle e_n, \varphi \rangle.$$

Applying this identity n times, we obtain

$$\langle f_n, \varphi \rangle + \sum_{s=0}^{n-1} a_s \langle e_s, \varphi \rangle = \sum_{s=0}^{n-1} b_s \langle \delta^{(s)}, \varphi \rangle + \langle f, \varphi \rangle,$$

which is exactly the equation in (3.7). To show that the coefficients b_n can be expressed as in (3.8), we first observe that equation (3.9) implies

$$-\int_t^\infty f_{n,n}(\tau)\, d\tau = f_{n+1,n+1}(t) + a_n E_n(t).$$

Letting $t \to 0$, we have

$$b_n = - \int_0^\infty f_{n,n}(t) \, dt. \tag{3.11}$$

Now insert (2.10) in (3.11) and interchange the order of integrations. The result is

$$b_n = \frac{(-1)^{n+1}}{n!} \int_0^\infty \tau^n f_n(\tau) \, d\tau, \tag{3.12}$$

which, by Lemma 1 in Chapter IV, is equivalent to

$$b_n = \frac{(-1)^{n+1}}{n!} \lim_{\varepsilon \to 0} \int_0^\infty \tau^n f_n(\tau) e^{-\varepsilon \tau} \, d\tau. \tag{3.13}$$

The desired result (3.8) now follows from substituting (3.4) into (3.13) and applying Ex. 8 in Chapter III and Lemma 2 in Chapter IV. ∎

Lemma 4. *Let $E_s(t)$ be the function given in (3.3). Then for $s = 0, 1, 2, \ldots$, we have*

$$(s+1)! \int_0^\infty \frac{E_s(t)}{(t+z)^{s+2}} \, dt = \frac{(-1)^s}{z^{s+1}} e^{-icz} E_0(z). \tag{3.14}$$

Proof.   The repeated integral in (3.3) can be written in the form

$$E_s(t) = \frac{(-1)^{s+1}}{s!} \int_t^\infty (\tau - t)^s e_s(\tau) \, d\tau, \tag{3.15}$$

as in the case of (2.10). Thus the integral on the left side of (3.14) is equal to

$$\frac{(-1)^{s+1}}{s!} \int_0^\infty \int_t^\infty \frac{(\tau - t)^s}{(t+z)^{s+2}} e_s(\tau) \, d\tau \, dt.$$

Interchanging the order of integration gives

$$\int_0^\infty \frac{E_s(t)}{(t+z)^{s+2}} \, dt = \frac{(-1)^{s+1}}{s!} \int_0^\infty e_s(\tau) \int_0^\tau \frac{(\tau - t)^s}{(t+z)^{s+2}} \, dt \, d\tau.$$

The inner integral on the right-hand side can be evaluated explicitly by integration by parts, and we have

$$\int_0^\tau \frac{(\tau - t)^s}{(t+z)^{s+2}} \, dt = \frac{1}{s+1} \frac{\tau^{s+1}}{z^{s+1}(\tau + z)}. \tag{3.16}$$

Therefore

$$\int_0^\infty \frac{E_s(t)}{(t+z)^{s+2}}\, dt = \frac{(-1)^{s+1}}{(s+1)!} \frac{1}{z^{s+1}} \int_0^\infty \frac{e^{ic\tau}}{\tau + z}\, d\tau$$

as required. ∎

The function $E_0(z)$ can be expressed in terms of the cosine and sine integrals:

$$E_0(z) = \operatorname{Ci}(zc) + i\operatorname{Si}(zc) - \frac{\pi i}{2}. \tag{3.17}$$

It is also related to the incomplete Gamma function by

$$E_0(z) = -\Gamma(0; -icz). \tag{3.18}$$

Asymptotic properties of these functions are well-known; see, for example, the book by Olver (1974a).

Theorem 3. *Let $f(t)$ be a locally integrable function on $[0, \infty)$ and suppose that $f(t)$ satisfies (3.1). Then for each $n \geq 1$*

$$S_f(z) = e^{-icz} E_0(z) \sum_{s=0}^{n-1} (-1)^s \frac{a_s}{z^{s+1}} - \sum_{s=0}^{n-1} b_s \frac{s!}{z^{s+1}} + R_n(z), \tag{3.19}$$

where b_s is given in (3.8) and $R_n(z)$ satisfies

$$R_n(z) = n! \int_0^\infty \frac{f_{n,n}(t)}{(t+z)^{n+1}}\, dt. \tag{3.20}$$

The function $f_{n,n}(t)$ has the same meaning as given in (2.10).

Proof. Let $\varphi_\eta(t)$ be the rapidly decreasing function in §2; see the equation preceding (2.23). By Lemma 4

$$\lim_{\eta \to 0^+} \langle e_s, \varphi_\eta \rangle = \frac{(-1)^s}{z^{s+1}} e^{-icz} E_0(z), \tag{3.21}$$

and by Lemma 3

$$\langle f, \varphi_\eta \rangle = \sum_{s=0}^{n-1} a_s \langle e_s, \varphi_\eta \rangle - \sum_{s=0}^{n-1} b_s \langle \delta^{(s)}, \varphi_\eta \rangle + \langle f_n, \varphi_\eta \rangle. \tag{3.22}$$

Taking the limit as $\eta \to 0^+$, we have from (2.30), (3.21), and (3.22)

$$S_f(z) = e^{-icz}E_0(z) \sum_{s=0}^{n-1} (-1)^s \frac{a_s}{z^{s+1}} - \sum_{s=0}^{n-1} b_s \frac{s!}{z^{s+1}} + \lim_{\eta \to 0^+} \langle f_n, \varphi_n \rangle.$$

Since

$$\lim_{\eta \to 0^+} \langle f_n, \varphi_n \rangle = n! \int_0^\infty \frac{f_{n,n}(t)}{(t+z)^{n+1}} \, dt,$$

the theorem is established. ∎

To obtain error bounds for the approximation (3.19), one may proceed as in §2. In fact, assuming (2.51), the estimates given in (2.52) and (2.53) hold also for the remainder terms in (3.20); see also Ex. 5. Thus Theorem 3 gives, in particular,

$$S_f(z) \sim e^{-icz}E_0(z) \sum_{s=0}^\infty (-1)^s \frac{a_s}{z^{s+1}} - \sum_{s=0}^\infty b_s \frac{s!}{z^{s+1}}, \tag{3.23}$$

as $|z| \to \infty$ in $|\arg z| \le \pi - \delta \, (<\pi)$.

The above method can readily be extended to functions $f(t)$ with an asymptotic expansion of the form

$$f(t) \sim e^{ict} \sum_{s=0}^\infty a_s t^{-s-\alpha}, \qquad \text{as } t \to \infty, \tag{3.24}$$

where $c \ne 0$, c is real, and $0 < \alpha < 1$. Since the argument here is essentially the same as that for Theorem 3, the following result is stated without proof. For convenience, we introduce the notation

$$-E_{1-\alpha}(z) = e^{icz} \int_0^\infty \frac{\tau^{1-\alpha}e^{ic\tau}}{\tau + z} \, d\tau, \tag{3.25}$$

provided the integral on the right exists.

Theorem 4. *Let $f(t)$ be locally integrable on $[0, \infty)$ and satisfy (3.24). Then for any $n \ge 1$*

$$S_f(z) = e^{-icz}E_{1-\alpha}(z) \sum_{s=0}^{n-1} (-1)^s \frac{a_s}{z^{s+1}} - \sum_{s=0}^{n-1} d_s \frac{s!}{z^{s+1}} + R_n(z), \tag{3.26}$$

where d_s is given by

$$d_s = \frac{(-1)^{s+1}}{s!} \left\{ M[f; s+1] - \sum_{j=0}^{s-1} a_j \exp\left[\frac{(s+1-j-\alpha)\pi i}{2} \right] \right.$$
$$\left. \times \frac{\Gamma(s-j-\alpha+1)}{c^{s-j-\alpha+1}} \right\} \tag{3.27}$$

and the remainder $R_n(z)$ has the same meaning as given in (3.20).

Although this result is stated for $0 < \alpha < 1$, it actually holds also for $\alpha = 1$, in which case Theorem 4 reduces to Theorem 3. The asymptotic expansion of $E_{1-\alpha}(z)$ can be obtained by repeated integration by parts, and the following result is left as an exercise.

Lemma 5. *Let $0 < \alpha \leq 1$, $c \neq 0$, and c be real. Then for any $n \geq 1$, we have*

$$E_{1-\alpha}(z) = e^{icz}\left[\sum_{s=1}^{n-1}(-1)^s \frac{\Gamma(s+1-\alpha)}{c^{s+1-\alpha}} \frac{e^{i\pi(s+1-\alpha)/2}}{z^s} + R_n(z) \right] \tag{3.28}$$

with

$$|R_n(z)| \leq \frac{n!}{c^n} \frac{C_{n,\alpha}(\theta)}{|z|^{n+\alpha-1}}, \tag{3.29}$$

where $C_{n,\alpha}(\theta)$ depends on $\theta = \arg z$, $-\pi < \theta < \pi$, but is independent of $|z|$.

Example 1. As an illustration of our error analysis, we return to the integral

$$I(x) = \int_0^\infty \frac{J_0^2(t)}{t+x}\, dt \tag{3.30}$$

considered in Chapter III, Example 2, where we derived the asymptotic expansion

$$I(x) \sim \sum_{k=0}^\infty [c_k \log x + d_k]x^{-2k-1}, \tag{3.31}$$

where c_k and d_k are explicitly determined constants. In particular, we have

$$I(x) = \frac{1}{\pi x}\log x + \frac{1}{\pi x}(\gamma + 3\log 2) + \varepsilon(x), \tag{3.32}$$

where $\varepsilon(x) = O(x^{-3} \log x)$. In this example, we shall show that

$$|\varepsilon(x)| \le [(0.1)\log(x+1) + 0.96]x^{-3}, \qquad x > 0. \tag{3.33}$$

For comparison, we note that in (3.31) we have $c_1 = -1/8\pi = -0.04$ and $d_1 = (\tfrac{5}{2} - 3\log 2 - \gamma)/8\pi = -0.006$.

To prove (3.33) we shall use, instead of (2.52), the error bound constructed in Ex. 5. First we recall the identity

$$J_0(t) = \tfrac{1}{2}\{H_0^{(1)}(t) + H_0^{(2)}(t)\}, \tag{3.34}$$

where $H_0^{(1)}$ and $H_0^{(2)}$ are the Hankel functions. Thus

$$I(x) = \tfrac{1}{4}[I_1(x) + I_2(x)] + \tfrac{1}{2}I_3(x), \tag{3.35}$$

where I_1, I_2, and I_3 are the Stieltjes transforms of $[H_0^{(1)}]^2$, $[H_0^{(2)}]^2$, and $H_0^{(1)}H_0^{(2)}$ respectively. From the well-known asymptotic expansion (Olver 1974a, pp. 266–269) of $H_0^{(1)}$, it is easily seen that

$$[H_0^{(1)}(t)]^2 = \left(\frac{2}{\pi t}\right)e^{2i(t - \pi/4)}\left(1 - \frac{i}{4t}\right) + \varepsilon_2^{(1)}(t)$$

and the remainder satisfies

$$|\varepsilon_2^{(1)}(t)| \le \frac{2}{\pi}\frac{93}{512}t^{-3}, \qquad t \ge 1,$$

and

$$\int_0^1 t^2 |\varepsilon_2^{(1)}(t)|\, dt \le \frac{5}{2\pi}.$$

From Theorem 3 and Ex. 5,

$$I_1(x) = e^{-i2x}E_0(x)\left[-\frac{2i}{\pi x} + \frac{1}{2\pi x^2}\right] - \left[b_0^{(1)}\frac{1}{x} + b_1^{(1)}\frac{1}{x^2}\right] + \delta_2^{(1)}(x) \tag{3.36}$$

with

$$|\delta_2^{(1)}(x)| \le \frac{1}{x^2}\int_0^\infty \frac{t^2\varepsilon_2^{(1)}(t)}{t + x}\, dt \le \left[\frac{186}{512\pi}\log(x+1) + \frac{5}{2\pi}\right]x^{-3}.$$

The coefficients $b_0^{(1)}$ and $b_1^{(1)}$ in (3.36) are given by

$$b_0^{(1)} = -M[(H_0^{(1)})^2; 1], \qquad b_1^{(1)} = M[(H_0^{(1)})^2; 2] - \frac{1}{\pi},$$

and the function $E_0(x)$ by

$$E_0(x) = -\int_x^\infty e^{i2t} t^{-1} \, dt.$$

It is easily shown that $|E_0(x)| \le x^{-1}$ and

$$E_0(x) = \frac{e^{i2x}}{2xi}\left[1 + \frac{\theta_1(x)}{2xi}\right],$$

where $|\theta_1(x)| \le 2$. Thus (3.36) reduces to

$$I_1(x) = -\frac{1}{\pi x^2} - \left(b_0^{(1)}\frac{1}{x} + b_1^{(1)}\frac{1}{x^2}\right) + \rho_2^{(1)}(x)$$

with

$$|\rho_2^{(1)}(x)| \le \left[\frac{186}{512\pi}\log(x+1) + \frac{4}{\pi}\right]x^{-3}.$$

Replacing i by $-i$, we obtain a similar expansion for $I_2(x)$. Adding the two gives

$$I_1(x) + I_2(x) = -\frac{2}{\pi x^2} - \left(\frac{b_0}{x} + \frac{b_1}{x^2}\right) + \rho_2(x), \qquad (3.37)$$

where

$$b_0 = -M[(H_0^{(1)})^2 + (H_0^{(2)})^2; 1], \qquad b_1 = M[(H_0^{(1)})^2 + (H_0^{(2)})^2; 2] - \frac{2}{\pi}$$

and

$$|\rho_2(x)| \le \frac{2}{x^3}\left[\frac{186}{512\pi}\log(x+1) + \frac{4}{\pi}\right]. \qquad (3.38)$$

To obtain a corresponding result for $I_3(x)$, we note that

$$H_0^{(1)}(t)H_0^{(2)}(t) = J_0^2(t) + Y_0^2(t),$$

where Y_0 is the Bessel function of the second kind. Also we recall the asymptotic expansion (Olver 1974a, p. 449)

$$J_0^2(t) + Y_0^2(t) \sim \frac{2}{\pi t}\sum_{s=0}^\infty [1\cdot3\cdots(2s-1)]\frac{A_s}{t^{2s}},$$

where

$$A_s = (-1)^s \frac{[1\cdot3\cdots(2s-1)]^2}{s!\,8^s}$$

and the remainder after n terms is of the same sign as, and numerically less than, the $(n + 1)$-th term. Coupling Theorem 2 and Ex. 5, we obtain

$$I_3(x) = \frac{2}{\pi x} \log x + \left[a_0^* \frac{1}{x} - a_1^* \frac{1}{x^2} \right] + R_2(x), \qquad (3.39)$$

where

$$|R_2(x)| \leq \left[\frac{1}{4\pi} \log(x + 1) + \frac{2}{\pi} \right] x^{-3}. \qquad (3.40)$$

With $a_0 = \dfrac{2}{\pi}$ and $a_1 = 0$, the coefficients a_0^* and a_1^* are given by

$$a_s^* = \lim_{z \to s+1} \left\{ M[J_0^2 + Y_0^2; z] + \frac{a_s}{z - s - 1} \right\}, \qquad s = 0, 1.$$

Since

$$\frac{1}{2} a_0^* - \frac{1}{4} b_0 = \lim_{z \to 1} \left\{ M[J_0^2; z] + \frac{1/\pi}{z - 1} \right\}$$

and

$$\frac{1}{2} a_1^* + \frac{1}{4} b_1 = \lim_{z \to 2} M[J_0^2; z] - \frac{1}{2\pi},$$

the approximation in (3.32) now follows from (3.35), (3.37), (3.39), and the Mellin transform (Chapter III)

$$M[J_0^2; z] = \frac{2^{z-1} \Gamma(\frac{1}{2}z) \Gamma(1 - z)}{\{\Gamma(1 - \frac{1}{2}z)\}^3}. \qquad (3.41)$$

Furthermore, since $\varepsilon(x) = \frac{1}{4} \rho_2(x) + \frac{1}{2} R_2(x)$, the desired estimate (3.33) is obtained from (3.38) and (3.40).

4.   Hilbert Transforms

Let f be a locally integrable function on $(-\infty, \infty)$. The Hilbert transform of f, when it exists, is defined by

$$H_f(x) = \frac{1}{\pi} \int_{-\infty}^{\infty} \frac{f(t)}{t - x} \, dt, \qquad x \in (-\infty, \infty), \qquad (4.1)$$

where the bar indicates that the integral is a Cauchy principal value at $t = x$. By subdividing the range of integration at the origin, we obtain

$$H_f(x) = \frac{1}{\pi}\{H_f^-(x) + H_f^+(x)\}, \tag{4.2}$$

where $H_f^-(x)$ and $H_f^+(x)$ denote the integrals corresponding to the intervals $(-\infty, 0)$ and $(0, \infty)$, respectively. For definiteness let us restrict x to be positive, in which case $H_f^-(x)$ is simply the Stieltjes transform of $-f(-t)$. Thus we may confine ourselves to the consideration of the one-sided Hilbert transform.

$$H_f^+(x) = \int_0^\infty \frac{f(t)}{t - x}\, dt. \tag{4.3}$$

Throughout this section we shall assume that $f(t)$ is locally integrable on $[0, \infty)$ and continuously differentiable in $(0, \infty)$, and has an asymptotic expansion of the form

$$f(t) \sim e^{ict} \sum_{s=0}^\infty a_s t^{-s-\alpha}, \qquad \text{as } t \to \infty, \tag{4.4}$$

where $0 < \alpha \le 1$ and c is real.

In view of the well-known formula of Plemelj (Bremermann 1965, p. 64)

$$H_f^+(x) = \frac{1}{2}\lim_{\varepsilon \to 0} \int_0^\infty \left[\frac{1}{t - x + i\varepsilon} + \frac{1}{t - x - i\varepsilon}\right] f(t)\, dt, \tag{4.5}$$

it is naturally tempting to use the results already established in §§2 and 3, since the integrals on the right-hand side of (4.5) are Stieltjes transforms of f. For instance, a combination of the results (2.27), (2.28), and (4.5) gives

$$H_f^+(x) = \pi \cot \alpha\pi \sum_{s=0}^{n-1} \frac{a_s}{x^{s+\alpha}} - \sum_{s=1}^{n} (s-1)! \frac{(-1)^s c_s}{x^s} + \varepsilon_n(x), \tag{4.6}$$

$$\varepsilon_n(x) = \frac{1}{2}\lim_{\varepsilon \to 0} [R_n(-x + i\varepsilon) + R_n(-x - i\varepsilon)]$$

$$= \frac{n!}{2}\lim_{\varepsilon \to 0} \int_0^\infty \left[\frac{1}{(t - x + i\varepsilon)^{n+1}} + \frac{1}{(t - x - i\varepsilon)^{n+1}}\right] f_{n,n}(t)\, dt. \tag{4.7}$$

This expression for the remainder is of no practical use, and would present a formidable difficulty if one attempted to use it to obtain an error bound. For this reason, we shall derive an alternative expression for the error $R_n(z)$ in (2.28), (2.33), and (3.20).

We recall from (2.28)

$$R_n(z) = n! \int_0^\infty \frac{f_{n,n}(t)}{(t+z)^{n+1}} \, dt, \tag{4.8}$$

and from (2.10)

$$f_{n,n}(t) = \frac{(-1)^n}{(n-1)!} \int_t^\infty (\tau - t)^{n-1} f_n(\tau) \, d\tau. \tag{4.9}$$

Substituting (4.9) into (4.8) and interchanging the order of integration gives

$$R_n(z) = (-1)^n n \int_0^\infty f_n(\tau) \int_0^\tau (\tau - t)^{n-1} (t+z)^{-n-1} \, dt \, d\tau.$$

The inner integral was evaluated in (3.16), from which we have

$$R_n(z) = \frac{(-1)^n}{z^n} \int_0^\infty \frac{\tau^n f_n(\tau)}{\tau + z} \, d\tau. \tag{4.10}$$

It now follows from the first equality in (4.7) and the Plemelj formula that

$$\varepsilon_n(x) = \frac{1}{x^n} \int_0^\infty \frac{\tau^n f_n(\tau)}{\tau - x} \, d\tau. \tag{4.11}$$

Here we have also used the fact that if φ is continuously differentiable on $[0, \infty)$ and $\varphi = O(t^{-\alpha})$ for some $\alpha > 0$, then

$$\lim_{\varepsilon \to 0} \int_0^\infty \frac{\varphi(t)}{t - x \pm i\varepsilon} \, dt$$

exists; see Bremermann (1965, p. 62). Coupling (4.6) and (4.11), we obtain the following result.

Theorem 5.   *If in (4.4) $c = 0$ and $0 < \alpha < 1$, then (4.6) and (4.11) hold. The coefficient c_s in (4.6) is given by*

$$c_s = \frac{(-1)^s}{(s-1)!} \, M[f; s]. \tag{4-12}$$

In exactly the same manner, we have from (2.31) the next result.

Theorem 6. *If in (4.4) $c = 0$ and $\alpha = 1$, then for any $n \geq 1$*

$$H_f^+(x) = \left(\log \frac{1}{x}\right) \sum_{s=0}^{n-1} \frac{a_s}{x^{s+1}} - \sum_{s=0}^{n-1} \frac{a_s^*}{x^{s+1}} + \varepsilon_n(x), \qquad (4.13)$$

where

$$a_s^* = \lim_{z \to s+1} \left\{ M[f; z] + \frac{a_s}{z - s - 1} \right\} \qquad (4.14)$$

and $\varepsilon_n(x)$ is as given in (4.11).

The cases in which $c \neq 0$ in (4.4) can be handled in a similar manner. For simplicity, we shall consider only the case $c > 0$. First we need the following lemma.

Lemma 6. *For $0 \leq \alpha < 1$ and $c > 0$, we have*

$$\lim_{\varepsilon \to 0^+} \int_0^\infty \frac{e^{ict}}{t^\alpha (t - x + i\varepsilon)} \, dt = \int_0^\infty \frac{e^{ict}}{t^\alpha (t - x)} \, dt - \frac{i\pi}{x^\alpha} e^{icx} \qquad (4.15)$$

$$\lim_{\varepsilon \to 0^+} \int_0^\infty \frac{e^{ict}}{t^\alpha (t - x - i\varepsilon)} \, dt = \int_0^\infty \frac{e^{ict}}{t^\alpha (t - x)} \, dt + \frac{i\pi}{x^\alpha} e^{icx}. \qquad (4.16)$$

Proof. By deforming the path of integration into the positive imaginary axis on which $t = i\tau$, it is easily shown that

$$\int_0^\infty \frac{e^{ict}}{t^\alpha (t - x + i\varepsilon)} \, dt = e^{-i\alpha\pi/2} \int_0^\infty \frac{e^{-c\tau}}{\tau^\alpha (\tau + ix + \varepsilon)} \, d\tau \qquad (4.17)$$

and

$$\int_0^\infty \frac{e^{ict}}{t^\alpha (t - x - i\varepsilon)} \, dt = \frac{2\pi i}{(x + i\varepsilon)^\alpha} e^{ic(x + i\varepsilon)} + e^{-i\alpha\pi/2} \int_0^\infty \frac{e^{-c\tau}}{\tau^\alpha (\tau + ix - \varepsilon)} \, d\tau. \qquad (4.18)$$

Similarly, we have

$$\int_0^\infty \frac{e^{ict}}{t^\alpha (t - x)} \, dt = \frac{i\pi}{x^\alpha} e^{icx} + e^{-i\alpha\pi/2} \int_0^\infty \frac{e^{-c\tau}}{\tau^\alpha (\tau + ix)} \, d\tau, \qquad (4.19)$$

where the integrated term comes from a small semi-circle centered at $t = x$ in the first quadrant of the complex t-plane. Thus, upon taking $\varepsilon \to 0$, (4.15) follows from (4.17) and (4.19), and (4.16) from (4.18) and (4.19). ∎

For convenience, let us set

$$E_\alpha^*(x) = \int_0^\infty \frac{e^{ict}}{t^\alpha(t-x)}\, dt, \qquad 0 \le \alpha < 1, \tag{4.20}$$

and recall the identity

$$\int_0^\infty t^{\mu-1} e^{i\lambda t}\, dt = \frac{\Gamma(\mu)}{\lambda^\mu} e^{i\mu\pi/2}, \qquad 0 < \mu < 1. \tag{4.21}$$

From the above lemma it follows that the integral $-E_{1-\alpha}(z)$ in (3.25) satisfies

$$\lim_{\varepsilon \to 0^+} \{e^{ic(x \mp i\varepsilon)}[-E_{1-\alpha}(-x \pm i\varepsilon)]\} = \frac{\Gamma(1-\alpha)}{c^{1-\alpha}} e^{(1-\alpha)\pi i/2}$$

$$+ x\left[E_\alpha^*(x) \mp \frac{i\pi}{x^\alpha} e^{icx}\right], \qquad 0 < \alpha < 1. \tag{4.22}$$

The following results are now consequences of Theorems 3 and 4.

Theorem 7.   *If in (4.4) $c > 0$ and $\alpha = 1$, then for any $n \ge 1$*

$$H_f^+(x) = E_0^*(x) \sum_{s=0}^{n-1} \frac{a_s}{x^{s+1}} - \sum_{s=0}^{n-1} (-1)^{s+1} b_s \frac{s!}{x^{s+1}} + \varepsilon_n(x), \tag{4.23}$$

where

$$(-1)^{s+1} s!\, b_s = M[f; s+1] - \sum_{j=0}^{s-1} a_j \exp\left\{\frac{(s-j)\pi i}{2}\right\} \frac{\Gamma(s-j)}{c^{s-j}} \tag{4.24}$$

and $\varepsilon_n(x)$ is given in (4.11).

Theorem 8.   *If in (4.4) $c > 0$ and $0 < \alpha < 1$, then for any $n \ge 1$*

$$H_f^+(x) = E_\alpha^*(x) \sum_{s=0}^{n-1} \frac{a_s}{x^s} - \sum_{s=0}^{n-1} \frac{d_s^*}{x^{s+1}} + \varepsilon_n(x), \tag{4.25}$$

where

$$d_s^* = M[f; s+1] - \sum_{j=0}^{s} a_j \exp\left\{\frac{(s+1-j-\alpha)\pi i}{2}\right\} \frac{\Gamma(s+1-j-\alpha)}{c^{s+1-j-\alpha}} \tag{4.26}$$

and $\varepsilon_n(x)$ is again given in (4.11).

Theorems 5, 6, 7, and 8 have been derived by a more elementary method in Wong (1980a). Here we take the opportunity to point out that the identity (2.14) in that paper is incorrect. The result quoted there is for the two-sided Hilbert transform. This error, however, does not affect the main result stated in Theorem 3 of that paper; cf. Theorem 7 given above.

To show that the expansions in Theorems 5, 6, 7, and 8 are indeed asymptotic in nature, we must prove that the remainder

$$\varepsilon_n(x) = \frac{1}{x^n} \int_0^\infty \frac{\tau^n f_n(\tau)}{\tau - x} \, d\tau$$

satisfies $\varepsilon_n(x) = o(x^{-n})$ as $x \to \infty$. We shall in fact show that there exists a positive number M_n such that for $0 < \alpha \le 1$,

$$\left| \int_0^\infty \frac{\tau^n f_n(\tau)}{\tau - x} \, d\tau \right| \le M_n \frac{\log x}{x^\alpha} \tag{4.27}$$

for all $x > e$.

Theorem 9. *Let $f(t)$ be a locally integrable function on $[0, \infty)$, which is continuously differentiable in $(0, \infty)$. If $f(t)$ has an asymptotic expansion of the form given in (4.4), and if $f'(t)$ has the asymptotic expansion obtained by termwise differentiation, then the estimate (4.27) holds.*

Proof. Let $\varphi_n(x)$ denote the integral in (4.27), and write

$$\varphi_n(x) = \varphi_{n,1}(x) + \varphi_{n,2}(x) + \varphi_{n,3}(x), \tag{4.28}$$

where the integrals $\varphi_{n,1}$, $\varphi_{n,2}$, $\varphi_{n,3}$ correspond respectively to the intervals $(0, x - 1)$, $(x - 1, x + 1)$, $(x + 1, \infty)$. Put

$$M_{n,0} = \int_0^1 t^n |f_n(t)| \, dt \tag{4.29}$$

$$M_{n,\infty} = \sup\{t^{\alpha + n} |f_n(t)| : t \ge 1\}. \tag{4.30}$$

Under the hypotheses, both numbers $M_{n,0}$ and $M_{n,\infty}$ are finite. To estimate $\varphi_{n,1}(x)$ we further divide the range of integration at $t = 1$. It is easy to see that

$$|\varphi_{n,1}(x)| \le \frac{M_{n,0}}{x - 1} + M_{n,\infty} \int_1^{x-1} \frac{1}{\tau^\alpha (x - \tau)} \, d\tau.$$

In the last integral we make a change of variables $\tau = xu$. The resulting integral is dominated by

$$\frac{1}{x^\alpha} \int_{1/x}^{(x-1)/x} \frac{1}{u(1-u)} \, du,$$

which in turn is dominated by $2 \log(x-1)/x^\alpha$. Therefore

$$|\varphi_{n,1}(x)| \le 2(M_{n,0} + M_{n,\infty}) \frac{\log x}{x^\alpha}. \tag{4.31}$$

The integral $\varphi_{n,3}(x)$ can be estimated similarly:

$$|\varphi_{n,3}(x)| \le \frac{1}{x^\alpha} M_{n,\infty} \int_{1/x}^{\infty} \frac{1}{(1+u)^\alpha u} \, du.$$

In the interval $(x^{-1}, 1)$, we use the bound $(1+u)^{-\alpha} \le 1$, whereas in the interval $(1, \infty)$, we use $(1+u)^{-\alpha} \le u^{-\alpha}$. Thus

$$|\varphi_{n,3}(x)| \le \left(1 + \frac{1}{\alpha}\right) M_{n,\infty} \frac{\log x}{x^\alpha}. \tag{4.32}$$

We now turn to the consideration of $\varphi_{n,2}(x)$. Let $h_n(\tau) = \tau^n f_n(\tau)$ and write

$$\varphi_{n,2}(x) = \int_{x-1}^{x+1} \frac{h_n(\tau) - h_n(x)}{\tau - x} \, d\tau = \int_{x-1}^{x+1} h_n'(\xi) \, d\tau, \tag{4.33}$$

where ξ is between τ and x. By hypotheses, as $t \to \infty$, $f_n(t) = O(t^{-n-\alpha})$ and $f_n'(t) = O(t^{-n-\alpha})$. Put

$$M_{n,\infty}' = \sup\{t^{n+\alpha} | f_n'(t)| : t \ge 1\}, \tag{4.34}$$

and note that, by assumption, $M_{n,\infty}' < \infty$. We then have

$$|h_n'(t)| \le (nM_{n,\infty} + M_{n,\infty}')t^{-\alpha}$$

for all $t \ge 1$. It now follows from (4.33) that

$$|\varphi_{n,2}(x)| \le 2\left(\frac{e}{e-1}\right)^\alpha (nM_{n,\infty} + M_{n,\infty}') \frac{\log x}{x^\alpha}. \tag{4.35}$$

Combination of this result together with (4.31) and (4.32) gives the desired conclusion (4.27). ■

If the explicit expression for the remainder $\varepsilon_n(x)$ is not needed, then Theorems 7 and 8 can be further simplified. First we note the following lemma.

Lemma 7. *Let $0 \le \alpha < 1$ and $c > 0$. Then for any $n \ge 1$ and $x > 0$*

$$E_\alpha^*(x) = \frac{i\pi}{x^\alpha} e^{icx} - \sum_{s=0}^{n-1} \frac{\Gamma(s+1-\alpha)}{c^{s+1-\alpha}} \frac{e^{i\pi(s+1-\alpha)/2}}{x^{s+1}} + \delta_n(x), \quad (4.36)$$

where

$$|\delta_n(x)| \le \frac{\Gamma(n+1-\alpha)}{c^{n+1-\alpha}} \frac{1}{x^{n+1}}. \quad (4.37)$$

Proof. Apply (4.19) and (2.3).

Theorem 10. *If in (4.4) $c > 0$ and $0 < \alpha \le 1$, then as $x \to \infty$*

$$H_f^+(x) \sim i\pi e^{icx} \sum_{s=0}^{\infty} \frac{a_s}{x^{s+\alpha}} - \sum_{s=0}^{\infty} \frac{M[f; s+1]}{x^{s+1}}. \quad (4.38)$$

Proof. If $\alpha = 1$, then this follows from Theorem 7 and Lemma 7. If $0 < \alpha < 1$, then it follows from Theorem 8 and Lemma 7. ∎

Ursell (1983) observed that the above results can be stated in more general terms. He showed that if in place of (4.4) we have

$$f(t) = \sum_{s=1}^{n} \frac{a_s}{t^s} + \cos wt \sum_{s=1}^{n} \frac{A_s}{t^s} + \sin wt \sum_{s=1}^{n} \frac{B_s}{t^s} + f_n(t), \quad (4.39)$$

where w is a positive constant and $f_n(t) = O(t^{-n-1})$ as $t \to \infty$, then

$$H_f^+(x) = -\sum_{s=1}^{n} \frac{c_s}{x^s} - \log x \sum_{s=1}^{n} \frac{a_s}{x^s} + \left(\sum_{s=1}^{n} \frac{A_s}{x^s} \right) \int_0^\infty \frac{\cos wt}{t-x} dt$$
$$+ \left(\sum_{s=1}^{n} \frac{B_s}{x^s} \right) \int_0^\infty \frac{\sin wt}{t-x} dt + \frac{1}{x^n} \int_0^\infty \frac{t^n f_n(t)}{t-x} dt, \quad (4.40)$$

with

$$c_s = \lim_{z \to s} \left\{ M[f; z] + \frac{a_s}{z-s} \right\}$$
$$- \sum_{j=1}^{s-1} \frac{\Gamma(s-j)}{w^{s-j}} \left[A_j \cos \frac{\pi}{2}(s-j) + B_j \sin \frac{\pi}{2}(s-j) \right]. \quad (4.41)$$

Substituting (4.41) in (4.40) and applying Lemma 7 with $\alpha = 0$, one obtains

Theorem 11.   *If (4.39) holds, then*

$$H_f^+(x) \sim - \sum_{s=1}^{\infty} \frac{d_s}{x^s} - \log x \sum_{s=1}^{\infty} \frac{a_s}{x^s}$$

$$- \pi \sin wx \sum_{s=1}^{\infty} \frac{A_s}{x^s} + \pi \cos wx \sum_{s=1}^{\infty} \frac{B_s}{x^s}, \tag{4.42}$$

where

$$d_s = \lim_{z \to s} \left\{ M[f; z] + \frac{a_s}{z - s} \right\}. \tag{4.43}$$

Similarly one also has the following generalization of Theorem 5.

Theorem 12.   *If, instead of (4.39),*

$$f(t) \sim \sum_{s=0}^{\infty} \frac{a_s}{t^{s+\alpha}} + \cos wt \sum_{s=0}^{\infty} \frac{A_s}{t^{s+\beta}} + \sin wt \sum_{s=0}^{\infty} \frac{B_s}{t^{s+\beta}} \tag{4.44}$$

where $0 < \alpha < 1$, $0 < \beta < 1$, and $w > 0$, then

$$H_f(x) \sim \pi \cot \pi\alpha \sum_{s=0}^{\infty} \frac{a_s}{x^{s+\alpha}} - \sum_{s=1}^{\infty} \frac{M[f; s]}{x^s}$$

$$- \pi \sin wx \sum_{s=0}^{\infty} \frac{A_s}{x^{s+\beta}} + \pi \cos wx \sum_{s=0}^{\infty} \frac{B_s}{x^{s+\beta}}. \tag{4.45}$$

Example 2.   A recent work on water waves involved the integral

$$I^*(x) = \int_0^{\infty} \frac{J_0^2(t)}{t - x} \, dt. \tag{4.46}$$

From Chapter I, equation (5.35), we have

$$J_0(t) = \left(\frac{2}{\pi t} \right)^{1/2} \left\{ \cos\left(t - \frac{\pi}{4} \right)\left(1 - \frac{9}{128t^2} \right) + \sin\left(t - \frac{\pi}{4} \right)\left(\frac{1}{8t} \right) + O\left(\frac{1}{t^3} \right) \right\}$$

as $t \to \infty$. Hence

$$J_0^2(t) = \frac{1}{\pi t} - \frac{1}{8\pi t^3} - \cos 2t \left(\frac{1}{4\pi t^2} \right) + \sin 2t \left(\frac{1}{\pi t} - \frac{5}{32\pi t^3} \right) + O\left(\frac{1}{t^4} \right). \tag{4.47}$$

In the notation of (4.39), $w = 2$ and

$$a_1 = \frac{1}{\pi}, \qquad a_2 = 0, \qquad a_3 = -\frac{1}{8\pi};$$

$$A_1 = 0, \qquad A_2 = -\frac{1}{4\pi}, \qquad A_3 = 0; \qquad\qquad (4.48)$$

$$B_1 = \frac{1}{\pi}, \qquad B_2 = 0, \qquad B_3 = -\frac{5}{32\pi}.$$

The Mellin transform of $J_0^2(t)$ is given in (3.41), from which and from the Laurent expansion (Erdélyi *et al.* 1953a, §1.17)

$$\Gamma(z) = \frac{(-1)^s}{s!}\left[\frac{1}{z+s} + \psi(s+1) + O(z+s)\right], \qquad (4.49)$$

it follows that

$$d_1 = \frac{1}{\pi}(\gamma + 3\log 2), \qquad d_2 = 0, \qquad d_3 = -\frac{1}{8\pi}\left(\gamma + 3\log 2 - \frac{5}{2}\right).$$

Therefore by Theorem 11,

$$I^*(x) \sim -\frac{1}{\pi x}\log x - \frac{1}{\pi x}(\gamma + 3\log 2) + \frac{1}{x}\cos 2x + \frac{1}{4x^2}\sin 2x$$

$$+ \frac{1}{8\pi x^3}\log x + \frac{1}{8\pi x^3}\left(\gamma + 3\log 2 - \frac{5}{2}\right) - \frac{5}{32x^3}\cos 2x + \cdots. \qquad (4.50)$$

This result, also, is due to Ursell (1983).

5. Laplace and Fourier Transforms Near the Origin

The results in Lemmas 1 and 2 can also be used to derive asymptotic expansions of the Laplace transform

$$L_f(z) = \int_0^\infty f(t)e^{-zt}\, dt, \qquad \text{Re } z > 0, \qquad (5.1)$$

as $z \to 0$. The test function $\varphi(t)$ in this case is chosen to be any function in $\mathcal{S}$ which satisfies

$$\varphi(t) = e^{-zt} \qquad \text{for } t \geq 0. \qquad (5.2)$$

From (2.8) and (2.9), it follows that

$$\langle t_+^{-s-\alpha}, \varphi \rangle = \Gamma(1 - s - \alpha)z^{\alpha + s - 1}, \qquad 0 < \alpha < 1, \qquad (5.3)$$

$$\langle t_+^{-s-1}, \varphi \rangle = \frac{(-1)^{s+1}}{s!}(\gamma + \log z)z^s, \qquad (5.4)$$

$s = 0, 1, 2, \ldots$, where γ is the Euler constant.

Theorem 13.   *Let $f(t)$ be a locally integrable function on $[0, \infty)$ and satisfy (2.5). (i) If $0 < \alpha < 1$ in (2.5) then for each $n \geq 1$*

$$L_f(z) = \sum_{s=0}^{n-1} a_s \Gamma(1 - s - \alpha)z^{\alpha + s - 1} - \sum_{s=1}^{n} c_s z^{s-1} + R_n(z), \qquad (5.5)$$

where the coefficients c_s are given by (2.13). (ii) If $\alpha = 1$ then for each $n \geq 1$

$$L_f(z) = \log z \sum_{s=0}^{n-1} \frac{(-1)^{s+1}}{s!} a_s z^s + \sum_{s=0}^{n-1} \gamma_s z^s + R_n(z), \qquad (5.6)$$

where

$$\gamma_s = \frac{(-1)^{s+1}}{s!} \gamma a_s - d_{s+1}, \qquad s = 0, 1, 2, \ldots, \qquad (5.7)$$

and d_{s+1} is given by (2.18) or (2.32) and (2.34). The remainders in both cases satisfy

$$R_n(z) = z^n \int_0^\infty f_{n,n}(t)e^{-zt}\, dt \qquad (5.8)$$

with $f_{n,n}(t)$ being defined as before.

Proof.   By Lemma 1

$$L_f(z) = \sum_{s=0}^{n-1} a_s \langle t_+^{-s-\alpha}, \varphi \rangle - \sum_{s=1}^{n} c_s \langle \delta^{(s-1)}, \varphi \rangle + \langle f_n, \varphi \rangle,$$

where φ is given by (5.2). Since $\langle \delta^{(s)}, \varphi \rangle = z^s$, (5.3) implies that

$$L_f(z) = \sum_{s=0}^{n-1} a_s \Gamma(1 - s - \alpha)z^{\alpha + s - 1} - \sum_{s=1}^{n} c_s z^{s-1} + \langle f_n, \varphi \rangle. \qquad (5.9)$$

The last term is, by definition, given by

$$\langle f_n, \varphi \rangle = z^n \int_0^\infty f_{n,n}(t) e^{-zt}\, dt,$$

thus proving (i). To prove (ii), we use Lemma 2 and (5.4) instead of Lemma 1 and (5.3). The argument is similar, however. ∎

Corresponding expansions can be obtained for the Fourier transform

$$F_f(x) = \int_0^\infty f(t) e^{ixt}\, dt. \tag{5.10}$$

To this end, we put $z = \varepsilon - ix$ in (5.5) and (5.6). Upon letting $\varepsilon \to 0$, the final results follow from Lemma 1 in Chapter IV.

Theorem 14. *Let f be a locally integrable function on $[0, \infty)$ and satisfy (2.5). (i) If $0 < \alpha < 1$ in (2.5) then*

$$F_f(x) = e^{-\alpha\pi i/2} \sum_{s=0}^{n-1} (-i)^{s-1} \Gamma(1 - s - \alpha) a_s x^{\alpha + s - 1}$$

$$- \sum_{s=1}^{n} c_s (-ix)^{s-1} + R_n(x), \tag{5.11}$$

where the coefficients c_s are given by (2.13). (ii) If $\alpha = 1$ then

$$F_f(x) = -\log x \sum_{s=0}^{n-1} \frac{a_s}{s!} (ix)^s + \sum_{s=0}^{n-1} \gamma_s^*(-ix)^s + R_n(x), \tag{5.12}$$

where

$$\gamma_s^* = \frac{(-1)^{s+1}}{s!}\left(\gamma - i\frac{\pi}{2}\right) a_s - d_{s+1} \tag{5.13}$$

and d_{s+1} has the same meaning as given in (2.18) or (2.32) and (2.34). The remainders in both cases satisfy

$$R_n(x) = (-ix)^n \int_0^\infty f_{n,n}(t) e^{ixt}\, dt \tag{5.14}$$

with $f_{n,n}(t)$ being defined as before.

Note that there is an error in equations (5.15) and (5.19) of McClure and Wong (1978). The quantity π there should be replaced by $\pi/2$.

Example 3.   As an application of the above result, we consider the Kontorovich–Lebedev transform

$$I(x) = \int_0^\infty K_{it}(x) f(t) \, dt, \qquad x > 0, \tag{5.15}$$

where $K_{it}(x)$ is the modified Bessel function of the third kind of imaginary order. This transform plays an important role in a number of problems concerning diffraction of waves by wedges. Here we shall assume that $f(t)$ is locally absolutely integrable on $(0, \infty)$ and satisfies.

$$f(t) = O(t^b), \qquad \text{as } t \to 0^+,$$

for some $b > -1$. Furthermore, we assume that $f(t)$ has an asymptotic expansion of the form (2.5):

$$f(t) \sim \sum_{s=0}^\infty a_s t^{-s-\alpha}, \qquad \text{as } t \to \infty, \tag{5.16}$$

where $0 < \alpha < 1$. To apply the result of Theorem 14, we first observe that the kernel function $K_{it}(x)$ has the integral representation (Oberhettinger 1972, p. 241)

$$K_{it}(x) = \int_0^\infty \cos ty \, e^{-x \cosh y} \, dy. \tag{5.17}$$

Substituting (5.17) in (5.15) and reversing the order of integration, we have

$$I(x) = \int_0^\infty F_c(y) e^{-x \cosh y} \, dy, \tag{5.18}$$

where $F_c(y)$ is the Fourier cosine transform of $f(t)$, i.e.,

$$F_c(y) = \int_0^\infty f(t) \cos ty \, dt. \tag{5.19}$$

The interchange of the order of integration is justified by uniform convergence; see Ex. 10.

Following the Laplace method in Chapter II, we set, in (5.18),

$$u^2 = \cosh y - 1 \tag{5.20}$$

and write $I(x)$ as

$$I(x) = e^{-x} \int_0^\infty \varphi(u) e^{-xu^2} \, du, \tag{5.21}$$

where

$$\varphi(u) = F_c[y(u)]\frac{dy}{du}. \tag{5.22}$$

Equating the real parts in (5.11), we obtain

$$F_c(y) \sim \sum_{s=0}^{\infty} a_s \Gamma(1 - s - \alpha) \cos\left[(1 - s - \alpha)\frac{\pi}{2}\right] y^{s + \alpha - 1}$$
$$+ \sum_{n=0}^{\infty} \frac{(-1)^n}{(2n)!} M[f; 2n + 1] y^{2n}, \tag{5.23}$$

and from (5.20) it follows that

$$\frac{dy}{du} = \sqrt{2}\left(1 + \frac{u^2}{2}\right)^{-1/2} \tag{5.24}$$

and hence

$$y \sim \sqrt{2}u - \frac{\sqrt{2}}{12}u^3 + \frac{3\sqrt{2}}{160}u^5 + \cdots, \qquad \text{as } u \to 0. \tag{5.25}$$

A combination of the results in (5.22)–(5.25) gives

$$\varphi(u) \sim \sqrt{2}\left[\sum_{s=0}^{\infty} \gamma_s u^{s + \alpha - 1} + \sum_{s=0}^{\infty} \delta_s u^{2s}\right], \qquad \text{as } u \to 0^+. \tag{5.26}$$

The coefficients γ_s and δ_s are expressible in terms of a_s and $M[f; 2s + 1]$, respectively; for example,

$$\gamma_0 = 2^{(\alpha - 1)/2} a_0 \Gamma(1 - \alpha) \sin\left(\frac{\alpha\pi}{2}\right),$$

$$\gamma_1 = 2^{\alpha/2} a_1 \Gamma(-\alpha) \cos\left(\frac{\alpha\pi}{2}\right),$$

$$\gamma_2 = -2^{(\alpha - 1)/2}\left(\frac{\alpha}{12} + \frac{1}{6}\right) a_0 \Gamma(1 - \alpha) \sin\left(\frac{\alpha\pi}{2}\right) \tag{5.27}$$

$$- 2^{(\alpha + 1)/2} a_2 \Gamma(-1 - \alpha) \sin\left(\frac{\alpha\pi}{2}\right),$$

and

$$\delta_0 = M[f; 1], \qquad \delta_1 = -M[f; 3] - \frac{1}{4} M[f; 1],$$

$$\delta_2 = \frac{1}{6} M[f; 5] + \frac{5}{12} M[f; 3] + \frac{3}{32} M[f; 1]. \tag{5.28}$$

Substituting (5.26) into (5.21) and integrating term by term, we obtain

$$I(x) \sim \frac{e^{-x}}{\sqrt{2}} \left[\sum_{s=0}^{\infty} \Gamma\left(\frac{s+1}{2}\right) \frac{\gamma_s}{x^{(s+1)/2}} + \sum_{s=0}^{\infty} \Gamma\left(s+\frac{1}{2}\right) \frac{\delta_s}{x^{s+1/2}} \right], \quad (5.29)$$

as $x \to +\infty$. The last step is justified by Watson's lemma; see Chapter I. For the case when $\alpha = 1$ in (5.16), we refer to Wong (1981), where the small-x behavior of $I(x)$ is also given.

6.   Fractional Integrals

The Riemann–Liouville fractional integral of order μ is defined by

$$I^{\mu} f(x) = \frac{1}{\Gamma(\mu)} \int_0^x (x - t)^{\mu-1} f(t) \, dt, \qquad \mu > 0. \tag{6.1}$$

Here we again assume that $f(t)$ is locally integrable on $[0, \infty)$ and satisfies (2.5).

In terms of the convolution product

$$(f * g)(x) = \int_0^x f(x - t) g(t) \, dt \tag{6.2}$$

of two locally integrable functions on $[0, \infty)$, (6.1) can be written as

$$I^{\mu} f(x) = \frac{1}{\Gamma(\mu)} (t^{\mu-1} * f)(x). \tag{6.3}$$

The replacement of $f(t)$ by its asymptotic expansion (2.5) and term-by-term integration results in convolution integrals of the form

$$(t^{\mu-1} * t^{-s-\alpha})(x) = \int_0^x (x - t)^{\mu-1} t^{-s-\alpha} \, dt, \qquad s = 0, 1, 2, \ldots . \tag{6.4}$$

Of course, except for $s = 0$ and $0 < \alpha < 1$, none of these integrals exists in the usual sense. However, as we have seen in §6 of Chapter V, the left-hand side of (6.4) can be regarded as the convolution of the distributions $t_+^{\mu-1}$ and $t_+^{-s-\alpha}$. Thus we shall proceed naturally to use the results given in that chapter and those in Lemmas 1 and 2 of the present chapter. Before proceeding, one further result is needed concerning differentiation under an integral sign. This result may be found in most books on Advanced Calculus, when "absolutely continuous" is replaced by "continuously differentiable".

Lemma 8. *Suppose that $f(t, x)$ is integrable in $[a, b] \times [c, d]$, and absolutely continuous as a function of x, for each $t \in [a, b]$. Suppose also that $\partial f(t, x)/\partial x$ is integrable in $[a, b] \times [c, d]$. Then the function $F(x)$ defined on $[c, d]$ by $F(x) = \int_a^b f(t, x)\, dt$ is absolutely continuous in $[c, d]$, and for almost all $x \in [c, d]$*

$$F'(x) = \int_a^b \frac{\partial}{\partial x} f(t, x)\, dt.$$

Proof. We first recall that a function $h(x)$ is absolutely continuous on an interval $[c, d]$ if and only if h' is absolutely integrable and

$$h(x) = \int_c^x h'(u)\, du + h(c)$$

for all $x \in [c, d]$. Thus

$$F(x_i') - F(x_i) = \int_a^b \int_{x_i}^{x_i'} \frac{\partial}{\partial \tau} f(t, \tau)\, d\tau\, dt. \tag{6.5}$$

Since $\partial f(t, x)/\partial x$ is integrable on $[a, b] \times [c, d]$, if $\sum_{i=1}^n |x_i' - x_i|$ is small then it follows from (6.5) that $\sum_{i=1}^n |F(x_i') - F(x_i)|$ is also small; see Rudin (1974, p. 33, Ex. 12). This proves that $F(x)$ is absolutely continuous. Now write

$$F(x) = \int_a^b \left[f(t, c) + \int_c^x \frac{\partial}{\partial \tau} f(t, \tau)\, d\tau \right] dt.$$

By Fubini's theorem,

$$F(x) = \int_a^b f(t, c)\, dt + \int_c^x \int_a^b \frac{\partial}{\partial \tau} f(t, \tau)\, dt\, d\tau$$

and also the inner integral on the right-hand side is an integrable function of τ. Thus

$$F'(x) = \int_a^b \frac{\partial}{\partial x} f(t, x)\, dt \quad \text{a.e. in } [c, d].$$

This proves the lemma. ∎

Theorem 15.   *Let f be a locally integrable function on $[0, \infty)$ and satisfy (2.5) with $0 < \alpha < 1$. Then for any $n \geq 1$*

$$I^\mu f(x) = \sum_{s=0}^{n-1} a_s \frac{\Gamma(1-s-\alpha)}{\Gamma(\mu+1-s-\alpha)} x^{\mu-s-\alpha}$$

$$- \sum_{s=1}^{n} \frac{c_s}{\Gamma(\mu+1-s)} x^{\mu-s} + \frac{1}{x^n} \delta_n(x), \tag{6.6}$$

where $c_s = (-1)^s M[f; s]/(s-1)!$ and the error term satisfies

$$\delta_n(x) = \sum_{j=0}^{n} \binom{n}{j} \frac{\Gamma(\mu+1)}{\Gamma(\mu+1-j)} I^\mu[t^{n-j}f_{n,j}](x), \tag{6.7}$$

$f_{n,j}(t)$ *being defined as in (2.10).*

Proof.   Since $\mathscr{D} \subset \mathscr{S}$, Lemma 1 holds with $\mathscr{S}$ replaced by $\mathscr{D}$. Thus we have, as distributions in $\mathscr{D}'$,

$$f = \sum_{s=0}^{n-1} a_s t_+^{-s-\alpha} - \sum_{s=1}^{n} c_s \delta^{(s-1)} + f_n.$$

This, in view of Theorem 3 of Chapter 5, gives

$$t_+^{\mu-1} * f = \sum_{s=0}^{n-1} a_s t_+^{\mu-1} * t_+^{-s-\alpha} - \sum_{s=1}^{n} c_s t_+^{\mu-1} * \delta^{(s-1)} + t_+^{\mu-1} * f_n. \tag{6.8}$$

Since $f_n = f_{n,n}^{(n)}$, by Definition 7 of Chapter V,

$$t_+^{\mu-1} * f_n = D^n(t_+^{\mu-1} * f_{n,n}), \tag{6.9}$$

where D denotes the distributional derivative. Now $t_+^{\mu-1} * f_{n,n}$ is a regular distribution in $(0, \infty)$ and is given by the locally integrable function

$$(t^{\mu-1} * f_{n,n})(x) = \int_0^x (x-t)^{\mu-1} f_{n,n}(t)\, dt = x^\mu \int_0^1 (1-u)^{\mu-1} f_{n,n}(xu)\, du.$$

By Lemma 8, the last integral is a locally absolutely continuous function in $(0, \infty)$, i.e., absolutely continuous on each compact subinterval of $(0, \infty)$; furthermore,

$$\left[x^\mu \int_0^1 (1-u)^{\mu-1} f_{n,n}(xu)\, du \right]'$$

$$= \mu x^{\mu-1} \int_0^1 (1-u)^{\mu-1} f_{n,n}(xu)\, du + x^\mu \int_0^1 (1-u)^{\mu-1} u f_{n,n-1}(xu)\, du$$

almost everywhere in $(0, \infty)$. Since $f_{n,j}(t)$ is locally absolutely continuous in $(0, \infty)$, for $j = 1, \ldots, n-1$, this process can be repeated. The result is

$$(t^{\mu-1} * f_{n,n})^{(n)}(x) = \sum_{j=0}^{n} \binom{n}{j} \frac{\Gamma(\mu+1)}{\Gamma(\mu+1-j)} x^{\mu-j}$$

$$\times \int_0^1 (1-u)^{\mu-1} u^{n-j} f_{n,j}(xu) \, du$$

$$= \Gamma(\mu) x^{-n} \delta_n(x)$$

almost everywhere in $(0, \infty)$. Since each function that has been differentiated is locally absolutely integrable, Lemma 2 of Chapter V implies

$$D^n(t_+^{\mu-1} * f_{n,n}) = (t^{\mu-1} * f_{n,n})^{(n)}(x) = \Gamma(\mu) x^{-n} \delta_n(x). \qquad (6.10)$$

Coupling (6.9) and (6.10) gives

$$t_+^{\mu-1} * f_n = \Gamma(\mu) x^{-n} \delta_n(x). \qquad (6.11)$$

We now return to (6.8) and suppose for a moment that $\mu \neq 1, 2, \ldots$ and $\mu - \alpha$ is not a nonnegative integer. From Chapter V, equations (6.10) and (6.13), we have

$$t_+^{\mu-1} * \delta^{(s-1)} = \frac{\Gamma(\mu)}{\Gamma(\mu+1-s)} t_+^{\mu-s} \qquad (6.12)$$

and

$$t_+^{\mu-1} * t_+^{-s-\alpha} = \frac{\Gamma(\mu)\Gamma(1-s-\alpha)}{\Gamma(\mu+1-s-\alpha)} t_+^{\mu-s-\alpha}. \qquad (6.13)$$

Inserting (6.12) and (6.13) into (6.8) yields

$$t_+^{\mu-1} * f = \sum_{s=0}^{n-1} a_s \frac{\Gamma(\mu)\Gamma(1-s-\alpha)}{\Gamma(\mu+1-s-\alpha)} t_+^{\mu-s-\alpha}$$

$$- \sum_{s=1}^{n} c_s \frac{\Gamma(\mu)}{\Gamma(\mu+1-s)} t_+^{\mu-s} + t_+^{\mu-1} * f_n. \qquad (6.14)$$

Note that each distribution in (6.14) is determined by a locally integrable function in $(0, \infty)$. By replacing these distributions by their corresponding functions, we have from (6.11)

$$(t^{\mu-1} * f)(x) = \sum_{s=0}^{n-1} a_s \frac{\Gamma(\mu)\Gamma(1-s-\alpha)}{\Gamma(\mu+1-s-\alpha)} x^{\mu-s-\alpha}$$

$$- \sum_{s=1}^{n} c_s \frac{\Gamma(\mu)}{\Gamma(\mu+1-s)} x^{\mu-s} + \Gamma(\mu) x^{-n} \delta_n(x) \qquad (6.15)$$

holding in the sense of distributions in $(0, \infty)$. It is straightforward to show that equation (6.15) in fact holds pointwise almost everywhere in $(0, \infty)$. Since every convolution in (6.14) has at least one factor which is continuous in $(0, \infty)$, we further conclude that (6.15), or equivalently (6.6), holds for each x in $(0, \infty)$. The assumption that $\mu \neq 1, 2, \ldots$ and $\mu - \alpha \neq 0, 1, \ldots$ can be removed by analytic continuation. This completes the proof of Theorem 15. ∎

Theorem 16.   *Let f be a locally integrable function on $[0, \infty)$ and let f satisfy (2.5) with $\alpha = 1$. Then for any $n \geq 1$*

$$
\begin{aligned}
I^\mu f(x) = \sum_{s=0}^{n-1} \frac{(-1)^s a_s}{\Gamma(\mu+1)s!} \{x^\mu[\log x - \gamma - \psi(\mu+1)]\}^{(s+1)} \\
- \sum_{s=1}^{n} \frac{d_s}{\Gamma(\mu-s+1)} x^{\mu-s} + \frac{1}{x^n} \delta_n(x),
\end{aligned}
\tag{6.16}
$$

where γ is the Euler-Mascheroni constant, ψ is the logarithmic derivative of the gamma function, and $\delta_n(x)$ satisfies (6.7). The coefficient d_s has the same meaning as in (2.18) or equivalently in (2.32) and (2.34).

Proof.   We again assume that $\mu \neq 1, 2, \ldots$. From Lemma 2, we have

$$
f = \sum_{s=0}^{n-1} a_s t_+^{-s-1} - \sum_{s=1}^{n} d_s \delta^{(s-1)} + f_n \qquad \text{in } \mathscr{D}'
$$

and hence

$$
t_+^{\mu-1} * f = \sum_{s=0}^{n-1} a_s t_+^{\mu-1} * t_+^{-s-1} - \sum_{s=1}^{n} d_s t_+^{\mu-1} * \delta^{(s-1)} + t_+^{\mu-1} * f_n. \tag{6.17}
$$

The results in Chapter V, equations (6.10) and (6.16), then give

$$
\begin{aligned}
t_+^{\mu-1} * f = \sum_{s=0}^{n-1} \frac{(-1)^s a_s}{\mu \cdot s!} D^{s+1}\{t_+^\mu [\log_+ t - \gamma - \psi(\mu+1)]\} \\
- \sum_{s=1}^{n} d_s \frac{\Gamma(\mu)}{\Gamma(\mu-s+1)} t_+^{\mu-s} + D^n(t_+^{\mu-1} * f_{n,n}).
\end{aligned}
\tag{6.18}
$$

Note that the function $t^\mu[\log t - \gamma - \psi(\mu+1)]$ and all its derivatives are locally absolutely continuous in $(0, \infty)$. Thus the distributional derivatives in the first sum can be replaced by the corresponding ordinary derivatives. A similar remark applies to the last term in (6.18);

see (6.10). Replacing each distribution in (6.18) by its corresponding function gives

$$(t^{\mu-1} * f)(x) = \sum_{s=0}^{n-1} \frac{(-1)^s a_s}{\mu \cdot s!} \{x^\mu [\log x - \gamma - \psi(\mu + 1)]\}^{(s+1)}$$

$$- \sum_{s=1}^{n} d_s \frac{\Gamma(\mu)}{\Gamma(\mu - s + 1)} x^{\mu-s} + (t^{\mu-1} * f_{n,n})^{(n)}(x),$$

valid for x in $(0, \infty)$. From here on the argument parallels that of the proof of Theorem 15, and Theorem 16 is thus proved. ∎

The error term $\delta_n(x)$ can be bounded as follows. Let f satisfy (2.5) with $0 < \alpha < 1$, and *assume*

$$M_n = \sup_{(0,\infty)} \{t^{n+\alpha} |f_n(t)|\} < \infty; \tag{6.19}$$

see the remark following (2.45). A typical term in the sum (6.7) is given by the integral

$$I^\mu[t^{n-j} f_{n,j}](x) = \frac{1}{\Gamma(\mu)} \int_0^x (x - t)^{\mu-1} t^{n-j} f_{n,j}(t) \, dt. \tag{6.20}$$

From (2.10) and (6.19) it is easy to see that the function $f_{n,j}(t)$ satisfies

$$|f_{n,j}(t)| \leq \frac{\Gamma(n + \alpha - j)}{\Gamma(n + \alpha)} M_n t^{j-n-\alpha},$$

and hence

$$|I^\mu[t^{n-j} f_{n,j}](x)| \leq \frac{\Gamma(n + \alpha - j)\Gamma(1 - \alpha)}{\Gamma(\mu + 1 - \alpha)\Gamma(n + \alpha)} M_n x^{\mu-\alpha}.$$

This together with (6.7) gives

$$|\delta_n(x)| \leq M_n \left[\frac{\Gamma(\mu + 1)\Gamma(1 - \alpha)}{\Gamma(\mu + 1 - \alpha)\Gamma(n + \alpha)} \sum_{j=0}^{n} \binom{n}{j} \frac{\Gamma(n + \alpha - j)}{|\Gamma(\mu + 1 - j)|} \right] x^{\mu-\alpha}. \tag{6.21}$$

The estimate (6.21) establishes the asymptotic nature of the expansion (6.6):

$$I^\mu f(x) \sim \sum_{s=0}^{\infty} a_s \frac{\Gamma(1 - s - \alpha)}{\Gamma(\mu + 1 - s - \alpha)} x^{\mu-s-\alpha} - \sum_{s=1}^{\infty} \frac{c_s}{\Gamma(\mu + 1 - s)} x^{\mu-s}, \tag{6.22}$$

as $x \to +\infty$.

In the case when f satisfies (2.5) with $\alpha = 1$, we assume, as in (2.51), that for some $\rho \in (0, 1)$

$$M_n(\rho) = \sup_{(0,\infty)} \{t^{n+\rho}|f_n(t)|\} < \infty. \tag{6.23}$$

The remainder in (6.16) then satisfies

$$|\delta_n(x)| \le M_n(\rho)\left[\frac{\Gamma(\mu+1)\Gamma(1-\rho)}{\Gamma(\mu+1-\rho)\Gamma(n+\rho)}\sum_{j=0}^{n}\binom{n}{j}\frac{\Gamma(n+\rho-j)}{|\Gamma(\mu+1-j)|}\right]x^{\mu-\rho}. \tag{6.24}$$

Example 4. We illustrate the above results with the integral

$$I(x) = \int_0^1 \frac{\tau}{\sqrt{1-\tau^2}} e^{-2ix\tau^2} F\left(2\tau\sqrt{\frac{x}{\pi}}\right) d\tau, \tag{6.25}$$

where F is the Fresnel integral defined by

$$F(t) = \int_t^\infty e^{i\pi\xi^2/2}\,d\xi. \tag{6.26}$$

The change of variable $x\tau^2 = t$ puts $I(x)$ into the form of a fractional integral:

$$I(x) = \frac{1}{2\sqrt{x}}\int_0^x (x-t)^{-1/2} f(t)\,dt, \tag{6.27}$$

where

$$f(t) = e^{-2it}F\left(2\sqrt{\frac{t}{\pi}}\right). \tag{6.28}$$

In terms of the complementary error function, $f(t)$ can be written as

$$f(t) = \frac{1}{\sqrt{2}} e^{-2it+i\pi/4}\,\mathrm{erfc}(e^{-i\pi/4}\sqrt{2t}). \tag{6.29}$$

Thus it follows from Ex. 1 of Chapter I that

$$f(t) = \frac{i}{2\sqrt{\pi}}t^{-1/2}\left[1+\sum_{s=1}^{n-1}(-i)^s\frac{1\cdot3\cdots(2s-1)}{2^{2s}t^s}+r_n(t)\right] \tag{6.30}$$

with

$$|r_n(t)| \le \sqrt{2}\frac{1\cdot3\cdots(2n-1)}{2^{2n}t^n}. \tag{6.31}$$

Let $a_0 = i/2\sqrt{\pi}$ and

$$a_s = (-1)^s i^{s+1} \frac{1 \cdot 3 \cdots (2s-1)}{2^{2s+1}\sqrt{\pi}}, \qquad s = 1, 2, \ldots. \qquad (6.32)$$

With $\mu = \alpha = \frac{1}{2}$, Theorem 15 gives at once

$$I(x) = \frac{\sqrt{\pi}}{2\sqrt{x}} \left[\sum_{s=0}^{n-1} a_s \frac{\Gamma(\frac{1}{2}-s)}{\Gamma(1-s)} x^{-s} - \sum_{s=1}^{n} \frac{c_s}{\Gamma(\frac{3}{2}-s)} x^{1/2-s} + \frac{1}{x^n} \delta_n(x) \right]. \quad (6.33)$$

The coefficient c_s can be calculated from the Mellin transform of f (see Erdélyi *et al.* (1954, p. 325)), and is given by

$$c_s = \frac{1}{\sqrt{2}} \left(\frac{i}{2} \right)^s e^{i\pi/4}. \qquad (6.34)$$

Since the terms in the first sum are all zero except for the first one, (6.33) reduces to

$$I(x) \sim \frac{i}{4} \left(\frac{\pi}{x} \right)^{1/2} + \frac{e^{-i\pi/4}}{4\sqrt{2}} \sum_{s=0}^{\infty} \frac{(\frac{1}{2})_s}{(2i)^s} x^{-s-1}, \qquad \text{as } x \to \infty, \qquad (6.35)$$

where $(\frac{1}{2})_s = \Gamma(s + \frac{1}{2})/\Gamma(\frac{1}{2})$.

The similarity of the two series in (6.30) and (6.35) indicates that the integral $I(x)$ can be evaluated in closed form in terms of the Fresnel integral; see Ex. 12.

7. The Method of Regularization

The idea of using distributions can further be extended to derive asymptotic expansions for the convolution integral

$$I(x) = \int_0^\infty f(t)h(xt)\,dt \qquad (7.1)$$

considered in Chapter III. It is easy to see that this integral is equivalent to the more symmetric convolution defined by

$$(f * g)(x) = \int_0^\infty f(t)g(xt^{-1})t^{-1}\,dt. \qquad (7.2)$$

Here we shall assume that both f and g are members of the following class of functions.

Let $\mathscr{F}$ be the family of locally integrable functions $f(t)$ on $(0, \infty)$, with asymptotic expansions of the forms

$$f(t) \sim \sum_{s=0}^{\infty} a_s t^{s+\alpha} \qquad \text{as } t \to 0^+ \tag{7.3}$$

and

$$f(t) \sim \sum_{s=0}^{\infty} b_s t^{-s-\beta} \qquad \text{as } t \to +\infty, \tag{7.4}$$

where α and β are complex constants. It is easily verified that (i) if f and g belong to $\mathscr{F}$ then fg belongs to $\mathscr{F}$, and (ii) if f belongs to $\mathscr{F}$ then so does $\tilde{f}$, where $\tilde{f}(t) = f(1/t)$.

Without restrictions on the exponents α and β in (7.3) and (7.4), the integral

$$\int_0^{\infty} f(t)\, dt \tag{7.5}$$

will normally diverge. We shall now use the concept of regularization discussed in Chapter V, Section 7, to give a meaningful definition to the infinite integral in (7.5). First we return to equations (7.10) and (7.13) in Chapter V, where we have assigned the meanings

$$\int_0^b t^{\lambda}\, dt = \frac{b^{\lambda+1}}{\lambda+1} \tag{7.6}$$

and

$$\int_b^{\infty} t^{\lambda}\, dt = -\frac{b^{\lambda+1}}{\lambda+1} \tag{7.7}$$

for all $\lambda \neq -1$. The case $\lambda = -1$ needs a separate treatment. Observe that for all test functions φ whose support does not contain 0, we have

$$\int_0^b t^{-1}\varphi(t)\, dt = \varphi(b) \log b - \int_0^b \varphi'(t) \log t\, dt. \tag{7.8}$$

This identity suggests that we may define the distribution $t_{0<t\le b}^{-1}$ by

$$\langle t_{0<t\le b}^{-1}, \varphi \rangle = \varphi(b) \log b - \int_0^b \varphi'(t) \log t\, dt, \tag{7.9}$$

that is

$$t_{0<t\le b}^{-1} = (\log b)\, \delta(t-b) + [(\log t)\chi_{[0,b]}]', \tag{7.10}$$

where $\chi_{[0,b]}$ is the characteristic function which is equal to 1 on $[0, b]$, and where the derivative on the right-hand side is taken in the distributional sense. As in (7.9) of Chapter V, we again let $\varphi^*(t)$ denote a test function in $\mathscr{D}(\mathbb{R})$ which equals "1" on $[0, b]$. From (7.9) it follows that

$$\langle t_{0<t\le b}^{-1}, \varphi^* \rangle = \log b. \tag{7.11}$$

We shall regard the integral $\int_0^b t^{-1}\,dt$ as the action of the distribution $t_{0<t\le b}^{-1}$ on φ^*. Thus (7.11) gives

$$\int_0^b t^{-1}\,dt = \log b. \tag{7.12}$$

By transforming the interval $[b, \infty)$ to $(0, 1/b]$ via the substitution $\tau = 1/t$, we also have

$$\int_b^\infty t^{-1}\,dt = -\log b; \tag{7.13}$$

cf. (7.12) and (7.13) in Chapter V. Formulas (7.12) and (7.13) suggest that equation (7.14) in Chapter V should remain valid when $\lambda = -1$. That is, we should assign the meaning

$$\int_0^\infty t^\lambda\,dt = \int_0^b t^\lambda\,dt + \int_b^\infty t^\lambda\,dt = 0 \tag{7.14}$$

for all complex λ.

Suggested by (7.14), we split the interval of integration in (7.5) at $t = b$ and consider first the integral $\int_0^b f(t)\,dt$. Choose n such that $n + \operatorname{Re}\alpha > -1$ and write (7.3) as

$$f(t) = \sum_{s=0}^{n-1} a_s t^{s+\alpha} + f_{0,n}(t) \tag{7.15}$$

with

$$f_{0,n}(t) = O(t^{n+\alpha}) \quad \text{as } t \to 0^+. \tag{7.16}$$

We define

$$\int_0^b f(t)\,dt \equiv \sum_{s=0}^{n-1} a_s \int_0^b t^{s+\alpha}\,dt + \int_0^b f_{0,n}(t)\,dt, \tag{7.17}$$

where the integrals under the summation sign are understood in the sense of either (7.6) or (7.12), depending upon whether $s + \alpha$ is not or is equal to -1, and where the last integral on the right exists and is hence taken in the ordinary sense.

Similarly, we choose m such that $m + \text{Re } \beta > 1$ and write (7.4) in the form

$$f(t) = \sum_{s=0}^{m-1} b_s t^{-s-\beta} + f_{\infty,m}(t) \tag{7.18}$$

with

$$f_{\infty,m}(t) = O(t^{-m-\beta}) \qquad \text{as } t \to +\infty. \tag{7.19}$$

The integral $\int_b^\infty f(t)\, dt$ is then defined by

$$\int_b^\infty f(t)\, dt \equiv \sum_{s=0}^{m-1} b_s \int_b^\infty t^{-s-\beta}\, dt + \int_b^\infty f_{\infty,m}(t)\, dt, \tag{7.20}$$

where the integrals in the sum are understood in the sense of either (7.7) or (7.13), depending upon whether $s + \beta$ is not or is equal to 1, and where the last integral on the right-hand side exists and is taken in the ordinary sense.

Definition 1. If f is in $\mathscr{F}$ then the integral $\int_0^\infty f(t)\, dt$ is given the meaning

$$\int_0^\infty f(t)\, dt = \int_0^b f(t)\, dt + \int_b^\infty f(t)\, dt, \tag{7.21}$$

where the two integrals on the right are defined by the formulas (7.17) and (7.20). The value of the right-hand side is called *the regularization of the (formal) integral* on the left.

We shall also call the right-hand sides of equations (7.6), (7.7), (7.12), (7.13), and (7.14), respectively, the regularizations of the integrals on the corresponding left-hand sides of these equations.

A few comments are now in order. First, it is easily shown that the definitions of $\int_0^b f(t)\, dt$ and $\int_b^\infty f(t)\, dt$ given in (7.17) and (7.20) are independent of the choices of n and m as long as $n + \text{Re } \alpha > -1$ and $m + \text{Re } \beta > 1$. Second, it can also be shown that the definition of $\int_0^\infty f(t)\, dt$ given in (7.21) is independent of the choice of b. Third, the new definition of $\int_0^\infty f(t)\, dt$ agrees with the usual one when the integral converges in the ordinary sense. Finally, for the functions in $\mathscr{F}$, we have the linearity property

$$\int_0^\infty [c_1 f_1(t) + c_2 f_2(t)]\, dt = c_1 \int_0^\infty f_1(t)\, dt + c_2 \int_0^\infty f_2(t)\, dt, \qquad (7.22)$$

where all three integrals are understood in the sense of Definition 1.

Definition 2. If $f \in \mathscr{F}$, then we define the (*generalized*) *Mellin transform* of f by

$$M[f; z] = \int_0^\infty t^{z-1} f(t)\, dt, \qquad (7.23)$$

where the integral on the right is understood in the sense of Definition 1.

The above definition is easily verified to be equivalent to the Mellin transform defined in Chapter III, Section 4, when $z \neq -s - \alpha$ and $z \neq s + \beta$, where $s + \alpha$ and $s + \beta$ are the exponents in the asymptotic expansions of f. Thus $M[f; z]$ determines a meromorphic function with poles at $-s - \alpha$ and $s + \beta$. Furthermore, the present definition allows one to assign values to $M[f; -s - \alpha]$ and $M[f; s + \beta]$. The following lemma shows that these values are closely related to the meromorphic function.

Lemma 9. *Let f be in $\mathscr{F}$, with asymptotic expansions (7.3) and (7.4), and let $M[f; z]$ be the Mellin transform defined in Definition 2.*

(i) *If $-i - \alpha \neq s + \beta$ for all s, then*

$$M[f; -i - \alpha] = \lim_{z \to -i-\alpha} \left\{ M[f; z] - \frac{a_i}{z + i + \alpha} \right\}.$$

(ii) *If $j + \beta \neq -s - \alpha$ for all s, then*

$$M[f; j + \beta] = \lim_{z \to j+\beta} \left\{ M[f; z] + \frac{b_j}{z - j - \beta} \right\}.$$

(iii) *If $-i - \alpha = j + \beta$ for some i and j, then*

$$M[f; -i - \alpha] = M[f; j + \beta]$$

$$= \lim_{z \to -i-\alpha} \left\{ M[f; z] - \frac{a_i}{z + i + \alpha} + \frac{b_j}{z - j - \beta} \right\}.$$

Proof.   (i) By Definitions 1 and 2

$$M[f; z] = \int_0^1 t^{z-1} f(t)\, dt + \int_1^\infty t^{z-1} f(t)\, dt. \qquad (7.24)$$

For convenience, we have taken $b = 1$ in (7.21). From (7.6), (7.7), (7.15), and (7.18), it follows that the first integral in (7.24) is given by

$$\sum_{s=0}^{n-1} \frac{a_s}{z + s + \alpha} + \int_0^1 t^{z-1} f_{0,n}(t)\, dt$$

and the second integral by

$$-\sum_{s=0}^{m-1} \frac{b_s}{z - s - \beta} + \int_1^\infty t^{z-1} f_{\infty,m}(t)\, dt.$$

The last two integrals converge, and each defines an analytic function of z in the strip $-\operatorname{Re}\alpha - n < \operatorname{Re} z < m + \operatorname{Re}\beta$. Choose $n > i$, and write

$$M[f; z] - \frac{a_i}{z + i + \alpha} = \sum_{s=0}^{n-1}{}' \frac{a_s}{z + s + \alpha} + \int_0^1 t^{z-1} f_{0,n}(t)\, dt$$

$$- \sum_{s=0}^{m-1} \frac{b_s}{z - s - \beta} + \int_1^\infty t^{z-1} f_{\infty,m}(t)\, dt, \qquad (7.25)$$

where $\sum'$ indicates that the ith term is missing in the sum. From this it follows that

$$\lim_{z \to -i-\alpha} \left\{ M[f; z] - \frac{a_i}{z + i + \alpha} \right\} = \sum_{s=0}^{n-1}{}' \frac{a_s}{s - i} + \int_0^1 t^{-i-\alpha-1} f_{0,n}(t)\, dt$$

$$+ \sum_{s=0}^{m-1} \frac{b_s}{i + \alpha + s + \beta} + \int_1^\infty t^{-i-\alpha-1} f_{\infty,m}(t)\, dt.$$

Direct calculation, using (7.12), shows that the right-hand side of the above equation is equal to $M[f; -i - \alpha]$, thus proving (i).

Statements (ii) and (iii) are proved in a similar manner, using (7.13) where necessary. ∎

As remarked earlier, divergent integrals must be handled with care since standard rules do not usually apply. The next result shows the change-of-variable formula is no longer legitimate.

Lemma 10. *Let f be in $\mathscr{F}$, with asymptotic expansions (7.3) and (7.4), and let $f_x(t) \equiv f(x/t)$. Then*

(i) $M[f_x; z] = x^z\, M[f; -z]$

if $z \neq s + \alpha$ and $z \neq -s - \beta$ for all s;

(ii) $M[f_x; i + \alpha] = x^{i+\alpha}\{M[f; -i - \alpha] - a_i \log x\}$

if $i + \alpha \neq -s - \beta$ for all s;

(iii) $M[f_x; -j - \beta] = x^{-j-\beta}\{M[f; j + \beta] + b_j \log x\}$

if $j + \beta \neq -s - \alpha$ for all s;

(iv) $M[f_x; z] = x^z\{M[f; -z] - (a_i - b_j) \log x\}$

if $z = i + \alpha = -j - \beta$ for some i and j.

Proof.   (i) Proceed as in the proof of Lemma 9, except that instead of (7.24) we begin with

$$M[f_x; z] = \int_0^x t^{z-1} f\left(\frac{x}{t}\right) dt + \int_x^\infty t^{z-1} f\left(\frac{x}{t}\right) dt. \qquad (7.26)$$

The two integrals on the right are given by

$$x^z \sum_{s=0}^{m-1} \frac{b_s}{z+s+\beta} + \int_0^x t^{z-1} f_{\infty,m}\left(\frac{x}{t}\right) dt$$

and

$$-x^z \sum_{s=0}^{n-1} \frac{a_s}{z-s-\alpha} + \int_x^\infty t^{z-1} f_{0,n}\left(\frac{x}{t}\right) dt,$$

respectively. For $-m - \operatorname{Re}\beta < \operatorname{Re} z < n + \operatorname{Re}\alpha$, the two remainder integrals exist as ordinary integrals, and hence by a change of variable they are equal to

$$x^z \int_1^\infty u^{-z-1} f_{\infty,m}(u)\, du \qquad \text{and} \qquad x^z \int_0^1 u^{-z-1} f_{0,n}(u)\, du.$$

A combination of these results gives

$$M[f_x; z] = x^z\Bigg[\sum_{s=0}^{m-1} \frac{b_s}{z+s+\beta} + \int_1^\infty u^{-z-1} f_{\infty,m}(u)\, du$$
$$- \sum_{s=0}^{n-1} \frac{a_s}{z-s-\alpha} + \int_0^1 u^{-z-1} f_{0,n}(u)\, du \Bigg].$$

The quantity inside the square brackets is exactly $M[f; -z]$; see (7.25). This completes the proof of statement (i).

Statements (ii), (iii), (iv): Proceed as in (i) and use (7.12) and (7.13) when it is appropriate. ∎

Returning to the convolution integral (7.2), we assume that $f(t)$ has the asymptotic expansions (7.3) and (7.4), and that $g(t)$ has the corresponding expansions

$$g(t) \sim \sum_{s=0}^{\infty} c_s t^{s+\mu} \qquad \text{as } t \to 0^+, \qquad (7.27)$$

$$g(t) \sim \sum_{s=0}^{\infty} d_s t^{-s-\nu} \qquad \text{as } t \to \infty, \qquad (7.28)$$

where μ and ν are complex numbers. For the convergence of the integral (7.2), we also impose the conditions that the exponents in these expansions satisfy

$$\text{Re}(\alpha + \nu) > 0 \qquad \text{and} \qquad \text{Re}(\beta + \mu) > 0. \qquad (7.29)$$

From (7.14) and (7.15) it is easily seen that

$$M[f; z] = M[f_{0,n}; z] \qquad (7.30)$$

for all values of z. Similarly, if (7.27) is written as

$$g(t) = \sum_{s=0}^{m-1} c_s t^{s+\mu} + g_{0,m}(t) \qquad (7.31)$$

with $g_{0,m}(t) = O(t^{m+\mu})$ as $t \to 0^+$, then we also have

$$M[g_{0,m}(x/t); z] = M[g_x; z], \qquad (7.32)$$

where $g_x(t) = g(x/t)$. Since $\text{Re}(\alpha + \nu) > 0$ by (7.29), Lemma 10 (i) gives

$$M[g_x; i + \alpha] = x^{i+\alpha} M[g; -i - \alpha] \qquad (7.33)$$

as long as $i + \alpha \neq s + \mu$ for all s, and Lemma 10 (ii) gives

$$M[g_x; i + \alpha] = x^{i+\alpha}\{M[g; -i - \alpha] - c_{i+\alpha-\mu} \log x\} \qquad (7.34)$$

if $i + \alpha = j + \mu$ for some j.

Multiplying (7.15) and (7.31) together gives

$$f(t)g(xt^{-1}) = \sum_{i=0}^{n-1}\sum_{j=0}^{m-1} a_i c_j x^{j+\mu} t^{i+\alpha-j-\mu}$$

$$+ \sum_{i=0}^{n-1} a_i t^{i+\alpha} g_{0,m}(xt^{-1}) \qquad (7.35)$$

$$+ \sum_{j=0}^{m-1} c_j x^{j+\mu} t^{-j-\mu} f_{0,n}(t) + f_{0,n}(t)g_{0,m}(xt^{-1}).$$

Inserting this into (7.2) and applying the linearity property mentioned above, we obtain from (7.14), (7.30), and (7.32)

$$(f*g)(x) = \sum_{i=0}^{n-1} a_i M[g_x; i+\alpha] + \sum_{j=0}^{m-1} c_j M[f; -j-\mu]x^{j+\mu} + \delta_{mn}(x) \quad (7.36)$$

where

$$\delta_{mn}(x) = \int_0^\infty f_{0,n}(t)g_{0,m}(xt^{-1})t^{-1}\,dt. \qquad (7.37)$$

Choose m and n so that

$$m \geq 1 - \text{Re}(\mu+v), \qquad n \geq 1 - \text{Re}(\alpha+\beta) \qquad (7.38)$$

and

$$m + \text{Re}(\mu-\alpha) - 1 < n < m + \text{Re}(\mu-\alpha) + 1. \qquad (7.39)$$

Under these conditions, the integral in (7.37) exists as an ordinary integral. If it happens that $0 < \alpha \leq 1$ and $0 < \mu \leq 1$, then (7.39) is satisfied with $n = m$.

Theorem 17. *Let f and g belong to $\mathcal{F}$, and let their asymptotic expansions be given by (7.3), (7.4), (7.27), and (7.28). Suppose that (7.29) holds and $s + \alpha \neq r + \mu$ for all r and s. Then*

$$(f*g)(x) = \sum_{i=0}^{n-1} a_i M[g; -i-\alpha]x^{i+\alpha}$$

$$+ \sum_{j=0}^{m-1} c_j M[f; -j-\mu]x^{j+\mu} + O(x^{\rho_{mn}}), \qquad (7.40)$$

as $x \to 0^+$, where $\rho_{mn} = \min\{n + \text{Re}\,\alpha, m + \text{Re}\,\mu\}$.

342   VI The Distributional Approach

Proof. Without loss of generality, we may assume that n and m are sufficiently large so that (7.38) and (7.39) hold. Then, from (7.33) and (7.36), it follows that

$$(f * g)(x) = \sum_{i=0}^{n-1} a_i \, M[g; -i - \alpha]x^{i+\alpha}$$

$$+ \sum_{j=0}^{m-1} c_j \, M[f; -j - \mu]x^{j+\mu} + \delta_{mn}(x),$$

where $\delta_{mn}(x)$ is given in (7.37). Hence it suffices to show

$$\delta_{mn}(x) = O(x^{\rho_{mn}}), \qquad \text{as } x \to 0^+. \tag{7.41}$$

For $0 < x < 1$ it is easy to see that

$$\int_0^x f_{0,n}(t)g_{0,m}(xt^{-1})t^{-1}\,dt = O(x^{n+\alpha})$$

$$\int_x^1 f_{0,n}(t)g_{0,m}(xt^{-1})t^{-1}\,dt = O(x^{\rho_{mn}}),$$

and

$$\int_1^\infty f_{0,n}(t)g_{0,m}(xt^{-1})t^{-1}\,dt = O(x^{m+\mu});$$

thus proving (7.41) and the theorem. ■

Theorem 18. *Let f and g be given as in Theorem 17, and let the conditions in (7.29) hold. If $s + \alpha = r + \mu$ for some s and r, then*

$$(f * g)(x) = \sum_{0 \le i < \mu - \alpha} a_i \, M[g; -i - \alpha]x^{i+\alpha}$$

$$+ \sum_{i \ge \mu - \alpha}^{n-1} a_i[c_i^* - c_{i+\alpha-\mu} \log x]x^{i+\alpha}$$

$$+ \sum_{0 \le j < \alpha - \mu} c_j \, M[f; -j - \mu]x^{j+\mu} \tag{7.42}$$

$$+ \sum_{j \ge \alpha - \mu}^{m-1} a_j^* c_j x^{j+\mu} + O(x^{\rho_{mn}} \log x),$$

as $x \to 0^+$, *where* $\rho_{mn} = \min\{n + \operatorname{Re} \alpha,\, m + \operatorname{Re} \mu\}$,

$$c_i^* = \lim_{z \to -i-\alpha} \left\{ M[g;\, z] - \frac{c_{i+\alpha-\mu}}{z+i+\alpha} \right\} \qquad (7.43)$$

and

$$a_j^* = \lim_{z \to -j-\mu} \left\{ M[f;\, z] - \frac{a_{j+\mu-\alpha}}{z+j+\mu} \right\}. \qquad (7.44)$$

The coefficients a_i and c_j with negative subscripts are understood to be zero.

Proof. We again assume without loss of generality that m and n are sufficiently large so that (7.38) and (7.39) hold. Observe that since $s + \alpha = r + \mu$, $\alpha - \mu$ is an integer. Hence, if $i \geq 0$ and $i \geq \mu - \alpha$ then $i = \mu - \alpha + j$ for some $j \geq 0$. Consequently, $i + \alpha = j + \mu$ for all $i \geq \mu - \alpha$ and $j \geq \alpha - \mu$. Therefore, from (7.33), (7.34), and (7.36), we obtain

$$(f * g)(x) = \sum_{0 \leq i < \mu - \alpha} a_i\, M[g;\, -i - \alpha] x^{i+\alpha}$$

$$+ \sum_{i \geq \mu - \alpha}^{n-1} a_i \left\{ M[g;\, -i - \alpha] - c_{i+\alpha-\mu} \log x \right\} x^{i+\alpha} \quad (7.45)$$

$$+ \sum_{j=0}^{m-1} c_j\, M[f;\, -j - \mu] x^{j+\mu} + \delta_{mn}(x).$$

In view of Lemma 9 (i) and (7.29),

$$M[g;\, -i - \alpha] = M[g;\, -j - \mu]$$

$$= \lim_{z \to -j-\mu} \left\{ M[g;\, z] - \frac{c_j}{z+j+\mu} \right\}$$

$$= \lim_{z \to -i-\alpha} \left\{ M[g;\, z] - \frac{c_{i+\alpha-\mu}}{z+i+\alpha} \right\} = c_i^*$$

for $i \geq \mu - \alpha$, and similarly $M[f;\, -j - \mu] = a_j^*$ for $j \geq \alpha - \mu$. Thus (7.42) will follow immediately from (7.45), provided that we can show

$$\delta_{mn}(x) = O(x^{\rho_{mn}} \log x) \qquad \text{as } x \to 0^+. \qquad (7.46)$$

For $0 < x < 1$, it is easily shown that

$$\int_0^x f_{0,n}(t)g_{0,m}(xt^{-1})t^{-1}\,dt = O(x^{n+\alpha})$$

and

$$\int_1^\infty f_{0,n}(t)g_{0,m}(xt^{-1})t^{-1}\,dt = O(x^{m+\mu}).$$

Since $n + \alpha = m + \mu$, we also have

$$\int_x^1 f_{0,n}(t)g_{0,m}(xt^{-1})t^{-1}\,dt = O(x^{m+\mu}\log x).$$

Therefore (7.46) holds and the theorem is proved. ■

Note that if $\alpha = \mu$ then (7.42) simplifies to

$$(f * g)(x) \sim \sum_{i=0}^\infty A_i x^{i+\alpha} - \sum_{i=0}^\infty a_i c_i x^{i+\alpha}\log x, \qquad (7.47)$$

as $x \to 0^+$, where $A_i = a_i c_i^* + a_i^* c_i$. Analogous results for large-x behavior can be found in Wong and McClure (1984).

We now return to the error term $\delta_{mn}(x)$ in (7.37), and assume that m and n satisfy (7.38) and (7.39). For simplicity, we also assume that the exponents α and μ are real. To construct numerical bounds for $\delta_{mn}(x)$, we set

$$M_n = \sup_{(0,\infty)} \{t^{-n-\alpha}|f_{0,n}(t)|\} \qquad (7.48)$$

and

$$N_m = \sup_{(0,\infty)} \{t^{-m-\mu}|g_{0,m}(t)|\}. \qquad (7.49)$$

If M_n is finite and $m + \mu - 1 < n + \alpha < m + \mu$ then from (7.37) it follows immediately that

$$|\delta_{mn}(x)| \le x^{n+\alpha}M_n \int_0^\infty \tau^{-n-\alpha-1}|g_{0,m}(\tau)|\,d\tau. \qquad (7.50)$$

On the other hand, if N_m is finite and $m + \mu < n + \alpha < m + \mu + 1$ then

$$|\delta_{mn}(x)| \le x^{m+\mu}N_m \int_0^\infty t^{-m-\mu-1}|f_{0,n}(t)|\,dt. \qquad (7.51)$$

In the case when both results (7.50) and (7.51) do not hold, then we may proceed as follows. In view of the assumptions that $f_{0,n}(t) = O(t^{n+\alpha})$ and $g_{0,m}(t) = O(t^{m+\mu})$ as $t \to 0^+$, it is not unreasonable to *suppose* that there exist numbers M_n^* and N_m^* such that

$$|f_{0,n}(t)| \leq M_n^* t^{n+\alpha} \qquad \text{for } 0 < t < 1 \tag{7.52}$$

and

$$|g_{0,m}(t)| \leq N_m^* t^{m+\mu} \qquad \text{for } 0 < t < 1. \tag{7.53}$$

Put

$$F_{mn} = \int_1^\infty t^{-m-\mu-1} |f_{0,n}(t)| \, dt \tag{7.54}$$

and

$$G_{mn} = \int_1^\infty t^{-n-\alpha-1} |g_{0,m}(t)| \, dt. \tag{7.55}$$

Under the conditions in (7.38) and (7.39), both integrals are convergent. A simple estimation then yields

$$
|\delta_{mn}(x)| \leq x^{n+\alpha} M_n^* \left[G_{mn} + \frac{N_m^*}{m+\mu-n-\alpha} \right]
$$
$$
+ x^{m+\mu} N_m^* \left[F_{mn} + \frac{M_n^*}{n+\alpha-m-\mu} \right]
\tag{7.56}
$$

if $n + \alpha \neq m + \mu$, and

$$|\delta_{mn}(x)| \leq x^{n+\alpha} [A_{mn}^* - M_n^* N_m^* \log x] \tag{7.57}$$

if $n + \alpha = m + \mu$, for $0 < x < 1$, where $A_{mn}^* = M_n^* G_{mn} + N_m^* F_{mn}$.

Example 5.   The above results can be applied to the function

$$R_F(x, y, z) = \frac{1}{2} \int_0^\infty \frac{d\tau}{\sqrt{(\tau + x)(\tau + y)(\tau + z)}}, \tag{7.58}$$

which is called by Carlson (1977, p. 264) the elliptic integral of the first kind. In our notations, this function can be written as

$$2\sqrt{z} R_F(x, y, z) = (f * g)\left(\frac{1}{z}\right) \tag{7.59}$$

with

$$f(t) = \frac{1}{\sqrt{(1 + xt)(1 + yt)}} \quad \text{and} \quad g(t) = \frac{1}{\sqrt{1 + t}}.$$

Clearly $f(t) \sim \sum_{s=0}^{\infty} a_s t^s$ as $t \to 0^+$, where a_s is a polynomial in x and y of degree s, and $g(t) \sim \sum_{s=0}^{\infty} c_s t^s$ as $t \to 0^+$, where $c_s = (-1)^s \Gamma(s + \frac{1}{2})/s! \sqrt{\pi}$. Thus, with $\alpha = \mu = 0$, Theorem 18 gives

$$2\sqrt{z} R_F(x, y, z) \sim \sum_{i=0}^{\infty} a_i c_i z^{-i} \log z + \sum_{i=0}^{\infty} A_i z^{-i}, \qquad (7.60)$$

as $z \to \infty$. It can be shown (Ex. 16) that

$$A_0 = 4 \log 2 - \log x - 2 \log\left(1 + \sqrt{\frac{y}{x}}\right), \qquad (7.61)$$

but higher coefficients are difficult to calculate. Since $a_0 = c_0 = 1$, coupling (7.60) and (7.61) gives

$$R_F(x, y, z) \sim \frac{1}{\sqrt{z}} \log \frac{4\sqrt{z}}{\sqrt{x} + \sqrt{y}}, \qquad \text{as } z \to \infty. \qquad (7.62)$$

For a more detailed discussion of this example, see the article by Carlson and Gustafson (1985).

Exercises

1. Prove the result stated in (2.3) and (2.4).

2. Show that the coefficient d_s in (2.18) has the alternative representation

    $$d_{s+1} = -\int_0^{\infty} \left[f_{s,s}(t) - \frac{(-1)^s}{s!} \frac{a_s}{1 + t} \right] dt, \qquad s = 0, 1, \dots.$$

3. For $n = 1$, derive (2.27) directly from (2.6) and (2.10). Use (2.15) to show that the remainder in (2.28) satisfies the recurrence relation

    $$R_n(z) = (-1)^n \frac{\pi}{\sin \alpha \pi} \frac{a_n}{z^{n+\alpha}} - n! \frac{c_{n+1}}{z^{n+1}} + R_{n+1}(z).$$

    Prove (2.27) by induction.

4. Derive the estimate in (2.48)–(2.49) by using the identity in (3.16).

5. Assume that the constants $M_{n,0}$ and $M_{n,\infty}$, given in (4.29) and (4.30), are finite. Show that the remainder in (2.33) satisfies

$$|R_n(x)| \le [M_{n,0} + M_{n,\infty} \log(x+1)]x^{-n-1} \qquad \text{for } x > 0.$$

6. Show that the function $E_0(x)$ in (3.3) has the asymptotic expansion

$$E_0(x) = e^{icx}\left[\sum_{s=0}^{n-1}\left(-\frac{i}{c}\right)^{s+1}\frac{s!}{x^{s+1}} + \delta_n(x)\right],$$

where $|\delta_n(x)| \le 2 \cdot n!/(cx)^{n+1}$ for $x > 0$. (Cf. Lemma 5.)

7. Let $f_n(t)$ denote, as before, the n-th remainder term in the asymptotic expansion (4.4), and assume that $0 < \alpha < 1$ and $c \ne 0$. Prove that

$$\int_0^\infty t^{n-1}f_n(t)\, dt = d_n^*,$$

where d_n^* is the constant defined in (4.26) with $s = n$. Put

$$\delta_n(x) = \int_0^\infty \frac{t^n f_n(t)}{t-x}\, dt,$$

and show that

$$\delta_n(x) = d_s^* + x\int_0^\infty \frac{t^{n-1}f_n(t)}{t-x}\, dt.$$

Derive the recurrence relation

$$\frac{1}{x^n}\delta_n(x) = -E_\alpha^*(x)\frac{a_{n-1}}{x^{n-1}} + \frac{d_n^*}{x^n} + \frac{1}{x^{n-1}}\delta_{n-1}(x).$$

Use this identity to prove Theorem 8.

8. Show that as $x \to +\infty$,

$$\int_0^\infty \frac{e^{-u}}{1-xu}\, du \sim \left(\log\frac{1}{x}\right)\sum_{s=0}^\infty \frac{(-1)^s}{s!\, x^{s+1}} - \sum_{s=1}^\infty \frac{(-1)^s\psi(s)}{s!\, x^s},$$

where $\psi(s)$ is the logarithmic derivative of $\Gamma(s)$.

9. Let $0 < \alpha < 1$, c be real, and $T > 0$. Justify that if y is real and $y \neq \pm c$ then

$$\frac{\partial}{\partial y} \int_T^\infty \frac{e^{ict}}{t^{\alpha+1}} \sin yt\, dt = \int_T^\infty \frac{e^{ict}}{t^\alpha} \cos yt\, dt.$$

By application of this result, or otherwise, show that if $x > 0$ then the order of integration in

$$\int_T^\infty \frac{e^{ict}}{t^\alpha}\left[\int_0^\infty e^{-x\cosh y} \cos ty\, dy\right] dt$$

can be reversed.

10. Let $h(t)$ be a locally integrable function in $[0, \infty)$ and $h(t) = e^{ict}t^{-\alpha} + O(t^{-\alpha-1})$ as $t \to \infty$, where $0 < \alpha < 1$ and c is real. Prove that for $x > 0$

$$\int_0^\infty h(t)\left[\int_0^\infty e^{-x\cosh y} \cos ty\, dy\right] dt$$

$$= \int_0^\infty e^{-x\cosh y}\left[\int_0^\infty h(t) \cos yt\, dt\right] dy.$$

11. Derive an asymptotic expansion for the integral

$$\frac{1}{\Gamma(\alpha)} \int_0^x (x-t)^{\alpha-1} e^{-t} I_0(t)\, dt, \qquad \alpha > 0, \quad \text{as } x \to +\infty.$$

12. Consider the integral $I(x)$ in (6.27) as a convolution in the Laplace transform theory, and let $f(t)$ denote the function defined in (6.28). Show by integration by parts that

$$L_f(p) = \frac{e^{\pi i/4}}{\sqrt{2}(p+2i)} - \frac{1}{p^{1/2}(p+2i)}.$$

By taking the inverse Laplace transform, prove that

$$I(x) = \frac{i}{4}\left(\frac{\pi}{x}\right)^{1/2} - \frac{e^{i\pi/4}}{2\sqrt{x}} e^{-i2x} \sqrt{\frac{\pi}{2}} F\left(2\sqrt{\frac{x}{\pi}}\right),$$

where F is the Fresnel integral defined in (6.26).

13. Show that in the sense of regularization

$$\int_0^\infty \frac{t^{-m}}{t+x} dt = (-1)^{m+1} \frac{\log x}{x}, \qquad m = 0, 1, 2, \ldots.$$

(Hint: Use induction.)

14. Put

$$I(x) = \int_0^\infty \frac{t}{t^2 + x^2} e^t K_0(t)\, dt.$$

Derive the asymptotic expansion

$$I(x) = \sum_{s=0}^{2m} a_s x^{-s-1/2} + \sum_{s=0}^{m-1} b_s x^{-2s-2} + \delta_{2m+1}(x),$$

as $x \to +\infty$, where

$$a_s = (-1)^s \left(\frac{\pi}{2}\right)^{3/2} \frac{[1 \cdot 3 \cdots (2s-1)]^2}{s!\, 8^s} \csc\left[\frac{\pi}{2}(s + \tfrac{1}{2})\right]$$

and

$$b_s = (-1)^s \frac{[(2s+1)!]^2}{3 \cdot 5 \cdots (4s+3)}.$$

Show that the remainder $\delta_{2m+1}(x)$ satisfies

$$|\delta_{2m+1}(x)| \le \frac{\pi^{3/2}}{2} \frac{[1 \cdot 3 \cdots (4m+1)]^2}{(2m+1)!\, 8^{2m+1}} x^{-2m-3/2}.$$

15. Construct an asymptotic expansion for the integral

$$\int_0^\infty t^n (\log t)^m \frac{e^{-t}}{t+x}\, dt$$

as $x \to 0^+$, where n and m are nonnegative integers.

16. Let $f(t) = 1/\sqrt{(1+xt)(1+yt)}$. Show that

$$M[f; z] = x^{-z} \Gamma(z)\Gamma(1-z) F\left(\frac{1}{2}, z; 1; 1 - \frac{y}{x}\right),$$

where $F(a, b; c; z)$ is the hypergeometric function. Prove that

$$a_0^* \equiv \lim_{z \to 0} \left\{ M[f; z] - \frac{1}{z} \right\} = \frac{\partial}{\partial b} F\left(\frac{1}{2}, b; 1; 1 - \frac{y}{x}\right)\Big|_{b=0} - \log x,$$

and

$$\frac{\partial}{\partial b} F\left(\frac{1}{2}, b; 1; z\right)\Big|_{b=0} = 2\log 2 - 2\log(1 + \sqrt{1-z}).$$

Use these to verify the result stated in (7.61).

17. The integral

$$I(x) = \int_0^1 \frac{J_\nu^2(xt)}{\sqrt{1-t^2}}\, dt, \qquad \nu > -\tfrac{1}{2},$$

can be written in the form $I(x) = \int_0^\infty f(t)h(xt)\, dt$, where $h(t) = J_\nu^2(t)$ and

$$f(t) = \begin{cases} \dfrac{1}{\sqrt{1-t^2}} & 0 < t < 1 \\[2mm] 0 & t \ge 1; \end{cases}$$

cf. Ex. 5, Chapter III. Recall from Example 1 that $h(t) = h_1(t) + h_2(t)$, where

$$h_1(t) = \tfrac{1}{4}\{[H_\nu^{(1)}(t)]^2 + [H_\nu^{(2)}(t)]^2\}, \qquad h_2(t) = \tfrac{1}{2}[J_\nu^2(t) + Y_\nu^2(t)].$$

Hence $I(x)$ can be also expressed as $I(x) = I_1(x) + I_2(x)$ with $I_i(x) = \int_0^\infty f(t)h_i(xt)\, dt$, $i = 1, 2$. Show that as $x \to \infty$,

$$I_2(x) = \frac{1}{\pi x}\left[\log x + 2\log 2 - \psi(\tfrac{1}{2} + \nu) + \frac{\pi}{2}\tan \nu\pi \right]$$

$$+ \frac{1}{16\pi x^3}\left\{ (4\nu^2 - 1)\left[\log x + 2\log 2 - \psi(\tfrac{1}{2} + \nu) + \frac{\pi}{2}\tan \nu\pi \right] \right.$$

$$\left. - (4\nu^2 - 3)\right\} + O\!\left(\frac{\log x}{x^5}\right).$$

Let $v_1(t)$ and $v_2(t)$ be C^∞-functions satisfying

$$v_1(t) = 1 \quad \text{for} \quad 0 \le t \le a, \quad v_1(t) = 0 \ \text{for} \ t \ge b, \quad 0 < a < b < 1,$$

$$v_2(t) = 0 \ \text{for} \ 0 \le t \le a, \qquad v_2(t) = 1 \ \text{for} \ t \ge b,$$

$$v_1^{(s)}(a^+) = v_2^{(s)}(b^-) = 0, \qquad s = 1, 2, \ldots,$$

$$v_1(t) + v_2(t) = 1 \quad \text{for} \quad a \le t \le b$$

(the existence of these functions is established in Lemma 1, Chapter V). Write

$$I_1(x) = \int_0^\infty f(t)v_1(t)h_1(xt)\, dt + \int_0^\infty f(t)v_2(t)h_1(xt)\, dt = I_{11}(x) + I_{21}(x).$$

Show that as $x \to \infty$,

$$I_{11}(x) \sim M[h_1; 1]x^{-1} + \frac{1}{2} M[h_1; 3]x^{-3} + \cdots,$$

where

$$M[h_1; 1] = -\frac{1}{2} \tan v\pi, \qquad M[h_1; 3] = -\frac{4v^2 - 1}{16} \tan v\pi,$$

and that

$$I_{21}(x) = \frac{1}{2\sqrt{\pi}} \sin\left(2x - v\pi - \frac{\pi}{4}\right) x^{-3/2} - \frac{5}{32\sqrt{\pi}} \cos\left(2x - v\pi - \frac{\pi}{4}\right) x^{-5/2}$$

$$+ \frac{1}{2\sqrt{\pi}}\left(v^2 - \frac{1}{4}\right) \cos\left(2x - v\pi - \frac{\pi}{4}\right) x^{-5/2} + O(x^{-7/2}).$$

Conclude that

$$I(x) = \frac{1}{\pi x} [\log x + 2 \log 2 - \psi(\tfrac{1}{2} + v)] + \frac{1}{2\sqrt{\pi} x^{3/2}} \sin\left(2x - v\pi - \frac{\pi}{4}\right)$$

$$+ \frac{1}{32\sqrt{\pi} x^{5/2}} (16v^2 - 9) \cos\left(2x - v\pi - \frac{\pi}{4}\right)$$

$$+ \frac{1}{16\pi x^3} \{(4v^2 - 1)[\log x + 2 \log 2 - \psi(\tfrac{1}{2} + v)] - (4v^2 - 3)\}$$

$$+ O(x^{-7/2}).$$

Supplementary Notes

§1. The expansion (1.7) can of course be obtained by the Mellin transform technique given in Chapter III. The idea of using distribution theory to construct asymptotic expansions has also been dealt with by Lighthill (1958), Jones (1969), Durbin (1979), and Zayed (1982).

§§2–3. The contents of these sections are taken from McClure and Wong (1978). An alternative derivation of the asymptotic expansions of the Stieltjes transform can be found in Lauwerier (1974). The estimate (3.33) is considerably better than that previously given by McClure and Wong (1978, Eq. (4.11)).

§4.   The argument leading to the explicit expression (4.11) for the error term is taken from Wong (1980c). Another way of deriving the expansion (4.6) with the error term (4.11) is to view the one-sided Hilbert transform as a Mellin convolution; for details of this approach, see Wong (1979).

§5.   Asymptotic expansions of the Fourier transform near the origin have been obtained earlier by Grosjean (1965) and Levey and Mahony (1968).

§6.   The material of this section is based on McClure and Wong (1979).

§7.   For a more thorough discussion of the regularization method, see Wong and McClure (1984).

Exercises.   Exercise 1 is taken from Olver's book (1974a, p. 92). The inductive method in Ex. 3 is also due to him (private communication). Exercise 7 is based on the article of Wong (1980a). The arguments in Exs. 9 and 10 were suggested by E. R. Love. The result in Ex. 12 was communicated by J. Boersma. The behavior of the integral in Ex. 15 has been studied by Kaper (1977). For a solution to Ex. 16, see Wong (1983). Exercise 17 is based on the paper of Wong (1988).

VII

Uniform Asymptotic Expansions

1. Introduction

In Chapter II, we have considered integrals of the form

$$I(\lambda) = \int_C g(t)e^{-\lambda f(t)}\, dt, \tag{1.1}$$

where λ is a large positive parameter and C is a continuous curve in the complex t-plane along which the functions $f(t)$ and $g(t)$ are defined. For simplicity, let us assume that $f(t)$ is an entire function and the real part of $f(t)$ tends to $+\infty$ as t approaches either end of the contour C. Some of the essential steps involved in the steepest descent method consist of (i) making $f(t)$ the variable of integration, (ii) tracing the new path of integration, and (iii) locating the singularities of the integrand. If we let $u = f(t)$, then the integral $I(\lambda)$ becomes

$$I(\lambda) = \int_{C'} \varphi(u)e^{-\lambda u}\, du, \tag{1.2}$$

where C' is the image of C and

$$\varphi(u) = g(t)\,\frac{dt}{du}.$$

The singularities of φ can arise either from the singularities of g as a function of t or from the saddle points of f, i.e., points at which

$$\frac{du}{dt} = f'(t) = 0.$$

It is envisaged that the singularities, say $u = \ldots, a_{-1}, a_0, a_1, \ldots$, are either poles or of branch-point type. The contour C' starts from infinity on the right-hand side of the u-plane, returns to infinity to the right, and is oriented in the positive direction. We can always order the singularities according to $\ldots \leq \operatorname{Re} a_{-1} \leq \operatorname{Re} a_0 \leq \operatorname{Re} a_1 \leq \ldots$, and, for simplicity, we shall assume that all inequalities are strict. Furthermore, we shall suppose that the situation is as illustrated in Figure 7.1, whereby a translation has been used to make $a_0 = 0$. Cuts are drawn in the u-plane so that $\varphi(u)$ is single-valued and analytic along the contour C'. By shifting C' to the right as far as possible, it is evident that C' can be replaced by a number of loops surrounding the branch points and poles (Figure 7.2). In the case of a branch point, the two straight-line parts of the loop are on different sides of the cut. In the case of a pole, the integrals over the two straight-line parts cancel and what is left is the residue at the pole.

One could now envisage making an asymptotic evaluation involving each loop around a singularity separately, and then adding them

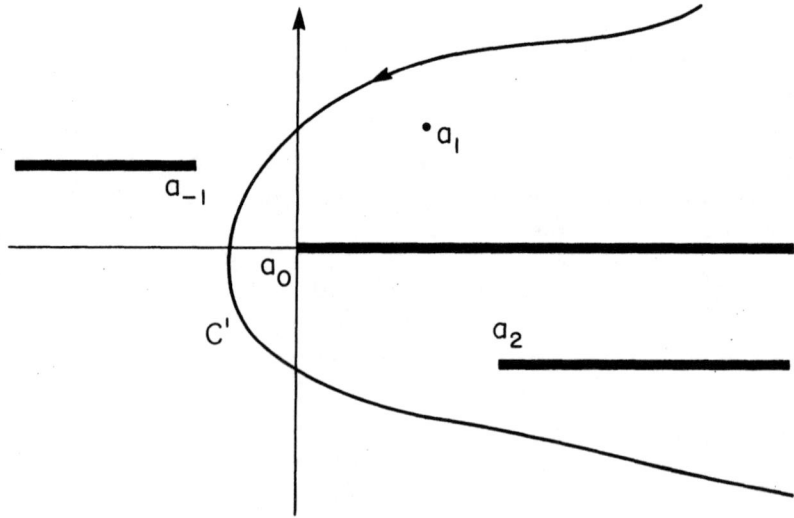

Fig. 7.1   Contour C' in the u-plane.

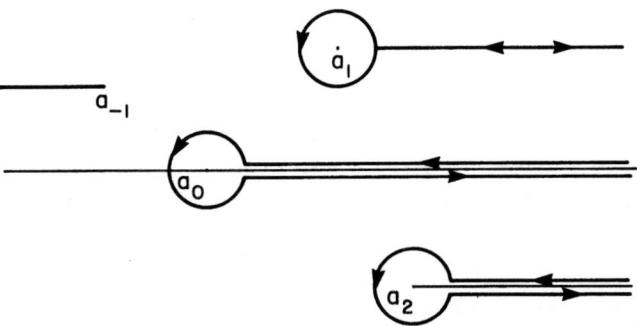

Fig. 7.2 Loop-paths of integration.

together. By Barnes' lemma (i.e., Watson's lemma for loop integrals), each loop integral is asymptotically equal to an exponential function multiplied by a series in descending powers of λ. Thus, except for the contribution from $u = a_0$, all contributions are exponentially small; i.e., only one singularity needs to be taken into account. (It should be noted that in some instances, exponentially small terms can be significant numerically as well as physically; see, for example, Olver (1974a, Chapter 3, §6) and Meyer (1980).)

The above situation is completely changed when the singularities are allowed to depend on some auxiliary parameters; the very form of the asymptotic expansion changes when two or more singularities coalesce. Furthermore, if the positions of the singularities are variable, it will no longer be possible to order them in a meaningful way like $\ldots < \operatorname{Re} a_{-1} < \operatorname{Re} a_0 < \operatorname{Re} a_1 < \ldots$. Let us suppose that the parameters are $\alpha_1, \alpha_2, \ldots, \alpha_m$, so that

$$I(\lambda) = \int_C g(t; \alpha_1, \ldots, \alpha_m) e^{-\lambda f(t; \alpha_1, \ldots, \alpha_m)} \, dt. \tag{1.3}$$

The inversion problem involves the solution of $f(t; \alpha_1, \ldots, \alpha_m) = u$ for t as a function of u. An extremely difficult problem is now envisaged if one seeks asymptotic expansions in λ which are uniform in the α's.

In this chapter, we shall study some simple situations in which the integration path C in (1.3) is either a fixed finite or infinite interval along the real-axis or a contour in the complex t-plane, and in which only two or three singularities are allowed to approach each other; some of these singularities may be located at the endpoints of integration. In §2 we consider the case where a pole lies near a saddle point.

The expansion obtained in this case has the form of an error function plus an asymptotic series in $\lambda^{-1/2}$. In §3, we derive a uniform asymptotic expansion of an integral with a saddle point near a branch-point singularity which is at the endpoint of the path of integration. The resulting expansion involves a parabolic cylinder function and its derivative. §4 is devoted to the famous cubic transformation used for the problem of two coalescing saddle points. The expansion in this case uses an Airy function and its derivative. An example involving the cubic transformation is provided by the Laguerre polynomial, and this is given in §5. An extension of the method in §4 to more than two nearby saddle points is presented in §6. The approximants used in this case are the "generalized" Airy function (eq. (6.13)) and its derivative. In §7, we introduce a rational transformation for a problem in which two saddle points coalesce onto a pole (of the phase function f). The expansion here involves Bessel functions. The final section deals with an integral in which two branch points coalesce with each other; the expansion obtained in this case is accompanied by computable error bounds.

We close this section with a definition of a uniform asymptotic expansion. Let $\{\varphi_n(\lambda, \alpha)\}$ be an asymptotic sequence as $\lambda \to +\infty$, i.e., $\varphi_{n+1}(\lambda, \alpha) = o\{\varphi_n(\lambda, \alpha)\}$ as $\lambda \to +\infty$ for every $n \geq 0$, where $\alpha = (\alpha_1, \ldots, \alpha_m)$ is a set of parameters, and suppose that $f(\lambda, \alpha)$ has the (generalized) asymptotic expansion

$$f(\lambda; \alpha) \sim \sum_{n=0}^{\infty} f_n(\lambda; \alpha); \qquad \{\varphi_n(\lambda; \alpha)\}, \qquad \text{as } \lambda \to +\infty,$$

i.e., for all $N \geq 1$,

$$f(\lambda, \alpha) = \sum_{n=0}^{N-1} f_n(\lambda; \alpha) + O\{\varphi_N(\lambda, \alpha)\}, \qquad \text{as } \lambda \to +\infty;$$

see Definition 3, Chapter I. If the o- and O-symbols hold independently of α in a domain Ω in $\mathbb{C}^m$, then we call the above expansion a *uniform asymptotic expansion* with respect to α.

2.   Saddle Point near a Pole

Returning to the contour integral

$$I(\lambda) = \int_C g(t) e^{-\lambda f(t)} \, dt \tag{2.1}$$

in (1.1), we now assume that $f(t)$ has a simple saddle point at $t = \alpha$ and $g(t)$ has a simple pole at $t = \beta$. We shall consider β near α, and in fact allow them to coalesce. In a neighborhood of α, we have

$$f(t) = f(\alpha) + \frac{1}{2} f''(\alpha)(t - \alpha)^2 + \frac{1}{3!} f'''(\alpha)(t - \alpha)^3 + \cdots, \qquad (2.2)$$

where $f''(\alpha) \neq 0$ for all α. If we set

$$u^2 = f(t) - f(\alpha), \qquad (2.3)$$

then $I(\lambda)$ becomes

$$I(\lambda) = e^{-\lambda f(\alpha)} \int_{-\infty}^{\infty} q(u) e^{-\lambda u^2}\, du. \qquad (2.4)$$

Here we have assumed that the image of C can be deformed into the real axis in the u-plane, and

$$q(u) = g(t) \frac{2u}{f'(t)}. \qquad (2.5)$$

Note that $q(u)$ has a pole at $u = ib$, where $-b^2 = f(\beta) - f(\alpha)$. We can extract the pole by writing

$$q(u) = \frac{c_{-1}}{u - ib} + \psi(u). \qquad (2.6)$$

The constant c_{-1} is given by

$$c_{-1} = \lim_{u \to ib} [(u - ib)q(u)] = \lim_{t \to \beta} \left[(u - ib)g(t) \frac{2u}{f'(t)} \right] = g_{-1}, \qquad (2.7)$$

where g_{-1} is the residue of $g(t)$ at $t = \beta$. The function $\psi(u)$ has no pole at $u = 0$ or $u = ib$, and its value at 0 is

$$
\begin{aligned}
\psi(0) &= g(\alpha) \sqrt{\frac{2}{f''(\alpha)}} + \frac{g_{-1}}{\sqrt{f(\beta) - f(\alpha)}} \\
&= g^*(\alpha) \sqrt{\frac{2}{f''(\alpha)}} - \frac{\sqrt{2}}{6} g_{-1} \frac{f'''(\alpha)}{[f''(\alpha)]^{3/2}} + O(\beta - \alpha),
\end{aligned}
\qquad (2.8)
$$

where $g^*(t)$ is the "regular" part of $g(t)$, i.e.,

$$g(t) = \frac{g_{-1}}{t - \beta} + g^*(t). \qquad (2.9)$$

To verify (2.8), we note that from (2.6) and (2.7), we have

$$\psi(0) = \lim_{u \to 0} \left[q(u) - \frac{c_{-1}}{u - ib} \right] = \lim_{t \to \alpha} \left[g(t) \frac{2u}{f'(t)} - \frac{g_{-1}}{u - ib} \right],$$

and from (2.2) and (2.3), we have

$$\lim_{t \to \alpha} \frac{2u}{f'(t)} = \sqrt{\frac{2}{f''(\alpha)}}.$$

Thus

$$\psi(0) = g(\alpha) \sqrt{\frac{2}{f''(\alpha)}} + \frac{g_{-1}}{ib},$$

which is equivalent to the first equality in (2.8). The second equality in (2.8) follows from (2.2). The asymptotic expansion of the integral

$$\int_{-\infty}^{\infty} \psi(u) e^{-\lambda u^2} \, du$$

can be obtained by Watson's lemma. We are thus left with the integral

$$K(\lambda) = \int_{-\infty}^{\infty} \frac{e^{-\lambda u^2}}{u - ib} \, du, \tag{2.10}$$

which can be expressed in terms of an incomplete gamma function or the complementary error function. To see this, we first observe that $K(\lambda)$ satisfies the first-order linear differential equation

$$K'(\lambda) - b^2 K(\lambda) = -ib \sqrt{\frac{\pi}{\lambda}}.$$

The solution to this equation satisfying $K(\infty) = 0$ is given by

$$K(\lambda) = i\sqrt{\pi} e^{\lambda b^2} \Gamma(\tfrac{1}{2}, \lambda b^2) = i\pi e^{\lambda b^2} \operatorname{erfc}(\lambda^{1/2} b). \tag{2.11}$$

Example 1.   As a simple illustration, we consider the integral

$$I = \int_C A(\theta) e^{ikR \cos(\theta - \alpha)} \, d\theta$$

encountered in the study of propagation of radio waves over a plane earth, where C is the contour shown in Figure 7.3 and α is a real

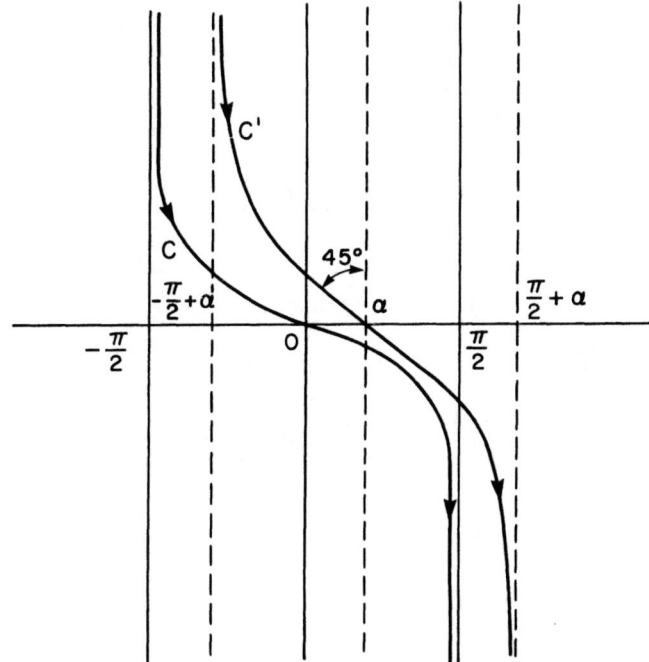

Fig. 7.3

parameter in the interval $(-\pi/2, \pi/2)$. In terms of our notations,

$$\lambda = kR, \qquad t = \theta, \qquad g(t) = A(\theta), \qquad f(t) = -i\cos(\theta - \alpha). \quad (2.12)$$

We now deform C into the steepest descent path C' for $-i\cos(\theta - \alpha)$ passing through $\theta = \alpha$. If $\theta - \alpha = \sigma + i\tau$, then this curve is given by

$$\cosh \tau \cos \sigma = 1.$$

It is easily verified that this curve begins at $\sigma = -\pi/2$ and $\tau = +\infty$, and ends at $\sigma = \pi/2$ and $\tau = -\infty$. Upon the change of variable

$$-i\cos(\theta - \alpha) = -i + u^2,$$

the integral I becomes

$$I = e^{i\lambda} \int_{-\infty}^{\infty} q(u)e^{-\lambda u^2}\, du, \qquad (2.13)$$

where

$$q(u) = A(\theta)\, \frac{2u}{i\sin(\theta - \alpha)}.$$

Now assume that $A(\theta)$ has a simple pole at $\theta = \beta$, β near α. Then $q(u)$ has a pole of order one at $u = ib$, where $b^2 = i\cos(\beta - \alpha) - i$. Note that b tends to zero as β approaches α. To write down the first two terms in the uniform expansion of I, we express $A(\theta)$ in the form

$$A(\theta) = \frac{A_{-1}}{\theta - \beta} + A^*(\theta),$$

where $A^*(\theta)$ is analytic in a neighborhood of $\theta = 0$. Inserting (2.6) in (2.13) with $c_{-1} = A_{-1}$ and integrating term by term, we obtain

$$
I = i\pi A_{-1} e^{ikR\cos(\beta - \alpha)} \operatorname{erfc}\{\sqrt{kR(i\cos(\beta - \alpha) - i)}\}
$$
$$
+ e^{ikR}\left[\psi(0)\sqrt{\frac{\pi}{kR}} + O\left(\frac{1}{(kR)^{3/2}}\right)\right], \tag{2.14}
$$

where

$$
\psi(0) = \left[\sqrt{2}A(\alpha) + \frac{A_{-1}}{\sqrt{1 - \cos(\alpha - \beta)}}\right]e^{-i\pi/4}
$$
$$
= \sqrt{2}e^{-i\pi/4}A^*(\alpha) + O(\beta - \alpha).
$$

Here use has been made of (2.8) and (2.11).

The above discussion is essentially that given by van der Waerden (1951).

3.   Saddle Point near an Endpoint

For simplicity, we suppose that the variables are real, the path of integration is the semi-infinite interval $(0, \infty)$, the integrand has a branch-point at the origin, and the phase function depends on a parameter α. Thus we consider here integrals of the form

$$
I(\lambda; \alpha) = \int_0^\infty t^{r-1}g(t)e^{-\lambda f(t;\alpha)} \, dt, \tag{3.1}
$$

where $r > 0$ is a fixed number and $g(t)$ is a sufficiently smooth function for $t \geq 0$. We assume that (i) $f(t; \alpha) \to +\infty$ as $t \to +\infty$; (ii) $f(t; \alpha)$ has a simple saddle point at α^* depending on α, $f_t(t; \alpha) > 0$ for $t > \alpha^*$ and $f_t(t; \alpha) < 0$ for $t < \alpha^*$ (if α^* is positive); (iii) $f(t; \alpha)$ is infinitely differentiable for $t \geq \min\{0, \alpha^*\}$.

If α is fixed, then so is the saddle point α^*, and hence, Laplace's method is applicable. However, this method gives different kinds of expansions for different values of α, depending on whether $\alpha^* < 0$, $\alpha^* = 0$, or $\alpha^* > 0$. If $\alpha^* < 0$, then we have, upon making the change of variable

$$f(t; \alpha) - f(0; \alpha) = u, \tag{3.2}$$

an asymptotic expansion of the form

$$I(\lambda; \alpha) \sim e^{-\lambda f(0;\alpha)} \sum_{m=0}^{\infty} \beta_m \frac{\Gamma(m+r)}{\lambda^{m+r}}; \tag{3.3}$$

if $\alpha^* \geq 0$ then the substitution

$$f(t; \alpha) - f(\alpha^*; \alpha) = u^2 \tag{3.4}$$

leads to expansions of the form

$$I(\lambda; \alpha) \sim e^{-\lambda f(\alpha^*;\alpha)} \sum_{m=0}^{\infty} \gamma_m \frac{\Gamma\{(m+r)/2\}}{\lambda^{(m+r)/2}} \tag{3.5}$$

when $\alpha^* = 0$ and

$$I(\lambda; \alpha) \sim e^{-\lambda f(\alpha^*;\alpha)} \sum_{m=0}^{\infty} \delta_m \frac{\Gamma(m+\frac{1}{2})}{\lambda^{m+1/2}} \tag{3.6}$$

when $\alpha^* > 0$. For the derivation of these expansions, we refer to Chapter II, §1.

If α is allowed to vary continuously, then the position of the saddle point α^* may move from outside the interval of integration to inside the interval, and, in the transition, may coincide with the endpoint $t = 0$ for a certain critical value of α, say α_0. Observe that as $\alpha^* \to 0$, there is a discontinuous change in the form of the expansions (3.3) and (3.5), from an expansion in powers of λ^{-1} to an expansion in powers of $\lambda^{-1/2}$. This occurs regardless of the fact that the integral $I(\lambda; \alpha)$ may be continuous in α. Furthermore, from equations (1.14) and (1.15) in Chapter II, it is easily seen that the coefficients β_m in (3.3) tend to infinity as $\alpha^* \to 0^-$, and that the same occurs with expansion (3.6) as $\alpha^* \to 0^+$ if $0 < r < 1$, although $I(\lambda; \alpha)$ may very well be an analytic function of α. It is therefore necessary to derive an asymptotic expansion for $I(\lambda; \alpha)$ as $\lambda \to +\infty$, which is uniformly valid for α^* varying in an interval containing $\alpha^* = 0$, or equivalently, for α varying in an interval containing $\alpha = \alpha_0$.

Examining the conditions (i)–(iii) imposed on $f(t; \alpha)$, we see that the simplest example of such a function is provided by the polynomial $\frac{1}{2}t^2 - \alpha^* t$. This suggests that instead of making the substitution (3.2) or (3.4), we let

$$f(t; \alpha) = \frac{u^2}{2} - au + b, \tag{3.7}$$

where a and b are parameters to be determined. We choose a and b so that the points $t = 0$ and $t = \alpha^*$ correspond to $u = 0$ and $u = a$. This leads to

$$b = f(0; \alpha) \quad \text{and} \quad a = \pm \{2f(0; \alpha) - 2f(\alpha^*; \alpha)\}^{1/2}, \tag{3.8}$$

where the upper or lower sign is taken according to $\alpha^* > 0$ or $\alpha^* < 0$. Solving (3.7) for u in terms of t gives

$$u = a \pm \{2f(t; \alpha) - 2f(\alpha^*; \alpha)\}^{1/2} \tag{3.9}$$

with the plus or minus sign chosen according as $t > \alpha^*$ or $t < \alpha^*$ and $\alpha^* > 0$. Equation (3.9) provides a one-to-one relationship between t and u, free from singularity at $t = \alpha^*$, since

$$\frac{du}{dt} = \pm \frac{f_t(t; \alpha)}{\sqrt{2f(t; \alpha) - 2f(\alpha^*; \alpha)}}. \tag{3.10}$$

In terms of u, (3.1) becomes

$$I(\lambda; \alpha) = e^{-\lambda f(0; \alpha)} J_0(\lambda; a), \tag{3.11}$$

where

$$J_0(\lambda; a) = \int_0^\infty u^{r-1} p_0(u) e^{-\lambda(u^2/2 - au)} \, du \tag{3.12}$$

and p_0 is defined by

$$p_0(u) = g(t)\left(\frac{t}{u}\right)^{r-1} \frac{dt}{du}. \tag{3.13}$$

Although the function $p_0(u)$ depends on the parameter a (and hence on α), we shall not indicate the dependence explicitly.

To derive a uniform expansion for $J_0(\lambda; a)$, we write

$$p_0(u) = a_0 + b_0 u + u(u - a) q_0(u), \tag{3.14}$$

where the constants a_0 and b_0 are defined by

$$a_0 = p_0(0) = g(0)\left[\frac{dt}{du}\bigg|_{u=0}\right]^r \tag{3.15}$$

and

$$b_0 = \frac{1}{a}\{p_0(a) - p_0(0)\} = \frac{1}{a}\left\{g(\alpha^*)\left(\frac{\alpha^*}{a}\right)^{r-1}\frac{dt}{du}\bigg|_{u=a} - g(0)\left[\frac{dt}{du}\bigg|_{u=0}\right]^r\right\}. \tag{3.16}$$

Inserting (3.14) into (3.12), we obtain

$$J_0(\lambda; a) = \frac{a_0}{\lambda^{r/2}}U_r(a\sqrt{\lambda}) + \frac{b_0}{\lambda^{(r+1)/2}}U_r'(a\sqrt{\lambda}) + J_1(\lambda; a), \tag{3.17}$$

where

$$\dot{U}_r(x) = \int_0^\infty u^{r-1}e^{-u^2/2 + xu}\,du \tag{3.18}$$

and

$$J_1(\lambda; a) = \int_0^\infty u^r(u-a)q_0(u)e^{-\lambda(u^2/2 - au)}\,du. \tag{3.19}$$

An integration by parts then yields

$$J_1(\lambda; a) = \frac{1}{\lambda}\int_0^\infty u^{r-1}p_1(u)e^{-\lambda(u^2/2 - au)}\,du \tag{3.20}$$

with

$$p_1(u) = uq_0'(u) + rq_0(u). \tag{3.21}$$

The same procedure can now be applied to the integral in (3.20). Thus, by writing

$$p_1(u) = a_1 + b_1u + u(u-a)q_1(u),$$

we obtain

$$J_0(\lambda; a) = \frac{U_r(\sqrt{\lambda}a)}{\lambda^{r/2}}\left(a_0 + \frac{a_1}{\lambda}\right) + \frac{U_r'(\sqrt{\lambda}a)}{\lambda^{(r+1)/2}}\left(b_0 + \frac{b_1}{\lambda}\right) + J_2(\lambda; a),$$

where

$$J_2(\lambda; a) = \frac{1}{\lambda^2}\int_0^\infty u^{r-1}p_2(u)e^{-\lambda(u^2/2 - au)}\,du, \qquad p_2(u) = uq_1'(u) + rq_1(u).$$

Continuing this process leads to an expansion of the form

$$J_0(\lambda; a) = \frac{U_r(\sqrt{\lambda a})}{\lambda^{r/2}} \sum_{m=0}^{n-1} \frac{a_m}{\lambda^m} + \frac{U_r'(\sqrt{\lambda a})}{\lambda^{(r+1)/2}} \sum_{m=0}^{n-1} \frac{b_m}{\lambda^m} + J_n(\lambda; a), \quad (3.22)$$

where we define inductively, for $m = 1, 2, \ldots$,

$$u^{r-1} p_m(u) = [u^r q_{m-1}(u)]' = u^{r-1}[a_m + b_m u + u(u-a)q_m(u)], \quad (3.23)$$

$$a_m = p_m(0), \qquad b_m = \frac{1}{a}\{p_m(a) - p_m(0)\}. \quad (3.24)$$

The remainder $J_n(\lambda; a)$ is explicitly given by

$$J_n(\lambda; a) = \frac{1}{\lambda^n} \int_0^{\infty} u^{r-1} p_n(u) e^{-\lambda(u^2/2 - au)} \, du. \quad (3.25)$$

We note that $J_n(\lambda; a)$ is of the same form as the integral $J_0(\lambda; a)$, except with p_n replacing p_0 and an extra factor λ^{-n}, which is important from an asymptotic point of view. Moreover, the function $U_r(x)$ can be expressed in terms of the parabolic cylinder function via

$$U_r(x) = \Gamma(r) \exp(\tfrac{1}{4}x^2) U(r - \tfrac{1}{2}, -x) = \Gamma(r) \exp(\tfrac{1}{4}x^2) D_{-r}(-x), \quad (3.26)$$

see Olver (1974a, pp. 207–208).

The above integration-by-parts technique is due to Bleistein (1966), and has since become a standard method for constructing uniform asymptotic expansions. In the following sections, we shall see several modifications of this technique.

To show that (3.22) is a uniform asymptotic expansion, we first note that without loss of generality we may assume $p_n(u)$ to be an infinitely differentiable function. From this it follows that it is also not unreasonable to assume there are nonnegative constants A_n, B_n, and c_n such that

$$|p_n(u)| \le (A_n + B_n u) \exp(c_n u^2) \qquad \text{for all } u \ge 0. \quad (3.27)$$

For simplicity, we suppose $c_n = 0$. Inserting (3.27) into (3.25) gives

$$|J_n(\lambda; a)| \le A_n \frac{U_r(\sqrt{\lambda a})}{\lambda^{n+r/2}} + B_n \frac{U_r'(\sqrt{\lambda a})}{\lambda^{n+(r+1)/2}}, \quad (3.28)$$

thus demonstrating that (3.22) is indeed a uniform compound asymptotic expansion (cf. Chapter I, Eq. (3.7)). (3.22) can also be viewed as a uniform expansion with respect to the asymptotic sequence

$$\varphi_n(\lambda; a) = \frac{1}{\lambda^{n+r/2}} U_r(\sqrt{\lambda}a) + \frac{1}{\lambda^{n+(r+1)/2}} U_r'(\sqrt{\lambda}a).$$

Example 2. An example which illustrates the coalescence of a saddle point with an endpoint is provided by the complementary incomplete Gamma function of large and nearly equal arguments. The starting point is the integral

$$y^{-1-x}e^y\Gamma(1+x, y) = \int_0^\infty e^{-x[\alpha t - \log(1+t)]} \, dt, \tag{3.29}$$

where $\alpha = y/x$. In our notation,

$$r = 1, \qquad g(t) = 1, \qquad f(t; \alpha) = \alpha t - \log(1+t), \qquad \lambda = x.$$

Note that the function $\alpha t - \log(1+t)$ has exactly one simple saddle point at $t = (1/\alpha) - 1$, which is inside the interval of integration when $\alpha < 1$ and outside the interval when $\alpha > 1$. As α approaches 1, the saddle point coalesces with the endpoint. This is exactly the situation described above. With $\alpha^* = 1/\alpha - 1$, it follows from (3.8) that

$$b = 0 \qquad \text{and} \qquad a = \pm[2(\alpha - 1 - \log \alpha)]^{1/2}, \tag{3.30}$$

where the upper or lower sign is taken according to $\alpha < 1$ or $\alpha > 1$. By (3.22), the first two terms of the uniform expansion are given by

$$y^{-1-x}e^y\Gamma(1+x, y) \sim \frac{a_0}{\sqrt{x}} U_1(\sqrt{x}a) + \frac{b_0}{x} U_1'(\sqrt{x}a). \tag{3.31}$$

Since $dt/du|_{t=0} = -a/f_t(0; \alpha)$ by (3.10) and (3.8), it is easily seen from (3.15) that the leading coefficient a_0 is given by

$$a_0 = \frac{\mp[2(\alpha - 1 - \log \alpha)]^{1/2}}{\alpha - 1} = \left[1 - \frac{1}{3}(\alpha - 1) + \cdots\right]. \tag{3.32}$$

To obtain the coefficient b_0, we observe from (3.10) that

$$f_t(t; \alpha)\frac{dt}{du} = \pm\sqrt{2f(t; \alpha) - 2f(\alpha^*; \alpha)}.$$

Differentiation of both sides gives

$$f_{tt}(t; \alpha)\left(\frac{dt}{du}\right)^2 + f_t(t; \alpha)\frac{d^2t}{du^2} = \pm\frac{f_t(t; \alpha)}{\sqrt{2f(t; \alpha) - 2f(\alpha^*; \alpha)}}\frac{dt}{du} = 1.$$

Since $f_t(\alpha^*; \alpha) = 0$, the last equation reduces to

$$\left(\frac{dt}{du}\bigg|_{u=a}\right)^2 = \frac{1}{f_{tt}(\alpha^*; \alpha)} = \alpha^2,$$

i.e.,

$$\left(\frac{dt}{du}\right)\bigg|_{u=a} = \alpha.$$

From (3.16) it follows that

$$b_0 = \frac{\alpha}{a} + \frac{1}{\alpha - 1} = \frac{\alpha}{\pm\{2(\alpha - 1 - \log\alpha)\}^{1/2}} + \frac{1}{\alpha - 1} = -\frac{4}{3} + O(\alpha - 1). \quad (3.33)$$

The above example was suggested by Olver (1975). For the derivation of an alternative expansion of the complementary incomplete Gamma function, see Temme (1975, 1979).

4.   Two Coalescing Saddle Points

Here we are concerned with contour integrals of the form

$$I(\lambda; \alpha) = \int_C g(t)e^{-\lambda f(t; \alpha)}\, dt, \quad (4.1)$$

where f and g are analytic functions of the complex variable t, and f is also an analytic function of the parameter α. Suppose that there exists a critical value of α, say $\alpha = \alpha_0$, such that for $\alpha \neq \alpha_0$ there are two distinct saddle points t_+ and t_- of multiplicity 1, but at $\alpha = \alpha_0$ these two points coincide and give a single saddle point t_0 of multiplicity 2. Thus

$$f_t(t_0; \alpha_0) = f_{tt}(t_0; \alpha_0) = 0, \qquad f_{ttt}(t_0; \alpha_0) \neq 0, \quad (4.2)$$

and

$$f_t(t_+; \alpha) = f_t(t_-; \alpha) = 0, \qquad f_{tt}(t_\pm; \alpha) \neq 0 \quad (4.3)$$

for $\alpha \neq \alpha_0$. By the saddle point method described in Chapter II, the asymptotic forms of $I(\lambda; \alpha)$ for large λ can be verified to be

$$I(\lambda; \alpha) \sim g(t_+)e^{-\lambda f(t_+; \alpha)}\left[\frac{2\pi}{\lambda f_{tt}(t_+; \alpha)}\right]^{1/2} + g(t_-)e^{-\lambda f(t_-; \alpha)}\left[\frac{2\pi}{\lambda f_{tt}(t_-; \alpha)}\right]^{1/2}$$

$$(4.4)$$

when $\alpha \neq \alpha_0$, and

$$I(\lambda; \alpha) \sim \mu g(t_0) e^{-\lambda f(t_0; \alpha_0)} \Gamma(\tfrac{4}{3}) \left[\frac{3!}{\lambda f_{ttt}(t_0; \alpha_0)} \right]^{1/3} \tag{4.5}$$

when $\alpha = \alpha_0$, μ being a constant whose value depends on the contour C. The behavior of these approximations differ radically, since (4.4) becomes singular, as $\alpha \to \alpha_0$, and the order of λ changes (discontinuously) from $\tfrac{1}{2}$ in (4.4) to $\tfrac{1}{3}$ in (4.5).

In view of the fact that the simplest phase function which exhibits two coalescing saddle points is a cubic polynomial, Chester, Friedman, and Ursell (1957) introduced in what is now regarded as a classic paper, a change of variable of the form

$$f(t; \alpha) = \tfrac{1}{3} u^3 - \zeta u + \eta, \tag{4.6}$$

where the coefficients ζ and η are to be determined. In order for (4.6) to result in a single-valued analytic function $t = t(u)$, neither dt/du nor du/dt can vanish in the relevant regions. But

$$\frac{dt}{du} = \frac{u^2 - \zeta}{f_t(t; \alpha)} \tag{4.7}$$

and $f_t(t; \alpha) = 0$ for $t = t_+$ and t_-. Therefore we must make t_- and t_+ correspond to $-\zeta^{1/2}$ and $\zeta^{1/2}$ respectively. This correspondence will make the derivative in (4.7) indeterminate. Accordingly, we find from (4.6) that

$$\zeta^{3/2} = \tfrac{3}{4}[f(t_-; \alpha) - f(t_+; \alpha)] \tag{4.8}$$

and

$$\eta = \tfrac{1}{2}[f(t_-; \alpha) + f(t_+; \alpha)]. \tag{4.9}$$

By rewriting (4.7) as

$$f_t(t; \alpha) \frac{dt}{du} = u^2 - \zeta \tag{4.10}$$

and differentiating with respect to u, we have

$$\left(\frac{dt}{du} \bigg|_{u = \pm \zeta^{1/2}} \right)^2 = \pm \frac{2\zeta^{1/2}}{f_{tt}(t_\pm; \alpha)} \tag{4.11}$$

if $\alpha \neq \alpha_0$. If $\alpha = \alpha_0$ then a further differentiation gives

$$\left(\frac{dt}{du} \bigg|_{u = 0} \right)^3 = \frac{2}{f_{ttt}(t_0; \alpha_0)}. \tag{4.12}$$

The transformation (4.6) was shown by Chester, Friedman, and Ursell to be one-to-one and analytic for all α in a neighborhood of α_0. More precisely, we have the following theorem. For a proof of this result, we refer to their paper (1957) and Friedman (1959).

Theorem 1.   *With the values given in* (4.8) *and* (4.9), *the transformation* (4.6) *has exactly one branch* $u = u(t, \alpha)$ *which can be expanded into a power series in* t, *with coefficients which are continuous in* α *for* α *near* α_0. *On this branch the points* $t = t_{\pm}$ *correspond to* $u = \pm\zeta^{1/2}$ *respectively. Furthermore, for* α *near* α_0, *the correspondence* $u \leftrightarrow t$ *is one-to-one.*

It is evident that the above theorem is only a local result. However, in practice, as we shall see in later examples, the transformation (4.6) often turns out to be one-to-one and analytic in a domain containing the path of integration. Assuming this to be the case, we substitute (4.6) in (4.1) and obtain

$$I(\lambda; \alpha) = e^{-\lambda\eta} \int_{C'} e^{-\lambda(u^3/3 - \zeta u)} \varphi_0(u) \, du, \qquad (4.13)$$

where C' is the image of C and

$$\varphi_0(u) = g(t) \frac{dt}{du}. \qquad (4.14)$$

Note that $\varphi_0(u)$ is an analytic function of u, and that it depends on ζ and hence on α. For simplicity of presentation, we shall not indicate the dependence on ζ explicitly. To derive an asymptotic expansion for the last integral, we follow Bleistein (1967) and write

$$\varphi_0(u) = a_0 + b_0 u + (u^2 - \zeta)\psi_0(u), \qquad (4.15)$$

where the coefficients a_0 and b_0 are easily determined to be

$$a_0 = \frac{1}{2}[\varphi_0(\zeta^{1/2}) + \varphi_0(-\zeta^{1/2})] \qquad (4.16)$$

$$b_0 = \frac{1}{2\zeta^{1/2}}[\varphi_0(\zeta^{1/2}) - \varphi_0(-\zeta^{1/2})]. \qquad (4.17)$$

Clearly,

$$\lim_{\sqrt{\zeta}\to 0} a_0 = \varphi_0(0), \qquad \lim_{\sqrt{\zeta}\to 0} b_0 = \varphi_0'(0). \qquad (4.18)$$

Thus, b_0 is an analytic function of $\zeta^{1/2}$ with a removable singularity at the origin. With a_0 and b_0 so chosen, it is easily seen that the function

$$\psi_0(u) = \frac{\varphi_0(u) - a_0 - b_0 u}{u^2 - \zeta} \tag{4.19}$$

is analytic in u and has removable singularities at $u = \pm\zeta^{1/2}$. Indeed, we have

$$\lim_{u \to \pm\sqrt{\zeta}} \psi_0(u) = \pm\frac{1}{2\zeta^{1/2}} [\varphi_0'(\pm\zeta^{1/2}) - b_0]. \tag{4.20}$$

Now we insert (4.15) in (4.13) and obtain

$$e^{\lambda\eta} I(\lambda; \alpha) = \frac{a_0}{\lambda^{1/3}} V(\lambda^{2/3}\zeta) + \frac{b_0}{\lambda^{2/3}} V'(\lambda^{2/3}\zeta) + I_1(\lambda; \alpha), \tag{4.21}$$

where

$$V(\lambda) = \int_{C'} e^{-v^3/3 + \lambda v} \, dv \tag{4.22}$$

and

$$I_1(\lambda; \alpha) = \int_{C'} (u^2 - \zeta) e^{-\lambda(u^3/3 - \zeta u)} \psi_0(u) \, du. \tag{4.23}$$

To the integral $I_1(\lambda; \alpha)$ we apply an integration by parts, and the result is

$$I_1(\lambda; \alpha) = \frac{1}{\lambda} \int_{C'} \varphi_1(u) e^{-\lambda(u^3/3 - \zeta u)} \, du, \tag{4.24}$$

where

$$\varphi_1(u) = \frac{d}{du} \psi_0(u). \tag{4.25}$$

Since $\varphi_1(u)$ is analytic in u, the last integral is exactly of the same form as the one on the right-hand side of (4.13). Hence the above procedure can be repeated. Defining inductively

$$\varphi_m(u) = a_m + b_m u + (u^2 - \zeta)\psi_m(u) \tag{4.26}$$

$$\varphi_{m+1}(u) = \psi_m'(u), \tag{4.27}$$

we obtain

$$e^{\lambda\eta}I(\lambda;\alpha) = V(\lambda^{2/3}\zeta) \sum_{m=0}^{n-1} \frac{a_m}{\lambda^{m+1/3}} + V'(\lambda^{2/3}\zeta) \sum_{m=0}^{n-1} \frac{b_m}{\lambda^{m+2/3}} + I_n(\lambda;\alpha), \quad (4.28)$$

where

$$I_n(\lambda;\alpha) = \frac{1}{\lambda^n} \int_{C'} \varphi_n(u) e^{-\lambda(u^3/3 - \zeta u)}\, du. \quad (4.29)$$

The integral $V(\lambda)$ in (4.22) can be expressed in terms of Airy functions, but which Airy function is appropriate depends on the path of integration C'. There are three choices for C': it can be any one of the contours γ_0, γ_1, and γ_2 shown in Figure 7.4. If C' is the contour γ_j, $j = 0, 1, 2$, then it can be easily verified that

$$V(\lambda) = 2\pi i \omega^j \mathrm{Ai}(\lambda\omega^j), \quad (4.30)$$

where ω is the cube root $e^{2\pi i/3}$ of unity. For definiteness, we suppose that C' is the contour γ_0. Thus (4.28) becomes

$$e^{\lambda\eta}I(\lambda;\alpha) = 2\pi i \left[\frac{\mathrm{Ai}(\lambda^{2/3}\zeta)}{\lambda^{1/3}} \sum_{m=0}^{n-1} \frac{a_m}{\lambda^m} + \frac{\mathrm{Ai}'(\lambda^{2/3}\zeta)}{\lambda^{2/3}} \sum_{m=0}^{n-1} \frac{b_m}{\lambda^m} \right] + I_n(\lambda;\alpha). \quad (4.31)$$

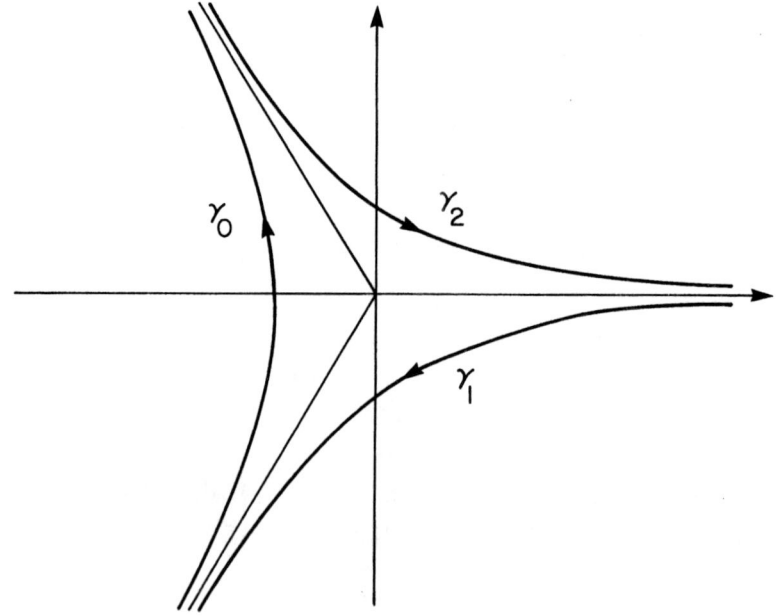

Fig. 7.4

For this to be a compound uniform asymptotic expansion, we must prove that there exist positive numbers C_n and D_n, independent of λ and α, such that

$$|I_n(\lambda; \alpha)| \leq \frac{C_n}{\lambda^{n+1/3}} |\text{Ai}(\lambda^{2/3}\zeta)| + \frac{D_n}{\lambda^{n+2/3}} |\text{Ai}'(\lambda^{2/3}\zeta)|. \qquad (4.32)$$

To achieve this, we first apply the recursive formulas (4.26)–(4.27) once more. This gives

$$I_n(\lambda; \alpha) = 2\pi i \left[\frac{a_n}{\lambda^{n+1/3}} \text{Ai}(\lambda^{2/3}\zeta) + \frac{b_n}{\lambda^{n+2/3}} \text{Ai}'(\lambda^{2/3}\zeta) \right]$$

$$+ \frac{1}{\lambda^{n+1}} \int_{C'} \varphi_{n+1}(u) e^{-\lambda(u^3/3 - \zeta u)} \, du.$$

Hence it suffices to show that the integral

$$I^* = \frac{1}{\lambda} \int_{C'} \varphi_{n+1}(u) e^{-\lambda(u^3/3 - \zeta u)} \, du \qquad (4.33)$$

satisfies

$$|I^*| \leq \frac{M_n}{\lambda^{1/3}} |\text{Ai}(\lambda^{2/3}\zeta)| + \frac{N_n}{\lambda^{2/3}} |\text{Ai}'(\lambda^{2/3}\zeta)| \qquad (4.34)$$

for some constants M_n and N_n independent of λ and α. If $\lambda^{2/3}\zeta$ is bounded, say $|\lambda^{2/3}\zeta| \leq \rho$, then we make the change of variable $u = \lambda^{-1/3}v$. This yields

$$I^* = \frac{1}{\lambda^{4/3}} \int_{C'} \varphi_{n+1}(\lambda^{-1/3}v) e^{-v^3/3 + \lambda^{2/3}\zeta v} \, dv.$$

Since φ_n is an analytic function of both u and α, it is evident that there is a constant B_n, independent of λ and α, such that

$$|I^*| \leq B_n \lambda^{-4/3}. \qquad (4.35)$$

Furthermore, from differential equation theory, none of the zeros of $\text{Ai}(z)$ and $\text{Ai}'(z)$ coincide; hence (4.34) follows immediately from (4.35). If $|\lambda^{2/3}\zeta| \geq \rho$ then we suppose without loss of generality that λ is positive, and consider two separate cases: (i) $|\arg \zeta| < 2\pi/3$ and (ii) $|\arg(-\zeta)| < 2\pi/3$. In case (i), we make the change of variable $u = \zeta^{1/2}v$; and in case (ii), we put $u = (-\zeta)^{1/2}v$. The analysis for the two cases is similar, and we first consider case (i). The substitution $u = \zeta^{1/2}v$ gives

$$I^* = \frac{\zeta^{1/2}}{\lambda} \int_{\tilde{C}} \varphi_{n+1}(\zeta^{1/2}v)e^{-\lambda\zeta^{3/2}(v^3/3 - v)}\, dv,$$

where $\tilde{C}$ is the image of C'. By temporarily assuming that ζ is positive, it is easily seen that $\tilde{C}$ retains the shape of C', i.e., that of γ_0 in Figure 7.4. The relevant saddle point for this integral is therefore at $v = -1$. The range of validity in ζ can now be extended by using analytic continuation. The ordinary saddle point method then yields

$$I^* \sim \frac{\pi^{1/2}}{\lambda^{3/2}\zeta^{1/4}} \varphi_{n+1}(-\zeta^{1/2}) \exp(-\tfrac{2}{3}\lambda\zeta^{3/2}).$$

Since

$$\mathrm{Ai}(z) \sim \frac{1}{2\sqrt{\pi}} z^{-1/4} \exp(-\tfrac{2}{3}z^{3/2}) \quad \text{and} \quad \mathrm{Ai}'(z) \sim -\frac{1}{2\sqrt{\pi}} z^{1/4} \exp(-\tfrac{2}{3}z^{3/2})$$

as $z \to \infty$ in $|\arg z| < 2\pi/3$ (see Olver (1974a, pp. 116 and 392)), (4.34) holds for large λ, bounded ζ, and $|\lambda^{2/3}\zeta| \geq \rho > 0$. In case (ii), we have

$$I^* = \frac{\eta^{1/2}}{\lambda} \int_{\tilde{C}} \varphi_{n+1}(\eta^{1/2}v)e^{-\lambda\eta^{3/2}(v^3/3 + v)}\, dv,$$

where $\eta = -\zeta$. We deform the contour $\tilde{C}$ to pass through the two saddle points $\pm i$. Using the saddle point method again and the fact that $\mathrm{Ai}(z)$ and $\mathrm{Ai}'(z)$ do not have the same zeros, it is easily seen that (4.34) also holds in this case, thus completing the proof of (4.32).

5. Laguerre Polynomials I

As an illustration of the method given in the preceding section, we consider the Laguerre polynomial $L_n^{(\alpha)}(x)$. We start with the integral representation (Erdélyi *et al.* 1953b, p. 190)

$$e^{-x/2}L_n^{(\alpha)}(x) = \frac{(-1)^n}{2^\alpha} \frac{1}{2\pi i} \int^{(1+)} e^{-xz/2}\left(\frac{1+z}{1-z}\right)^{v/4}(1-z^2)^{(\alpha-1)/2}\, dz, \quad (5.1)$$

where $v = 4n + 2\alpha + 2$. The path of integration encircles $z = 1$ in the positive direction, and closes at $\mathrm{Re}\, z = \infty$, $\mathrm{Im}\, z = \text{constant}$ (see Figure 7.5). Putting $x = vt$, we can rewrite (5.1) as

$$e^{-vt/2}L_n^{(\alpha)}(vt) = \frac{(-1)^n}{2^\alpha} \frac{1}{2\pi i} \int^{(1+)} e^{vf(z,t)}(1-z^2)^{(\alpha-1)/2}\, dz, \quad (5.2)$$

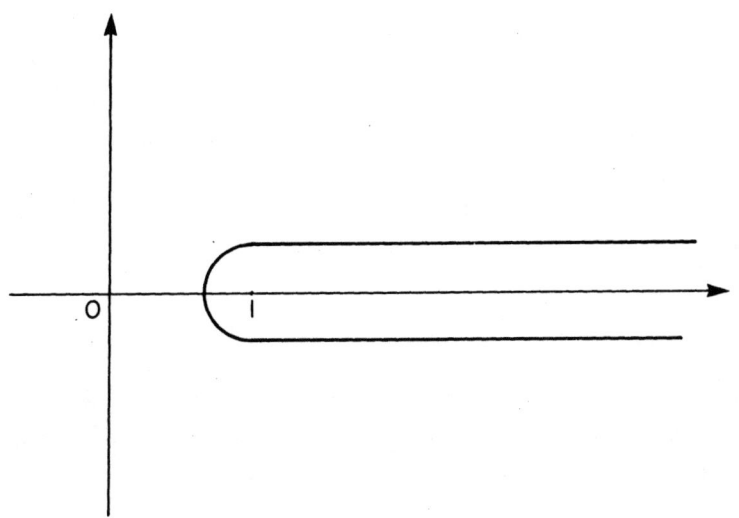

Fig. 7.5

where

$$f(z, t) = \frac{1}{4} \log\left(\frac{1+z}{1-z}\right) - \frac{1}{2} zt. \tag{5.3}$$

The saddle points of $f(z, t)$ are located at $z_+ = \sqrt{1 - 1/t}$ and $z_- = -\sqrt{1 - 1/t}$ if $t > 1$, and at $z_+ = i\sqrt{1/t - 1}$ and $z_- = -i\sqrt{1/t - 1}$ if $0 < t \le 1$. As $t \to 1$, the saddle points z_+ and z_- coalesce at $z = 0$. This is exactly the situation described in §4. Therefore we shall use the cubic transformation given in (4.6) with (t, α) replaced by (z, t). Since the phase function $f(z, t)$ in this case is an odd function of z, the coefficient η of the constant term on the right-hand side of (4.6) is zero. Thus we set

$$\frac{1}{4} \log\left(\frac{1+z}{1-z}\right) - \frac{1}{2} zt = \frac{u^3}{3} - B^2(t)u, \tag{5.4}$$

$B^2(t)$ being the coefficient ζ of the linear term in (4.6). We must choose $B(t)$ so that (5.4) gives a single-valued analytic function $z = z(u)$. Differentiating (5.4) with respect to u, we find that we must make the correspondence $u = B(t)$ with $z = z_+$ and $u = -B(t)$ with $z = z_-$. This gives

$$B(t) = \begin{cases} i[\frac{3}{2}\beta(t)]^{1/3}, & 0 < t \le 1, \\ [\frac{3}{2}\gamma(t)]^{1/3}, & t > 1, \end{cases} \tag{5.5}$$

where

$$\beta(t) = \tfrac{1}{2}(\cos^{-1}\sqrt{t} - \sqrt{t - t^2}), \qquad \gamma(t) = \tfrac{1}{2}(\sqrt{t^2 - t} - \cosh^{-1}\sqrt{t}). \quad (5.6)$$

Note that the mapping $z \leftrightarrow u$ in (5.4) depends only on $B^2(t)$ rather than $B(t)$. Furthermore, it is evident that $B(t)$ is continuous for $0 < t < \infty$ and $B^2(t)$ is analytic in a neighborhood of $t = 1$. The Taylor-series expansion of $B^2(t)$ near $t = 1$ is of the form

$$B^2(t) = \sum_{n=1}^{\infty} \alpha_n(t - 1)^n,$$

where $\alpha_1 = 2^{-2/3}$, $\alpha_2 = -1/(2^{2/3} \cdot 5)$, etc.

The cubic equation (5.4) for u can be solved explicitly in terms of trigonometric functions. For $t \neq 1$, the three solutions are given by

$$u_1(z, t) = 2B(t) \sin \tfrac{1}{3}\varphi$$

$$u_2(z, t) = 2B(t) \sin \tfrac{1}{3}(\varphi + 2\pi) \qquad\qquad (5.7)$$

$$u_3(z, t) = 2B(t) \sin \tfrac{1}{3}(\varphi + 4\pi),$$

where φ is the solution of

$$\frac{2}{3} B^3(t) \sin \varphi = \frac{1}{2} zt - \frac{1}{4} \log\left(\frac{1 + z}{1 - z}\right); \qquad\qquad (5.8)$$

see Ex. 9. For $t = 1$, we have

$$u(z, t) = \left[\frac{3}{4} \log\left(\frac{1 + z}{1 - z}\right) - \frac{3}{2} z\right]^{1/3}$$

and again, there are three branches.

Since $z = z_+$ corresponds to $u = B(t)$, from (5.4) and (5.8) it follows that $\sin \varphi = 1$ when $z = z_+$. Similarly, we have $\sin \varphi = -1$ when $z = z_-$. Thus,

$$u_1(z_+, t) = B(t), \qquad\qquad u_1(z_-, t) = -B(t),$$

$$u_2(z_+, t) = B(t), \qquad\qquad u_2(z_-, t) = 2B(t),$$

$$u_3(z_+, t) = -2B(t), \qquad\qquad u_3(z_-, t) = -B(t).$$

Therefore, $u_1(z, t)$ is the desired solution of (5.4), and, from here on, we set $u(z, t) \equiv u_1(z, t)$.

The easiest way to study the mapping from z-plane to u-plane is to introduce intermediate variables Z and φ defined by

$$Z = \frac{1}{4} \log\left(\frac{1+z}{1-z}\right) - \frac{1}{2} zt, \qquad \frac{2}{3} B^3(t) \sin\varphi = -Z, \qquad u = 2B(t) \sin\frac{1}{3}\varphi.$$
$$(5.9)$$

In the subsequent analysis, we shall temporarily assume $t > 1$. The first quadrant of the z-plane is shown in Figure 7.6; the saddle point $z = z_+$ and the singularity $z = 1$ are excluded by small indentations and the quadrant is closed by a large circular arc at ∞. As z travels along the boundary of the region ABCC'DEFA once, its image point Z travels once along the corresponding boundary in Figure 7.7. The line segments AB and BC in Figure 7.7 are considered as distinct parts of the boundary. Hence the mapping $\xi : z \to Z$ defined by $\xi(z) = \frac{1}{4}\log[(1+z)/(1-z)] - \frac{1}{2}zt$ is one-to-one also in the interior of the region; see Titchmarsh (1939, §6.45). By the same argument, as u describes once the closed curved BCC'DEB in Figure 7.9, φ and Z

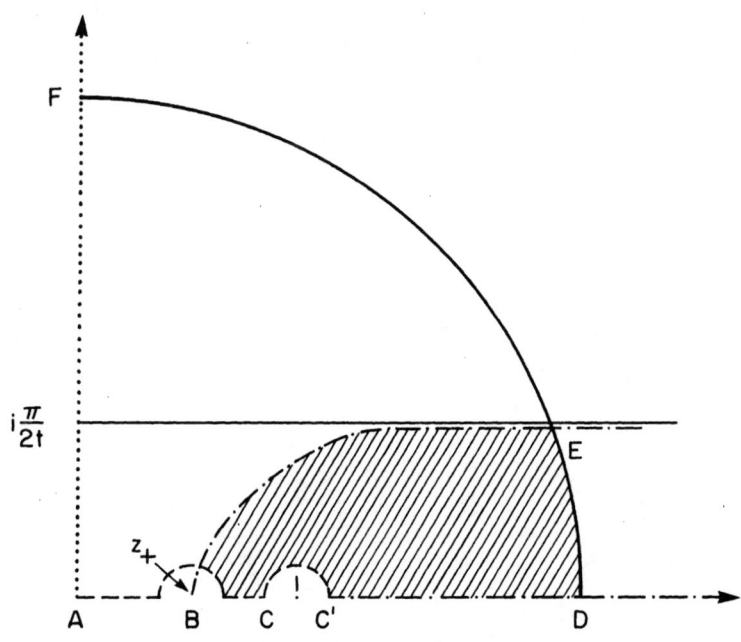

Fig. 7.6 z-plane ($t > 1$).

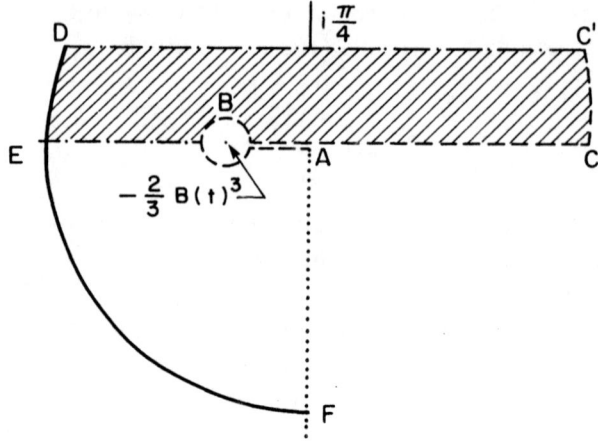

Fig. 7.7 Z-plane ($t > 1$).

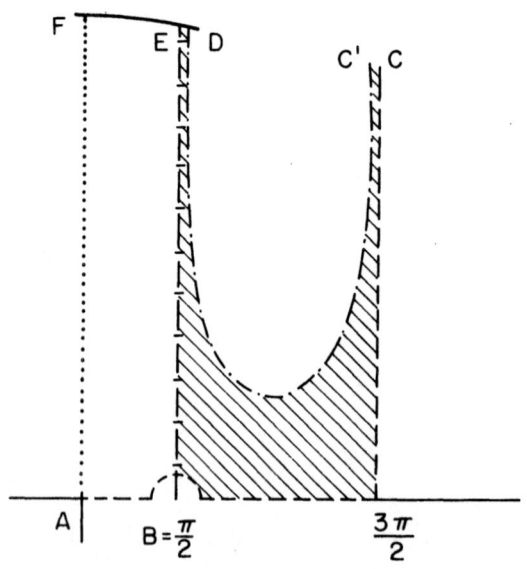

Fig. 7.8 φ-plane ($t > 1$).

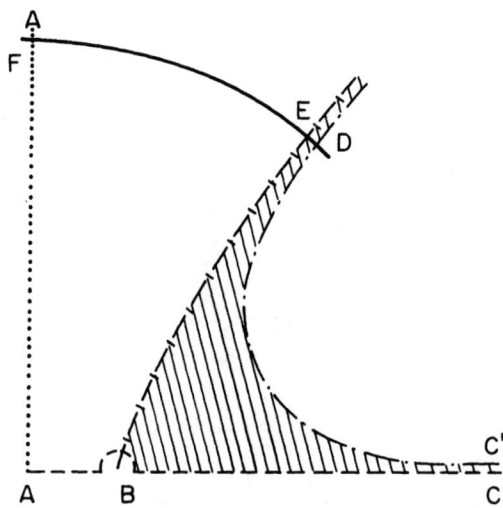

Fig. 7.9 u-plane $(t > 1)$.

describe the corresponding curve in Figures 7.8 and 7.7 once, respectively. Hence the mappings $u \leftrightarrow \varphi$ and $\varphi \leftrightarrow Z$, and consequently the resultant $u \leftrightarrow Z$, are all one-to-one in the interior of the region bounded by this curve. That is, the function $\eta(u) = \frac{1}{3}u^3 - B^2(t)u$ is univalent in the shaded region in Figure 7.9. The parametric equation of the boundary curve BE in Figure 7.9 is given by

$$s^2 - 3r^2 + 3B^2(t) = 0, \qquad s > 0, \quad r > 0, \tag{5.10}$$

where $s = \operatorname{Im} u$ and $r = \operatorname{Re} u$. The transformation $z \leftrightarrow u$ is obtained by composing $\xi^{-1} \colon Z \to z$ and $\eta \colon u \to Z$. Since $z \leftrightarrow Z$ and $u \leftrightarrow Z$ are both one-to-one in the region BCC'DEB, so is $z \leftrightarrow u$. If $z = x + iy$, then it is readily verified that the real parts of $\frac{1}{4}\log[(1 + z)/(1 - z)] - \frac{1}{2}zt$ and $\frac{1}{3}u^3 - B^2(t)u$ are even in y and s, respectively, and that the imaginary parts of these functions are odd in y and s, respectively. Hence, the mapping of the fourth quadrant is deducible from Figures 7.6 and 7.9 by reflection in the real axis. Fitting the results together, we conclude that the function $u(z, t)$ is analytic and one-to-one in the shaded region in Figure 7.10 and on its boundary. From Figures 7.6 and 7.9, it is also clear that a neighborhood of $z = z_+$ is mapped into a neighborhood of its image $u = B(t)$. Hence, $u(z, t)$ is bounded at that point.

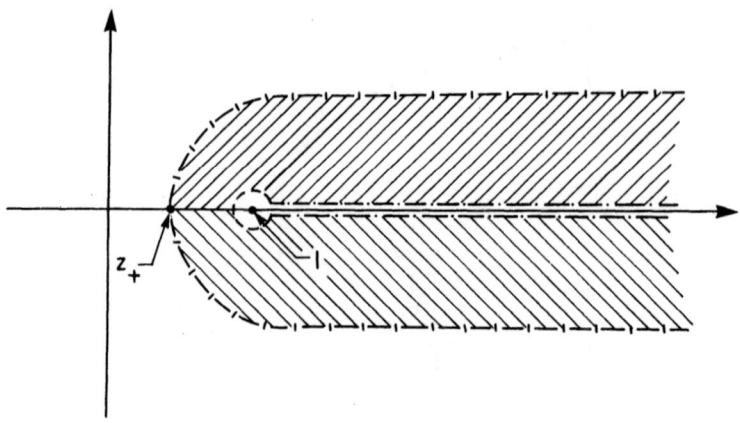

Fig. 7.10

We now return to the integral (5.2). With the transformation (5.4) this integral becomes

$$e^{-vt/2}L_n^{(\alpha)}(vt) = \frac{(-1)^n}{2^\alpha}\frac{1}{2\pi i}\int_{\mathscr{L}} h(u)\exp\left\{v\left[\frac{u^3}{3} - B^2(t)u\right]\right\} du, \quad (5.11)$$

where

$$h(u) = [1 - z^2(u)]^{(\alpha - 1)/2}\frac{dz}{du} \qquad (5.12)$$

and $\mathscr{L}$ is the branch of the hyperbolic curve in the right-half plane, half of which is given by (5.10). Note that (5.11) is established under the condition $t > 1$. The validity of this result for $0 < t < 1$ can be established in a similar manner. The region bounded by ACC'DEGBA in the z-, Z-, φ-, and u-planes is shown in Figures 7.11, 7.12, 7.13, and 7.14, respectively. By continuity, (5.11) holds also for $t = 1$.

To derive the asymptotic expansion of the integral in (5.11), we adopt the integration-by-parts technique introduced in §4. Thus we write

$$h(u) = h_0(u) = \alpha_0 + \beta_0 u + [u^2 - B^2(t)]g_0(u), \qquad (5.13)$$

where α_0, β_0, and $g_0(u)$ are to be determined; cf. (4.15). From (5.4), it is easily seen that $z(u)$ is an odd function of u. Hence, $h(u)$ is an even function of u, and $h(B) = h(-B)$. From this and (5.13), it follows that $\beta_0 = 0$, $\alpha_0 = h_0(B)$, and

$$g_0(u) = \frac{h_0(u) - h_0(B)}{u^2 - B^2}.$$

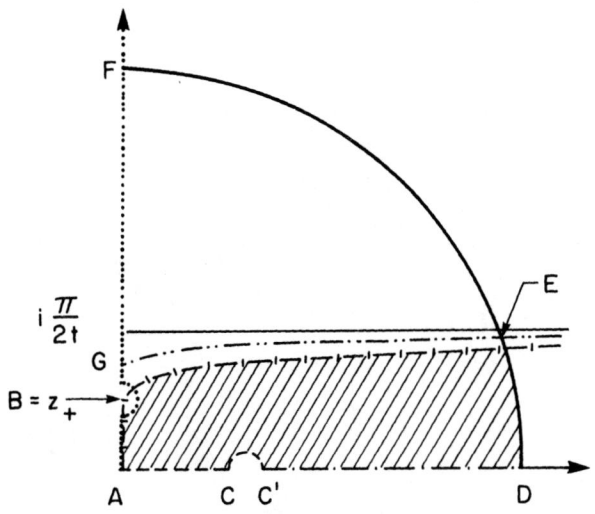

Fig. 7.11 z-plane $(0 < t < 1)$.

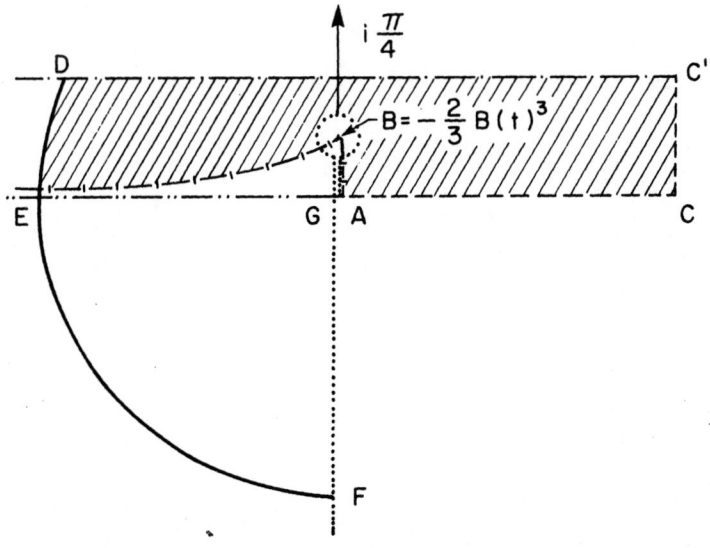

Fig. 7.12 Z-plane $(0 < t < 1)$.

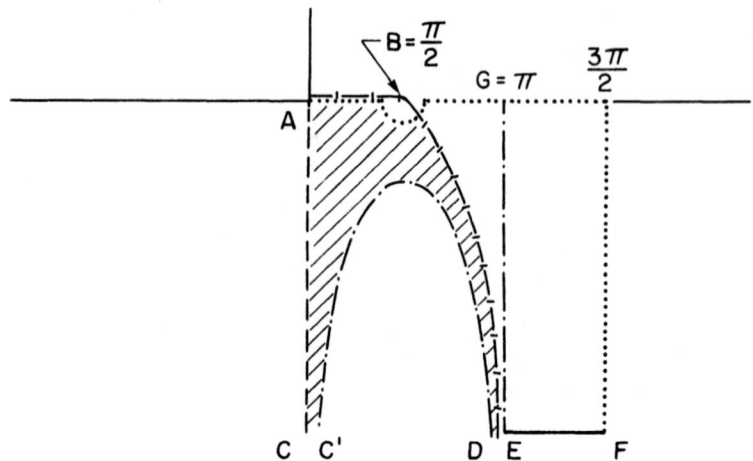

Fig. 7.13 φ-plane ($0 < t < 1$).

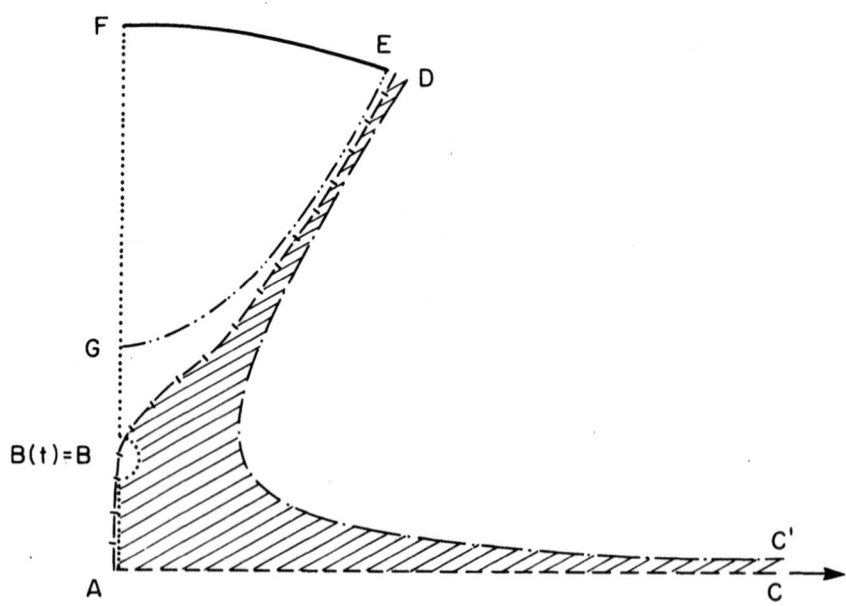

Fig. 7.14 u-plane ($0 < t < 1$).

Therefore, $g_0(u)$ is analytic in the domain of $h_0(u)$ with a removable singularity at $u = B$. We now substitute (5.13) in (5.11), and express the first integral in terms of the Airy function. The second integral can be integrated by parts; the integrated term vanishes, since $g_0(u)$ has at most algebraic growth at infinity. The result is

$$(-1)^n 2^\alpha e^{-vt/2} L_n^{(\alpha)}(vt) = \text{Ai}(v^{2/3}B^2) \frac{\alpha_0}{v^{1/3}}$$
$$- \left(\frac{1}{v}\right) \frac{1}{2\pi i} \int_{\mathcal{L}} g_0'(u) \exp\left[v\left(\frac{u^3}{3} - B^2 u\right)\right] du. \tag{5.14}$$

Repeating this procedure as explained in §4, we obtain

$$(-1)^n 2^\alpha e^{-vt/2} L_n^{(\alpha)}(vt) = \text{Ai}(v^{2/3}B^2) \sum_{k=0}^{[(p-1)/2]} \alpha_{2k} v^{-2k-1/3}$$
$$- \text{Ai}'(v^{2/3}B^2) \sum_{k=0}^{[p/2]-1} \beta_{2k+1} v^{-2k-5/3} + \varepsilon_p, \tag{5.15}$$

where we define inductively, for $m = 0, 1, 2, \ldots$,

$$h_m(u) = \alpha_m + \beta_m u + (u^2 - B^2)g_m(u), \qquad h_{m+1}(u) = g_m'(u). \tag{5.16}$$

It can be shown by induction that for $n \geq 0$, h_{2n} and g_{2n} are even analytic functions, and that h_{2n+1} and g_{2n+1} are odd analytic functions. Consequently,

$$\alpha_{2n} = h_{2n}(B), \qquad \alpha_{2n+1} = 0,$$
$$\beta_{2n} = 0, \qquad \beta_{2n+1} = \frac{h_{2n+1}(B)}{B}. \tag{5.17}$$

For each n, both α_n and β_n are continuous in t for $0 < t < \infty$. The remainder ε_p is explicitly given by

$$\varepsilon_p = \frac{1}{v^p} \frac{1}{2\pi i} \int_{\mathcal{L}} h_p(u) \exp\left[v\left(\frac{u^3}{3} - B^2 u\right)\right] du. \tag{5.18}$$

To estimate ε_p, we introduce the auxiliary functions

$$\check{\text{A}}\text{i}(z) = \begin{cases} \text{Ai}(z) & \text{if } z \geq 0 \\ [\text{Ai}^2(z) + \text{Bi}^2(z)]^{1/2} & \text{if } z < 0, \end{cases} \tag{5.19}$$

and

$$\check{\text{A}}\text{i}(z) = \begin{cases} \text{Ai}'(z) & \text{if } z \geq 0 \\ [\text{Ai}'^2(z) + \text{Bi}'^2(z)]^{1/2} & \text{if } z < 0. \end{cases} \tag{5.20}$$

Furthermore, we define

$$\tilde{\alpha}_p(t) = \begin{cases} 1 & \text{if } 0 < t < \xi \\ |\alpha_p| & \text{if } t > \xi, \end{cases}$$

and define $\tilde{\beta}_p(t)$ in a similar manner, where α_p and β_p are the coefficients in the expansion (5.15), and ξ is a positive number. In (5.19), Bi(z) denotes the second solution of the Airy equation $d^2W/dz^2 = zW$, related to the Airy function Ai(z) via the identity

$$\text{Bi}(z) = i\omega^2 \, \text{Ai}(\omega^2 z) - i\omega \, \text{Ai}(\omega z), \tag{5.21}$$

ω being the cube root $e^{i2\pi/3}$ of unity; cf. Copson (1965, p. 99). $\tilde{\text{Ai}}(z)$ and $\tilde{\tilde{\text{Ai}}}(z)$ mimic the behavior of Ai(z) and Ai'(z), respectively. Moreover, $\tilde{\alpha}_p$ and $\tilde{\beta}_p$ have the same behavior as α_p and β_p, respectively. For large values of v and for $0 < b \le t < \infty$, it can be shown that

$$|\varepsilon_p| \le \frac{N_p}{v^{p+2/3}} \, |\tilde{\beta}_p(t)| \, |\tilde{\text{Ai}}(v^{2/3}B^2)| \qquad \text{if } p \text{ is odd}, \tag{5.22}$$

and

$$|\varepsilon_p| \le \frac{M_p}{v^{p+1/3}} \, |\tilde{\alpha}_p(t)| \, |\tilde{\text{Ai}}(v^{2/3}B^2)| \qquad \text{if } p \text{ is even}, \tag{5.23}$$

M_p and N_p being some constants independent of v and t. These estimates of course imply that (5.15) is a compound uniform asymptotic expansion; cf. (4.32).

We conclude this section with an evaluation of the leading coefficient α_0. From (5.17) and (5.12), it follows that

$$\alpha_0 = [1 - z^2(u)]^{(\alpha-1)/2} \frac{dz}{du}\bigg|_{u=B}. \tag{5.24}$$

Since $u = B$ corresponds to $z = z_+$ and $1 - z_+^2 = 1/t$, upon differentiating (5.4) twice, we obtain

$$\left(\frac{dz}{du}\right)^2\bigg|_{u=B} = \frac{2B}{z_+ t^2},$$

which in turn gives

$$\alpha_0 = \begin{cases} t^{(1-\alpha)/2} \dfrac{\sqrt{2B}}{(t-1)^{1/4}t^{3/4}} & \text{if } t > 1 \\[3mm] t^{(1-\alpha)/2} \dfrac{\sqrt{2|B|}}{(1-t)^{1/4}t^{3/4}} & \text{if } 0 < t < 1. \end{cases} \tag{5.25}$$

For completeness, we also record the formula

$$\beta_1 = \frac{\alpha_0}{2B}\left[\frac{5}{24}B^{-3} - \frac{3}{4}\frac{\sqrt{t-1}}{\sqrt{t}} + \frac{1}{2}\frac{\sqrt{t}}{\sqrt{t-1}} - \frac{5}{12}\frac{t^{3/2}}{(t-1)^{3/2}}\right.$$
$$\left. - (\alpha^2 - 1)\frac{\sqrt{t-1}}{\sqrt{t}}\right] \tag{5.26}$$

if $t > 1$. A corresponding formula exists for the case $0 < t < 1$, and can be formally obtained from (5.26) by simply replacing $\sqrt{t-1}$ by $i\sqrt{1-t}$. To calculate the coefficients of higher terms, one can use a matching procedure similar to that described in Olver (1974a, Chap. 11, Ex. 7.4); that is, we re-expand the Airy functions in (5.15) and compare the results with the expansions obtained by using the ordinary method of steepest descent for t bounded away from 1.

6.   Many Coalescing Saddle Points

We now return to the contour integral (4.1):

$$I(\lambda; \alpha) = \int_C g(t)e^{-\lambda f(t,\alpha)}\, dt, \tag{6.1}$$

and assume that the parameter α is a vector $\alpha = (\alpha_1, \ldots, \alpha_l)$ and the phase function $f(t, \alpha)$ has m saddle points $t_1(\alpha), \ldots, t_m(\alpha)$. Thus,

$$\frac{\partial f}{\partial t}(t_i(\alpha), \alpha) = 0, \qquad i = 1, \ldots, m. \tag{6.2}$$

Also, we assume that as $\alpha \to 0$, the saddle points all approach the origin. Hence, when $\alpha = 0$, $t = 0$ is a saddle point of order at least m. We shall assume that this order is exactly m, i.e.,

$$\frac{\partial^{m+1} f}{\partial t^{m+1}}(0, 0) \neq 0.$$

The simplest function with these properties is a polynomial of degree $m + 1$, and the following result of Levinson (1961) shows that a locally one-to-one analytic transformation exists which will reduce $f(t, \alpha)$ to such a polynomial.

Theorem 2. *Let $f(t, \alpha)$ be analytic in (t, α) for small $|t|$ and $|\alpha|$, and let*

$$\frac{\partial^i f}{\partial t^i}(0, 0) = 0, \qquad 1 \leq i \leq m,$$

and

$$\frac{\partial^{m+1} f}{\partial t^{m+1}}(0, 0) \neq 0.$$

Then there is an analytic function φ of (z, α) for small $|z|$ and $|\alpha|$ such that setting

$$t = z + z^2 \varphi(z, \alpha) \tag{6.3}$$

in $f(t, \alpha)$ gives

$$f(t, \alpha) = P(z, \alpha) = \sum_{s=1}^{m+1} a_s(\alpha) \frac{z^s}{s} + a_0(\alpha), \tag{6.4}$$

where the $a_s(\alpha)$ are analytic for small $|\alpha|$, $a_s(0) = 0$ for $1 \leq s \leq m$, and $a_{m+1}(0) \neq 0$. Condition (6.3) implies $z = t + t^2 \psi(t, \alpha)$ where ψ is analytic for small $|t|$ and $|\alpha|$.

The above result establishes the existence of a one-to-one analytic correspondence between t and z for any small $|\alpha|$, but this result is only *local*. Thus, to apply Theorem 2, one must show that the contribution of the infinite integral from the part of the contour outside a neighborhood of the origin is negligibly small. In specific examples, as we have shown in §5, the change of variable (6.4) may actually be one-to-one and analytic along the entire path of integration.

By a trivial change of scale, we can always let $a_{m+1}(\alpha) = 1$, and by a linear change of variable in the z-plane, we can also choose $a_m(\alpha) = 0$. Thus (6.4) becomes

$$f(t, \alpha) = P(z, \alpha) = \frac{z^{m+1}}{m+1} + \sum_{s=1}^{m-1} a_s(\alpha) \frac{z^s}{s} + a_0(\alpha). \tag{6.5}$$

To determine the remaining m coefficients $a_0, a_1, \ldots, a_{m-1}$, we note that for the transformation (6.5) to be one-to-one analytic, the saddle points on either side of (6.5) must correspond;

$$t = t_i(\alpha) \qquad \leftrightarrow \qquad z = z_i(\alpha),$$

$z_i(\alpha)$ being the zeros of $\partial P/\partial z$. The coefficients $a_s(\alpha)$ are then determined by the set of equations

$$f(t_i, \alpha) = \frac{z_i^{m+1}}{m+1} + \sum_{s=1}^{m-1} \frac{a_s(\alpha)}{s} z_i^s + a_0(\alpha), \qquad i = 1, \ldots, m. \qquad (6.6)$$

Formally, these equations are linear in $a_0(\alpha), \ldots, a_{m-1}(\alpha)$. However, being the zeros of $\partial P/\partial z$, the z_i's themselves depend on these coefficients. Thus (6.6) is actually a set of nonlinear equations. To calculate the $a_i(\alpha)$'s, this procedure is obviously impractical. A more practical approach is to make use of the elementary symmetric functions in the classical theory of equations. The details of the two special cases $m = 2$ and $m = 3$ are given in the thesis of Kaminski (1987).

Assuming that the change of variable (6.5) is one-to-one and analytic along the entire path of integration, the integral (6.1) then becomes

$$I(\lambda; \alpha) = \int_{C'} G(z, \alpha)e^{-\lambda P(z,\alpha)}\, dz, \qquad (6.7)$$

where C' is the image of C in the z-plane and

$$G(z, \alpha) = g(t)\frac{dt}{dz}. \qquad (6.8)$$

Extending the integration-by-parts technique of §4, we now set

$$G(z, \alpha) = H_0(z, \alpha) + P_z(z, \alpha)F_0(z, \alpha), \qquad (6.9)$$

where $P_z = \partial P/\partial z$ and $H_0(z, \alpha)$ is a polynomial of degree $m - 1$,

$$H_0(z, \alpha) = \sum_{s=0}^{m-1} h_s^{(0)}z^s. \qquad (6.10)$$

The coefficients $h_s^{(0)} = h_s^{(0)}(\alpha)$, $s = 0, \ldots, m - 1$, are analytic functions of α for small values of $|\alpha|$, and can be calculated as follows: If we set $z = z_i$ in (6.9), $i = 1, \ldots, m$, then the second term on the right vanishes; this together with (6.10) gives a set of m linear equations in $h_0^{(0)}, \ldots, h_{m-1}^{(0)}$. Since the coefficients z_i^s ($1 \le i \le m$, $0 \le s \le m - 1$) of this system of m equations are analytic functions of α, the solution $(h_0^{(0)}, \ldots, h_{m-1}^{(0)})$ is also analytic in α. By using l'Hôpital's rule, it can be shown that the function $F_0(z, \alpha)$ in (6.9) is analytic in z along the entire path of integration with removable singularities at $z_1, \ldots, z_m$, the zeros of $P_z(z, \alpha)$. An explicit expression for the function F_0 in the form of a contour integral can be found in Ex. 16.

Inserting (6.9) in (6.7) gives

$$I(\lambda; \alpha) = \sum_{s=0}^{m-1} h_s^{(0)} \int_{C'} z^s e^{-\lambda P(z,\alpha)} \, dz + R_0(\lambda; \alpha), \qquad (6.11)$$

where the remainder is given by

$$R_0(\lambda; \alpha) = \int_{C'} P_z(z, \alpha) F_0(z, \alpha) e^{-\lambda P(z,\alpha)} \, dz. \qquad (6.12)$$

The integrals under the summation sign can be expressed in terms of the "generalized" Airy function

$$Y_0(\mathbf{a}) = \int_{\Gamma} e^{-Q(w,\mathbf{a})} \, dw, \qquad \mathbf{a} = (a_0, a_1, \ldots, a_{m-1}) \qquad (6.13)$$

and its derivatives

$$Y_s(\mathbf{a}) = \left(-\frac{\partial}{\partial a_1}\right)^s Y_0(\mathbf{a}) = \int_{\Gamma} w^s e^{-Q(w,\mathbf{a})} \, dw, \qquad (6.14)$$

where

$$Q(w, \mathbf{a}) = \frac{w^{m+1}}{m+1} + \sum_{s=1}^{m-1} a_s \frac{w^s}{s} + a_0, \qquad (6.15)$$

and the path of integration Γ can be any contour in the complex w-plane which has endpoints at ∞ and along which $\operatorname{Re} Q(w, \mathbf{a}) \to +\infty$ as $|w| \to +\infty$. Indeed, making the change of variable $\lambda z^{m+1} = w^{m+1}$, we have

$$\int_{C'} z^s e^{-\lambda P(z,\alpha)} \, dz = \lambda^{-(s+1)/(m+1)} Y_s(\mathbf{a}'), \qquad (6.16)$$

where $\mathbf{a}' = (a_0', \ldots, a_{m-1}')$ is given by

$$a_s' = \lambda^{1-s/(m+1)} a_s, \qquad s = 0, 1, \ldots, m-1. \qquad (6.17)$$

Thus (6.11) becomes

$$I(\lambda; \alpha) = \sum_{s=0}^{m-1} \frac{h_s^{(0)}}{\lambda^{(s+1)/(m+1)}} Y_s(\mathbf{a}') + R_1(\lambda; \alpha). \qquad (6.18)$$

To the remainder term $R_0(\lambda; \alpha)$, we now apply an integration by parts. From (6.12), it follows that

$$R_1(\lambda; \alpha) = \frac{1}{\lambda} \int_{C'} G_1(z, \alpha) e^{-\lambda P(z,\alpha)} \, dz, \qquad (6.19)$$

where

$$G_1(z, \alpha) = \frac{\partial}{\partial z} F_0(z, \alpha).$$ (6.20)

Note that the integral on the right-hand side of (6.19) is of the same form as the original integral (6.7), except that G is now being replaced by G_1. Hence the above procedure can be repeated. Let $G_0(z, \alpha) = G(z, \alpha)$, and define inductively

$$G_n(z, \alpha) = H_n(z, \alpha) + P_z(z, \alpha)F_n(z, \alpha)$$ (6.21)

$$H_n(z, \alpha) = \sum_{s=0}^{m-1} h_s^{(n)} z^s$$ (6.22)

and

$$G_{n+1}(z, \alpha) = \frac{\partial}{\partial z} F_n(z, \alpha)$$ (6.23)

for $n \geq 0$. Integration by parts m times yields

$$I(\lambda; \alpha) = \frac{1}{\lambda^{1/(m+1)}} Y_0(\mathbf{a}') \sum_{v=0}^{N} \frac{h_0^{(v)}}{\lambda^v} + \frac{1}{\lambda^{2/(m+1)}} Y_1(\mathbf{a}') \sum_{v=0}^{N} \frac{h_1^{(v)}}{\lambda^v}$$
$$+ \cdots + \frac{1}{\lambda^{m/(m+1)}} Y_{m-1}(\mathbf{a}') \sum_{v=0}^{N} \frac{h_{m-1}^{(v)}}{\lambda^v} + R_{N+1}(\lambda; \alpha),$$ (6.24)

where

$$R_{N+1}(\lambda; \alpha) = \frac{1}{\lambda^{N+1}} \int_{C'} G_{N+1}(z, \alpha) e^{-\lambda P(z, \alpha)} \, dz.$$ (6.25)

To show that (6.24) is an asymptotic expansion, uniformly valid for α in some domain containing $\mathbf{0}$, one must establish the existence of positive constants $M_0^{(N+1)}, M_1^{(N+1)}, \ldots, M_{m-1}^{(N+1)}$, independent of λ and α, such that for sufficiently large λ

$$|R_{N+1}(\lambda; \alpha)| \leq \frac{1}{\lambda^{N+1}} \sum_{s=0}^{m-1} \frac{M_s^{(N+1)}}{\lambda^{(s+1)/(m+1)}} |Y_s(\mathbf{a}')|.$$ (6.26)

A proof of this result has been given by Ursell (1972).

In order for the expansion (6.24) to be of any practical use, behavior of the functions $Y_s(\mathbf{a})$, $s = 0, 1, \ldots, m - 1$, in (6.14) must first be sought.

Since $Y_s(\mathbf{a})$ is just the s-th derivative of $Y_0(\mathbf{a})$, one need consider only the function

$$Y_0(\mathbf{a}) = \int_\Gamma \exp\left(-\frac{w^{m+1}}{m+1} - a_{m-1}\frac{w^{m-1}}{m-1} - \cdots - a_1 w - a_0\right) dw; \qquad (6.27)$$

cf. (6.13)–(6.15). If $m = 2$, then Y_0 can be expressed in terms of the Airy integral

$$\mathrm{Ai}(x) = \frac{1}{2\pi i}\int_{\gamma_0} \exp\left(-\frac{w^3}{3} + xw\right) dw, \qquad (6.28)$$

where γ_0 is any path from $\infty\omega^2$ to 0 and then to $\infty\omega$, ω being the cube root $e^{2\pi i/3}$ of unity; see Figure 7.4. Indeed, if the contour Γ in (6.27) can be deformed into γ_0, then $Y_0(a_0, a_1) = 2\pi i e^{-a_0}\,\mathrm{Ai}(-a_1)$. Similarly, by deforming the path of integration and making a change of variable, Y_0 can be expressed in terms of the Pearcey integral

$$P(x, y) = \int_{-\infty}^{\infty} \exp\left[i\left(\frac{t^4}{4} + x\frac{t^2}{2} + yt\right)\right] dt \qquad (6.29)$$

and the "swallowtail" integral

$$S(x, y, z) = \int_{-\infty}^{\infty} \exp\left[i\left(\frac{t^5}{5} + x\frac{t^3}{3} + y\frac{t^2}{2} + zt\right)\right] dt \qquad (6.30)$$

in the cases $m = 3$ and $m = 4$ respectively. The integral $P(x, y)$ arose in Pearcey's investigation (1946) of electromagnetic fields near a "cusp", and hence is named after him. Both integrals $P(x, y)$ and $S(x, y, z)$ play an important role in geometric optics and Catastrophe theory; see Gilmore (1981).

Analytical properties of the Airy integral are well known; see, for example, Copson (1965) and Olver (1974a), and numerical tables have been constructed for this integral; see Miller (1946). The same, however, cannot be said about the Pearcey and swallowtail integrals. It is only recently that attempts have been made by Connor and his associates (1981 and 1984) to calculate these integrals numerically. However, their work covers only small or moderate values of the variables. For large values of the variables, one naturally turns to asymptotic methods for approximation. Now, the Pearcey integral depends on two variables, and the swallowtail integral on three. Asymptotic expansions of these integrals, when one of the variables is large while the others are fixed, can be obtained by the method of steepest descent. If all variables are allowed to grow without bound, then the problem becomes extremely complicated, and itself requires the uniform treatment discussed in this chapter. The construction of the uniform asymptotic expansion of the Pearcey integral in terms of the Airy function is outlined in Exs. 10–12. In the following example, we shall consider a special case of the swallowtail integral. Our discussion will be brief, and we refer to Kaminski (1987) for more detailed analysis.

Example 3. The derivative of the phase function

$$F(t) = \frac{t^5}{5} + x\frac{t^3}{3} + y\frac{t^2}{2} + zt \qquad (6.31)$$

in (6.30) is $F'(t) = t^4 + xt^2 + yt + z$. For t_0 to be a zero of $F'(t)$ of order two, $F''(t) = 4t^3 + 2xt + y$ must also vanish there. Similarly, if t_0 is a zero of order three, then $F'(t_0) = F''(t_0) = F'''(t_0) = 0$. Since $F'''(t) = 12t^2 + 2x$, the phase function $F(t)$ can have saddle points of order three only when $x \leq 0$. If x is positive, then $F(t)$ can have saddle points of order at most two, in which case the method of Chester–Friedman–Ursell can be used. Here we are primarily interested in the confluence of three saddle points. Thus we shall assume that x is negative; furthermore, for notational convenience, we shall replace x by $-x$ and assume $x > 0$. That is, we consider the integral

$$S(-x, y, z) = \int_{-\infty}^{\infty} \exp\left[i\left(\frac{t^5}{5} - x\frac{t^3}{3} + yt^2 + zt\right)\right] dt, \qquad x > 0. \quad (6.32)$$

First, we deform the path of integration so that it begins at $\infty e^{9\pi i/10}$ and ends at $\infty e^{\pi i/10}$. Along the new path of integration, the integrand in (6.32) decays rapidly to zero as $|t| \to \infty$. Next, we make the substitution $t = x^{1/2}u$. This gives us

$$S(-x, y, z) = x^{1/2} \int_{\infty e^{9\pi i/10}}^{\infty e^{\pi i/10}} \exp[ix^{5/2}f(u; yx^{-3/2}, zx^{-2})] \, du, \quad (6.33)$$

where

$$f(t; b, c) = \frac{t^5}{5} - \frac{t^3}{3} + b\frac{t^2}{2} + ct. \qquad (6.34)$$

It is now evident that the asymptotic behavior of (6.32), in the case when x and one or both of y and z are large, can be deduced from that of

$$I(\lambda; b, c) = \int_{\infty e^{9\pi i/10}}^{\infty e^{\pi i/10}} \exp[i\lambda f(t; b, c)]\, dt. \tag{6.35}$$

For this reason, we shall restrict ourselves to the last integral.

To examine the saddle-point structure of the phase function (6.34), we note that $\partial f/\partial t = t^4 - t^2 + bt + c$ and $\partial^2 f/\partial t^2 = 4t^3 - 2t + b$. Hence, if $\partial f/\partial t$ and $\partial^2 f/\partial t^2$ have simultaneous real zeros, then we have $(b, c) = (2t - 4t^3, -t^2 + 3t^4)$. This provides parametric equations of the curve in the bc-plane along which f has saddle points of order two or higher. These curves are plotted in Figure 7.15; the two cusps $(\pm 4/3\sqrt{6}, -1/12)$ of the curves correspond to values of b and c for which f has saddle points of order 3, and the other points on the curves correspond to saddle points of order 2. The white region corresponds to those points (b, c) for which f has exactly two distinct real saddle points, the shaded region gives no real zeros of f', and the dotted region yields four

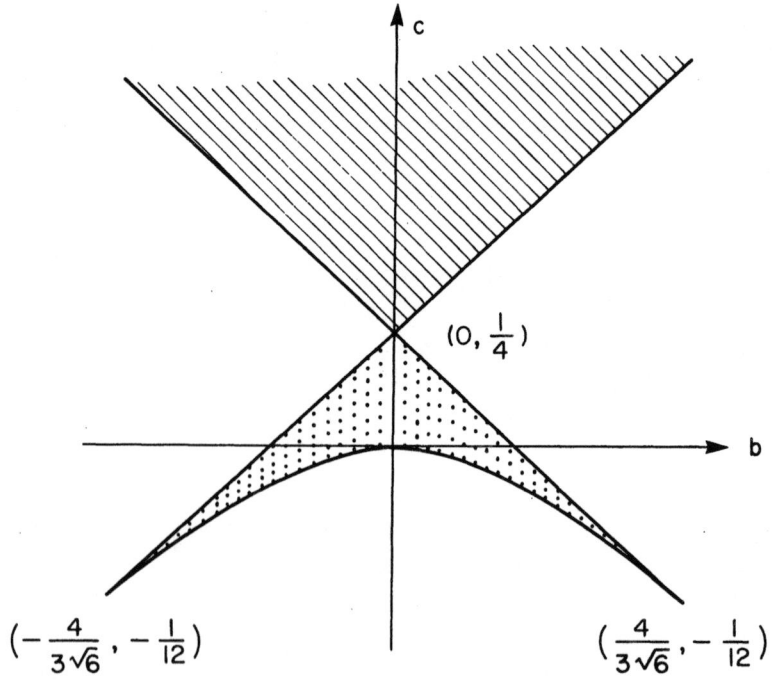

Fig. 7.15

distinct real zeros. In view of the fact that $S(x, y, z)$ and $S(x, -y, z)$ are complex conjugates, we need only consider the case $y \geq 0$. The case $y < 0$ follows by conjugation. Hence, when examining the behavior of $I(\lambda; b, c)$, we may restrict ourselves to the case $b \geq 0$. We are now ready to derive the asymptotic expansion of $I(\lambda; b, c)$, as $\lambda \to +\infty$, uniformly valid in a neighborhood of the cusp point $(4/3\sqrt{6}, -1/12)$.

The saddle points of $f(t; b, c)$ in (6.34) are the zeros of a quartic polynomial, and explicit formulas for the zeros of quartics can be found in the theory of equations using Lagrange resolvents (Uspensky 1948, p. 274) or in the work of Greenhill (1886, 1892) using the Weierstrass elliptic function. With proper choice of branches, it has been shown by Kaminski (1987) that the zeros of

$$\frac{\partial f}{\partial t} = t^4 - t^2 + bt + c \tag{6.36}$$

are given by

$$
\begin{aligned}
t_0 &= \sqrt{\tfrac{1}{6} - e_1} + \sqrt{\tfrac{1}{6} - e_2} - \sqrt{\tfrac{1}{6} - e_3}, \\
t_1 &= \sqrt{\tfrac{1}{6} - e_1} - \sqrt{\tfrac{1}{6} - e_2} + \sqrt{\tfrac{1}{6} - e_3}, \\
t_2 &= -\sqrt{\tfrac{1}{6} - e_1} + \sqrt{\tfrac{1}{6} - e_2} + \sqrt{\tfrac{1}{6} - e_3}, \\
t_3 &= -\sqrt{\tfrac{1}{6} - e_1} - \sqrt{\tfrac{1}{6} - e_2} - \sqrt{\tfrac{1}{6} - e_3},
\end{aligned}
\tag{6.37}
$$

where all square roots have their principal branches and e_i, $i = 1, 2, 3$, are the zeros of the associated cubic

$$S = 4s^3 - \left(c + \frac{1}{12}\right)s + \left(\frac{c}{6} + \frac{b^2}{16} - \frac{1}{216}\right).$$

By using the trigonometric solution of cubics, the e_i's are given by

$$e_1 = \left(\frac{c}{3} + \frac{1}{36}\right)^{1/2} \sin\left(\frac{\pi}{3} - \psi\right)$$

$$e_2 = \left(\frac{c}{3} + \frac{1}{36}\right)^{1/2} \sin\psi$$

$$e_3 = -\left(\frac{c}{3} + \frac{1}{36}\right)^{1/2} \sin\left(\frac{\pi}{3} + \psi\right),$$

where

$$\sin 3\psi = \left(\frac{c}{6} + \frac{b^2}{16} - \frac{1}{216}\right) \Big/ \left(\frac{c}{3} + \frac{1}{36}\right)^{3/2}.$$

The curves in Figure 7.15, along which the phase function f has saddle points of order ≥ 2, are known as the *caustics* associated with f. Kaminski has also shown that for values of (b, c) inside the caustic, the zeros t_i, $i = 0, 1, 2, 3$, in (6.37) satisfy the inequalities

$$t_3 < t_0 < t_1 < t_2. \tag{6.38}$$

Furthermore, on the arch joining the cusp $(4/3\sqrt{6}, -1/12)$ to $(0, 0)$, we have $t_0 = t_1 < t_2$, and on the arch joining the cusp to $(0, 1/4)$, we have $t_0 < t_1 = t_2$. As (b, c) tends to the cusp, t_0, t_1, and t_2 all approach $1/\sqrt{6}$. The fourth zero t_3, however, remains bounded away from them. Thus, we are in the situation of a confluence of three saddle points and the appropriate change of variable is therefore the quartic transformation

$$f(t; b, c) = \frac{z^4}{4} - \zeta \frac{z^2}{2} + \eta z + \theta; \tag{6.39}$$

cf. (6.5) and Ursell (1972). By using elementary symmetric functions, it was found (Kaminski 1987) that the parameters ζ, η, and θ can be expressed in terms of the σ_i's defined by

$$\sigma_1 = f(t_0) + f(t_1) + f(t_2),$$
$$\sigma_2 = f(t_0)f(t_1) + f(t_0)f(t_2) + f(t_1)f(t_2), \tag{6.40}$$
$$\sigma_3 = f(t_0)f(t_1)f(t_2).$$

Here we have suppressed the dependence of f on (b, c). More explicitly, we have

$$\zeta = (\tfrac{4}{3})^{1/2}(\sigma_1^2 - 3\sigma_2)^{1/4}\sqrt{-(1 - \chi)^{1/2} + (1 - \omega\chi)^{1/2} + (1 - \omega^2\chi)^{1/2}}, \tag{6.41}$$

where ω is the cube root of unity,

$$\chi = \left[1 - \frac{(9\sigma_1\sigma_2 - 2\sigma_1^3 - 27\sigma_3)^2}{4(\sigma_1^2 - 3\sigma_2)^3}\right]^{1/3}, \tag{6.42}$$

and all roots are taken to have their principal value. The quantity χ tends to zero when (b, c) approaches the caustic; see Kaminski (1987,

p. 83). The values of η and θ are given by

$$\eta^2 = -\frac{\zeta^3}{54} + \frac{8(\sigma_1^2 - 3\sigma_2)}{27\zeta} \tag{6.43}$$

and

$$\theta = \frac{\sigma_1}{3} + \frac{\zeta^2}{6}. \tag{6.44}$$

We are now ready to derive the asymptotic expansion of $I(\lambda; b, c)$ in (6.35). Since t_3 always stays away from t_0, t_1, t_2 for $b > 0$, we rewrite $I(\lambda)$ as

$$I(\lambda; b, c) = \int_{\Gamma_1} e^{i\lambda f(t; b, c)}\, dt + \int_{\Gamma_2} e^{i\lambda f(t; b, c)}\, dt$$

$$\equiv I_1(\lambda; b, c) + I_2(\lambda; b, c), \tag{6.45}$$

where $I_i(\lambda; b, c)$ denotes the integral of $\exp[i\lambda f(t; b, c)]$ over Γ_i and the contours Γ_i are as shown in Figure 7.16.

By taking Γ_1 to be the steepest descent curve through t_3 beginning at $\infty e^{9\pi i/10}$ and ending at $\infty e^{13\pi i/10}$, one can show that $I_1(\lambda; b, c)$ has an asymptotic expansion of the form

$$I_1(\lambda; b, c) \sim e^{i\lambda f(t_3; b, c) - i\pi/4} \sum_{s=1}^{\infty} a_s(b, c)\lambda^{-s/2}, \tag{6.46}$$

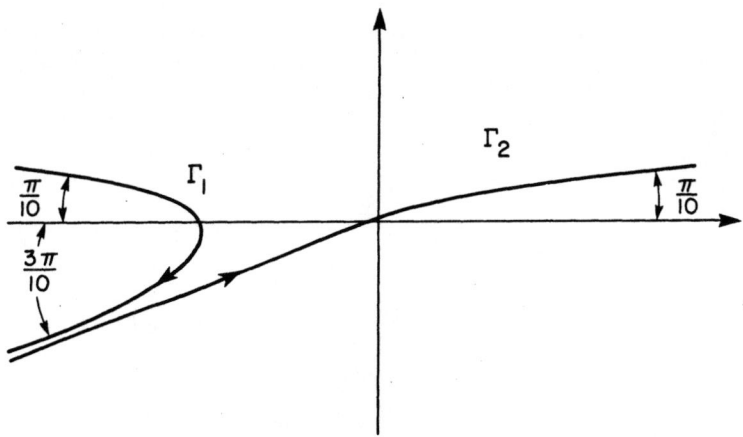

Fig. 7.16 t-plane.

as $\lambda \to +\infty$. The leading coefficient $a_1(b, c)$ is given by

$$a_1(b, c) = \sqrt{\frac{2\pi}{-f''(t_3; b, c)}}, \tag{6.47}$$

which is well behaved for $b \geq b_0 > 0$. Note that since t_3 is a local maximum of the quintic $f(t; b, c)$, $f''(t_3; b, c) < 0$.

In the integral $I_2(\lambda; b, c)$, we make the quartic change of variable (6.39). This yields

$$e^{-i\lambda\theta}I_2(\lambda; b, c) = \int_C g_0(z; \zeta, \eta) \exp\left[i\lambda\left(\frac{z^4}{4} - \zeta\frac{z^2}{2} + \eta z\right)\right] dz,$$

where $g_0(z; \zeta, \eta) = dt/dz$; cf. equation (6.7). The new path of integration C is the image of Γ_2, which begins at $\infty e^{i9\pi/8}$ and ends at $\infty e^{i\pi/8}$. The transformation (6.39) has been shown by Kaminski (1987) to be one-to-one analytic in a region containing the entire path of integration. Now we define successively

$$g_0(z; \zeta, \eta) = p_0 + q_0 z + r_0 z^2 + (z^3 - \zeta z + \eta)h_0(z; \zeta, \eta)$$

$$\frac{\partial h_k}{\partial z} = g_{k+1}(z; \zeta, \eta) = p_{k+1} + q_{k+1}z + r_{k+1}z^2 \tag{6.48}$$

$$+ (z^3 - \zeta z + \eta)h_{k+1}(z; \zeta, \eta),$$

$k = 0, 1, 2, \ldots$. The coefficients p_k, q_k, and r_k clearly depend on ζ and η. In view of our formulas for ζ, η, and θ in terms of b and c, they are also functions of b and c. By repeated integration by parts (cf. equation (6.24)), we obtain

$$e^{-i\lambda\theta}I_2(\lambda; b, c) = p_0 F_0(\lambda; \zeta, \eta) + q_0 F_1(\lambda; \zeta, \eta) + r_0 F_2(\lambda; \zeta, \eta)$$

$$+ \frac{i}{\lambda}\int_C g_1(z; \zeta, \eta)e^{i\lambda(z^4/4 - \zeta z^2/2 + \eta z)}\, dz$$

$$= \cdots = \sum_{s=0}^{n}\left(\frac{i}{\lambda}\right)^s [p_s F_0(\lambda; \zeta, \eta) + q_s F_1(\lambda; \zeta, \eta) + r_2 F_2(\lambda;\zeta,\eta)]$$

$$+ \left(\frac{i}{\lambda}\right)^{n+1}\int_C g_{n+1}(z; \zeta, \eta)e^{i\lambda(z^4/4 - \zeta z^2/2 + \eta z)}\, dz,$$

$$\tag{6.49}$$

where

$$F_k(\lambda; \zeta, \eta) = \int_C z^k e^{i\lambda(z^4/4 - \zeta z^2/2 + \eta z)} \, dz. \tag{6.50}$$

The functions $F_k(\lambda; \zeta, \eta)$ can be expressed in terms of the Pearcey integral (6.29). Indeed, putting $\lambda z^4 = u^4$, we have

$$F_0(\lambda; \zeta, \eta) = \lambda^{-1/4} P(-\lambda^{1/2}\zeta, \lambda^{3/4}\eta)$$
$$F_1(\lambda; \zeta, \eta) = -i\lambda^{-1/2} P_y(-\lambda^{1/2}\zeta, \lambda^{3/4}\eta) \tag{6.51}$$
$$F_2(\lambda; \zeta, \eta) = -2i\lambda^{-3/4} P_x(-\lambda^{1/2}\zeta, \lambda^{3/4}\eta),$$

P_x and P_y being the first order partial derivatives of $P(x, y)$ with respect to x and y respectively. Equation (6.49) thus gives

$$I_2(\lambda; b, c) \sim e^{i\lambda\theta} \sum_{s=0}^{\infty} \left(\frac{i}{\lambda}\right)^s [p_s \lambda^{-1/4} P(-\lambda^{-1/2}\zeta, \lambda^{3/4}\eta)$$
$$- iq_s \lambda^{-1/2} P_y(-\lambda^{1/2}\zeta, \lambda^{3/4}\eta) - 2ir_s \lambda^{-3/4} P_x(-\lambda^{1/2}\zeta, \lambda^{3/4}\eta)] \tag{6.52}$$

as $\lambda \to +\infty$, uniformly for (b, c) in a band of fixed width from the caustic, with $b \geq b_0 > 0$ for some fixed b_0.

In view of (6.45), the asymptotic expansion of $I(\lambda; b, c)$ is obtained by adding (6.46) and (6.52). In particular, we have

$$I(\lambda; b, c) = e^{i\lambda f(t_3; b, c) - i\pi/4} \sqrt{\frac{2\pi}{-\lambda f''(t_3; b, c)}} \left[1 + O\left(\frac{1}{\sqrt{\lambda}}\right)\right]$$
$$+ e^{i\lambda\theta}\{p_0(b, c)\lambda^{-1/4} P(-\lambda^{1/2}\zeta, \lambda^{3/4}\eta)$$
$$- iq_0(b, c)\lambda^{-1/2} P_y(-\lambda^{1/2}\zeta, \lambda^{3/4}\eta) \tag{6.53}$$
$$- 2ir_0(b, c)\lambda^{-3/4} P_x(-\lambda^{1/2}\zeta, \lambda^{3/4}\eta)\} \left[1 + O\left(\frac{1}{\lambda}\right)\right],$$

where the O-symbols are uniform with respect to b and c.

The asymptotic expansion of $S(x, y, z)$ now follows directly from (6.53) by using the relation

$$S(-x, y, z) = x^{1/2} I(x^{5/2}; yx^{-3/2}, zx^{-2}), \qquad x > 0; \tag{6.54}$$

see equation (6.33).

7.   Laguerre Polynomials II

In §5, we have derived an asymptotic expansion for the Laguerre polynomials $L_n^{(\alpha)}(vt)$, as $n \to \infty$, where $v = 4n + 2\alpha + 2$. This expansion holds uniformly for $t \geq b > 0$. To cover the entire real axis, we must also consider the case $t \leq a$ with $0 < b < a < 1$. The starting point in this case is the integral representation (Szegö 1967, p. 384)

$$e^{-x/2}L_n^{(\alpha)}(x) = \frac{1}{2\pi i} \int_{-\infty}^{(0+)} \exp\left(-\frac{x}{2}\frac{1+e^{-z}}{1-e^{-z}}\right)(1 - e^{-z})^{-\alpha-1}e^{nz}\, dz, \quad (7.1)$$

where the path of integration is the usual loop which begins and ends at $-\infty$ and encircles the origin in the positive direction. *Throughout this section we shall assume that* $\alpha > -1$. Clearly (7.1) can be written as

$$e^{-x/2}L_n^{(\alpha)}(x) = \frac{1}{2\pi i} \int_{-\infty}^{(0+)} \exp\left(-\frac{x}{2}\coth\frac{z}{2} + Nz\right)\left(\frac{\sinh\frac{1}{2}z}{\frac{1}{2}z}\right)^{-\alpha-1} z^{-\alpha-1}\, dz$$

$$(7.2)$$

with $N = n + \frac{1}{2}(\alpha + 1)$. If we replace z by $2z$ and let $v = 4N$ and $x = vt$, $-\infty < t < 1$, then (7.2) becomes

$$e^{-vt/2}L_n^{(\alpha)}(vt) = \frac{2^{-\alpha}}{2\pi i} \int_{-\infty}^{(0+)} \exp\left(\frac{v}{2}f(z, t)\right)\left(\frac{\sinh z}{z}\right)^{-\alpha-1} z^{-\alpha-1}\, dz, \quad (7.3)$$

where $f(z, t)$ is given by

$$f(z, t) = z - t \coth z. \tag{7.4}$$

Note that this function has two symmetrically located saddle points $z_\pm = \pm i \sin^{-1}\sqrt{t}$ and, in addition, a simple pole at $z = 0$. As t tends to zero, the saddle points coalesce with each other and also with the pole. The simplest function which also possesses these essential features is provided by the rational function $u - A^2(t)/u$. Thus the appropriate transformation is

$$f(z, t) = u - \frac{A^2(t)}{u}, \tag{7.5}$$

where $A(t)$ is to be determined. For this to be an analytic transformation, we must have $dz/du \neq 0$ or ∞. Now

$$f_z(z, t)\frac{dz}{du} = 1 + \frac{A^2(t)}{u^2} \tag{7.6}$$

and $f_z(z, t)$ vanishes at $z = z_+$ and z_-, where

$$z_\pm = \begin{cases} \pm i \sin^{-1} \sqrt{t}, & 0 \le t < 1, \\ \mp \sinh^{-1} \sqrt{-t}, & t < 0. \end{cases}$$

Since the right-hand side of (7.6) vanishes at $u = \pm iA(t)$, we must make $z = z_+$ correspond to $u = iA(t)$, and $z = z_-$ to $u = -iA(t)$. This gives

$$A(t) = \begin{cases} \dfrac{1}{2} [\sin^{-1} \sqrt{t} + \sqrt{t(1-t)}], & 0 \le t < 1 \\ \dfrac{i}{2} [\sinh^{-1} \sqrt{-t} + \sqrt{t(t-1)}], & t < 0. \end{cases} \qquad (7.7)$$

It will be shown that with this choice, the transformation (7.5) is one-to-one and analytic along the whole infinite loop given in (7.3). Thus, changing the variable to u, we obtain

$$e^{-vt/2} L_n^{(\alpha)}(vt) = \frac{2^{-\alpha}}{2\pi i} \int_{-\infty}^{(0+)} u^{-\alpha-1} h(u) \exp\left[\frac{v}{2}\left(u - \frac{A^2(t)}{u}\right)\right] du, \qquad (7.8)$$

where

$$h(u) = \left[\frac{\sinh z(u)}{z(u)}\right]^{-\alpha-1} \left[\frac{z(u)}{u}\right]^{-\alpha-1} \frac{dz}{du}. \qquad (7.9)$$

Here we have temporarily assumed that the mapping $z \leftrightarrow u$ preserves the loop nature of the path of integration.

The properties of the mapping between z and u are best seen by introducing an intermediate variable Z defined by

$$z - t \coth z = Z = u - \frac{A^2(t)}{u}. \qquad (7.10)$$

For our purpose, it suffices to consider only the strip $|\operatorname{Im} z| \le \pi/2$. We first restrict ourselves to the case $0 < t < 1$. The half-strip $\{z : \operatorname{Re} z \le 0, 0 \le \operatorname{Im} z \le \pi/2\}$ is shown in Figure 7.17. Its image in the Z-plane is depicted in Figure 7.18. Note that as z traverses once along the idented boundary ABCC'DEFA in Figure 7.17, Z also traverses exactly once along the corresponding curve in Figure 7.18. Here we treat the straight lines C'D and DE in Figure 7.18 as distinct parts of the boundary; see Ursell (1970, p. 375, lines 22–25). Hence, by Theorem 4.5 in Markushevich (1977, vol. 2, p. 118), $\varphi = z - t \coth z$ is univalent in

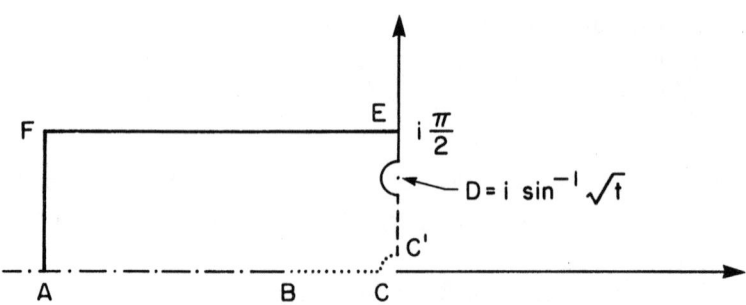

Fig. 7.17 z-plane $(0 < t < 1)$.

the interior of the region bounded by this curve; see also Titchmarsh (1939, §§6.45 and 6.46).

We next consider the mapping $\psi: u \to Z$ defined by $\psi(u) = u - A^2(t)/u$. By the same argument as above, when u traverses once along the boundary of the region ABCC′DEFA in Figure 7.19, Z goes once around the corresponding curve in the Z-plane. Hence ψ is univalent in the interior of the region ABCC′DEFA in Figure 7.19. The equation of the boundary curve EF in Figure 7.19 is given implicitly by

$$\frac{\pi}{2} = s + \frac{A^2(t)s}{r^2 + s^2}, \qquad \text{Re } Z = r - \frac{A^2(t)r}{r^2 + s^2},$$

where $u = r + is$ and $r < 0$. Since $r \leq 0$ and Re $Z \leq 0$, we have, from the second equation above, $A^2(t) \leq r^2 + s^2$. This together with the first

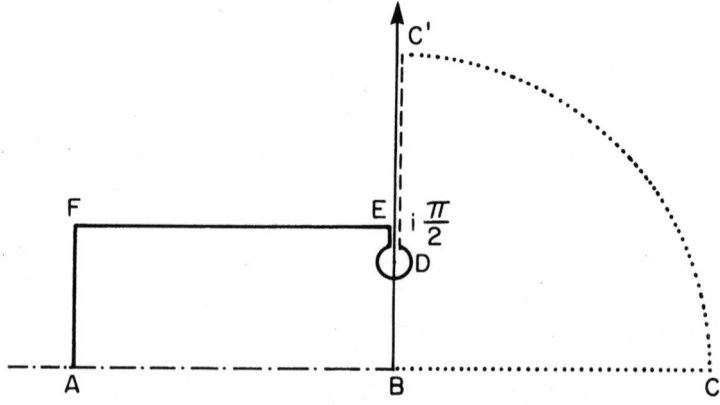

Fig. 7.18 Z-plane.

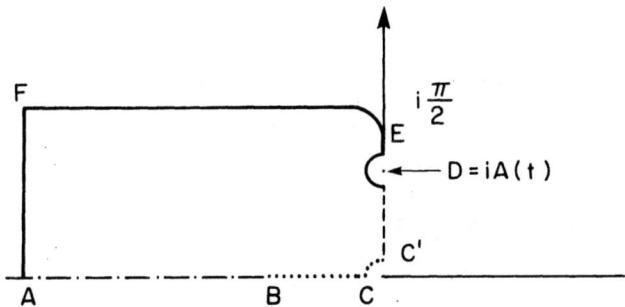

Fig. 7.19 *u*-plane.

equation implies that $s \geq \pi/4$, i.e., the curve EF in Figure 7.19 remains in the region Im $u \geq \pi/4$. Also, from (7.7), we have $A(0) = 0$, $A'(t) > 0$ for $0 < t < 1$, and $A(1^-) = \pi/4$. Hence, $0 < A(t) < \pi/4$ for $0 < t < 1$. Therefore, for $0 < t \leq a < 1$, the point D in Figure 7.19 is at a nonzero distance away from the curve EF.

The transformation $z \leftrightarrow u$ is obtained by composing $\varphi^{-1} : Z \to z$ and $\psi : u \to Z$. Since the transformations $z \leftrightarrow Z$ and $u \leftrightarrow Z$ are one-to-one within the boundary ABCC'DEFA, so is $z \leftrightarrow u$. Let $z = x + iy$ and $u = r + is$. It can be shown by direct computation that the real parts of $z - t \coth z$ and $u - A^2(t)/u$ are odd in x and r, and even in y and s, respectively, and that the imaginary parts of these functions are odd in y and s, and even in x and r, respectively (Ex. 17). Hence, the mapping of the rest of the strip $|\text{Im } z| \leq \pi/2$ is deducible from Figures 7.17 and 7.19 by reflection in the real and imaginary axes. This establishes the one-to-one and analytic nature of the function $u(z, t)$ in $|\text{Im } z| \leq \pi/2$, except possibly at $z = z_+$ and $z = 0$. From the above argument (cf. Figures 7.17 and 7.19), it is evident also that neighborhoods of these points are mapped into neighborhoods of their corresponding images. Consequently, $u(z, t)$ is bounded and analytic at these points. (Note also that near $z = 0$ and $t = 0$, we have $u \sim A^2(t)z/t$ and $A^2(t) \sim t$, respectively.)

We have therefore proved (7.8) for the case $0 < t < 1$. The validity of (7.8) for $-\infty < t < 0$ can be established in a similar manner. The region bounded by ABCC'DEFA in the z-, Z- and u-planes is shown in Figures 7.20, 7.21, and 7.22, respectively. By continuity, (7.8) holds also for $t = 0$. To show that $u(z, t)$ is continuous in t, we note that when $z = 0$, u is identically zero and hence is continuous in t. For $z \neq 0$, we have

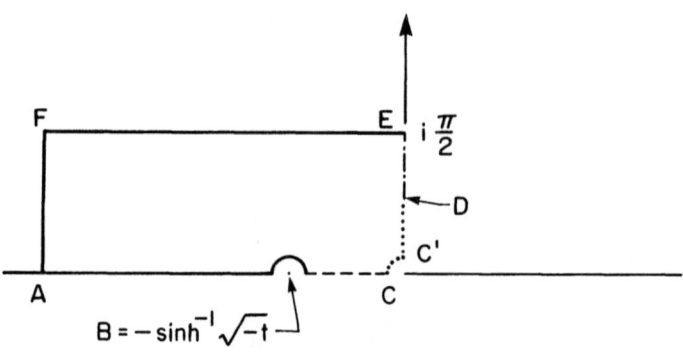

Fig. 7.20 z-plane $(t < 0)$.

$u \neq 0$, as there is a one-to-one correspondence between z and u. Thus, for $z \neq 0$, equation (7.10) is equivalent to the quadratic equation $u^2 - (z - t \coth z)u - A^2(t) = 0$. Since $A^2(t)$ is continuous in t and the solution of the quadratic equation depends continuously on its coefficients, u is continuous in t for $-\infty < t < 1$. We now return to the integral in (7.8), and deform the loop path of integration so that it consists of two straight lines along the negative real axis and the circle centered at the origin with radius ρ. The radius will be specified later, and depends on whether t is close to, or bounded away from, the origin.

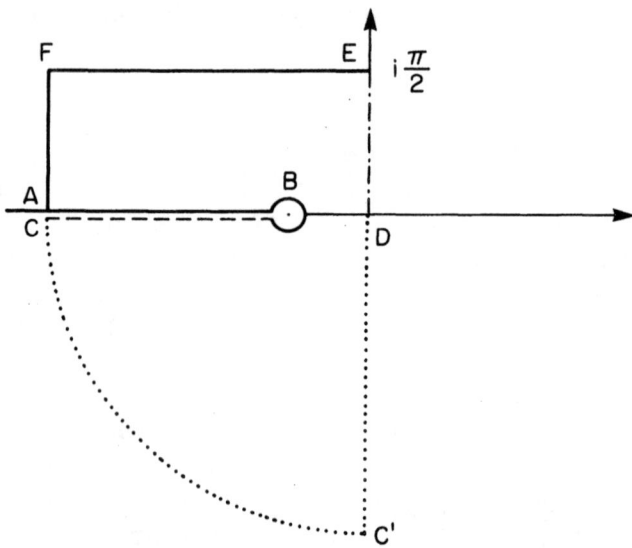

Fig. 7.21 Z-plane $(t < 0)$.

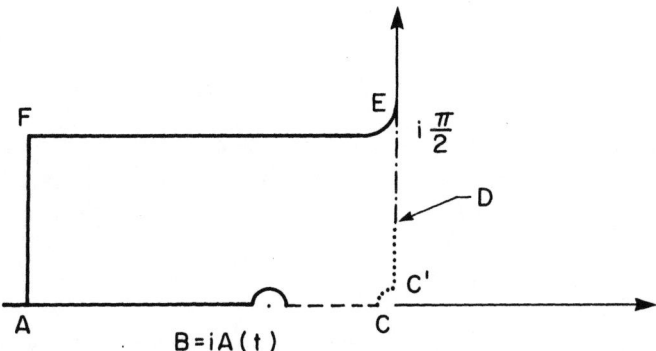

Fig. 7.22 u-plane $(t < 0)$.

From (7.10) it can be shown that $z(u)$ is an odd function of u. Thus, (7.9) implies that $h(u)$ is an even function. Put $h_0(u) = h(u)$ and write

$$h_0(u) = \alpha_0 + \frac{\beta_0}{u} + \left(1 + \frac{A^2(t)}{u^2}\right)g_0(u), \qquad (7.11)$$

where α_0, β_0, and $g_0(u)$ are to be determined. Since $h(u)$ is even, $h_0(iA) = h_0(-iA)$ and it follows that $\beta_0 = 0$, $\alpha_0 = h_0(iA)$. Thus

$$g_0(u) = u^2 \frac{h_0(u) - h_0(iA)}{u^2 - (iA)^2}. \qquad (7.12)$$

Clearly, $g_0(u)$ is analytic for $u^2 \neq (iA)^2$ and has removable singularities at $u = \pm iA$. Near $u = \pm iA$,

$$g_0(u) = \frac{u^2}{u + (\pm iA)} \frac{h_0(u) - h_0(\pm iA)}{u - (\pm iA)},$$

where either all "$+$" or all "$-$" signs are to be taken. From this, we conclude that $g_0(u)$ is analytic everywhere in the domain of $h_0(u)$. Now substitute (7.11) in (7.8), and express the first integral in terms of the Bessel function J_α. To the second integral, we apply an integration by parts. The integrated term vanishes, since $z(u) \sim u$ and $dz/du = O(1)$ as $u \to -\infty$, for fixed t. The final result is

$$2^\alpha e^{-vt/2}L_n^{(\alpha)}(vt) = \frac{\alpha_0}{A^\alpha} J_\alpha(vA)$$

$$- \left(\frac{2}{v}\right)\frac{1}{2\pi i}\int_{-\infty}^{(0+)} u^{-\alpha-1}h_1(u)\exp\left[\frac{v}{2}\left(u - \frac{A^2(t)}{u}\right)\right]du, \qquad (7.13)$$

where

$$h_1(u) = g_0'(u) - \frac{\alpha + 1}{u} g_0(u).$$

We now digress to briefly discuss the asymptotic behavior of $z^{(n)}(u)$, for fixed t, as $u \to -\infty$, $n = 0, 1, 2, \ldots$. From (7.10), it is easily seen that $z(u) \sim u$ as $u \to -\infty$. Differentiating (7.10) with respect to u gives

$$(1 + t \operatorname{csch}^2 z)z'(u) = 1 + \frac{A^2(t)}{u^2}, \tag{7.14}$$

which in turn yields $z'(u) = O(1)$ as $u \to -\infty$. From (7.14), we also have by Leibniz's rule

$$(1 + t \operatorname{csch}^2 z)z^{(n+1)}(u) + \sum_{j=1}^{n} \binom{n}{j-1}(1 + t \operatorname{csch}^2 z(u))^{(n-j+1)}z^{(j)}(u) \tag{7.15}$$

$$= (-1)^n A^2(t) \frac{(n+1)!}{u^{n+2}}.$$

Since every term in the above sum is exponentially decaying, (7.15) implies, by induction, that for fixed t,

$$\frac{d^{n+1}z}{du^{n+1}} = O\left(\frac{1}{u^{n+2}}\right), \qquad n \geq 1, \qquad \text{as } n \to -\infty. \tag{7.16}$$

To derive the full asymptotic expansion of $L_n^{(\alpha)}(vt)$, we repeat the procedures indicated in (7.11) and (7.13). Thus, we define recursively: $h_0(u) = h(u)$,

$$h_n(u) = \alpha_n + \frac{\beta_n}{u} + \left(1 + \frac{A^2}{u^2}\right)g_n(u), \qquad h_{n+1}(u) = g_n'(u) - \frac{\alpha + 1}{u} g_n(u).$$

Using induction, it can be shown that $h_n(u) = O(1)$, as $u \to -\infty$, for fixed t and for all $n \geq 0$. From this, it follows that we also have $g_n(u) = O(1)$, as $u \to -\infty$, $n \geq 0$. Furthermore, we can show that for $n \geq 0$, $g_{2n}(u)$ and $h_{2n}(u)$ are even analytic functions, and that $g_{2n+1}(u)$ and $h_{2n+1}(u)$ are odd analytic functions. Consequently, we have $\alpha_{2n} = h_{2n}(iA)$, $\alpha_{2n+1} = 0$, and $\beta_{2n} = 0$, $\beta_{2n+1} = iAh_{2n+1}(iA)$. Thus, for each n, both α_n and β_n are continuous in t for $-\infty < t < 1$. Repeated application of integration by parts then gives

$$2^\alpha e^{-vt/2}L_n^{(\alpha)}(vt) = \frac{J_\alpha(vA)}{A^\alpha} \sum_{k=0}^{[(p-1)/2]} \alpha_{2k}\left(\frac{2}{v}\right)^{2k}$$

$$- \frac{J_{\alpha+1}(vA)}{A^{\alpha+1}} \sum_{k=0}^{[p/2]-1} \beta_{2k+1}\left(\frac{2}{v}\right)^{2k+1} + \varepsilon_p, \tag{7.17}$$

where

$$\varepsilon_p = \left(-\frac{2}{v}\right)^p \frac{1}{2\pi i} \int_{-\infty}^{(0+)} u^{-\alpha-1} h_p(u) \exp\left\{\frac{v}{2}\left[u - \frac{A^2(t)}{u}\right]\right\} du. \quad (7.18)$$

To estimate ε_p, we introduce the auxiliary function

$$\tilde{J}_\alpha(z) = \begin{cases} J_\alpha(z) & \text{if } z \text{ is imaginary or } 0 \le z \le \delta \\ (|J_\alpha(z)|^2 + |Y_\alpha(z)|^2)^{1/2} & \text{if } z > \delta, \end{cases} \quad (7.19)$$

where δ is chosen so that $J_\alpha(z) \ne 0$ when $0 < |z| \le \delta$ and $\delta > 0$. Furthermore, we define

$$\tilde{\beta}_p(t) = \begin{cases} 1 & \text{for } -\eta \le t \le a < 1 \\ |\beta_p| & \text{for } -\infty < t < -\eta, \end{cases} \quad (7.20)$$

and define $\tilde{\alpha}_p(t)$ in a similar manner, where α_p and β_p are the coefficients in the expansion (7.17), and η is a positive number to be chosen later. We shall now show that for $-\infty < t \le a < 1$,

$$|\varepsilon_p| \le \frac{M_p}{v^p} \tilde{\beta}_p(t) \left|\frac{J_{\alpha+1}(vA(t))}{A(t)^{\alpha+1}}\right| \qquad \text{if } p \text{ is odd,}$$

$$|\varepsilon_p| \le \frac{N_p}{v^p} \tilde{\alpha}_p(t) \left|\frac{J_\alpha(vA(t))}{A(t)^\alpha}\right| \qquad \text{if } p \text{ is even,}$$
(7.21)

where M_p and N_p are constants depending only on p. Recall that $vA(t)$ in our case is either positive or imaginary. Since α_p and β_p are continuous at $t = 0$, the estimates in (7.21) show that the error term ε_p has the same behavior as the first neglected term in the expansion (7.17) near both $t = 0$ and $t = -\infty$.

An alternative bound for the remainder ε_p could be expressed also in terms of the auxiliary functions $E_\alpha(x)$ and $M_\alpha(x)$ given in Olver (1974a, Chapter 12, §1.3) for real argument or $\mathfrak{E}_\alpha(z)$ and $\mathfrak{M}_\alpha(z)$ given in Olver (1974a, Chapter 12, §8) for complex argument.

The proof of (7.21) is divided into two separate cases: (i) $0 \le t \le a < 1$ and (ii) $-\infty < t < 0$. We first consider case (i). Here, $0 \le A(t) < \pi/4$. We shall subdivide this case into two subcases: (ia) $0 \le vA(t) \le \delta$ and (ib) $vA(t) > \delta > 0$. In subcase (ia), we make the change of variable $vu = w$ in (7.18). The result is

$$\varepsilon_p = \left(-\frac{2}{v}\right)^p \frac{v^\alpha}{2\pi i} \int_{-\infty}^{(0+)} w^{-\alpha-1} h_p\left(\frac{w}{v}\right) \exp\left[\frac{1}{2}\left(w - \frac{v^2 A^2}{w}\right)\right] dw. \quad (7.22)$$

Since the last integral is bounded uniformly with respect to v and t, (7.22) gives

$$\varepsilon_p = O(v^{-p+\alpha}), \tag{7.23}$$

the O-symbol being independent of t and v. The estimate (7.21) now follows, taking into account the small-z behavior of $\tilde{J}_{\alpha+1}(z)$. In subcase (ib), we recall that the curve EF in Figure 7.19 is bounded away from the point D. Thus, we may take the radius ρ of the circle in the loop path of integration in (7.18) to be equal to $|A(t)|$. Consequently, we can write (7.18) as

$$\varepsilon_p = \left(-\frac{2}{v}\right)^p A(t)^{-\alpha} \left\{ \frac{\sin \alpha\pi}{\pi} \int_1^\infty r^{-\alpha-1} h_p(-A(t)r) \exp\left[\frac{vA(t)}{2}\left(\frac{1}{r}-r\right)\right] dr \right.$$
$$\left. - \frac{1}{2\pi} \int_{-\pi}^{\pi} \exp(ivA(t)\sin\theta) h_p(A(t)e^{i\theta}) e^{-i\alpha\theta}\, d\theta \right\}. \tag{7.24}$$

By Laplace's method (Chapter II, §1), the first integral is asymptotically equal to $h_p(-A(t))/vA(t)$; and by the method of stationary phase (Chapter II, §3), we see that the second integral behaves like

$$-2i\sqrt{\frac{2\pi}{vA}}\, h_p(iA(t)) \sin\left(vA - \frac{\pi}{4} - \frac{\alpha\pi}{2}\right).$$

Therefore we conclude

$$\varepsilon_p = \left(\frac{2}{v}\right)^p A(t)^{-\alpha} h_p(iA(t)) O\left(\frac{1}{\sqrt{vA(t)}}\right). \tag{7.25}$$

(Note that the contribution from the first integral in (7.24) actually cancels with the endpoint contribution from the second integral there.) The result (7.21) now follows from (7.25), in view of the large-z behavior of $\tilde{J}_\alpha(z)$.

We next consider case (ii). Here $A(t)$ is purely imaginary and $iA(t)$ is negative. We shall divide this case also into two subcases: (iia) $-\eta \le t < 0$ and (iib) $-\infty < t < -\eta$, where $\eta > 0$ is chosen so that $|A(t)| < \pi/2$ for $-\eta \le t < 0$; cf. (7.20). Since the argument for case (iia) is similar to that of case (i), it will be omitted. However, we point out that in this case $vA(t)$ is purely imaginary and positive, and hence that the roles of Laplace's method and the method of stationary phase in case (i) must be interchanged. In case (iib), we choose the radius ρ of the circle in the loop integral (7.18) to be any fixed positive number less than $\pi/2$, and make the substitution $u = -iA(t)v$. The result is

$$\varepsilon_p = \left(-\frac{2}{v}\right)^p [-iA(t)]^{-\alpha} \frac{1}{2\pi i} \int_{-\infty}^{(0+)} v^{-\alpha-1} h_p(-iA(t)v)$$

$$\times \exp\left[-\frac{ivA(t)}{2}\left(v + \frac{1}{v}\right)\right] dv. \tag{7.26}$$

Observe that in our present case, $iA(t)$ is large and negative, and that the saddle points are at $v = \pm 1$. By using Cauchy's theorem, we may make the circular portion of the contour in (7.26) have radius 1. An argument similar to that leading to (7.24) and (7.25), or an application of the Perron method (Chapter II, §5), then gives

$$\varepsilon_p \sim \left(-\frac{2}{v}\right)^p [-iA(t)]^{-\alpha} \frac{1}{2\pi i} h_p(-iA(t)) \exp[-ivA(t)] \sqrt{\frac{-2\pi}{ivA(t)}}. \tag{7.27}$$

Note that $h_p(-iA(t))$ can be expressed in terms of the coefficients α_{2k} and β_{2k+1}, depending on whether p is even or odd. In view of the large-z behavior of $J_\alpha(z)$, z and i being real and positive, the result in (7.21) now follows immediately from (7.27). This completes our proof of (7.21).

The leading coefficient α_0 in (7.17) can be calculated as follows. Since $\alpha_0 = h_0(iA) = h(iA)$, from (7.9) we have

$$\alpha_0 = \left[\frac{\sinh z(u)}{u}\right]^{-\alpha-1} \frac{dz}{du}\bigg|_{u=iA}.$$

Since $u = iA$ corresponds to $z = z_+$ and $\sinh^2 z_+ = -t$, differentiating (7.10) twice with respect to u gives

$$\left(\frac{dz}{du}\right)^2\bigg|_{u=iA} = \frac{i \sinh^3 z_+}{tA \cosh z_+}.$$

Thus

$$\alpha_0 = \left(\frac{A}{\sqrt{t}}\right)^{\alpha+1} \left(\frac{t}{1-t}\right)^{1/4} A^{-1/2} \qquad \text{if } 0 \le t < 1$$

$$= \left(\frac{|A|}{\sqrt{-t}}\right)^{\alpha+1} \left(\frac{-t}{1-t}\right)^{1/4} |A|^{-1/2} \qquad \text{if } t < 0. \tag{7.28}$$

The second coefficient β_1 in (7.17) can also be computed, but the work is overwhelming. After much computation, we find that for $0 \le t < 1$,

$$\beta_1 = \frac{\alpha_0 A}{2}\left[\frac{1 - 4\alpha^2}{8}A^{-1} + \frac{\sqrt{1 - t}}{\sqrt{t}}\left\{\frac{4\alpha^2 - 1}{8} + \frac{1}{4}\frac{t}{1 - t} + \frac{5}{24}\left(\frac{t}{1 - t}\right)^2\right\}\right],$$

(7.29)

and for $t < 0$,

$$\beta_1 = \frac{\alpha_0 A}{2}\left[\frac{1 - 4\alpha^2}{8}A^{-1} - i\frac{\sqrt{1 - t}}{\sqrt{-t}}\left\{\frac{4\alpha^2 - 1}{8} + \frac{1}{4}\frac{t}{1 - t} + \frac{5}{24}\left(\frac{t}{1 - t}\right)^2\right\}\right],$$

(7.30)

where A is given by (7.7). Note that (7.30) can be formally obtained from (7.29) by simply replacing $\sqrt{t}$ by $i\sqrt{-t}$.

By using a matching procedure, it is possible to obtain recursive formulas for the coefficients α_n and β_n. This procedure consists of re-expanding the Bessel functions in (7.17) and comparing the results with expansions obtained by using the ordinary steepest descent method for t bounded away from 0; see Olver (1974a, Chapter 11, Ex. 7.4, and Chapter 12, Ex. 5.2).

To conclude this section, we wish to remark that the method presented here is quite general. It can be applied to many other integrals whose phase function has two symmetrically located saddle points and a simple pole. We have considered only the case of Laguerre polynomials, and in this case we can compare our result with that given by Erdélyi (1960).

8. Legendre Function $P_n^{-m}(\cosh z)$

So far in this chapter, we have discussed uniform expansions for integrals that involve only the confluence of either saddle points with each other or with an endpoint or with a pole. In this section, we shall consider a different type of problem. Here we have no saddle point, but two branch-point singularities located at the endpoints of integration. As the auxiliary (uniformity) parameter tends to the critical value 0, the two singularities approach each other. The function which we consider is the Legendre function $P_n^{-m}(\cosh z)$, and our starting point is the integral representation (Erdélyi *et al.* 1953a, Vol. 1, p. 156, (8))

$$P_n^{-m}(\cosh z) = \left(\frac{1}{2\pi}\right)^{1/2}\frac{(\sinh z)^{-m}}{\Gamma(m+\frac{1}{2})}\int_{-z}^{z} e^{-ut}(\cosh z - \cosh t)^{m-1/2}\,dt,$$

(8.1)

where $u = n + \frac{1}{2}$. In this section, we shall assume that $m + \frac{1}{2} > 0$ and $0 < z \le \rho < \infty$. Since $\cosh z - \cosh t$ behaves like $(z^2 - t^2)/2!$ as $t \to \pm z$, we first wish to express this function in a series involving $z^2 - t^2$. To do this, we begin with the result (Watson 1944, p. 140, (3))

$$\left(\frac{2}{\pi z}\right)^{1/2}\cos\sqrt{z^2 - 2z\theta} = \sum_{m=0}^{\infty}\frac{\theta^m}{m!}J_{m-1/2}(z),$$

(8.2)

which holds for all complex values of θ, where $J_\nu(z)$ is the Bessel function of the first kind. Replacing z by iz and θ by $i\theta$, (8.2) becomes

$$\left(\frac{2}{\pi z}\right)^{1/2}\cosh\sqrt{z^2 - 2z\theta} = \sum_{m=0}^{\infty}\frac{(-1)^m}{m!}\theta^m I_{m-1/2}(z).$$

(8.3)

Here use has been made of the identity $J_\nu(iz) = i^\nu I_\nu(z)$. Now put $z^2 - 2z\theta = t^2$, and recall the special cases

$$I_{-1/2}(z) = \left(\frac{2}{\pi z}\right)^{1/2}\cosh z, \qquad I_{1/2}(z) = \left(\frac{2}{\pi z}\right)^{1/2}\sinh z.$$

Inserting these in (8.3) gives

$$2(\cosh z - \cosh t) = \frac{\sinh z}{z}(z^2 - t^2)\left[1 + \sum_{\nu=1}^{\infty}(-1)^\nu\varphi_\nu(z)(z^2 - t^2)^\nu\right],$$ (8.4)

where

$$\varphi_\nu(z) = \frac{1}{2^\nu(\nu+1)!}\frac{I_{\nu+1/2}(z)}{z^\nu I_{1/2}(z)}.$$

(8.5)

The desired expansion

$$(\cosh z - \cosh t)^{m-1/2} = \frac{1}{2^{m-1/2}}\left(\frac{\sinh z}{z}\right)^{m-1/2}(z^2 - t^2)^{m-1/2}$$

$$\times \sum_{\nu=0}^{\infty}(-1)^\nu\psi_\nu(z)(z^2 - t^2)^\nu$$

(8.6)

now follows immediately, where the coefficients $\psi_\nu(z)$ satisfy the recurrence relation

$$\psi_{\nu+1}(z) = \sum_{j=0}^{\nu}\left[\left(m - \frac{1}{2}\right) - \frac{j}{\nu+1}\left(m + \frac{1}{2}\right)\right]\varphi_{\nu+1-j}(z)\psi_j(z), \quad (8.7)$$

$\nu = 0, 1, 2, \ldots$, with $\psi_0(z) = 1$; see Pourahmadi (1984). Simple calculation gives

$$\psi_1(z) = -\frac{1}{4}\left(m - \frac{1}{2}\right)\frac{1 - z\coth z}{z^2} \qquad (8.8)$$

and

$$\psi_2(z) = \frac{1}{24}\left(m - \frac{1}{2}\right)\left[\frac{1}{z^2}\left(1 + \frac{3}{z^2}\right) - \frac{3}{z^3}\coth z\right]$$
$$+ \frac{1}{32}\left(m - \frac{1}{2}\right)\left(m - \frac{3}{2}\right)\left(\frac{1 - z\coth z}{z^2}\right)^2. \qquad (8.9)$$

We now introduce the remainder $\Delta_p(z, t)$ defined by

$$\left[\frac{2(\cosh z - \cosh t)}{z^2 - t^2}\frac{z}{\sinh z}\right]^{m-1/2} = \sum_{\nu=0}^{p-1}(-1)^\nu\psi_\nu(z)(z^2 - t^2)^\nu$$
$$+ (-1)^p(z^2 - t^2)^p\Delta_p(z, t). \qquad (8.10)$$

On account of the well-known formula

$$I_\nu(uz) = \frac{2^{-\nu}u^\nu z^{-\nu}}{\pi^{1/2}\Gamma(\nu + \frac{1}{2})}\int_{-z}^{z}(z^2 - t^2)^{\nu - 1/2}e^{-ut}\,dt, \qquad (8.11)$$

we have, upon inserting (8.10) in (8.1),

$$P_n^{-m}(\cosh z) = \left(\frac{z}{\sinh z}\right)^{1/2}\left[\sum_{\nu=0}^{p-1}c_\nu(z)\frac{I_{m+\nu}(uz)}{u^{m+\nu}} + \varepsilon_p(z, u)\right], \qquad (8.12)$$

where

$$\varepsilon_p(z, u) = \frac{(-1)^p}{\sqrt{\pi}(2z)^m\Gamma(m + \frac{1}{2})}\int_{-z}^{z}e^{-ut}(z^2 - t^2)^{p+m-1/2}\Delta_p(z, t)\,dt. \qquad (8.13)$$

The coefficient $c_\nu(z)$ is given explicitly by

$$c_\nu(z) = (-1)^\nu\frac{\Gamma(m + \nu + \frac{1}{2})}{\Gamma(m + \frac{1}{2})}(2z)^\nu\psi_\nu(z). \qquad (8.14)$$

In particular, we have from (8.8) and (8.9)

$$c_0(z) = 1, \qquad c_1(z) = \frac{1}{2}\left(m^2 - \frac{1}{4}\right)\left(\frac{1 - z\coth z}{z}\right),$$

$$c_2(z) = \left(m^2 - \frac{1}{4}\right)\left\{\frac{1}{6}\left(m + \frac{3}{2}\right)\left[1 + \frac{3}{z^2} - \frac{3}{z}\coth z\right]\right. \tag{8.15}$$

$$\left. + \frac{1}{8}\left(m^2 - \frac{9}{4}\right)\frac{(1 - z\coth z)^2}{z^2}\right\}.$$

Since the series in (8.6) is convergent for $-z \le t \le z$, it is easily seen that the remainder $\Delta_p(z, t)$ in (8.10) is $O(1)$ for *bounded* z. Hence, from (8.11), it follows that

$$\varepsilon_p(z, u) = O\left(\frac{I_{m+p}(zu)}{u^{m+p}}\right) \tag{8.16}$$

uniformly for $0 \le z \le \rho < \infty$. This result in fact holds for z in the *unbounded* interval $0 \le z < \infty$, but its proof is very complicated. Shivakumar and Wong (1989) have shown that

$$|\Delta_p(z, t)| \le \frac{M_p}{(1 + z)^p}, \qquad -z \le t \le z, \tag{8.17}$$

for some numerical quantity M_p. The first three M_p's are given by

$$M_1 = |m - \tfrac{1}{2}|, \qquad M_2 = \tfrac{7}{6}|m - \tfrac{1}{2}| + |(m - \tfrac{1}{2})(m - \tfrac{3}{2})|, \tag{8.18}$$

$$M_3 = \tfrac{5}{8}|m - \tfrac{1}{2}| + \tfrac{19}{8}|(m - \tfrac{3}{2})(m - \tfrac{1}{2})| + \tfrac{1}{2}|(m - \tfrac{5}{2})(m - \tfrac{3}{2})(m - \tfrac{1}{2})|.$$

Thus, from (8.13) and (8.11), we have

$$|\varepsilon_p(z, u)| \le \frac{\Gamma(m + p + \tfrac{1}{2})}{\Gamma(m + \tfrac{1}{2})} \frac{(2z)^p}{(1 + z)^p} M_p \frac{I_{m+p}(uz)}{u^{m+p}} \tag{8.19}$$

for all z in $[0, \infty)$.

Exercises

1. Let $f(t)$ be an analytic function in $|t| \le |\alpha|$, and let α be a pole of f of order p, which is not on the positive real t-axis. Suppose that

$$f(t) = \frac{b_{-p}}{(t - \alpha)^p} + \cdots + \frac{b_{-1}}{t - \alpha} + f^*(t),$$

where $f^*(t) = \sum_{n=0}^{\infty} a_n t^n$, for $|t| < \delta$ and $\delta > |\alpha|$. Let

$$I(\lambda) = \int_0^{\infty} t^{\mu} f(t) e^{-\lambda t}\, dt$$

be absolutely convergent for all large values of λ. Show that

$$I(\lambda) = \sum_{n=0}^{\infty} a_n \Gamma(\mu + n + 1) \lambda^{-\mu-n-1}$$

$$+ \Gamma(\mu + 1)(-\alpha)^{\mu/2^{\prime\prime}} \lambda^{-\mu/2-1^{\prime\prime}} e^{-\lambda\alpha/2}$$

$$\times \sum_{k=1}^{p} b_{-k}\left(\frac{\lambda}{-\alpha}\right)^{k/2} W_{-k/2-\mu/2,\,\mu/2-k/2+1/2}(-\lambda\alpha),$$

as $\lambda \to \infty$, uniformly for α in a neighborhood of the origin, where $W_{k,m}(z)$ is the Whittaker function (Erdélyi et al. 1953a, p. 274).

2. Consider the integral

$$I(\lambda) = \int_0^{\infty} g(t) e^{i\lambda f(t)}\, dt,$$

where λ is a large positive parameter, $f(t)$ is a real-valued function, and $g(t)$ vanishes identically in the neighborhood of infinity. Assume that $f'(\alpha) = 0$, $f''(\alpha) = \pm 2a^2$, $a > 0$, and that $f'(t) \neq 0$ for $t \neq \alpha$, where α is some real number. Let $\alpha_- = \min(\alpha, 0)$, and suppose that both f and g are C^{∞} in $[\alpha_-, \infty)$. Note that as $\alpha \to 0$, the saddle point $t = \alpha$ coalesces with the endpoint $t = 0$. Put

$$u^2 = \pm[f(t) - f(\alpha)], \qquad \operatorname{sgn} u = \operatorname{sgn}(t - \alpha).$$

Then

$$I(\lambda) = e^{i\lambda f(\alpha)} \int_{\eta}^{\infty} e^{\pm i\lambda u^2} h(u)\, du$$

where

$$h(u) = \frac{2u}{\pm f'(t)}\, g(t)$$

and $\eta = \operatorname{sgn}(-\alpha)|f(0) - f(\alpha)|^{1/2}$. Define $h_0(u) = h(u)$ and

$$h_{s+1}(u) = \frac{d}{du}\left[\frac{h_s(u) - h_s(0)}{u}\right] = \int_0^1 v h_s''(uv)\, dv, \qquad s \geq 0.$$

Derive the asymptotic expansion

$$I(\lambda) = e^{i\lambda f(0)} \sum_{s=0}^{N} \left(\frac{\pm i}{2\lambda}\right)^s \left[\frac{h_s(0)}{\sqrt{\lambda}} F_{\pm}(\sqrt{\lambda}\eta) \pm \frac{i}{2\lambda} \frac{h_s(\eta) - h_s(0)}{\eta}\right]$$

$$+ \left(\frac{\pm i}{2\lambda}\right)^{N+1} \int_{\eta}^{\infty} e^{\pm i\lambda u^2} h_{N+1}(u)\, du,$$

where $F_{\pm}(x)$ is the Fresnel integral given by $F_{\pm}(x) = e^{\mp ix^2} \int_x^{\infty} e^{\pm iu^2}\, du$.

3. Define $S_n(x)$ by

$$e^{nx} = \sum_{r=0}^{n} \frac{(nx)^r}{r!} + \frac{(nx)^n}{n!} S_n(x).$$

Show that

$$(nx)^{nx} S_n(x) = e^{nx} \int_0^{nx} (nx - t)^n e^{-(nx-t)}\, dt.$$

Deduce from this the alternative representation

$$S_n(x) = (nx) \int_0^1 \exp\{-n[-xt - \log(1 - t)]\}\, dt.$$

Note that the saddle point $t = (x - 1)/x$ coincides with the end-point $t = 0$ when $x = 1$. Let $a^2 = 2[(x - 1) - \log x]$ and prove that

$$S_n(x) \sim (nx)e^{na^2/4} \frac{a}{1 - x} \frac{D_{-1}(\sqrt{na})}{\sqrt{n}}, \qquad \text{as } n \to \infty,$$

for $0 < \delta \le x \le 1$, D_{-1} being the parabolic cylinder function of order -1.

4. For $0 \le \alpha < \pi/2$ and λ positive, let

$$I(\alpha, \lambda) = \int_0^{\pi/2} e^{\lambda(\cos t + t \sin \alpha)}\, dt.$$

Use the argument in §3 to show that

$$e^{-\lambda} I(\alpha, \lambda) = \frac{U_1(\sqrt{\lambda}a)}{\lambda^{1/2}} \left[a_0 + O\left(\frac{1}{\lambda}\right)\right] + \frac{U_1'(\sqrt{\lambda}a)}{\lambda} \left[b_0 + O\left(\frac{1}{\lambda}\right)\right],$$

as $\lambda \to \infty$, uniformly with respect to $\alpha \in [0, \alpha_0]$, where α_0 is any constant in the interval $(0, \pi/2)$,

$$a = \sqrt{2(\cos \alpha + \alpha \sin \alpha - 1)}, \qquad a_0 = \frac{a}{\sin \alpha},$$

$$b_0 = \frac{1}{a}\left[\frac{1}{\sqrt{\cos \alpha}} - \frac{a}{\sin \alpha}\right],$$

and $U_1(x)$ is defined by (3.18).

Use the relation $U_1'(x) = 1 + xU_1(x)$ to deduce the uniform asymptotic approximation

$$I(\alpha, \lambda) = \frac{e^{\lambda(\cos \alpha + \alpha \sin \alpha)}}{\sqrt{2\lambda}}\left[\frac{1}{\sqrt{\cos \alpha}}X_0(\alpha, \lambda) + O\left(\frac{1}{\sqrt{\lambda}}\right)\right],$$

where

$$X_0(\alpha, \lambda) = \sqrt{2}\, e^{\lambda(1 - \cos \alpha - \alpha \sin \alpha)}U_1(\sqrt{\lambda}a).$$

(Note: $\sqrt{\pi} \le X_0(\alpha, \lambda) < 2\sqrt{\pi}$.)

5. Let $f \in C^\infty[0, \infty)$, and assume that f and all its derivatives are bounded on $[0, \infty)$. Put

$$t^{\alpha - 1}f(t) = \sum_{s=0}^{n-1} c_s t^{s+\alpha-1} + \varphi_n(t), \qquad n = 1, 2, \ldots,$$

and assume that $\varphi_n^{(k)}(t) = O(t^{n-k+\alpha-1})$ as $t \to 0^+$, $k = 0, 1, \ldots, n$. Use Taylor's theorem with integral remainder to show that for $k = 1, 2, \ldots,$

$$\varphi_k^{(k-1)}(t) = t^\alpha \sum_{l=0}^{k-1} K_l \int_0^1 (1 - u)^l f^{(k)}(tu)\, du,$$

where

$$K_l = \binom{k-1}{l}\frac{\Gamma(\alpha)}{\Gamma(\alpha - l)\Gamma(l + 1)}.$$

6. Consider the integral

$$F(x, a) = \int_a^\infty t^{\alpha - 1}f(t)e^{-x(t - a)}\, dt,$$

where $a \ge 0$, $0 < \alpha < 1$, and $x > 0$. Note that as $a \to 0^+$, the endpoint $t = a$ approaches the singularity $t = 0$. Let $\varphi_n(t)$ be defined as in Exercise 5, and assume that $f(t)$ satisfies all the conditions imposed there. Show, by integration by parts, that for $k = 1, 2, \ldots$,

$$\int_a^\infty e^{-xt} \varphi_k^{(k)}(t)\, dt = c_k \frac{\Gamma(\alpha + k)}{\Gamma(\alpha)} \int_a^\infty e^{-xt} t^{\alpha-1}\, dt$$

$$+ e^{-ax} \varphi_{k+1}^{(k)}(a) x^{-1} + x^{-1} \int_a^\infty e^{-xt} \varphi_{k+1}^{(k+1)}(t)\, dt.$$

Deduce from this the expansion

$$F(x, a) = I(x, a) \sum_{k=0}^{n-1} \frac{\Gamma(k + \alpha)}{\Gamma(\alpha)} \frac{c_k}{x^k} + \sum_{k=1}^{n} \varphi_k^{(k-1)}(a) x^{-k} + E_n,$$

$$n = 1, 2, \ldots,$$

where

$$I(x, a) = \int_a^\infty e^{-x(t-a)} t^{\alpha-1}\, dt,$$

$$E_n = x^{-n} \int_a^\infty e^{-x(t-a)} \varphi_n^{(n)}(t)\, dt.$$

Show also that there exists a positive constant M, independent of a, such that

$$|E_n| \leq M_n x^{-n} I(x, a) \qquad \text{for } x \geq a.$$

Note that for small values of a, the coefficients $\varphi_k^{(k-1)}(a)$ in the above expansion can be computed by using its integral expression in Exercise 5.

7. The modified Bessel function $K_\nu(z)$ has the integral representation

$$\int_1^\infty (t^2 - 1)^\mu e^{-zt}\, dt = \frac{\Gamma(\mu + 1)}{\sqrt{\pi} (\tfrac{1}{2} z)^{\mu + 1/2}} K_{-\mu-1/2}(z),$$

where $\operatorname{Re} z > 0$ and $\mu > 1$; see Watson (1944, p. 82). Use connection formulas for various Bessel functions to show that

$$\lim_{\varepsilon \to 0^+} \int_1^\infty e^{-\varepsilon t} (t^2 - 1)^\mu \sin xt\, dt = \frac{\sqrt{\pi}\, \Gamma(\mu + 1)}{2(\tfrac{1}{2} x)^{\mu + 1/2}} J_{-\mu-1/2}(x).$$

This result is used in the following exercise.

8.   Consider the integral

$$I(\lambda) = \int_a^z g(\sqrt{y^2 - a^2}) \sin \lambda y \, dy,$$

where $0 \leq a < z < \infty$ and $\lambda \to +\infty$. Assume that the following conditions hold: (i) $g(x)$ is n times continuously differentiable in the interval $0 < x < (z^2 - a^2)^{1/2}$. (ii) As $x \to 0^+$,

$$g(x) \sim \sum_{s=0}^{\infty} a_s x^{s + \alpha - 1}, \qquad 0 \leq \alpha < 1,$$

and the asymptotic expansions of the derivatives of g, up to and including order n, can be obtained by differentiating this series term by term. (iii) If $G(y) = g(\sqrt{y^2 - a^2})$ then $G^{(s)}(z^-)$ exists for each s, $s = 0, 1, \ldots, n$. Show that for each $n \geq 0$,

$$I(\lambda) = \sum_{s=0}^{2n-1} J_{-(s+\alpha)/2}(a\lambda) \frac{b_s(a)}{\lambda^{(s+\alpha)/2}} + \delta_n(\lambda)$$

$$- \sum_{s=0}^{n-1} \sin\left(\frac{s+1}{2}\pi + \lambda z\right) \frac{G^{(s)}(z^-)}{\lambda^{s+1}} - \varepsilon_n(\lambda),$$

where $\delta_n(\lambda) = o(\lambda^{-n})$ and $\varepsilon_n(\lambda) = O(\lambda^{-n-1})$, as $\lambda \to +\infty$, uniformly for $0 \leq a \leq a_0 < z$. The coefficients $b_s(a)$ are given by

$$b_s(a) = \frac{\sqrt{\pi}}{2} a_s \Gamma\left(\frac{s+\alpha+1}{2}\right)(2a)^{(s+\alpha)/2}, \quad s = 0, 1, 2, \ldots.$$

(Hint: Use Lemma 1 in Chapter IV and Exercise 7 above.)

9.   Consider the polynomial equation $u^3 - 3\zeta u + A = 0$. Establish the identity $4 \sin^3 \theta - 3 \sin \theta + \sin 3\theta = 0$ and use it to show that the zeros of the cubic in u can be expressed in the form

$$u_1 = -2\sqrt{\zeta} \sin\left(\frac{\pi}{3} + \theta\right), \quad u_2 = 2\sqrt{\zeta} \sin \theta, \quad u_3 = 2\sqrt{\zeta} \sin\left(\frac{\pi}{3} - \theta\right),$$

where θ is given by

$$\theta = \frac{1}{3} \arc \sin\left(\frac{A}{2\zeta^{3/2}}\right)$$

for nonzero ζ.

10. Let

$$f(t; \mu) = \frac{1}{4} t^4 - \frac{1}{2} t^2 + \mu t \quad \text{and} \quad J(\lambda; \mu) = \int_\Gamma e^{i\lambda f(t; \mu)} \, dt,$$

where Γ is a contour beginning at $\infty e^{9\pi i/8}$ and ending at $\infty e^{5\pi i/8}$, λ is a large positive parameter, and μ is a positive number in some interval containing $2/\sqrt{27}$.

(a) Show that f has three saddle points given by

$$t_1(\mu) = -\frac{2}{\sqrt{3}} \sin\left(\frac{\pi}{3} + \varphi\right), \qquad t_2(\mu) = \frac{2}{\sqrt{3}} \sin\varphi,$$

$$t_3(\mu) = \frac{2}{\sqrt{3}} \sin\left(\frac{\pi}{3} - \varphi\right),$$

where

$$3\varphi = \text{arc sin}\left(\frac{\mu\sqrt{27}}{2}\right).$$

Further, show that if δ is a positive constant, then for all μ in the interval $(2/\sqrt{27} - \delta, \ 2/\sqrt{27} + \delta)$, $t_1(\mu)$ is real and negative, and $\text{Re } t_2(\mu)$, $\text{Re } t_3(\mu) > 0$.

(b) Determine the steepest descent path through t_1 and show that Γ may be deformed into it.

(c) Deduce the asymptotic approximation

$$J(\lambda) = e^{i\lambda f(t_1; \mu)} \left[\left(\frac{\pi}{3t_1^2 - 1}\right)^{1/2} \frac{1 + i}{\lambda^{1/2}} + O\left(\frac{1}{\lambda^{3/2}}\right) \right]$$

as $\lambda \to +\infty$, for $\mu \in (2/\sqrt{27} - \delta, \ 2/\sqrt{27} + \delta)$.

11. Set $I(\lambda; \mu) = \int_\gamma e^{i\lambda f(t; \mu)} \, dt$ with f, λ, and μ as in Exercise 10. Here, γ is a contour beginning at $\infty e^{5\pi i/8}$ and ending at $\infty e^{\pi i/8}$.

(a) Refer to Exercise 10. Show that the saddle points of f with positive real part are complex conjugates when $\mu > 2/\sqrt{27}$, are equal to $1/\sqrt{3}$ when $\mu = 2/\sqrt{27}$, and satisfy $t_2(\mu) < 1/\sqrt{3} < t_3(\mu)$ when $\mu < 2/\sqrt{27}$.

(b)   Introduce the cubic change of variables of Chester–Friedman–Ursell and conclude that the uniformly analytic univalent solution of

$$f(t; \mu) = \frac{u^3}{3} - \zeta u + \eta$$

that preserves the correspondences $t_2 \leftrightarrow -\sqrt{\zeta}$, $t_3 \leftrightarrow +\sqrt{\zeta}$, for $\zeta > 0$, is given by

$$u = 2\sqrt{\zeta} \sin \psi, \qquad \psi = \frac{1}{3} \arc \sin\left\{ \frac{3[\eta - f(t; \mu)]}{2\zeta^{3/2}} \right\}.$$

Provide formulas for η and ζ.

(c)   Deduce the expansion

$$I(\lambda; \mu) \sim e^{i\lambda\eta} \frac{2\pi}{\lambda^{1/3}} \sum_{j=0}^{\infty} \left(\frac{i}{\lambda}\right)^j \left[p_j(\mu)\, \text{Ai}(-\lambda^{2/3}\zeta) + \frac{q_j(\mu)}{i\lambda^{1/3}}\, \text{Ai}'(-\lambda^{2/3}\zeta) \right]$$

as $\lambda \to \infty$, with μ in an open interval containing $2/\sqrt{27}$.

(d)   For the change of variables in (b), show that, for $\zeta > 0$,

$$\left.\frac{dt}{du}\right|_{u=\sqrt{\zeta}} = \sqrt{\frac{2\zeta^{1/2}}{3t_3^2 - 1}},$$

$$\left.\frac{dt}{du}\right|_{u=-\sqrt{\zeta}} = \sqrt{\frac{-2\zeta^{1/2}}{3t_2^2 - 1}},$$

all square roots having their principal branches. Use these formulas, the expressions for $t_2(\mu)$ and $t_3(\mu)$ in Exercise 10(a), and the expression ζ given in (b) above to obtain

$$p_0(\mu) = 3^{-1/6}\left[1 + O\left\{ \left(\frac{2}{\sqrt{27}} - \mu\right)^{1/2} \right\} \right],$$

$$q_0(\mu) = -\frac{3^{-5/6}}{2}\left[1 + O\left\{ \left(\frac{2}{\sqrt{27}} - \mu\right)^{1/2} \right\} \right]$$

as $\mu \to 2/\sqrt{27}$. (Warning: There is extensive computation involved in deriving these limiting forms for $p_0(\mu)$ and $q_0(\mu)$.)

12.   Put $y = \mu x^{3/2}$ in the Pearcey integral

$$P(-x, y) = \int_{-\infty}^{\infty} \exp\left(\frac{t^4}{4} - x\frac{t^2}{2} + yt \right) dt,$$

and show that

$$P(-x, \mu x^{3/2}) = x^{1/2}I(x^2; \mu) + x^{1/2}J(x^2; \mu),$$

where I and J are the integrals in Exercises 10 and 11. Using the results of these exercises, show that

$$P(-x, \mu x^{3/2}) = e^{ix^2[f(t_2; \mu) + f(t_3; \mu)]/2}$$

$$\times \left\{ p_0(\mu) \frac{2\pi}{x^{1/6}} \operatorname{Ai}(-x^{4/3}\zeta)\left[1 + O\left(\frac{1}{x^2}\right)\right] \right.$$

$$+ q_0(\mu) \frac{2\pi}{ix^{5/6}} \operatorname{Ai}'(-x^{4/3}\zeta)\left[1 + O\left(\frac{1}{x^2}\right)\right] \right\}$$

$$+ e^{ix^2 f(t_1; \mu)}\left(\frac{\pi}{3t_1^2 - 1}\right)^{1/2} \frac{1 + i}{x^{1/2}}\left[1 + O\left(\frac{1}{x^2}\right)\right]$$

as $x \to +\infty$, uniformly for μ in an interval containing $2/\sqrt{27}$. (f, t_1, t_2, t_3, and ζ are given in Exercises 10 and 11). Discuss the behavior of this approximation when ζ is bounded away from 0 (i.e., μ is bounded away from $2/\sqrt{27}$) and x is large. Compare your conclusions with (non-uniform) expansions of $P(-x; \mu x^{3/2})$ obtained by applying the method of stationary phase to each of the cases $\mu > 2/\sqrt{27}$, $\mu = 2/\sqrt{27}$, and $0 \le \mu < 2/\sqrt{27}$.

13. Consider the Bessel function (Watson 1944, §6.2, equation (3))

$$J_\nu(\nu a) = \frac{1}{2\pi i} \int_{\infty - \pi i}^{\infty + \pi i} e^{N(\sinh z - z \cosh \alpha)}\, dz,$$

where $N = \nu a$, $a = \operatorname{sech} \alpha$, and α is a small nonnegative parameter.

(a) Find the relevant saddle points of this integral. Set

$$\sinh z - z \cosh \alpha = \tfrac{1}{3}u^3 - b^2 u + c,$$

and find the coefficients b and c so that the mapping from the z-plane to the u-plane is analytic in the relevant regions. Show that the coefficient b so chosen is an analytic function of α in a neighborhood of $\alpha = 0$.

(b) With the choice of b and c made in (a), show by using Ex. 9 that the appropriate solution to the above cubic equation is given by

$$u = 2b \sin \tfrac{1}{3}\varphi,$$

where φ is the solution of

$$\frac{2}{3} b^3 \sin \varphi = z \cosh \alpha - \sinh z.$$

Show also that u is an odd function of z and expressible as

$$u(z) = \sum_{n=0}^{\infty} b_n(\alpha) z^{2n+1}.$$

Calculate the first two coefficients $b_0(\alpha)$ and $b_1(\alpha)$.

(c)   Consider the sequence of mappings $z \leftrightarrow Z$, $Z \leftrightarrow \varphi$, and $\varphi \leftrightarrow u$ defined by

$$Z = z \cosh \alpha - \sinh z, \qquad \frac{2}{3} b^3 \sin \varphi = Z, \qquad u = 2b \sin \frac{1}{3}\varphi.$$

Find the images of the half-strip $\{z : \operatorname{Re} z > 0, 0 \le \operatorname{Im} z \le \pi\}$ in the Z-, φ-, and u-planes. Show that the mapping $z \leftrightarrow u$ is analytic and one-to-one from the region $|\operatorname{Im} z| \le \pi$ to its image in the u-plane.

(d)   Find the steepest descent curve in the u-plane, and derive the uniform asymptotic formula

$$J_\nu(\nu \operatorname{sech} \alpha) = \sqrt{\frac{2b}{\sinh \alpha}} \frac{\operatorname{Ai}(N^{2/3} b^2)}{N^{1/3}} \left[1 + O\!\left(\frac{1}{N}\right) \right], \qquad \alpha \in [0, \infty).$$

14.   Consider the Anger function

$$\mathbf{A}_{-\nu}(\nu \operatorname{sech} \alpha) = \int_0^\infty e^{-\nu f(t;\alpha)} \, dt$$

where $\alpha \ge 0$ and $f(t; \alpha) = \operatorname{sech} \alpha \sinh t - t$. Define ζ by

$$\frac{2}{3} \zeta^{3/2} = \alpha - \tanh \alpha,$$

and the sequences $\{a_n\}$, $\{b_n\}$, $\{\varphi_n(u)\}$, and $\{\psi_n(u)\}$ as in (4.26) and (4.27). By modifying the argument in §4, show that

$$\mathbf{A}_{-\nu}(\nu \operatorname{sech} \alpha) = W(\nu^{2/3}\zeta) \sum_{s=0}^{n-1} \frac{a_s}{\nu^{s+(1/3)}} + W'(\nu^{2/3}\zeta) \sum_{s=0}^{n-1} \frac{b_s}{\nu^{s+(2/3)}}$$

$$+ \sum_{s=0}^{n-1} \frac{\psi_s(0)}{\nu^{s+1}} + E_n(\zeta, \nu),$$

where

$$W(x) = \int_0^\infty e^{-v^3/3 + xv}\, dv, \qquad E_n(\zeta, v) = \frac{1}{v^n} \int_0^\infty \varphi_n(u) e^{-v[u^3/3 - \zeta u]}\, du.$$

Calculate a_0, b_0, and $\psi_0(0)$. Show also that

$$E_n(\zeta, v) = W(v^{2/3}\zeta)O(v^{-n-(1/3)})$$

uniformly for $\alpha \geq 0$. (Note: $W(x)$ can be expressed in terms of Scorer's function (Olver 1974a, p. 332).)

15. In Chapter II, Exercise 20, we have considered the integral

$$I(N, \theta) = \int_{-\infty - i\pi/4}^{\infty - i\pi/4} \cosh^3 u \; e^{Nf(u,\theta)}\, du,$$

where $f(u, \theta) = i(\cos\theta \cosh u - \frac{1}{2}\sin\theta \sinh 2u)$ and θ is a real parameter. It was stated that there exists a critical value θ_c such that for $\theta \neq \theta_c$ there are two (relevant) saddle points of order 1, and for $\theta = \theta_c$ there is only one (relevant) saddle point (but) of order 2. Two asymptotic approximations were also given, one for $0 < \theta < \theta_c$ and one for $\theta = \theta_c$. It is easily seen that the amplitude of the approximation for $\theta < \theta_c$ approaches infinity when θ tends to θ_c from below. Thus two approximations do not join smoothly. A similar abrupt change occurs when θ tends to θ_c from above. To derive an approximation which is uniform with respect to θ in a neighborhood of θ_c, we make the cubic change of variable

$$if(u, \theta) = \tfrac{1}{3}v^3 - \mu(\theta)v + v(\theta),$$

as suggested in §4.

(a) Determine the coefficients $\mu(\theta)$ and $v(\theta)$ so that the transformation $u \leftrightarrow v$ is analytic in a neighborhood of θ_c.

(b) Show that near $\theta = \theta_c$, $I(N, \theta)$ has an expansion of the form

$$e^{iNv(\theta)}I(N, \theta) \sim \frac{2\pi i\, \mathrm{Ai}(-N^{2/3}\mu(\theta))}{N^{1/3}} \sum_{m=0}^\infty \frac{A_m(\theta)}{N^m}$$

$$+ \frac{2\pi i\, \mathrm{Ai}'(-N^{2/3}\mu(\theta))}{N^{2/3}} \sum_{m=0}^\infty \frac{B_m(\theta)}{N^m}, \qquad \text{as } N \to \infty.$$

(c) Calculate the leading coefficients $A_0(\theta)$ and $B_0(\theta)$.

16. Let $P(z, \alpha)$ be the polynomial of degree $m + 1$ given in (6.5), and let $H_0(z, \alpha)$ be a polynomial of degree $m - 1$. Show that the function F_0 in equation (6.9),

$$G(z, \alpha) = H_0(z, \alpha) + P_z(z, \alpha)F_0(z, \alpha),$$

can be represented by the contour integral

$$F_0(z, \alpha) = \frac{1}{2\pi i} \int_C \frac{G(\xi, \alpha)}{(\xi - z)P_z(\xi, \alpha)} \, d\xi,$$

where C is any closed contour enclosing z and all the zeros of P_z.

17. Consider the mapping $Z = z - t \cosh z$, t real, and write $Z = X + iY$ and $z = x + iy$. Show that X is odd in x and even in y, and that Y is odd in y and even in x. Similarly, if $Z = u - A^2/u$, A^2 real, and $u = r + is$, show that X is odd in r and even in s, whereas Y is odd in s and even in r.

18. Consider the mapping $Z = u - A^2/u$, and write $u = \rho e^{i\theta}$ and $Z = X + iY$. Show that (i) for $\rho \neq A$, the circle $u = \rho e^{i\theta}$, $0 \leq \theta \leq 2\pi$, is mapped onto the ellipse $X^2/a^2 + Y^2/b^2 = 1$, where $a = \rho - A^2/\rho$ and $b = \rho + A^2/\rho$; (ii) the image of the circle $u = Ae^{i\theta}$, $0 \leq \theta \leq 2\pi$, is the slit $S = \{(X, Y) : X = 0, Y = 2A \sin \theta, 0 \leq \theta \leq 2\pi\}$, which is a degenerate ellipse; and (iii) both the interior and the exterior of the circle $\rho = A$ in the u-plane are mapped onto $Z \backslash S$ in a one-to-one manner. Is this inconsistent with what has been said in §7 about the mapping $\psi: u \to Z$ defined by $\psi(u) = u - A^2(t)/u$, $A^2(t)$ being given by (7.7)?

19. Use the Dirichlet–Mehler integral (Szegö 1967, §4.8, equation (4.8.6))

$$P_n(\cos \theta) = \frac{2}{\pi} \int_0^\theta \frac{\cos[(n + \frac{1}{2})\varphi]}{(2 \cos \varphi - 2 \cos \theta)^{1/2}} \, d\varphi$$

to show that there exist analytic functions $a_\nu(\theta)$ in the interval $[0, \pi)$ such that

$$P_n(\cos \theta) = \sum_{\nu = 0}^{m - 1} a_\nu(\theta) \frac{J_\nu[(n + \frac{1}{2})\theta]}{(n + \frac{1}{2})^\nu} + O(n^{-m - 1/2})$$

for every fixed $m \geq 1$, where the O-term is uniform with respect to $\theta \in [0, \pi - \varepsilon]$, $\varepsilon > 0$. Calculate the first two coefficients $a_0(\theta)$ and $a_1(\theta)$.

Supplementary Notes

§1. Problems on uniform asymptotic expansions have been discussed
in several survey articles, see, for example, Erdélyi (1970), Jones
(1972), Olver (1975), Wong (1980c), and Temme (1985). As a general
guide in deriving such expansions, one may use the following
steps. (i) Look for points which contribute to the asymptotic
expansion of the integral; these are the so-called critical points of
van der Corput (1948), and they include the endpoints of the
interval of integration, singularities of the integrand and the
saddle points of the phase function. (ii) Examine the possibility of
coalescence of these critical points, and then reduce the integral
to a canonical form by using an appropriate transformation. (iii)
Construct a formal uniform expansion and prove its validity. The
derivations in §§3–7 all follow this pattern.

§2. For a related method of evaluating the integral in Example 1, see
Ott (1943).

§§3 and 4. In constructing successive terms in the asymptotic expansions, we have used exclusively the technique of integration by
parts. There is, however, an alternative approach to this construction; that is, expand the integrand in a suitable series and then
integrate term by term. For this approach, see Chester, Friedman,
and Ursell (1957), Olver (1974a, Chapter 9, §§9, 10 & 12), and Wong
(1973).

§5. The material in this section, and also in §7, is taken from Frenzen
and Wong (1988). The leading terms of the expansions in (5.15) and
(7.17) agree with the two asymptotic formulas given by Erdélyi
(1960) who used the theory of differential equations. Since Laguerre polynomials can be expressed in terms of confluent hypergeometric functions, it is also possible to derive their infinite
asymptotic expansions with error bounds directly from those for
Whittaker functions given by Olver (1974a, p. 412 and pp. 446–447).

§6. The general discussion given at the beginning of this section is
based on Bleistein (1967).

§7. A transformation similar to (7.5) has also been suggested by
Temme (1985), but his argument is only formal. An integration-by-
parts technique similar to that presented in this section has also
been used by him.

§8.   Other uniform treatments of the Legendre functions include Olver
(1974a, Chapter 12, §12) and Ursell (1984).

Exercises.   Exercise 1 is based on the paper by Oberhettinger (1959).
The material in Ex. 2 is taken from Lewis (1967). Exercise 4 is related to
Example 10.1 in Olver (1974a, pp. 346–351). For Exs. 5 and 6, see Soni
(1983). The integral in Ex. 6 was first studied by Erdélyi (1974); see also
Temme (1976). The results of Exs. 7 and 8 can be found in Wong (1980b).
The contents of Exs. 10–12 are taken from Kaminski (1989). For Ex. 13,
see Copson (1965). Exercise 15 is based on §13 of Ursell (1972). The
result in Ex. 16 is taken from Bleistein (1967). The expansion in Ex. 19
is due to Szegö (1932).

VIII

Double Integrals

1. Introduction

In most of this chapter we shall make a systematic study of the asymptotic behavior of double integrals of the form

$$I(\lambda) = \iint_D g(x, y)e^{i\lambda f(x,y)}\, dx\, dy, \qquad (1.1)$$

where D is a bounded domain, $f(x, y)$ is a real-valued function, and λ is a large positive parameter. For simplicity of presentation, we shall assume that both f and g are infinitely differentiable in the closure of D. Integrals of this type arise frequently in problems of the diffraction theory of optics, and the behavior of $I(\lambda)$, as $\lambda \to \infty$, has been studied by a multitude of people using a variety of methods, see, for example, van Kampen (1949), Focke (1954), Chako (1965), and Bleistein and Handelsman (1975b). The method we shall use is based on the paper by Jones and Kline (1958). It consists of a sequence of coordinate changes and an application of the method of resolution of multiple integrals mentioned in Chapter V.

In §2 it will be shown that, as in the one-dimensional case, the main contributions to the asymptotic expansion of $I(\lambda)$ again come only from certain *critical points*. These include (i) *stationary points* of $f(x, y)$, i.e., points in D or on its boundary Γ which satisfy $f_x = f_y = 0$, (ii) points on the boundary at which a level curve of f is tangential to Γ, and (iii) points where Γ has a discontinuously turning tangent. Interior stationary points are treated in §§3 and 4. A degenerate case is considered in §5. In §6 we discuss boundary stationary points. Points of tangential contact and corner points are the subjects in §§7 and 8, respectively. §9 concludes our investigation of the integral (1.1) with a discussion of a curve of stationary points inside the domain D.

The final two sections of this chapter (§§10, 11) contain analogous results for the Laplace-type integral

$$J(\lambda) = \iint\limits_{D} g(x, y)e^{-\lambda f(x,y)} \, dx \, dy. \tag{1.2}$$

Included are the derivation of a two-dimensional analogue of Laplace's approximation for $J(\lambda)$ and a discussion of boundary extrema.

2.   Classification of Critical Points

To indicate how the critical points are determined, we shall apply the divergence theorem to the integral $I(\lambda)$. The resulting expression is the higher dimensional analogue of the integration-by-parts formula

$$\int_a^b ge^{i\lambda f} \, dx = \frac{1}{i\lambda}\left(\frac{g}{f'}\right)e^{i\lambda f}\Big|_a^b + \frac{i}{\lambda}\int_a^b \left(\frac{g}{f'}\right)' e^{i\lambda f} \, dx, \tag{2.1}$$

where $f'(x) \neq 0$ in $[a, b]$. Equation (2.1) is obtained from the product rule $d(uv) = u \, dv + v \, du$ with $u = g/f'$ and $v = e^{i\lambda f}$. We therefore first look for a vector field $\mathbf{u}$ such that

$$\nabla \cdot (\mathbf{u}e^{i\lambda f}) = (\nabla \cdot \mathbf{u})e^{i\lambda f} + i\lambda ge^{i\lambda f}. \tag{2.2}$$

The two-dimensional analogue of the condition "$f'(x) \neq 0$ in $[a, b]$" is the nonvanishing of the gradient $\nabla f(x, y)$ in the closure of D. Under this condition, a simple calculation shows that a suitable choice for $\mathbf{u}$ is given by

$$\mathbf{u} = \frac{\nabla f}{|\nabla f|^2}g. \tag{2.3}$$

Inserting (2.2) in (1.1), we have from the divergence theorem

$$I(\lambda) = \frac{1}{i\lambda} \iint\limits_{D} \nabla \cdot (\mathbf{u}e^{i\lambda f}) \, dx \, dy + \frac{i}{\lambda} \iint\limits_{D} (\nabla \cdot \mathbf{u})e^{i\lambda f} \, dx \, dy$$

$$= -\frac{i}{\lambda} \int_{\Gamma} (\mathbf{u} \cdot \mathbf{n})e^{i\lambda f} \, ds + \frac{i}{\lambda} \iint\limits_{D} (\nabla \cdot \mathbf{u})e^{i\lambda f} \, dx \, dy, \qquad (2.4)$$

where Γ is the positively oriented boundary of D, s is the arc length of Γ, and $\mathbf{n}$ is the unit outward normal to Γ. The last integral in (2.4) is of the same form as the original integral $I(\lambda)$. Thus we may repeat the process by letting $g_1 = \nabla \cdot \mathbf{u}$ and

$$\mathbf{u}_1 = \frac{\nabla f}{|\nabla f|^2} g_1. \qquad (2.5)$$

Replacing g and $\mathbf{u}$ by g_1 and $\mathbf{u}_1$, equation (2.4) becomes

$$\iint\limits_{D} g_1 e^{i\lambda f} \, dx \, dy = -\frac{i}{\lambda} \int_{\Gamma} (\mathbf{u}_1 \cdot \mathbf{n})e^{i\lambda f} \, ds + \frac{i}{\lambda} \iint\limits_{D} (\nabla \cdot \mathbf{u}_1)e^{i\lambda f} \, dx \, dy. \quad (2.6)$$

Coupling (2.4) and (2.6), we obtain

$$I(\lambda) = -\frac{i}{\lambda} \int_{\Gamma} (\mathbf{u} \cdot \mathbf{n})e^{i\lambda f} \, ds - \left(\frac{i}{\lambda}\right)^2 \int_{\Gamma} (\mathbf{u}_1 \cdot \mathbf{n})e^{i\lambda f} \, ds + \left(\frac{i}{\lambda}\right)^2 \iint\limits_{D} g_2 e^{i\lambda f} \, dx \, dy,$$

where $g_2 = \nabla \cdot \mathbf{u}_1$. Repeating this procedure n times gives

$$I(\lambda) = -\sum_{s=0}^{n-1} \left(\frac{i}{\lambda}\right)^{s+1} \int_{\Gamma} (\mathbf{u}_s \cdot \mathbf{n})e^{i\lambda f} \, ds + \left(\frac{i}{\lambda}\right)^n \iint\limits_{D} g_n e^{i\lambda f} \, dx \, dy, \quad (2.7)$$

where $\mathbf{u}_0 = \mathbf{u}$ and

$$g_{s+1} = \nabla \cdot \mathbf{u}_s, \qquad \mathbf{u}_{s+1} = \frac{\nabla f}{|\nabla f|^2} g_{s+1}, \qquad (2.8)$$

for $s = 0, 1, 2, \ldots$. Since the double integral on the right of (2.7) is $O(1)$ as $\lambda \to \infty$, we have

$$I(\lambda) = -\sum_{s=0}^{n-1} \left(\frac{i}{\lambda}\right)^{s+1} \int_{\Gamma} (\mathbf{u}_s \cdot \mathbf{n})e^{i\lambda f} \, ds + O(\lambda^{-n}). \qquad (2.9)$$

An immediate consequence of (2.9) is the following result.

Lemma 1. *If* $\nabla f \neq 0$ *in* $\bar{D}$, *and if* g *vanishes* C^∞-*smoothly on the boundary of* D, *i.e.*, g *together with all its derivatives vanish on* Γ, *then*

$$I(\lambda) = O(\lambda^{-N}), \qquad as \ \lambda \to +\infty, \qquad (2.10)$$

for all $N \geq 1$.

From this lemma it is clear that a contribution to the asymptotic expansion of $I(\lambda)$ as $\lambda \to \infty$ may come from points where the phase function f is stationary. Often they are called *critical points of the first kind*.

Critical points of other kinds are on the boundary of D. To see this, we note that each integral on the right-hand side of (2.9) is a line integral. Hence, representing Γ parametrically, they are just one-dimensional integrals, which we have considered in Chapter II, Section 3. From this we know that contributions to the asymptotic expansion of these integrals come only from (i) points on the boundary at which $\nabla f \neq (0, 0)$, but at which f, considered along Γ as a function of a single variable, has a stationary point, and (ii) points on the boundary at which some parametrization of Γ has a discontinuous derivative of some order. The former are known as *critical points of the second kind* and the latter as *critical points of the third kind*. Critical points of the second kind are also points on the boundary where a level curve $f(x, y) = c$ is tangent to Γ for some c. To prove this, we suppose that Γ is given by $h(x, y) = 0$, where h is a smooth function with $(x(t), y(t))$ as a solution to the equation. We also suppose that $\nabla h \neq (0, 0)$, so that Γ is locally a smooth curve. Stationary points of $f|_\Gamma$, the restriction of f to Γ, are exactly those points where

$$f_x \dot{x}(t) + f_y \dot{y}(t) = 0.$$

Since

$$h_x \dot{x} + h_y \dot{y} = \frac{d}{dt} h(x, y) = 0,$$

a stationary point (x_0, y_0) of $f|_\Gamma$ is one for which

$$\begin{vmatrix} f_x & f_y \\ h_x & h_y \end{vmatrix} = f_x h_y - f_y h_x = 0. \qquad (2.11)$$

Thus it follows that $\nabla f(x_0, y_0)$ and $\nabla h(x_0, y_0)$ are linearly dependent. Geometrically, this means that these two vectors are parallel, which in turn implies that the level curve $f(x, y) = f(x_0, y_0)$ is tangent to the boundary $h(x, y) = 0$ at (x_0, y_0). This completes our discussion on the classification of critical points.

In the following sections, except §9 and a special case in §7, we shall assume that there are only a finite number of critical points in D and on ∂D. Let them be labelled (x_1, y_1), (x_2, y_2), ..., (x_s, y_s). Now take pairwise disjoint neighborhoods Ω_i' of (x_i, y_i), and let Ω_i be a small neighborhood of (x_i, y_i) such that $\bar{\Omega}_i \subset \Omega_i'$; see Figure 8.1. For each i, construct a neutralizer $\psi_i(x, y)$ (Chapter V, Ex. 7) which is 1 on Ω_i and 0 outside Ω_i'. Next write

$$I(\lambda) = \sum_{i=1}^{s} \iint_{D} \psi_i(x, y) g(x, y) e^{i\lambda f(x,y)} \, dx \, dy + E(\lambda),$$

where

$$E(\lambda) = \iint_{D} \left[1 - \sum_{i=1}^{s} \psi_i(x, y) \right] g(x, y) e^{i\lambda f(x,y)} \, dx \, dy.$$

If $\tilde{D}$ denotes the region obtained from D by removing the shaded neighborhoods $\Omega_i \cap D$, $i = 1, 2, \ldots, s$, then $E(\lambda)$ can be expressed as

$$E(\lambda) = \iint_{\tilde{D}} \tilde{g}(x, y) e^{i\lambda f(x,y)} \, dx \, dy,$$

where $\tilde{g}$ is a C^∞-function which vanishes to infinite order at every point on ∂D at which ∂D intersects the boundary of Ω_i for some i. Since $\nabla f \neq 0$ in the closure of $\tilde{D}$, the result in (2.7) applies with g and D

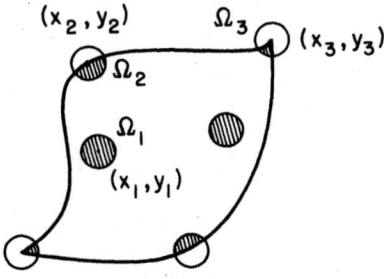

Fig. 8.1

replaced by $\tilde{g}$ and $\tilde{D}$. By using integration by parts, it can be shown that each line integral in (2.7) is $O(\lambda^{-n})$ for all $n \geq 1$. This is due to the fact that $\tilde{g}$ vanishes identically at every point where ∂D intersects $\partial \Omega_i$ for some i. Therefore, $E(\lambda) = O(\lambda^{-n})$ and

$$I(\lambda) = \sum_{i=1}^{s} \iint_D \psi_i(x, y) g(x, y) e^{i\lambda f(x,y)} \, dx \, dy + O(\lambda^{-n}) \qquad (2.12)$$

for all $n \geq 1$.

We have thus demonstrated that it suffices to consider the canonical situation in which the integral $I(\lambda)$ in (1.1) has only one critical point in $\overline{D}$. Without loss of generality, we may suppose that the critical point is located at $(0, 0)$ and that the support of $g(x, y)$ can be as small as necessary.

3.   Local Extrema

Let $(0, 0)$ be a stationary point of $f(x, y)$. Then the initial terms in the Maclaurin expansion of $f(x, y)$ have the form

$$f(x, y) = f_{00} + f_{20} x^2 + f_{11} xy + f_{02} y^2 + \cdots .$$

Since the constant term f_{00} merely contributes a factor $\exp(i\lambda f_{00})$ to $I(\lambda)$, we can take $f_{00} = 0$ without loss of generality. Also, the cross-product term xy can always be eliminated by using a linear transformation. Indeed, the change of variables

$$\xi = x + \frac{b}{a} y \qquad \text{and} \qquad \eta = y \qquad (3.1)$$

gives

$$ax^2 + 2bxy + cy^2 = \lambda \xi^2 + \mu \eta^2, \qquad (3.2)$$

where $\lambda = a$ and $\mu = (ac - b^2)/a$. (The Jacobian of this transformation is 1.) Thus we may assume, without loss of generality, that in a neighborhood of the stationary point $(0, 0)$, the Maclaurin expansion of f takes the form

$$f(x, y) = f_{20} x^2 + f_{02} y^2 + \cdots . \qquad (3.3)$$

The theory to be presented, however, applies for expansions of the form (3.3) that may be asymptotic rather than convergent power series. This remark applies equally well to other Maclaurin series in this and subsequent sections.

Equation (3.3) can be written as

$$f(x, y) = f_{20}x^2[1 + P(x, y)] + f_{02}y^2[1 + Q(x, y)], \qquad (3.4)$$

where P and Q are power series in x and y satisfying $P(0, 0) = Q(0, 0) = 0$. In the integral (1.1), we shall make the change of variables

$$u = x[1 + P(x, y)]^{1/2} \qquad \text{and} \qquad v = y[1 + Q(x, y)]^{1/2} \qquad (3.5)$$

so that $f(x, y) = f_{20}u^2 + f_{02}v^2$. Let D' denote the image of D under this change of variables. From (3.5), it is clear that

$$\left.\frac{\partial(x, y)}{\partial(u, v)}\right|_{(0,0)} = 1.$$

Hence, by shrinking the support of g if necessary, we may assume that $\partial(x, y)/\partial(u, v)$ is positive. Put

$$G(u, v) = g(x, y)\frac{\partial(x, y)}{\partial(u, v)} \qquad \text{and} \qquad F(u, v) = f_{20}u^2 + f_{02}v^2. \quad (3.6)$$

The double integral $I(\lambda)$ then becomes

$$I(\lambda) = \iint\limits_{D'} G(u, v)e^{i\lambda F(u, v)}\, du\, dv. \qquad (3.7)$$

The new amplitude function $G(u, v)$ has the Maclaurin series

$$G(u, v) = \sum G_{ij}u^i v^j, \qquad G_{00} = g_{00}, \qquad (3.8)$$

which is obtained by inverting the variables in (3.5). The coefficients G_{ij} in (3.8) can be expressed in terms of the derivatives of f and g at $(0, 0)$. Let m and M denote the infimum and supremum of F in D' (or equivalently, of f in D), respectively. By the method of resolution of multiple integrals (Theorem 9, Chapter V), the double integral in (3.7) can be written as

$$I(\lambda) = \int_m^M h(t)e^{i\lambda t}\, dt, \qquad (3.9)$$

where

$$h(t) = \int_{F(u, v) = t} \frac{G(u, v)}{\sqrt{F_u^2 + F_v^2}}\, d\sigma, \qquad (3.10)$$

σ being the arc length of the curve $F(u, v) = t$.

If f_{20} and f_{02} are both positive, then $(0, 0)$ is a local minimum and $m = 0$ in (3.9). Since the support of $g(x, y)$ can be made arbitrarily small, so can the support of $G(u, v)$. Hence $h(t)$ and all its derivatives are zero at the upper limit $t = M$. From the one-dimensional theory, it follows that the asymptotic behavior of the Fourier integral (3.9) is completely determined by the behavior of the integrand in a neighborhood of $t = 0$. We shall use (3.10) to calculate $h(t)$ in D'; the level curves $F(u, v) = t$ are ellipses (Figure 8.2). Let

$$u = \left(\frac{\xi}{f_{20}}\right)^{1/2} \cos \eta \qquad \text{and} \qquad v = \left(\frac{\xi}{f_{02}}\right)^{1/2} \sin \eta. \qquad (3.11)$$

Clearly $\xi = f_{20} u^2 + f_{02} v^2 = F(u, v)$ and

$$\frac{1}{|\nabla F|} \, d\sigma = \frac{1}{2\sqrt{f_{20} f_{02}}} \, d\eta.$$

Thus from (3.10) we have

$$h(t) = \int_{\xi = t} \Phi(\xi, \eta) \, d\eta = \int_0^{2\pi} \Phi(t, \eta) \, d\eta, \qquad (3.12)$$

where

$$\Phi(\xi, \eta) = \frac{1}{2\sqrt{f_{20} f_{02}}} \, G(u, v).$$

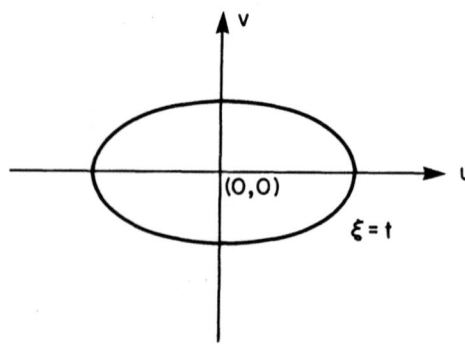

Fig. 8.2

The second equality in (3.12) holds when t is sufficiently small. Put

$$G_{ij}^* = \frac{G_{ij}}{f_{20}^{i/2} f_{02}^{j/2}}.$$

Then $\Phi(\xi, \eta)$ has the asymptotic representation

$$\Phi(\xi, \eta) \sim \frac{1}{2\sqrt{f_{20}f_{02}}} \sum G_{ij}^* \xi^{(i+j)/2} \cos^i \eta \sin^j \eta, \tag{3.13}$$

as $\xi \to 0^+$. Since

$$\int_0^{2\pi} \cos^i \eta \sin^j \eta \, d\eta = 0 \qquad \text{for} \quad i, j = 1, 3, 5, \ldots$$

and

$$\int_0^{2\pi} \cos^{2m} \eta \sin^{2n} \eta \, d\eta = 4 \int_0^{\pi/2} \cos^{2m} \eta \sin^{2n} \eta \, d\eta = 2 \frac{\Gamma(m + \frac{1}{2})\Gamma(n + \frac{1}{2})}{(m + n)!},$$

we find, on inserting (3.13) in (3.12),

$$h(t) \sim \frac{1}{\sqrt{f_{20}f_{02}}} \sum \frac{\Gamma(m + \frac{1}{2})\Gamma(n + \frac{1}{2})}{(m + n)!} G_{2m,2n}^* t^{m+n}, \tag{3.14}$$

as $t \to 0^+$. The final result

$$I(\lambda) \sim \frac{1}{\sqrt{f_{20}f_{02}}} \sum G_{2m,2n}^* e^{i\pi(m+n+1)/2} \frac{\Gamma(m + \frac{1}{2})\Gamma(n + \frac{1}{2})}{\lambda^{m+n+1}}, \tag{3.15}$$

as $\lambda \to \infty$, now follows from termwise integration in the sense of Abel summability (Chapter IV); cf. Ex. 29, Chapter V.

Taking into account the elimination of the cross product xy (cf. (3.2)) and the constant term $f(0, 0)$ in the Maclaurin expansion of $f(x, y)$, the leading term in (3.15) gives

$$I(\lambda) \sim \frac{2\pi i}{\lambda} g(0, 0)[\det f''(0, 0)]^{-1/2} \exp[i\lambda f(0, 0)], \tag{3.16}$$

where

$$\det f''(0, 0) = f_{xx}(0, 0)f_{yy}(0, 0) - f_{xy}^2(0, 0). \tag{3.17}$$

Since f_{20} and f_{02} are assumed to be positive in this case, (3.16) holds under the condition that $f_{xx}(0, 0) > 0$ and $\det f''(0, 0) > 0$.

If f_{20} and f_{02} are both negative, then $(0,0)$ is a local maximum and $M = 0$ in (3.9). The function $h(t)$ in (3.10) now vanishes to infinite order at the lower limit $t = m$. The result (3.14) however remains the same, except that "$t \to 0^+$" is replaced by "$t \to 0^-$"; and, instead of (3.15), the asymptotic expansion of $I(\lambda)$, as $\lambda \to +\infty$, is given by

$$I(\lambda) \sim \frac{1}{\sqrt{f_{20}f_{02}}} \sum G^*_{2m,2n} e^{i\pi(m+n-1)/2} \frac{\Gamma(m+\frac{1}{2})\Gamma(n+\frac{1}{2})}{\lambda^{m+n+1}}. \qquad (3.18)$$

4.   Saddle Points

If, in (3.3), f_{20} and f_{02} have opposite signs, say $f_{20} > 0$ and $f_{02} < 0$, then $(0,0)$ is a saddle point of $f(x,y)$. In this case, the one-dimensional Fourier integral representation (3.9)–(3.10) is still valid, except that the level curves

$$F(u,v) = f_{20}u^2 + f_{02}v^2 = t$$

are now hyperbolas. We shall evaluate the line integral $h(t)$ along the hyperbolas which intersect the domain D' and are bounded between $u = d$ and $u = -d$, d being a small positive number; see Figure 8.3. Note that these paths of integration are symmetric with respect to both u- and v-axes. Thus in the expansion (3.8) we need retain only those

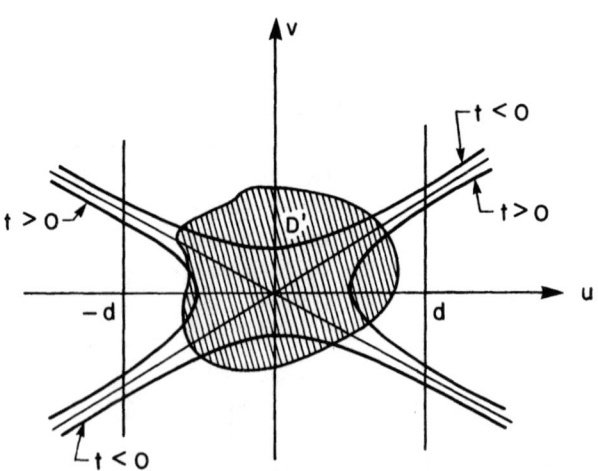

Fig. 8.3

terms that contain even powers of both u and v. Moreover, for these terms the integral is equal to four times the corresponding integral taken along the part of the hyperbolas that lies in the first quadrant.

Now, in place of (3.11), we let

$$u = \left(\frac{\xi}{f_{20}}\right)^{1/2} \cosh \eta, \qquad v = \left(\frac{\xi}{-f_{02}}\right)^{1/2} \sinh \eta, \qquad (4.1)$$

for $\xi > 0$. Clearly, we again have $\xi = f_{20}u^2 + f_{02}v^2 = F(u, v)$ and

$$\frac{1}{|\nabla F|} d\sigma = \frac{1}{2\sqrt{|f_{20}f_{02}|}} d\eta.$$

Thus

$$h(t) = \int_{\xi = t} \Phi(\xi, \eta) \, d\eta, \qquad (4.2)$$

where

$$\Phi(\xi, \eta) = \frac{1}{2\sqrt{|f_{20}f_{02}|}} G(u, v).$$

If we define

$$G_{ij}^* \equiv \frac{G_{ij}}{f_{20}^{i/2}(-f_{02})^{j/2}},$$

then as $\xi \to 0^+$,

$$\Phi(\xi, \eta) \sim \frac{1}{2\sqrt{|f_{20}f_{02}|}} \sum G_{ij}^* \xi^{(i+j)/2} \cosh^i \eta \sinh^j \eta. \qquad (4.3)$$

In view of the earlier remark on the symmetry of the paths of integration, we obtain, upon inserting (4.3) in (4.2),

$$h(t) \sim \frac{2}{\sqrt{|f_{20}f_{02}|}} \sum G_{2m,2n}^* t^{m+n} \int_0^{\eta_1(t)} \cosh^{2m} \eta \sinh^{2n} \eta \, d\eta, \qquad (4.4)$$

as $t \to 0^+$, where $\eta_1(t) = \cosh^{-1}[d(f_{20}/t)^{1/2}]$.

Now we recall from the one-dimensional theory that the contribution to the asymptotic expansion of the Fourier integral $I(\lambda)$ in (3.9) comes from the endpoints $t = M$ and $t = m$, and also from points where a derivative of $h(t)$ is discontinuous. Since $g(x, y)$ vanishes

C^∞-smoothly outside a small neighborhood of the origin, so does $G(u, v)$. Hence there is no endpoint contribution to $I(\lambda)$. The integrals

$$I_{mn} = t^{m+n} \int_0^{\eta_1(t)} \cosh^{2m} \eta \sinh^{2n} \eta \, d\eta$$

in (4.4) can be evaluated by elementary methods. By repeated application of formula (2.413) in Gradshteyn and Ryzhik (1980, p. 94), the exponents $2m$ and $2n$ can be successively reduced to 0. The integrated terms are C^∞-functions of t near the origin, and hence do not contribute to the asymptotic expansion of $I(\lambda)$. Using $\sigma(t)$ as a generic symbol for such a function, we have

$$I_{mn} = \frac{(-1)^n}{2^{2m}} \frac{1 \cdot 3 \cdot 5 \cdots (2n-1)}{(2n+2m)(2n+2m-2)\cdots(2m+2)} \binom{2m}{m} t^{m+n} \eta_1(t) + \sigma(t),$$

σ being dependent on m and n. Since

$$\eta_1(t) = \log(d\sqrt{f_{20}} + \sqrt{d^2 f_{20} - t}) - \tfrac{1}{2} \log t$$

and the first logarithmic term on the right-hand side is again C^∞ near $t = 0$, I_{mn} can further be written as

$$I_{mn} = C_{mn} t^{m+n} \log t + \sigma(t), \tag{4.5}$$

where

$$C_{mn} = \frac{(-1)^{n+1}}{2^{2m+1}} \binom{2m}{m} \frac{1 \cdot 3 \cdot 5 \cdots (2n-1)}{(2n+2m)(2n+2m-2)\cdots(2m+2)}.$$

Inserting (4.5) in (4.4) and disregarding terms that do not contribute to the asymptotic expansion of $I(\lambda)$, we obtain

$$h(t) \sim \frac{2}{\sqrt{|f_{20} f_{02}|}} \sum c_\nu^{(+)} t^\nu \log t, \qquad t \to 0^+, \tag{4.6}$$

where $c_\nu^{(+)} = \sum G_{2m,2n}^* C_{mn}$, the summation being over all m and n satisfying $m + n = \nu$.

For $\xi < 0$, the argument proceeds in a similar manner, except that here we set

$$u = \left(-\frac{\xi}{f_{20}}\right)^{1/2} \sinh \eta, \qquad v = \left(\frac{\xi}{f_{02}}\right)^{1/2} \cosh \eta.$$

The end result is given by

$$h(t) \sim \frac{2}{\sqrt{|f_{20}f_{02}|}} \sum c_\nu^{(-)} t^\nu \log(-t), \qquad t \to 0^-, \tag{4.7}$$

where the coefficients $c_\nu^{(-)}$ satisfy $c_\nu^{(-)} = c_\nu^{(+)}$ for all $\nu \geq 0$.

The summability method for one-dimensional Fourier integrals (Ex. 11, Chapter IV) then gives

$$I(\lambda) \sim \frac{2\pi i}{\sqrt{|f_{20}f_{02}|}} \sum c_\nu^{(+)} e^{i(\nu+1)\pi/2} \frac{\nu!}{\lambda^{\nu+1}}. \tag{4.8}$$

Note that the logarithmic terms are all absent in this expansion, in view of the fact that $c_\nu^{(+)} = c_\nu^{(-)}$ for all $\nu \geq 0$. Furthermore, the leading term

$$I(\lambda) \sim \frac{g_{00}}{\sqrt{|f_{20}f_{02}|}} \frac{\pi}{\lambda} \tag{4.9}$$

differs from the corresponding term in (3.15) only by a factor i.

5. A Degenerate Case

A stationary point (x_0, y_0) of $f(x, y)$ is said to be *degenerate* if the Hessian of f,

$$\det f''(x, y) = f_{xx}(x, y) f_{yy}(x, y) - f_{xy}^2(x, y), \tag{5.1}$$

vanishes there, cf. (3.17). In terms of the expansion (3.3), this would mean either f_{20} or f_{02} is zero. We shall assume that $f_{20} \neq 0$ and $f_{02} = 0$. Thus (3.3) becomes

$$f(x, y) = f_{20} x^2 + f_{30} x^3 + f_{21} x^2 y + f_{03} y^3 + \cdots. \tag{5.2}$$

Here we have taken f_{12} to be zero. If it is not, then the linear change of variables

$$x = \bar{x}, \qquad y = \bar{y} - \frac{f_{12}}{3f_{03}} \bar{x},$$

can always be used to eliminate this term. (The Jacobian of this transformation is 1.) We shall suppose that $f_{03} \neq 0$. Then, suggested by (3.4), we write

$$f(x, y) = f_{20} x^2 [1 + P(x, y)] + f_{03} y^3 [1 + Q(x, y)] \tag{5.3}$$

with $P(x, y)$ and $Q(x, y)$ being power series in x and y satisfying $P(0, 0) = Q(0, 0) = 0$. Furthermore, we set

$$u = x[1 + P(x, y)]^{1/2}, \qquad v = y[1 + Q(x, y)]^{1/3}$$

with sgn $u =$ sgn x. Clearly,

$$\left.\frac{\partial(x, y)}{\partial(u, v)}\right|_{(0,0)} = 1$$

and

$$f(x, y) = f_{20}u^2 + f_{03}v^3. \tag{5.4}$$

As in the case of local extrema, the double integral $I(\lambda)$ is first expressed in terms of the variables u and v, and then written as a one-dimensional Fourier integral. We again let m and M denote the infimum and supremum of $f(x, y)$ in D, and put $G(u, v) = g(x, y)|\partial(x, y)/\partial(u, v)|$ and $F(u, v) = f_{20}u^2 + f_{03}v^3$. The integral $I(\lambda)$ then becomes

$$I(\lambda) = \int_m^M h(t)e^{i\lambda t}\, dt, \tag{5.5}$$

where

$$h(t) = \int_{F(u,v)\,=\,t} \frac{G(u, v)}{|\nabla F|}\, d\sigma.$$

We shall evaluate $h(t)$ along the curves $F(u, v) = f_{20}u^2 + f_{03}v^3 = t$, which are bounded between the vertical lines $u = -d$ and $u = d$ (Figure 8.4).

First we assume that $f_{20} > 0$ and that $f_{03} > 0$. Set

$$\xi = f_{20}u^2 + f_{03}v^3, \qquad \eta = u, \tag{5.6}$$

and define

$$\Phi(\xi, \eta) = G(u, v)\left|\frac{\partial(u, v)}{\partial(\xi, \eta)}\right|.$$

Again as in the case of local extrema, the line integral $h(t)$ is reduced to

$$h(t) = \int_{\xi\,=\,t} \Phi(\xi, \eta)\, d\eta.$$

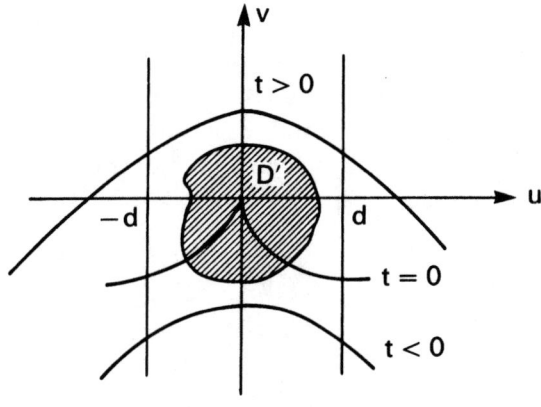

Fig. 8.4

Since the path of integration is symmetric with respect to the v-axis (recall that $G(u, v)$ vanishes outside D'), we need retain only terms involving even powers of u in the Maclaurin expansion of $G(u, v)$ given in (3.8), and multiply the result by 2; see a similar argument in §4. Clearly

$$\frac{\partial(u, v)}{\partial(\xi, \eta)} = -\frac{1}{3f_{03}^{1/3}}(\xi - f_{20}\eta^2)^{-2/3}$$

and hence

$$\Phi(\xi, \eta) = \sum \Phi_{ij}\eta^i(\xi - f_{20}\eta^2)^{(j-2)/3}, \tag{5.7}$$

where

$$\Phi_{ij} = \frac{G_{ij}}{3f_{03}^{(j+1)/3}}. \tag{5.8}$$

The expansion corresponding to (4.4) is

$$h(t) = 2\sum \Phi_{2m,n}\int_0^d \eta^{2m}(t - f_{20}\eta^2)^{(n-2)/3}\, d\eta. \tag{5.9}$$

The integrals

$$J_{m,n} = \int_0^d \eta^{2m}(t - f_{20}\eta^2)^{(n-2)/3}\, d\eta \tag{5.10}$$

can be evaluated as follows. First, the exponent $2m$ can be successively brought down to 0 by using formula (1) in Gradshteyn and Ryzhik

(1980, p. 70, §2.18). Next, consider separately the three cases $n = 3s$, $n = 3s + 1$, and $n = 3s + 2$. In each case, the exponent s can be brought down to zero by using formula (2) in the above mentioned reference. The integrated terms are C^∞-functions of t near $t = 0$ and, therefore, do not contribute to the asymptotic expansion of the Fourier integral $I(\lambda)$ in (5.5). Using $\sigma(t)$ again as a generic symbol for such a function, we have

$$J_{m,3s} = \frac{\Gamma(m + \frac{1}{2})\Gamma(s + \frac{1}{3})\Gamma(\frac{5}{6})t^{m+s}}{\sqrt{\pi}\,\Gamma(m + s + \frac{5}{6})\Gamma(\frac{1}{3})f_{20}^m} \int_0^d (t - f_{20}\eta^2)^{-2/3}\, d\eta + \sigma(t)$$

$$J_{m,3s+1} = \frac{\Gamma(m + \frac{1}{2})\Gamma(s + \frac{2}{3})\Gamma(\frac{7}{6})t^{m+s}}{\sqrt{\pi}\,\Gamma(m + s + \frac{7}{6})\Gamma(\frac{2}{3})f_{20}^m} \int_0^d (t - f_{20}\eta^2)^{-1/3}\, d\eta + \sigma(t)$$

$$J_{2m,3s+2} = 0.$$

Note that d is independent of t. Hence, we may write

$$\int_0^d (t - f_{20}\eta^2)^{-2/3}\, d\eta = \int_0^\infty (t - f_{20}\eta^2)^{-2/3}\, d\eta + \sigma(t).$$

If $t > 0$ then we break the interval of integration on the right-hand side at $\eta = \eta^* \equiv \sqrt{t/f_{20}}$. On the finite interval $(0, \eta^*)$, we make the substitution $\eta = \sqrt{t/f_{20}}\,\sin\theta$, and on the infinite interval (η^*, ∞), we put $\eta = \sqrt{t/f_{20}}\,\sec\theta$. The resulting integrals can be expressed in terms of the Gamma function, and we obtain

$$\int_0^\infty (t - f_{20}\eta^2)^{-2/3}\, d\eta = \frac{3\sqrt{\pi}\,\Gamma(\frac{1}{3})t^{-1/6}}{2\Gamma(\frac{5}{6})f_{20}^{1/2}}.$$

If $t < 0$, then the substitution $\eta = \sqrt{-t/f_{20}}\,\tan\theta$ gives

$$\int_0^\infty (t - f_{20}\eta^2)^{-2/3}\, d\eta = \frac{\sqrt{3\pi}\,\Gamma(\frac{1}{3})(-t)^{-1/6}}{2\Gamma(\frac{5}{6})f_{20}^{1/2}}.$$

Similarly, we have for $t < 0$

$$\int_0^d (t - f_{20}\eta^2)^{-1/3}\, d\eta = \int_0^\infty [(t - f_{20}\eta^2)^{-1/3} - (-f_{20}\eta^2)^{-1/3}]\, d\eta + \sigma(t)$$

$$= \frac{3\sqrt{\pi}\,\Gamma(\frac{5}{6})}{\Gamma(\frac{1}{3})f_{20}^{1/2}}(-t)^{1/6} + \sigma(t),$$

by using an integral formula in Gradshteyn and Ryzhik (1980, p. 293, §3.245). (The condition "$\mu > v > 0$" there can be weakened to "$\mu > v > -1$" by analytic continuation.) For $t > 0$, it is found by again breaking the interval of integration at $\eta = \sqrt{t/f_{20}}$ that

$$\int_0^\infty (t - f_{20}\eta^2)^{-1/3} \, d\eta = \frac{\sqrt{3\pi}\,\Gamma(\frac{5}{6})}{\Gamma(\frac{4}{3})f_{20}^{1/2}} \, t^{1/6}.$$

Now put $C_{mn}^{(+)} = 0$ for $n = 3s + 2$,

$$C_{mn}^{(+)} = \frac{3\Gamma\left(m + \frac{1}{2}\right)\Gamma\left(\frac{n+1}{3}\right)}{2\Gamma\left(m + \frac{n}{3} + \frac{5}{6}\right)f_{20}^{m+1/2}} \qquad \text{for } n \neq 3s + 2$$

and

$$C_{mn}^{(-)} = \frac{1}{\sqrt{3}}(-1)^{m + [n/3]}C_{mn}^{(+)},$$

where $[n/3]$ denotes the largest integer $\leq n/3$. Then

$$J_{m,n}(t) = \begin{cases} C_{mn}^{(+)}t^{m + n/3 - 1/6} + \sigma(t), & t > 0, \\ C_{mn}^{(-)}(-t)^{m + n/3 - 1/6} + \sigma(t), & t < 0. \end{cases}$$

Thus, disregarding terms that are C^∞ near the origin, we have

$$h(t) \sim 2 \sum \Phi_{2m,n} C_{mn}^{(\pm)} |t|^{m + n/3 - 1/6}.$$

Consequently it follows from (5.5) that

$$I(\lambda) \sim \frac{a_0}{\lambda^{5/6}} + \frac{a_1}{\lambda^{7/6}} + \frac{a_2}{\lambda^{11/6}} + \cdots \qquad (5.11)$$

as $\lambda \to +\infty$, where

$$a_0 = \frac{1}{\sqrt{3}} \frac{\Gamma(\frac{1}{2})\Gamma(\frac{1}{3})}{f_{20}^{1/2}f_{03}^{1/3}} g_{00} e^{i\pi/4}$$

and

$$a_1 = \frac{1}{\sqrt{3}} \frac{\Gamma(\frac{1}{2})\Gamma(\frac{2}{3})}{f_{03}^{4/3}f_{20}^{1/2}} G_{01} e^{3\pi i/4}.$$

The arguments are similar for the other three cases: $f_{20} < 0$ and $f_{03} > 0$, $f_{20} > 0$ and $f_{03} < 0$, $f_{02} < 0$ and $f_{03} < 0$, and only the coefficients in the expansion (5.11) are slightly altered; see Ex. 6.

6.   Boundary Stationary Points

We now consider the case where the stationary point $(0, 0)$ is on the boundary of D. We shall assume that the boundary of D is an infinitely smooth curve in the neighborhood of $(0, 0)$ such that if $h(x, y) = 0$ is the equation of the curve then h_x and h_y are not both zero at any point there. Furthermore, we will again assume that the Maclaurin expansion of $f(x, y)$ is given by (3.3)

$$f(x, y) = f_{20}x^2 + f_{02}y^2 + \text{cubic terms} + \cdots. \tag{6.1}$$

Our first objective is to introduce a new system of coordinates (X, Y) in which $h = 0$ becomes $X = 0$. Write

$$h(x, y) = h_{10}x + h_{01}y + h_{20}x^2 + h_{11}xy + h_{02}y^2 + \cdots. \tag{6.2}$$

Without loss of generality we may assume that the sign of h_{10} is the sign of $h(x, y)$ in D so that the positive x-axis points into D. Otherwise, we can simply change the variable x to $-x$. Let $d = h_{10}^2 f_{02} + h_{01}^2 f_{20}$, and assume that $d \neq 0$ (see Ex. 7). Set

$$\begin{pmatrix} x \\ y \end{pmatrix} = \frac{1}{d} \begin{pmatrix} h_{10}f_{02} & -h_{01} \\ h_{01}f_{20} & h_{10} \end{pmatrix} \begin{pmatrix} \bar{x} \\ \bar{y} \end{pmatrix}. \tag{6.3}$$

If $\bar{h}(\bar{x}, \bar{y})$ and $\bar{f}(\bar{x}, \bar{y})$ denote respectively the transform of $h(x, y)$ and $f(x, y)$, then

$$\bar{h}(\bar{x}, \bar{y}) = \bar{x} + \bar{h}_{20}\bar{x}^2 + \bar{h}_{11}\bar{x}\bar{y} + \bar{h}_{02}\bar{y}^2 + \cdots,$$
$$\bar{f}(\bar{x}, \bar{y}) = \bar{f}_{20}\bar{x}^2 + \bar{f}_{02}\bar{y}^2 + \cdots.$$

Note that this linear change of variable eliminates the term $h_{01}y$ while retaining the form of $f(x, y)$. Now we define

$$X = \bar{h}(\bar{x}, \bar{y}) \qquad \text{and} \qquad Y = \bar{y}. \tag{6.4}$$

Solving $\bar{x}$ and $\bar{y}$ in terms of X and Y, we obtain

$$\bar{x} = X + b_{20}X^2 + b_{11}XY + b_{02}Y^2 + \cdots,$$
$$\bar{y} = Y,$$

and hence

$$\bar{f}(\bar{x}, \bar{y}) = \bar{f}_{20}X^2 + \bar{f}_{02}Y^2 + \text{higher order terms}.$$

The original boundary $h(x, y) = 0$ is now mapped into $X = 0$, but the form of $f(x, y)$ remains unchanged. Let $F(X, Y)$ denote the transform of $\bar{f}(\bar{x}, \bar{y})$ and put

$$G(X, Y) = \frac{1}{d} g(x, y) \left| \frac{\partial(\bar{x}, \bar{y})}{\partial(X, Y)} \right|.$$

(Note: $\partial(x, y)/\partial(\bar{x}, \bar{y}) = 1/d$.) The double integral (1.1) then becomes

$$I(\lambda) = \iint_{D'} G(X, Y)e^{i\lambda F(X, Y)} \, dX \, dY, \tag{6.5}$$

where D' is the image of D under the above sequence of transformations (Figure 8.5). The analysis of §3 can now be applied with x, y, f, g, and D replaced by X, Y, F, G, and D'. The only essential difference here is that the limits of integration in (3.12) are now $-\pi/2$ and $\pi/2$ instead of 0 and 2π. As a result of this, (3.14) no longer holds and the asymptotic expansion of $h(t)$ takes the form

$$h(t) \sim \sum_{s=0}^{\infty} b_s t^{s/2}, \qquad \text{as } t \to 0, \tag{6.6}$$

where $b_0 = \pi g_{00}/2d\sqrt{\bar{f}_{20}\bar{f}_{02}}$.

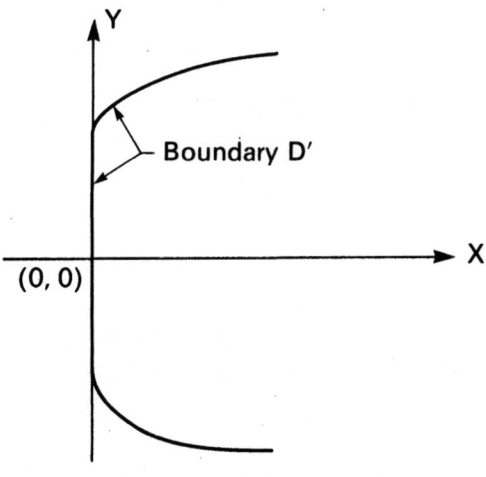

Fig. 8.5

If both $\bar{f}_{20}$ and $\bar{f}_{02}$ are positive then, as in §3, $m = 0$ in (3.9). The asymptotic expansion of $I(\lambda)$ is given by

$$I(\lambda) \sim \sum_{s=0}^{\infty} b_s \exp\left[i\frac{\pi}{2}\left(1 + \frac{s}{2}\right)\right] \frac{\Gamma\left(1 + \frac{s}{2}\right)}{\lambda^{1+s/2}}. \tag{6.7}$$

The first term in (6.7) will yield

$$I(\lambda) \sim \frac{\pi i}{\lambda} g(0, 0)[\det f''(0, 0)]^{-1/2} \exp[i\lambda f(0, 0)], \tag{6.8}$$

which is one-half of the stationary phase approximation (3.16).

If both $\bar{f}_{20}$ and $\bar{f}_{02}$ are negative, then $M = 0$ in (3.9), and (6.6) holds with "$t \to 0^+$" replaced by "$t \to 0^-$". Thus we have

$$I(\lambda) \sim \sum_{s=0}^{\infty} b_s \exp\left[i\frac{\pi}{2}\left(\frac{s}{2} - 1\right)\right] \frac{\Gamma\left(1 + \frac{s}{2}\right)}{\lambda^{1+s/2}}. \tag{6.9}$$

If $\bar{f}_{20}$ and $\bar{f}_{02}$ are of opposite sign, then we can repeat the argument given in §4 for the interior saddle point except that the path of integration is symmetric only with respect to the u-axis and only those terms containing odd powers of v in the expansion (3.8) can be dropped. Hence, instead of (4.4), we have as $t \to 0^+$

$$h(t) \sim \frac{1}{\sqrt{|\bar{f}_{20}\bar{f}_{02}|}} \sum G^*_{m,2n} t^{m/2+n} \int_0^{\eta_1(t)} \cosh^m \eta \sinh^{2n} \eta \, d\eta. \tag{6.10}$$

The integrals in (6.10) have already been considered in §4, and from here on the analysis parallels that given there; see Ex. 8.

7.   Critical Points of the Second Kind

These are points where a level curve of the phase function $f(x, y)$ is tangent to the boundary of the integration domain D. Let $(0, 0)$ be a critical point of the second kind, and suppose that $f_{10} = f_x(0, 0) \neq 0$. Let the boundary curve be given by $h(x, y) = 0$. We shall first show that by a change of coordinates, we may assume that the portion of the boundary curve $h = 0$ which passes through $(0, 0)$ actually lies on the vertical line

$x = 0$, and that the positive x-axis points into D. Furthermore, the y-term is absent in the Maclaurin expansion of $f(x, y)$.

Since the level curve $f(x, y) = 0$ is tangent to the boundary $h(x, y) = 0$ at $(0, 0)$, we have at $(0, 0)$

$$f_x h_y - f_y h_x = 0; \qquad (7.1)$$

cf. (2.11). Let us write

$$h(x, y) = h_{10}x + h_{01}y + h_{20}x^2 + h_{11}xy + \cdots, \qquad (7.2)$$

and make the linear change of variables

$$\begin{pmatrix} x \\ y \end{pmatrix} = \frac{1}{d} \begin{pmatrix} a & -h_{01} \\ c & h_{10} \end{pmatrix} \begin{pmatrix} \bar{x} \\ \bar{y} \end{pmatrix},$$

where $d = ah_{10} + ch_{01}$ and a, c are chosen so that the positive $\bar{x}$-axis points into D; cf. (6.3). Note that the Jacobian of this transformation is 1. If $\bar{f}(\bar{x}, \bar{y})$ and $\bar{h}(\bar{x}, \bar{y})$ denote respectively the transform of $f(x, y)$ and $h(x, y)$, then in view of (7.1)

$$\bar{f}(\bar{x}, \bar{y}) = \bar{f}_{10}\bar{x} + \bar{f}_{20}\bar{x}^2 + \bar{f}_{11}\bar{x}\bar{y} + \cdots$$

and

$$\bar{h}(\bar{x}, \bar{y}) = \bar{h}_{10}\bar{x} + \bar{h}_{20}\bar{x}^2 + \bar{h}_{11}\bar{x}\bar{y} + \cdots.$$

The boundary $\bar{h}(\bar{x}, \bar{y}) = 0$ of D becomes $X = 0$ if we make the further change of variables

$$\bar{h}_{10}X = \bar{h}(\bar{x}, \bar{y}), \qquad Y = \bar{y}.$$

Solving for $\bar{x}$ and $\bar{y}$ in terms of X and Y, we obtain

$$\bar{x} = X + b_{20}X^2 + b_{11}XY + b_{02}Y^2 + \cdots,$$

$\bar{y} = Y$, and hence

$$F(X, Y) = F_{10}X + F_{20}X^2 + F_{11}XY + F_{02}Y^2 + \cdots. \qquad (7.3)$$

Therefore, we may assume without loss of generality that the original phase function has such a Maclaurin series expansion, i.e.,

$$f(x, y) = f_{10}x + f_{20}x^2 + f_{11}xy + f_{02}y^2 + \cdots, \qquad (7.4)$$

and that a part of the boundary of D lies on the y-axis. As in (3.4) and (5.3), we now write

$$f(x, y) = f_{10}x[1 + P(x, y)] + f_{02}y^2[1 + Q(x, y)],$$

where $P(x, y)$ and $Q(x, y)$ are power series in x and y satisfying $P(0, 0) = Q(0, 0) = 0$. Put

$$u = x[1 + P(x, y)] \qquad \text{and} \qquad v = y[1 + Q(x, y)]^{1/2}. \tag{7.5}$$

Clearly

$$\left. \frac{\partial(x, y)}{\partial(u, v)} \right|_{(0,0)} = 1 \qquad \text{and} \qquad f(x, y) = f_{10}u + f_{02}v^2 = F(u, v).$$

(Note that this F differs from that given in (7.3).)

We first suppose that $f_{10} > 0$ and $f_{02} > 0$. The transformation from (x, y) to (ξ, η), corresponding to (3.11) and (4.1), is given by

$$u = \frac{\xi}{f_{10}} \cos^2 \eta, \qquad v = \left(\frac{\xi}{f_{02}} \right)^{1/2} \sin \eta,$$

from which it follows that

$$F(u, v) = f_{10}u + f_{02}v^2 = \xi \tag{7.6}$$

and

$$\frac{\partial(u, v)}{\partial(\xi, \eta)} = \frac{\xi^{1/2}}{f_{10}f_{02}^{1/2}} \cos \eta.$$

Put

$$\Phi(\xi, \eta) = G(u, v) \left| \frac{\partial(u, v)}{\partial(\xi, \eta)} \right|,$$

where $G(u, v) = g(x, y)|\partial(x, y)/\partial(u, v)|$. Expanding $G(u, v)$ as in (3.8), we have

$$\Phi(\xi, \eta) = \sum \Phi_{ij} \xi^{i + (j + 1)/2} \cos^{2i + 1} \eta \sin^j \eta, \tag{7.7}$$

where

$$\Phi_{ij} = \frac{G_{ij}}{f_{10}^{i + 1} f_{02}^{(j + 1)/2}}.$$

The line integral (3.10) in this case is given by

$$h(t) = \int_{F(u,v) = t} \frac{G(u, v)}{|\nabla F|} \, d\sigma = \int_{-\pi/2}^{\pi/2} \Phi(t, \eta) \, d\eta. \tag{7.8}$$

Note that the curves $F = t$ in (7.6) are parabolas (Figure 8.6). For $t < 0$, the curves lie entirely on the left half-plane and hence completely

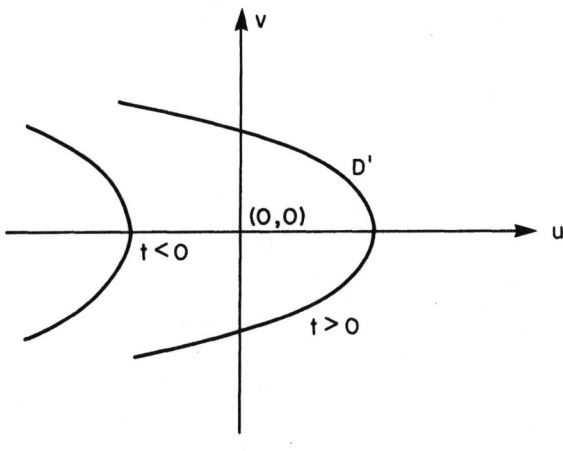

Fig. 8.6

outside the domain D'. Therefore, $h(t) = 0$ for $t < 0$. For $t > 0$, we have from (7.8)

$$h(t) = \sum \Phi_{i,j} t^{i + (j + 1)/2} \int_{-\pi/2}^{\pi/2} \cos^{2i+1} \eta \sin^j \eta \, d\eta.$$

The integrals on the right vanish for odd j, and hence

$$h(t) \sim \sum b_{ij} t^{i + j + 1/2}, \qquad \text{as } t \to 0^+,$$

where

$$b_{ij} = \Phi_{i,2j} \frac{i! \, \Gamma(j + \tfrac{1}{2})}{\Gamma(i + j + \tfrac{3}{2})}.$$

The asymptotic expansion of the double integral $I(\lambda)$ now follows from its Fourier integral representation (3.9) with $m = 0$. The result is

$$I(\lambda) \sim \sum b_{ij} e^{i\pi(i + j + 3/2)/2} \frac{\Gamma(i + j + \tfrac{3}{2})}{\lambda^{i + j + 3/2}}. \tag{7.9}$$

We next suppose that $f_{10} < 0$ and $f_{02} > 0$. In this case, the curves $F(u, v) = t$ in (7.6) are as shown in Figure 8.7. The domain D' still lies in the right half-plane $u \geq 0$. However, we may consider a larger domain D^* which contains $(0, 0)$ in its interior, e.g., the domain bounded by ABCDEA. It is easily seen that in this domain the line integral

$$h(t) = \frac{1}{|f_{10}|} \int_{-v_1(t)}^{v_1(t)} G(u, v) \, dv, \qquad v_1(t) = \sqrt{(t - f_{10}d)/f_{02}}, \tag{7.10}$$

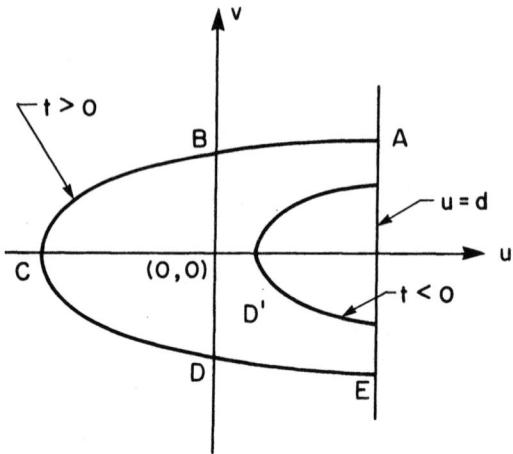

Fig. 8.7

is a C^∞-function of t near $t = 0$; cf. (7.8). Hence the $h(t)$ corresponding to the domain D' is just the difference between a C^∞-function and the value of $h(t)$ for the paths in the left half-plane $u \leq 0$. Note that the domain in the left half-plane is symmetric to the domain in the right half-plane for the previous case. Thus the final result is the same as (7.9).

The arguments for the other two cases ($f_{10} > 0$ and $f_{02} < 0$, $f_{10} < 0$ and $f_{02} < 0$) are similar and need only slight modifications. The final form of the expansions is again given by (7.9).

Finally we consider the case of a curve of critical points of the second kind, that is, the boundary of D coincides with a level curve of $f(x, y)$ over a finite length. In this case, equation (7.1) holds for every point on the curve of coincidence. We make the same transformations as above, first from (x, y) to $(\bar{x}, \bar{y})$ and then to (X, Y). Since $F_x(0, 0) \neq 0$, the level curve $F(X, Y) = 0$ defines X as a function of Y in a neighborhood of $(0, 0)$. Furthermore, since $F(X, Y) = 0$ lies on the Y-axis over an interval of finite length, say $a \leq Y \leq b$, $dX/dY = 0$ for $a \leq Y \leq b$; see Figure 8.8. Thus, by using implicit differentiation, it is easily seen that the partial derivatives F_y, F_{yy}, F_{yyy} all vanish at $(0, 0)$. Hence, instead of (7.3), we have

$$F(X, Y) = F_{10}X + F_{20}X^2 + F_{11}XY + F_{30}X^3 + \cdots . \qquad (7.11)$$

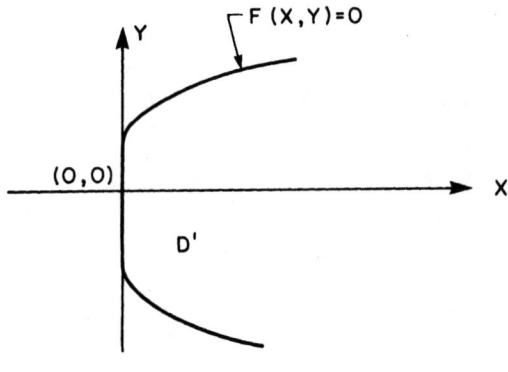

Fig. 8.8

Without loss of generality, we may therefore assume that the original phase function $f(x, y)$ is already in this form, i.e.,

$$f(x, y) = f_{10}x + f_{20}x^2 + f_{11}xy + f_{30}x^3 + \cdots$$
$$= f_{10}x[1 + P(x, y)], \qquad P(0, 0) = 0. \tag{7.12}$$

The variables u and v, corresponding to (7.5), are defined by

$$u = x[1 + P(x, y)], \qquad v = y \tag{7.13}$$

and it is clear that

$$\frac{\partial(x, y)}{\partial(u, v)}\bigg|_{(0,0)} = 1 \qquad \text{and} \qquad f(x, y) = f_{10}u \equiv F(u, v).$$

Suppose that $f_{10} > 0$. Then the level curve $F(u, v) = t$ will lie in the left half-plane $u < 0$ if $t < 0$, while the domain D' is in the right half-plane $u \geq 0$. Thus, the line integral $h(t)$ given in (3.10) is zero for $t < 0$. For $t > 0$, $h(t)$ is given by

$$h(t) = \frac{1}{f_{10}} \int_{F(u, v) = t} G(u, v)\, dv = \frac{1}{f_{10}} \int_{v_1(t)}^{v_2(t)} G\left(\frac{t}{f_{10}}, v\right) dv, \tag{7.14}$$

where $v_1(t)$ and $v_2(t)$ are some C^∞-functions in a neighborhood of 0. (Actually v_1 and v_2 are the least and greatest values of v in all (u, v) belonging to the domain D' for which $f_{10}u = t$.) From this, it follows that $h(t)$ is an infinitely differentiable function at $t = 0$, and that we have

$$h(t) \sim \sum_{s=0}^{\infty} h_s t^s, \qquad \text{as } t \to 0^+.$$

(The coefficients h_s of course depend on the length of the arc on which the boundary of D coincides with the level curve $f(x, y) = 0$.) The asymptotic expansion of $I(\lambda)$ is therefore of the form

$$I(\lambda) \sim \sum_{s=0}^{\infty} h_s e^{i\pi(s+1)/2} \frac{s!}{\lambda^{s+1}}, \qquad \lambda \to +\infty. \tag{7.15}$$

8.   Critical Points of the Third Kind

These are the points where the boundary has two intersecting tangent lines (Figure 8.9), but neither tangent line coincides with that of the level curve $f(x, y) = f(0, 0)$ at $(0, 0)$.

We shall first make a change of coordinates from (x, y) to (X, Y) so that the boundary of D lies along the positive X- and Y-axes.

Let $h(x, y) = 0$ and $k(x, y) = 0$ represent the boundary curves on the two sides of $(0, 0)$. Since the direction of the tangent to the curve jumps at $(0, 0)$, we have

$$h_x k_y - h_y k_x|_{(0,0)} \neq 0.$$

Moreover, since the tangent to the level curve $f(x, y) = 0$ does not coincide with either of the above tangents (Figure 8.9), it also follows that

$$h_x f_y - h_y f_x|_{(0,0)} \neq 0 \qquad \text{and} \qquad k_x f_y - k_y f_x|_{(0,0)} \neq 0. \tag{8.1}$$

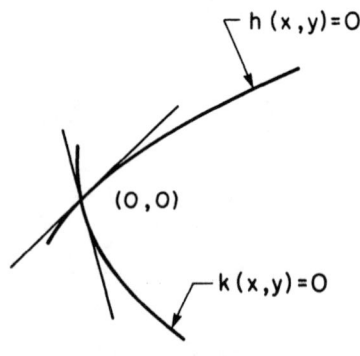

Fig. 8.9

Now consider the linear transformation

$$\begin{pmatrix} x \\ y \end{pmatrix} = \begin{pmatrix} a & b \\ c & d \end{pmatrix} \begin{pmatrix} \bar{x} \\ \bar{y} \end{pmatrix}, \qquad \begin{vmatrix} a & b \\ c & d \end{vmatrix} = 1,$$

where a, b, c, and d are chosen so that $ak_x + ck_y = 0$ and $bh_x + dh_y = 0$, and the positive $\bar{x}$- and $\bar{y}$-axes lie along the tangents to D at $(0, 0)$. If $\bar{h}$ and $\bar{k}$ denote the transforms of h and k, then

$$\bar{h}(\bar{x}, \bar{y}) = \bar{h}_{10}\bar{x} + \bar{h}_{20}\bar{x}^2 + \bar{h}_{11}\bar{x}\bar{y} + \bar{h}_{02}\bar{y}^2 + \cdots$$

$$\bar{k}(\bar{u}, \bar{v}) = \bar{k}_{01}\bar{y} + \bar{k}_{20}\bar{x}^2 + \bar{k}_{11}\bar{x}\bar{y} + \bar{k}_{02}\bar{y}^2 + \cdots.$$

The expansion of $f(x, y)$ becomes

$$\bar{f}(\bar{x}, \bar{y}) = \bar{f}_{10}\bar{x} + \bar{f}_{01}\bar{y} + \bar{f}_{20}\bar{x}^2 + \cdots, \tag{8.2}$$

where $\bar{f}_{10} \neq 0$ and $\bar{f}_{01} \neq 0$ in view of (8.1).

The boundary of D at $(0, 0)$ coincides with the coordinate axes in the XY-plane if we make the further change of variables

$$\bar{h}_{10}X = \bar{h}_{10}\bar{x} + \bar{h}_{20}\bar{x}^2 + \bar{h}_{11}\bar{x}\bar{y} + \bar{h}_{02}\bar{y}^2 + \cdots,$$

$$\bar{k}_{01}Y = \bar{k}_{01}\bar{y} + \bar{k}_{20}\bar{x}^2 + \bar{k}_{11}\bar{x}\bar{y} + \bar{k}_{02}\bar{y}^2 + \cdots.$$

Inverting the two power series and substituting the results in (8.2) gives

$$F(X, Y) = F_{10}X + F_{01}Y + \cdots$$

with $F_{10} \neq 0$ and $F_{01} \neq 0$. Thus we may assume, without loss of generality, that the original phase function $f(x, y)$ has the expansion

$$f(x, y) = f_{10}x + f_{01}y + \cdots, \qquad f_{10} \neq 0, \quad f_{01} \neq 0. \tag{8.3}$$

We shall again write (8.3) in the form

$$f(x, y) = f_{10}x[1 + P(x, y)] + f_{01}y[1 + Q(x, y)],$$

where $P(0, 0) = Q(0, 0) = 0$, and make the substitutions $u = x[1 + P(x, y)]$ and $v = y[1 + Q(x, y)]$. Then $f(x, y) = f_{10}u + f_{01}v$. The final transformation from (u, v) to (ξ, η) is given by

$$u = \frac{\xi}{f_{10}}\cos^2\eta \quad \text{and} \quad v = \frac{\xi}{f_{01}}\sin^2\eta, \tag{8.4}$$

which gives

$$\xi = f_{10}u + f_{01}v \equiv F(u, v). \tag{8.5}$$

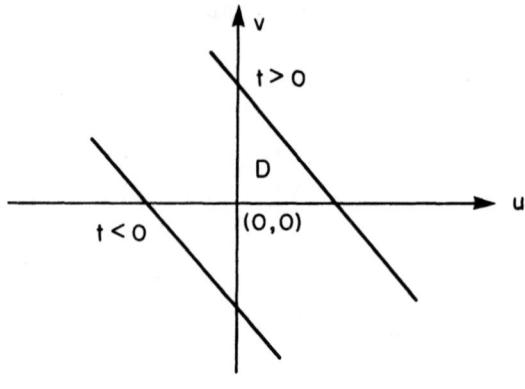

Fig. 8.10

Suppose that $f_{10} > 0$ and $f_{01} > 0$. Then, for $t < 0$, the line $F(u, v) = t$ lies outside of D (Figure 8.10), and hence $h(t) = 0$. For $t > 0$, we use (3.10), (8.4), and (8.5) to obtain

$$h(t) = \int_0^{\pi/2} \Phi(t, \eta) \, d\eta, \qquad (8.6)$$

where

$$\Phi(t, \eta) = \frac{2}{f_{10} f_{01}} \sum \Phi_{ij} t^{i+j+1} \cos^{2i+1} \eta \sin^{2j+1} \eta$$

with $\Phi_{ij} = G_{ij}/f_{10}^i f_{01}^j$ and G_{ij} being given as in (3.8). Thus we have

$$h(t) \sim \frac{1}{f_{10} f_{01}} \sum \Phi_{ij} \frac{i! \, j!}{(i+j+1)!} t^{i+j+1}, \qquad (8.7)$$

as $t \to 0^+$. The contribution from the critical point $(0, 0)$ to the integral is given by

$$I(\lambda) \sim \sum_{m=0}^{\infty} h_m e^{i\pi(m+2)/2} \lambda^{-m-2}, \qquad \lambda \to +\infty, \qquad (8.8)$$

where

$$h_m = \frac{1}{f_{10} f_{01}} \sum_{i+j=m} \Phi_{ij} i! \, j!.$$

If $f_{10} < 0$ and $f_{01} > 0$, then the level curves $F(u, v) = t$ are as shown in Figure 8.11. In this case, we add to the domain D the domain D' in the

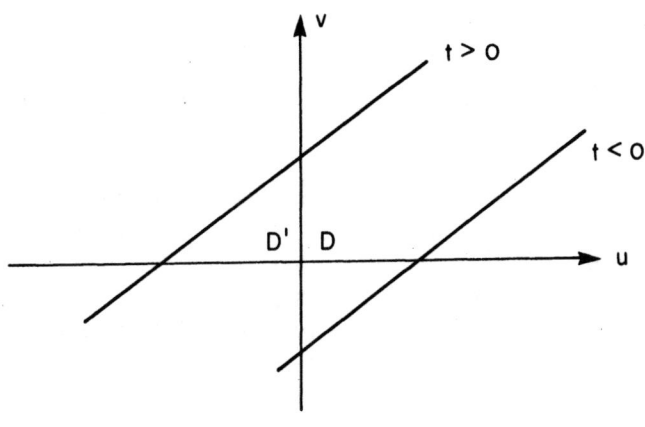

Fig. 8.11

second quadrant (see Figure 8.11) and extend the functions F and G to $D \cup D'$. The behavior of the double integral $I(\lambda)$ over the original domain D is obtained by subtracting from the integral over $D \cup D'$ the same integral over D'. Since the portion of the boundary of $D \cup D'$ containing the origin is the u-axis, which is a smooth curve, and since the level curve $F(u, v) = t$ cuts the boundary (and does not touch it), $(0, 0)$ is not a critical point for the double integral over $D \cup D'$ and hence does not contribute to the asymptotic expansion of the integral. The treatment of $I(\lambda)$ over D' is similar to that given for the case $f_{10} > 0$ and $f_{01} > 0$. Instead of (8.6), here we have

$$h(t) = \int_{\pi/2}^{\pi} \Phi(t, \eta) \, d\eta,$$

$h(t)$ being taken over D'. The expansion (8.7) is therefore replaced by

$$h(t) \sim -\frac{1}{f_{10} f_{01}} \sum \Phi_{ij} \frac{i! \, j!}{(i + j + 1)!} t^{i+j+1}.$$

The final result (8.8), however, remains the same.

If $f_{10} < 0$ and $f_{01} < 0$, then the curves $t > 0$ and $t < 0$ in Figure 8.10 are interchanged. Hence, (8.7) holds as $t \to 0^-$, instead of $t \to 0^+$. The final result is given by the negative of (8.8), which also holds in the case where $f_{10} > 0$ and $f_{01} < 0$.

Finally we discuss briefly the case of a corner point where one of the two intersecting boundary curves *coincides* with the level curve $f(x, y) = f(0, 0)$. Here we first proceed as in the beginning of this section to arrive at equations (8.3), except that in the present case the coefficient f_{01} in (8.3) is zero. In fact, by the argument in §7 leading to (7.11), the coefficients of y^i, $i = 1, 2, \ldots$, all vanish. Thus, instead of (8.3), we have

$$f(x, y) = f_{10}x + f_{20}x^2 + f_{11}xy + f_{30}x^3 + \cdots. \qquad (8.9)$$

The remaining analysis parallels that following (7.11) in Section 7. Since the domain D now lies only in the first quadrant, the lower limit $v_1(t)$ in (7.14) is zero. The final expansion for $I(\lambda)$ is again in the form (7.15)

$$I(\lambda) \sim \sum_{s=0}^{\infty} h_s e^{i\pi(s+1)/2} \frac{s!}{\lambda^{s+1}}, \qquad \lambda \to \infty,$$

where the h_s are constants.

9.   A Curve of Stationary Points

So far in our discussion, we have assumed that the stationary points of $f(x, y)$ are always isolated, whether they are interior or boundary points of the domain D. In this section, we shall consider a case in which D contains a curve of stationary points. This type of problem occurs in the calculation of decoupling between rectangular antennas with parallel sides, and has been studied previously by Kontorovich, Karatygin, and Rozov (1970). To formulate our problem more precisely, throughout this section we assume that the following conditions hold.

(C_1) D contains a C^∞-curve γ such that $\nabla f = (0, 0)$ on γ and $\nabla f \neq (0, 0)$ on $D\backslash\gamma$; furthermore, γ is simple, i.e., no loops, and $f_{xx} + f_{yy} \neq 0$ on γ.
(C_2) Let γ be parametrized by s, the arc length of γ, and let L denote the total length of γ. Write $x = \xi(s)$ and $y = \eta(s)$, for $0 \leq s \leq L$. If $A = (\xi(0), \eta(0))$ and $B = (\xi(L), \eta(L))$, then $A, B \in \Gamma$, $A \neq B$, $(\xi(s), \eta(s)) \notin \Gamma$ for $0 < s < L$, Γ being the boundary of D.
(C_3) The curves γ and Γ are not tangent to each other at A and B.

Under condition (C_1), $f(x, y) = f_0 = $ constant for $(x, y) \in \gamma$. Without loss of generality, we suppose that $f_0 = 0$. It will be seen later that the condition $f_{xx} + f_{yy} \neq 0$ on γ forces $f(x, y)$ to have a fixed sign for (x, y) close to, but not on, γ; see (9.11) and (9.13). We shall assume that $f(x, y) > 0$ for such (x, y); equivalently, we shall assume that $f_{xx} + f_{yy} > 0$ on γ.

Condition (C_2) says that the stationary curve γ cuts the boundary Γ at, and only at, A and B. Eventually we shall require analytical versions of condition (C_3). If Γ is represented by $h(x, y) = 0$ and $k(x, y) = 0$ in the neighborhoods of A and B respectively, where h and k are smooth functions with nonvanishing gradients, then (C_3) is equivalent to

$$h_x[\xi(0), \eta(0)] \, \xi'(0) + h_y[\xi(0), \eta(0)] \, \eta'(0) \neq 0 \qquad (9.1)$$

and

$$k_x[\xi(L), \eta(L)] \, \xi'(L) + k_y \, [\xi(L), \eta(L)] \, \eta'(L) \neq 0. \qquad (9.2)$$

(We can actually allow Γ to have corners at A and B, but this will require two one-sided versions of (9.1) and (9.2); see Ex. 12.)

For $\delta > 0$, let D_δ denote the set of points in D whose distance from the stationary curve γ is less than δ, and let E_δ denote the set of points in D whose distance from γ is δ; see Figure 8.12. Since we are interested in the contribution to $I(\lambda)$ from the stationary curve γ, we may use a partition of unity to restrict our attention to the integral

$$I_0(\lambda) = \iint\limits_{D_\delta} g(x, y) e^{i\lambda f(x,y)} \, dx \, dy, \qquad (9.3)$$

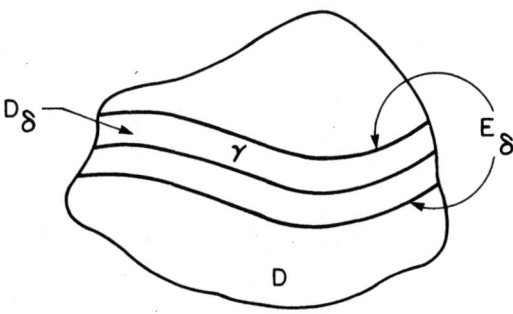

Fig. 8.12

where g vanishes to infinite order on the "edges" E_δ of D_δ. The main result of this section is the following

Theorem 1.   *Under the above conditions, we have*

$$I_0(\lambda) \sim \sum_{s=0}^{\infty} b_s \lambda^{-(s+1)/2}, \qquad as \ \lambda \to \infty, \tag{9.4}$$

where the coefficients b_s are independent of λ; in particular

$$b_0 = \sqrt{2\pi}\, e^{i\pi/4} \int_\gamma \frac{g(x,y)}{\sqrt{f_{xx} + f_{yy}}}\, ds. \tag{9.5}$$

Before proceeding to the derivation of (9.4), we first remark that since f and g are infinitely differentiable in $\overline{D}$ and γ is a C^∞-curve, by Theorem 4.1 in Malgrange's book (1966, p. 10), we can assume that f and g are extended to C^∞-functions in some open neighborhood of $\overline{D}$, and that the parametrizing functions ζ and η of the curve γ are extended to C^∞-functions in some open neighborhood $(-\varepsilon_0, L + \varepsilon_0)$ of $[0, L]$. Although we shall not actually use the values of these extensions outside their original domains, the existence of these extensions will be a technical convenience.

With ζ and η as in (C_2), and extended to a neighborhood of $[0, L]$ as mentioned above, we now define a transformation **M**: $(s, t) \to (x, y)$ by

$$x = \zeta(s) - t\eta'(s), \qquad y = \eta(s) + t\zeta'(s). \tag{9.6}$$

Initially, **M** can be defined in a strip $N \times \mathbb{R}$ of $\mathbb{R}^2$, where N is a neighborhood of $[0, L]$. The motivation behind the mapping **M** is to "straighten" the stationary curve γ to a segment $[0, L]$, and the band D_δ to a more rectangular region R_δ; see Figure 8.13. The value $|t|$ represents the distance from a point (x, y) to γ. Since s is the arc length of γ, it follows that $\zeta'(s)^2 + \eta'(s)^2 = 1$, and the Jacobian of the transformation **M** is

$$\frac{\partial(x, y)}{\partial(s, t)} = 1 + t[\zeta''(s)\eta'(s) - \zeta'(s)\eta''(s)]. \tag{9.7}$$

Since the Jacobian is 1 at each point $(s, 0)$, the inverse function theorem implies that **M** is one-to-one in a neighborhood of each point of $[0, L]$. By a fairly routine compactness argument, it can be shown that there exist positive numbers ε and δ such that **M** is one-to-one in the rectangle $Q_\delta = (-\varepsilon, L + \varepsilon) \times (-\delta, \delta)$; see Ex. 13.

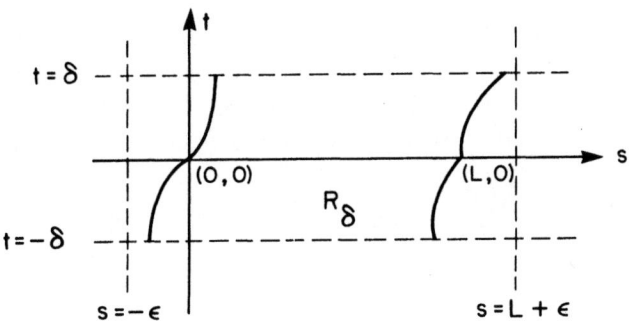

Fig. 8.13

Now let $R_\delta = \mathbf{M}^{-1}(D_\delta)$, as in Figure 8.13. Changing variables in (9.3) from (x, y) to (s, t), we obtain

$$I_0(\lambda) = \iint\limits_{R_\delta} G(s, t) e^{i\lambda F(s,t)} \, ds \, dt, \qquad (9.8)$$

where $F(s, t) = f(x, y)$ and $G(s, t) = g(x, y) \, \partial(x, y)/\partial(s, t)$. Near the end-points A and B of the curve γ, the boundary Γ of D is represented by $h(x, y) = 0$ and $k(x, y) = 0$ respectively; see the second paragraph following (C_3). Put $H(s, t) = h(\mathbf{M}(s, t))$ and $K(s, t) = k(\mathbf{M}(s, t))$. From (9.1) and (9.2), it follows immediately that $H_s(0, 0) \neq 0$ and $K_s(L, 0) \neq 0$. By the implicit function theorem, there exist functions $s = a(t)$, with $a(0) = 0$, and $s = b(t)$, with $b(0) = L$ such that near $(0, 0)$ and $(L, 0)$ the boundary of R_δ is given by $s = a(t)$ and $s = b(t)$, respectively. We shall assume that δ has been chosen sufficiently small so that the boundary of R_δ in Q_δ is completely determined by the equations $s = a(t)$ and $s = b(t)$; see, again, Figure 8.13. Then the integral $I_0(\lambda)$ can be expressed as an iterated integral

$$I_0(\lambda) = \int_{-\delta}^{\delta} \int_{a(t)}^{b(t)} G(s, t) e^{i\lambda F(s,t)} \, ds \, dt. \qquad (9.9)$$

The transformation $\mathbf{M}$ of (9.6) has other useful aspects, which we shall now explore. From the condition (C_1), we have $\nabla F(s, 0) = (0, 0)$ for $0 \leq s \leq L$, and therefore

$$\frac{\partial^{k+1} F}{\partial s^{k+1}}(s, 0) = \frac{\partial^{k+1} F}{\partial s^k \, \partial t}(s, 0) = 0 \qquad (9.10)$$

for any $k \geq 0$ and any s in $[0, L]$. This, of course, shows that $F(s, 0)$ is constant on the segment $0 \leq s \leq L$; we suppose without loss of generality that $F(s, 0) = 0$. At each point $(s, 0)$, we also obtain from (9.10) the following expansions for $F(s, t)$ and its derivatives

$$F(s, t) = F_{tt}(s, 0)\frac{t^2}{2} + F_{ttt}(s, 0)\frac{t^3}{6} + O(t^4),$$

$$F_s(s, t) = F_{tts}(s, 0)\frac{t^2}{2} + F_{ttts}(s, 0)\frac{t^3}{6} + O(t^4),$$

$$\qquad\qquad (9.11)$$

$$F_t(s, t) = F_{tt}(s, 0)t + F_{ttt}(s, 0)\frac{t^2}{2} + O(t^3),$$

$$F_{tt}(s, t) = F_{tt}(s, 0) + F_{ttt}(s, 0)t + O(t^2).$$

The O-estimates in (9.11) apply when $t \to 0$ uniformly for s in $[0, L]$.

From the definition $F(s, t) = f(x, y)$ and the chain rule, we have

$$F_{tt} = f_{xx}(\eta')^2 - 2f_{xy}\xi'\eta' + f_{yy}(\xi')^2.$$

Since s is the arc length of the stationary curve γ, it follows that $\xi'(s)^2 + \eta'(s)^2 = 1$ for $0 \leq s \leq L$. This gives

$$F_{tt} = f_{xx} + f_{yy} - [f_{xx}(\xi')^2 + 2f_{xy}\xi'\eta' + f_{yy}(\eta')^2],$$

and therefore

$$F_{tt}(s, 0) = f_{xx}(\xi(s), \eta(s)) + f_{yy}(\xi(s), \eta(s)) - \left(\frac{d}{ds}\right)^2 f(\xi(s), \eta(s)) \quad (9.12)$$

for $0 \leq s \leq L$. Since f is constant on γ, this implies

$$F_{tt}(s, 0) = f_{xx}(\xi(s), \eta(s)) + f_{yy}(\xi(s), \eta(s)) \qquad (9.13)$$

for $0 \leq s \leq L$. By (C_1), $F_{tt}(s, 0) \neq 0$ for s in $[0, L]$. Without loss of generality, we can suppose $F_{tt}(s, 0) > 0$. Then, by choosing δ sufficiently small, we have $F_{tt}(s, t) > 0$ in the region R_δ. From this and (9.11), one can deduce

$$F(s, t) > 0 \quad \text{and} \quad \text{sgn } F_t(s, t) = \text{sgn } t \quad \text{for } (s, t) \text{ in } R_\delta \text{ with } t \neq 0,$$

$$\qquad\qquad (9.14)$$

shrinking δ again if necessary. For points (s, t) in R_δ with $s < 0$, the justification of (9.14) is slightly different. Here we use the two-variable Taylor expansion of $F(s, t)$ around $(0, 0)$, and note that if (s, t) is in R_δ

with $s < 0$ then by (C_3), $s = O(t)$ as $(s, t) \to (0, 0)$. A similar remark applies in the case $s > L$. From (9.14), we have, in particular, $F_t(s, t) \neq 0$ if (s, t) is in R_δ with $t \neq 0$. Thus in R_δ, the equation $F(s, t) = c$ can be solved locally for t as a function of s. The solution will be in two parts, say $t = t_c^+(s)$ and $t = t_c^-(s)$, corresponding to $t > 0$ and $t < 0$. For small c, the curves $t = t_c^\pm(s)$ will extend right across R_δ, from a point $(\alpha_\pm(c), t_c^\pm(\alpha_\pm(c)))$ on $s = a(t)$ to a point $(\beta_\pm(c), t_c^\pm(\beta_\pm(c)))$ on $s = b(t)$; see Figure 8.14. As c increases, the curves $t = t_c^\pm(s)$ may become disconnected, and for $c > \sup\{F(s, t) : (s, t) \text{ in } R_\delta\}$, the curves will be empty.

Following our ideas in earlier sections, we now wish to transform the integral (9.9) to a one-dimensional Fourier integral. To do this, we define a second transformation $N: (s, t) \to (w, z)$ by

$$w = s, \qquad z^2 = F(s, t) \quad \text{with} \quad \operatorname{sgn} z = \operatorname{sgn} t. \tag{9.15}$$

Reasoning similar to that used to justify (9.14) shows that this mapping is one-to-one on R_δ. The Jacobian is

$$\frac{\partial(w, z)}{\partial(s, t)} = \frac{F_t}{2z} = \frac{F_t(s, t)}{2\sqrt{F(s, t)} \operatorname{sgn} t}. \tag{9.16}$$

As $t \to 0$ with $0 \leq s \leq L$, we have, from (9.11),

$$\frac{F_t(s, t)}{2\sqrt{F(s, t)} \operatorname{sgn} t} \to \sqrt{\frac{F_{tt}(s, 0)}{2}}. \tag{9.17}$$

Thus, we can suppose the Jacobian is nonzero in R_δ. Changing variables in (9.9), we obtain

$$I_0(\lambda) = \int_{-\rho}^{\rho} e^{i\lambda z^2} \Phi(z) \, dz, \tag{9.18}$$

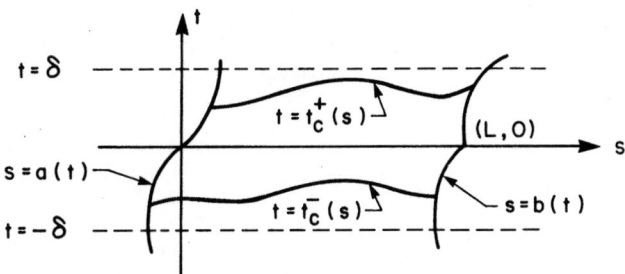

Fig. 8.14

where

$$\rho^2 = \sup\{F(s, t): (s, t) \in R_\delta\}, \tag{9.19}$$

$$\Phi(z) = \int_{\gamma(z)} G(s, t) \frac{2z}{F_t(s, t)} \, dw, \tag{9.20}$$

and

$$\gamma(z) = \{w: (w, z) = N(s, t) \quad \text{for some } (s, t) \text{ in } R_\delta\}. \tag{9.21}$$

In (9.20), $s = w$ and $t = t(w, z)$ represent the inversion of (9.15), which exists and is infinitely differentiable in $N(R_\delta)$, by the inverse function theorem. Comparing (9.21) with the discussion of the level sets of F given after (9.14), and setting $c = z^2$, we see that $t(w, z) = t_c^{\pm}(s)$, where $w = s$ and $+$ or $-$ is chosen according to whether z is positive or negative. By the comment following (9.3), and the definition of $G(s, t)$, we see that $\Phi(z)$ vanishes to infinite order as $z \to \rho^-$ and as $z \to (-\rho)^+$. Thus, the asymptotic behavior of $I_0(\lambda)$ is determined by the behavior of $\Phi(z)$ for z near 0. Put

$$I_0(\lambda) = I_0^+(\lambda) + I_0^-(\lambda), \tag{9.22}$$

where

$$I_0^{\pm}(\lambda) = \int_0^\rho e^{i\lambda z^2} \Phi(\pm z) \, dz. \tag{9.23}$$

Referring again to the discussion given after (9.14), and putting $c = z^2$, we see that for sufficiently small z,

$$\Phi(z) = \int_{\alpha(c)}^{\beta(c)} G(s, t) \frac{2z}{F_t(s, t)} \, dw, \tag{9.24}$$

where, as in (9.20), $s = w$ and $t = t(w, z)$, and α and β are taken to be α_+ or α_- and β_+ or β_- depending on whether z is positive or negative. It will be shown later that $\Phi(z)$ is infinitely differentiable in each of $[0, \sigma]$ and $[-\sigma, 0]$, for sufficiently small σ (Φ will always be continuous at 0, but may fail to be infinitely differentiable there, if Γ has a corner at A or B; see Ex. 16). Now, by repeated integration by parts, we obtain

$$I_0^{\pm}(\lambda) = \frac{1}{2} \sum_{s=0}^{n-1} \frac{1}{s!} \Gamma\left(\frac{s+1}{2}\right) e^{i(s+1)\pi/4} (\pm 1)^s \Phi^{(s)}(0^{\pm}) \lambda^{-(s+1)/2}$$

$$+ O(\lambda^{-(n+1)/2}). \tag{9.25}$$

The desired expansion of $I_0(\lambda)$ is obtained by adding these two results; compare (9.22).

To make this expansion useful, the coefficients must be expressed in terms of the original data of the problem. Since the computation of the $\Phi^{(s)}(0\pm)$ becomes quite complicated as s increases, we shall carry out only the case $s = 0$. The case $s = 1$ is given in Ex. 16.

We work with the expression for $\Phi(z)$ given in (9.24), which is valid for sufficiently small z. It will be enough to consider the case $z > 0$; the analysis in the case $z < 0$ is entirely similar. By the inverse function theorem, $s = w$, $t = t(w, z)$, and $\partial(s, t)/\partial(w, z) = 2z/F_t(s, t)$ are infinitely differentiable for small z and $0 \le s \le L$. The lower limit of integration $w = \alpha(c) = \alpha(z^2)$ is determined by the equations $w = s$, $z^2 = F(s, t)$, and $s = a(t)$, or by $z^2 = F(a(t(w, z)), t(w, z))$, or by $H(w, t(w, z)) = 0$. From (9.15), it follows that $\partial t/\partial w = -F_s/F_t$. This, together with (9.11), gives $t_w(0, 0) = 0$; since $H_w(0, 0) \ne 0$ (see the remark following (9.8)), the implicit function theorem implies that $w = \alpha(z^2)$ is infinitely differentiable for small nonzero z. Clearly, $w(0^+) = 0$, and we can show that $w'(0^+)$ exists (Ex. 15). Similar calculations, which we shall not carry out, show that $w^{(n)}(0^+)$ exists for all n, and thus the claim that the lower limit of integration is infinitely differentiable in $[0, \sigma]$, for small positive σ, is justified. Similarly, the upper limit of integration $w = \beta(z^2)$ is infinitely differentiable in $[0, \sigma]$. Taken together, these results imply that $\Phi(z)$ is infinitely differentiable in each of $[0, \sigma]$ and $[-\sigma, 0]$. Since $\alpha(0) = 0$ and $\beta(0) = L$, we now have, using (9.16), (9.17), and (9.13),

$$
\begin{aligned}
\Phi(0^+) = \Phi(0^-) &= \int_0^L G(w, 0) \sqrt{\frac{2}{F_{tt}(w, 0)}}\, dw \\
&= \sqrt{2} \int_0^L \frac{g(\xi(s), \eta(s))}{\sqrt{f_{xx}(\xi(s), \eta(s)) + f_{yy}(\xi(s), \eta(s))}}\, ds.
\end{aligned}
\tag{9.26}
$$

Inserting (9.26) in (9.25), we obtain the leading coefficient given in (9.5).

10. Laplace's Approximation

We now turn to the Laplace-type integral

$$
J(\lambda) = \iint_D g(x, y) e^{-\lambda f(x, y)}\, dx\, dy,
\tag{10.1}
$$

where, as before, λ is a large positive parameter, D is a domain, and f and g are real-valued C^∞-functions in the closure of D. We shall allow D to be unbounded, but will assume that the integral (10.1) converges absolutely for all large values of λ.

Our first objective is to show that the major contribution to the asymptotic expansion of $J(\lambda)$ comes from points where $f(x, y)$ attains its absolute minimum. Without loss of generality, we may assume that the minimum value of f is zero and occurs at, and only at, $(0, 0)$. Let N_δ be a circular neighborhood of the point with radius $\delta > 0$, and put $D_\delta = N_\delta \cap D$. Clearly

$$J(\lambda) = \iint_{D_\delta} + \iint_{D \setminus D_\delta} = J_1(\lambda) + J_2(\lambda). \tag{10.2}$$

Let λ_0 be a constant such that $J(\lambda)$ converges absolutely for all $\lambda \geq \lambda_0$, and set

$$K = \iint_D |g(x, y)| e^{-\lambda_0 f(x,y)} \, dx \, dy. \tag{10.3}$$

Then it is evident that for $\lambda \geq \lambda_0$

$$|J_2(\lambda)| \leq K e^{-(\lambda - \lambda_0)c}, \tag{10.4}$$

where

$$c = \inf\{f(x, y): (x, y) \in D \setminus D_\delta\}. \tag{10.5}$$

We will show that $J_1(\lambda) = O(\lambda^{-1})$; then it will follow that the estimate (10.4) is exponentially small compared to $J_1(\lambda)$ and hence the contribution from the region $D \setminus D_\delta$ is negligible. Since δ is an arbitrary positive number, our claim that the principal contribution to $J(\lambda)$ comes from the immediate neighborhood of the origin will be justified.

In this section, we consider only the case in which $(0, 0)$ is an interior point of D. Since $f(x, y)$ is infinitely differentiable, $(0, 0)$ must be a critical point of $f(x, y)$, i.e., $f_x(0, 0) = f_y(0, 0) = 0$. By eliminating the cross-product term xy by a transformation as in (3.1), we can assume without loss of generality that the Maclaurin expansion of $f(x, y)$ takes the form

$$f(x, y) = f_{20} x^2 + f_{02} y^2 + \sum_{i+j=3} f_{ij} x^i y^j + \cdots, \tag{10.6}$$

where f_{20} and f_{02} are both positive. From here on the analysis parallels that given in §3, except that instead of the one-dimensional Fourier integral (3.9), we now have the Laplace integral

$$J_1(\lambda) = \int_0^M h(t)e^{-\lambda t}\, dt, \qquad (10.7)$$

where $M = \max\{f(x, y): (x, y) \in D_\delta\}$ and

$$h(t) \sim \frac{1}{\sqrt{f_{20}f_{02}}} \sum b_{mn} t^{m+n}, \qquad \text{as } t \to 0^+; \qquad (10.8)$$

see (3.14). The coefficients b_{mn} can be expressed in terms of the derivatives of f and g at $(0, 0)$. In particular, $b_{00} = \pi g(0, 0)$. To the integral (10.7), we now apply Watson's lemma. In view of (10.2) and (10.4), we obtain

$$J(\lambda) \sim \frac{1}{\sqrt{f_{20}f_{02}}} \sum b_{mn} \frac{\Gamma(m + n + 1)}{\lambda^{m+n+1}}, \qquad \text{as } \lambda \to +\infty. \qquad (10.9)$$

Here we wish to point out that the small-t expansion of $h(t)$ given in (10.8) is derived under the condition that both $f(x, y)$ and $g(x, y)$ are C^∞-functions in $\mathbb{R}^2$. If f and g have only a finite number of continuous derivatives, then a finite (instead of an infinite) asymptotic expansion of the form (10.8) can still be obtained for $h(t)$ and hence for the double integral $J(\lambda)$ in (10.1). Also, it is easily seen that this weaker assumption will not cause any complication in the derivation of the final result.

Transforming back into the original variables of $f(x, y)$ and taking into account the constant term $f(0, 0)$, the leading term in (10.9) becomes

$$J(\lambda) \sim \frac{2\pi}{\lambda} g(0, 0)[\det f''(0, 0)]^{-1/2} \exp\{-\lambda f(0, 0)\}, \qquad (10.10)$$

where $\det f''(0, 0) = f_{xx}(0, 0)f_{yy}(0, 0) - f_{xy}^2(0, 0)$. Formula (10.10) is the two-dimensional analogue of the Laplace approximation given in Chapter II.

Example. Consider the double integral

$$S_n = \iint_{D'} [\cos u \cos v \cos(u + v)]^n \, du \, dv, \qquad (10.11)$$

where D' is the hexagonal region given by $D' = \{(u, v): |u| < \pi/2,$ $|v| < \pi/2, |u + v| < \pi/2\}$. This integral arose in the asymptotic evaluation of the sum

$$S(3, n) = \sum_{k=0}^{n} \binom{n}{k}^3;$$

see Henrici (1974, p. 413). The integrand can be written in the form $\exp[-nh(u, v)]$ with

$$h(u, v) = -\log \cos u - \log \cos v - \log \cos(u + v) = u^2 + uv + v^2 + \cdots.$$

$$(10.12)$$

By (10.10), we have

$$S_n \sim \frac{2\pi}{n\sqrt{3}}, \qquad \text{as } n \to \infty. \tag{10.13}$$

The derivation of formula (10.10) does not lend itself readily to the construction of error bounds. This is mainly due to the fact that it is difficult to determine the exact region in which the Jacobian of the transformation (3.5) is nonvanishing. For this reason, we present in the following a brief description of a modification of the above derivation, by which we can, for instance, show that the asymptotic formula in (10.13) can be sharpened to

$$S_n = \frac{2}{\sqrt{3}}\left[\frac{\pi}{n} + \delta_1(n)\right], \tag{10.14}$$

where

$$|\delta_1(n)| \le \frac{\pi}{(n-1)^2}. \tag{10.15}$$

Let ∂D denote the boundary of D, and put

$$\rho = \inf\{f(x, y): (x, y) \in \partial D\} \tag{10.16}$$

and

$$\Gamma_\rho = \{(x, y) \in \bar{D}: f(x, y) = \rho\}.$$

Note that ρ might be infinite, in which case we take $\Gamma_\rho = \partial D$. If D^* denotes the region bounded by Γ_ρ (see Figure 8.15), then clearly

$$J(\lambda) = \iint_{D^*} + \iint_{D/D^*} = J_1(\lambda) + J_2(\lambda). \tag{10.17}$$

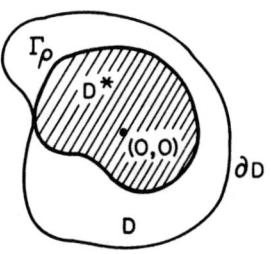

Fig. 8.15

By the argument leading to (10.4), we have the explicit bound

$$|J_2(\lambda)| \le Ke^{-(\lambda - \lambda_0)\rho}, \qquad \lambda \ge \lambda_0,$$

where the constant K is given by (10.3). Thus our problem is reduced to that of finding an asymptotic approximation for the integral $J_1(\lambda)$.

By the nesting property given in Theorem 10, Chapter V, the region D^* can be covered by the family of simple curves

$$\Gamma_t = \{(x, y): f(x, y) = t\}, \qquad 0 < t < \rho. \tag{10.18}$$

Moreover, by the method of resolution of multiple integrals given in Theorem 9, Chapter V, the double integral $J_1(\lambda)$ can be reduced to the single integral

$$J_1(\lambda) = \int_0^\rho h(t)e^{-\lambda t}\, dt, \tag{10.19}$$

where

$$h(t) = \int_{\Gamma_t} \frac{g(x, y)}{|\nabla f|}\, d\sigma, \tag{10.20}$$

and σ is the arc length of the curve Γ_t. The essential difference between this modification and the original derivation is that here we have avoided the use of the transformation $(x, y) \to (u, v)$ given in (3.5). The disadvantage of the present approach is that we now have to work with the more complicated level curves Γ_t, instead of the simple ellipses $f_{20}u^2 + f_{02}v^2 = t$ shown in Figure 8.2.

To obtain a more workable expression for the function $h(t)$ in (10.20), we make the change of variables

$$x = \left(\frac{\xi}{f_{20}}\right)^{1/2} \cos\eta, \qquad y = \left(\frac{\xi}{f_{02}}\right)^{1/2} \sin\eta.$$

Clearly

$$\xi = f_{20}x^2 + f_{02}y^2 \qquad \text{and} \qquad \frac{\partial(x, y)}{\partial(\xi, \eta)} = \frac{1}{2\sqrt{f_{20}f_{02}}}.$$

Equation (10.6) suggests that we write

$$f(x, y) = \xi + F(\xi, \eta). \tag{10.21}$$

Put

$$\Phi(\xi, \eta) = \frac{1}{2\sqrt{f_{20}f_{02}}} g(x, y).$$

In view of (13.6) and (13.7) in Chapter V, the line integral (10.20) can be expressed as

$$h(t) = \lim_{c \to 0} \frac{1}{c} \iint_{-c \le f - t < 0} g(x, y) \, dx \, dy.$$

Thus, by the familiar change of variables formula,

$$h(t) = \lim_{c \to 0} \frac{1}{c} \iint_{-c \le \xi + F - t < 0} \Phi(\xi, \eta) \, d\xi \, d\eta.$$

The same argument then gives

$$h(t) = \int_{t = \xi + F} \frac{\Phi(\xi, \eta)}{|\nabla(\xi + F)|} \, d\sigma, \tag{10.22}$$

where $d\sigma$ denotes the length of the curve $t = \xi + F(\xi, \eta)$, and the gradient ∇ is taken with respect to ξ and η.

We now impose the additional condition that there exists a positive number δ such that

$$xf_x + yf_y \ge 2\delta(f_{20}x^2 + f_{02}y^2) \tag{10.23}$$

for all $(x, y) \in D$. The left-hand side of (10.23) is almost the directional derivative of $f(x, y)$ along the direction of the line segment joining $(0, 0)$ to (x, y). Thus, (10.23) compares it with the directional derivative of the function $f_{20}x^2 + f_{02}y^2$. In a sense, (10.23) is the two-dimensional analogue of the condition $f'(x) \ge \delta > 0$ in the one-dimensional case.

From (10.21) and (10.23), we have $1 + F_\xi(\xi, \eta) \ge \delta > 0$. This ensures that the solution $\xi_t(\eta)$ to

$$t = \xi + F(\xi, \eta) \tag{10.24}$$

exists, and that the arc length $d\sigma$ can be explicitly given by

$$d\sigma = \sqrt{1 + \left(\frac{\partial \xi}{\partial \eta}\right)^2} \, d\eta = \frac{|\nabla(\xi + F)|}{1 + F_\xi(\xi_t, \eta)} \, d\eta. \qquad (10.25)$$

Inserting (10.25) in (10.22) gives

$$h(t) = \int_0^{2\pi} \frac{\Phi(\xi_t, \eta)}{1 + F_\xi(\xi_t, \eta)} \, d\eta. \qquad (10.26)$$

Here we have used the fact that for each $t \in (0, \rho)$, the solution $\xi_t(\eta)$ to (10.24) is a simple closed C^1-curve which stays entirely within D^* for all η in $[0, 2\pi]$; see the nesting property stated in Theorem 10, Chapter V.

We now return to the example (10.11), and make the change of variables

$$u = x - \tfrac{1}{2}y, \qquad v = y; \qquad (10.27)$$

compare (3.1). This transformation eliminates the cross product uv in (10.12), and maps the region D' into

$$D = \left\{(x, y): \left| x \pm \frac{1}{2} y \right| < \frac{\pi}{2}, |y| < \frac{\pi}{2} \right\}. \qquad (10.28)$$

The integral (10.11) becomes

$$S_n = \iint\limits_D e^{-\lambda f(x,y)} \, dx \, dy,$$

where

$$f(x, y) = -\log \cos(x - \tfrac{1}{2}y) - \log \cos y - \log \cos(x + \tfrac{1}{2}y). \quad (10.29)$$

The result which we have just developed is now immediately applicable. In the present case, the constant ρ in (10.16) is $+\infty$, and hence $J_2(\lambda)$ in (10.17) is absent. On account of (10.19) and (10.26), we obtain

$$S_n = \int_0^\infty h(t) e^{-nt} \, dt, \qquad (10.30)$$

where

$$h(t) = \frac{1}{\sqrt{3}} \int_0^{2\pi} \frac{1}{1 + F_\xi(\xi_t, \eta)} \, d\eta. \qquad (10.31)$$

From (10.21), it is easily seen that $F(\xi, \eta) = O(\xi^{3/2})$ and $F_\xi(\xi, \eta) = O(\xi^{1/2})$ as $\xi \to 0$. Hence the solution $\xi_t(\eta)$ to (10.24) is $O(t)$, and $F_\xi(\xi_t, \eta) = O(t^{1/2})$ as $t \to 0^+$. Therefore (10.31) gives $h(t) \sim 2\pi/\sqrt{3}$ as $t \to 0^+$. Set

$$h(t) = \frac{2\pi}{\sqrt{3}} - h_1(t) \tag{10.32}$$

with

$$h_1(t) = \frac{1}{\sqrt{3}} \int_0^{2\pi} \frac{F_\xi(\xi_t, \eta)}{1 + F_\xi(\xi_t, \eta)} \, d\eta. \tag{10.33}$$

By using the inequality $\theta \tan \theta \geq \theta^2$ for all $\theta \in (-\pi/2, \pi/2)$ and the identity

$$(x - \tfrac{1}{2}y)^2 + y^2 + (x + \tfrac{1}{2}y)^2 = 2(x^2 + \tfrac{3}{4}y^2),$$

the constant δ in (10.23) can easily be shown to be 1. Hence

$$1 + F_\xi(\xi_t, \eta) \geq 1. \tag{10.34}$$

Furthermore, since each term on the right-hand side of (10.29) is positive for all (x, y) lying in the hexagon given by (10.28), from the equation $t = \xi + F(\xi, \eta) = f(x, y)$ it follows that the reciprocals of the quantities $\cos(x - \tfrac{1}{2}y)$, $\cos y$, and $\cos(x + \tfrac{1}{2}y)$ are all bounded by e^t. Therefore

$$xf_x + yf_y \leq 2(x^2 + \tfrac{3}{4}y^2)e^t, \qquad \text{and} \qquad 1 + F_\xi(\xi, \eta) \leq e^t.$$

Since $F_\xi \geq 0$ by (10.34), we also have

$$0 \leq F_\xi(\xi_t, \eta) \leq e^t - 1 \leq te^t. \tag{10.35}$$

Applying the inequalities (10.34) and (10.35) to (10.33) gives

$$|h_1(t)| \leq \frac{2\pi}{\sqrt{3}} te^t. \tag{10.36}$$

The desired result (10.14)–(10.15) now follows from (10.30), (10.32), and (10.36).

The success of (10.15) depends heavily on the inequality (10.35). In general, these types of inequalities are not available. For a discussion of the construction of an error bound associated with the Laplace approximation (10.10), see McClure and Wong (1983).

11. Boundary Extrema

We next consider the case where the critical point $(0, 0)$ of $f(x, y)$ is on the boundary of D. The analysis here follows that of §6. First, we apply the transformations $(x, y) \to (\bar{x}, \bar{y}) \to (X, Y)$ given in (6.3) and (6.4). This maps the part of the boundary containing $(0, 0)$ into the Y-axis. Let $F(X, Y)$ denote the transform of $f(x, y)$, and put $G(X, Y) = g(x, y)| \partial(x, y)/\partial(X, Y)|$. The double integral then can be written as

$$J(\lambda) = \iint_{D'} G(X, Y) e^{-\lambda F(X, Y)} \, dX \, dY,$$

where D' is the image of D. Next, we use the method of resolution of multiple integrals to reduce the last integral to a one-dimensional Laplace transform

$$J(\lambda) = \int_0^M h(t) e^{-\lambda t} \, dt,$$

where $M = \max\{f(x, y) : (x, y) \in D\}$, and

$$h(t) \sim \sum_{s=0}^{\infty} b_s t^{s/2}, \qquad \text{as } t \to 0^+,$$

with $b_0 = \pi g_{00}/2\sqrt{\bar{f}_{20}\bar{f}_{02}}$; cf. (6.6). Thus, by Watson's lemma, we have

$$J(\lambda) \sim \sum_{s=0}^{\infty} b_s \Gamma\left(1 + \frac{s}{2}\right) \lambda^{-s/2 - 1} \qquad \text{as } \lambda \to +\infty. \tag{11.1}$$

By the reasoning used for (10.10), the first term of the series gives

$$J(\lambda) \sim \frac{\pi}{\lambda} g(0, 0)[\det f''(0, 0)]^{-1/2} \exp[-\lambda f(0, 0)]. \tag{11.2}$$

As expected, this is one-half of the Laplace approximation (10.10).

We finally consider the case in which the function $f(x, y)$ attains its minimum at $(0, 0)$, which is on the boundary of D but is not a critical point, i.e., $\nabla f(0, 0) \neq (0, 0)$. Here we proceed as in §7. First, we show that without loss of generality, we may assume that the portion of the boundary curve which passes through $(0, 0)$ actually lies on the vertical axis, and that the function $f(x, y)$ has the Maclaurin expansion (7.4)

$$f(x, y) = f_{10}x + f_{20}x^2 + f_{11}xy + f_{02}y^2 + \cdots.$$

Next, we use the transformation $(x, y) \to (u, v)$, given in (7.5), to map $f(x, y)$ into its canonical form $f_{10}u + f_{02}v^2 = F(u, v)$. Finally, we derive the asymptotic expansion of the line integral

$$h(t) = \int_{F(u,v) = t} \frac{G(u, v)}{|\nabla F|} \, d\sigma,$$

as $t \to 0^+$, where $G(u, v) = g(x, y)| \partial(x, y)/\partial(u, v)|$. Since

$$h(t) \sim \sum b_{ij} t^{i + j + 1/2}, \qquad \text{as } t \to 0^+,$$

we have by Watson's lemma

$$J(\lambda) = \int_0^M h(t) e^{-\lambda t} \, dt \sim \sum b_{ij} \frac{\Gamma(i + j + \frac{3}{2})}{\lambda^{i + j + 3/2}}, \qquad \text{as } \lambda \to +\infty, \quad (11.3)$$

where $b_{00} = 2g_{00}/f_{10}\sqrt{f_{02}}$.

In terms of the original function $f(x, y)$ before any coordinate changes are made, the first term in (11.3) leads to

$$J(\lambda) \sim \left(\frac{2\pi}{\kappa_0}\right)^{1/2} g(0, 0) \exp[-\lambda f(0, 0)] \lambda^{-3/2}, \qquad (11.4)$$

as $\lambda \to +\infty$, where

$$\kappa_0 = \{[f_{xx}(f_y)^2 - 2f_{xy}(f_x f_y) + f_{yy}(f_x)^2] \pm \kappa |\nabla f|^3\}(0, 0), \qquad (11.5)$$

and κ is the curvature of the boundary curve at $(0, 0)$; see Ex. 18. An n-dimensional generalization of (11.4) is given in Chapter IX.

Exercises

1. Find the leading term in the asymptotic expansion of the integral

$$\iint_D \exp(i\lambda \cos x \cos y) \, dx \, dy,$$

as $\lambda \to +\infty$, where D is the unit disk $\{(x, y): x^2 + y^2 \le 1\}$.

2. Derive the asymptotic expansion of the integral

$$\iint_D \exp[i\lambda(\cos x + \cos y)] \, dx \, dy,$$

as $\lambda \to +\infty$, up to the order $O(\lambda^{-2})$, where D is the square $\{(x, y): |x| < \pi/2, |y| < \pi/2\}$.

3. Consider the double integral

$$I(\lambda) = \int_{-1}^{1} \int_{-1/2}^{1/2} \exp[i\lambda(x^2 + xy + y^2)(1 - \tfrac{1}{2}x)] \, dx \, dy.$$

Show that there are three critical points of the second kind, and that they are located at $(-\tfrac{1}{3}, 1)$, $(\tfrac{1}{2}, -\tfrac{1}{4})$, and $(-\tfrac{1}{2}, \tfrac{1}{4})$. Write down the leading-term contribution from each of these points.

4. Show that $(0, 0)$ is the only critical point of the first kind for the integral in Ex. 3, and that its contribution is

$$\frac{2\pi}{\sqrt{3}} \left[\frac{i}{\lambda} - \frac{261}{512} \frac{1}{\lambda^2} + \cdots \right].$$

5. Show that $(0, 0)$ is a boundary stationary point for the integral

$$\iint_{D} \exp[i\lambda(x^2 + xy + y^2)(1 - \tfrac{1}{2}x)] \, dx \, dy, \qquad \lambda > 0,$$

where D is the region bounded by the circle $(x - 1)^2 + (y - 3)^2 = 10$. Also show that the first two terms in the asymptotic expansion from this critical point are given by

$$\frac{\pi i}{\sqrt{3\lambda}} - \frac{i17\sqrt{2\pi}e^{i\pi/4}}{2(14\lambda)^{3/2}}.$$

6. Consider the degenerate case

$$f(x, y) = f_{20}x^2 + f_{30}x^3 + f_{21}x^2y + f_{03}y^3 + \cdots;$$

cf. Eq. (5.2). Show that (i) if $f_{20} < 0$ and $f_{03} > 0$, then the coefficients a_0 and a_1 in (5.11) are given by

$$a_0 = \frac{1}{\sqrt{3}} \frac{\Gamma(\tfrac{1}{2})\Gamma(\tfrac{1}{3})}{(-f_{20})^{1/2} f_{03}^{1/3}} g_{00} e^{-i\pi/4}, \qquad a_1 = \frac{1}{\sqrt{3}} \frac{\Gamma(\tfrac{1}{2})\Gamma(\tfrac{2}{3})}{(-f_{20})^{1/2} f_{03}^{2/3}} G_{01} e^{i\pi/4};$$

(ii) if $f_{20} > 0$ and $f_{03} < 0$, then

$$a_0 = \frac{1}{\sqrt{3}} \frac{\Gamma(\tfrac{1}{2})\Gamma(\tfrac{1}{3})}{f_{20}^{1/2}(-f_{03})^{1/3}} g_{00} e^{i\pi/4}, \qquad a_1 = \frac{1}{\sqrt{3}} \frac{\Gamma(\tfrac{1}{2})\Gamma(\tfrac{2}{3})}{f_{20}^{1/2}(-f_{03})^{2/3}} G_{01} e^{-i\pi/4};$$

and (iii) if $f_{20} < 0$ and $f_{03} < 0$, then

$$a_0 = \frac{1}{\sqrt{3}} \frac{\Gamma(\tfrac{1}{2})\Gamma(\tfrac{1}{3})}{(-f_{20})^{1/2}(-f_{03})^{1/3}} g_{00} e^{-i\pi/4},$$

$$a_1 = \frac{1}{\sqrt{3}} \frac{\Gamma(\tfrac{1}{2})\Gamma(\tfrac{2}{3})}{(-f_{20})^{1/2}(-f_{03})^{2/3}} G_{01} e^{-3\pi i/4}.$$

7. Let h_{10}, h_{01}, f_{20}, and f_{02} be nonzero constants, and consider a nonsingular linear transformation

$$\begin{pmatrix} x \\ y \end{pmatrix} = \begin{pmatrix} \alpha & \beta \\ \gamma & \delta \end{pmatrix} \begin{pmatrix} \bar{x} \\ \bar{y} \end{pmatrix}.$$

Show that a necessary and sufficient condition for the existence of nonzero constants $\bar{f}_{20}$ and $\bar{f}_{02}$ such that $f_{20}x^2 + f_{02}y^2 = \bar{f}_{20}\bar{x}^2 + \bar{f}_{02}\bar{y}^2$ and $h_{10}x + h_{01}y = \bar{x}$ is $h_{10}^2 f_{02} + h_{01}^2 f_{20} \neq 0$; see §6.

8. Consider the case of a saddle point on the boundary of the domain D discussed in §6, and suppose that the coefficients $\bar{f}_{20}$ and $\bar{f}_{02}$ in the expansion of $\bar{f}(\bar{x}, \bar{y})$ are of opposite sign, say $\bar{f}_{20} > 0$ and $\bar{f}_{02} < 0$. Show that, for $t < 0$, equation (6.10) is replaced by

$$h(t) \sim \frac{1}{\sqrt{|\bar{f}_{20}\bar{f}_{02}|}} \sum G^*_{m,2n}(-t)^{m/2+n} \int_0^{\eta_2(t)} \sinh^m \eta \cosh^{2n} \eta \, d\eta,$$

where

$$\eta_2(t) = \cosh^{-1}[d_1(\bar{f}_{02}/t)^{1/2}]$$

$$= \log(d_1\sqrt{-\bar{f}_{02}} + \sqrt{-\bar{f}_{02}d_1^2 + t}) - \frac{1}{2}\log(-t),$$

and d_1 is a positive constant. The integrals under the summation sign can be evaluated as in §4. Let $d_1 = d[\bar{f}_{20}/(-\bar{f}_{02})]^{1/2}$, where d is as given in Figure 8.3, and use $\sigma(t)$ as a generic symbol for a C^∞-function near $t = 0$. Show that

$$(-t)^{m+n} \int_0^{\eta_2(t)} \sinh^{2m} \eta \cosh^{2n} \eta \, d\eta = C_{mn}(-t)^{m+n} \log(-t) + \sigma(t),$$

and

$$(-t)^{m+n+1/2} \int_0^{\eta_2(t)} \sinh^{2m+1} \eta \cosh^{2n} \eta \, d\eta$$

$$= D_{mn}(-t)^{m+n+1/2} + \sigma(t),$$

where C_{mn} is as given in (4.5),

$$C_{mn} = \frac{(-1)^{m+1}}{2^{m+1}} \frac{1 \cdot 3 \cdot 5 \cdots (2n-1)}{(2n+2m)(2n+2m-2) \cdots (2m+2)} \binom{2m}{m},$$

and D_{mn} is given by

$$D_{mn} = \frac{1 \cdot 3 \cdot 5 \cdots (2n-1)}{(2n+2m+1)(2n+2m-1) \cdots (2m+3)} \frac{1}{2^{2m}}$$

$$\times \sum_{k=0}^{m} (-1)^k \binom{2m+1}{k} \frac{1}{2m-2k+1}.$$

Deduce from the asymptotic behavior of $h(t)$, as $t \to 0^+$ and as $t \to 0^-$, that the double integral $I(\lambda)$ in (6.5) has an expansion of the form

$$I(\lambda) \sim \sum_{s=0}^{\infty} c_s \lambda^{-1-s/2}, \qquad \text{as } \lambda \to +\infty,$$

where

$$c_0 = \frac{g_{00}}{\sqrt{|\bar{f}_{20}\bar{f}_{02}|}} \frac{\pi}{2}.$$

9. Let $G(u, v)$ be a C^∞-function with compact support, and suppose that its support contains the point $(0, 0)$. Consider the integral

$$I_0(\lambda) = \int_0^\infty \int_{-\infty}^\infty G(u, v) e^{i\lambda(u^2 - v^2)} \, dv \, du.$$

Note that here we have a saddle point on the boundary of the domain of integration, which is essentially the case discussed in §6 and Ex. 8. Let $G_0 = G$, and define

$$H_0^{(1)} = \frac{1}{u} [G_0(u, v) - G_0(0, v)], \qquad H_0^{(2)} = -\frac{1}{v} [G_0(0, v) - G_0(0, 0)].$$

Put $\mathbf{H}_0 = (H_0^{(1)}, H_0^{(2)})$. Show that $G_0(u, v) = G_0(0, 0) + (u, -v) \cdot \mathbf{H}_0$ and

$$\nabla \cdot (\tfrac{1}{2}\mathbf{H}_0 e^{i\lambda(u^2-v^2)}) = (\nabla \cdot \tfrac{1}{2}\mathbf{H}_0) e^{i\lambda(u^2-v^2)} + i\lambda(u, -v) \cdot \mathbf{H}_0 e^{i\lambda(u^2-v^2)},$$

where "$\cdot$" means inner product. Furthermore, use the divergence theorem to show that

$$I_0(\lambda) = G_0(0, 0) \int_0^\infty \int_{-\infty}^\infty e^{i\lambda(u^2-v^2)} \, dv \, du - \frac{i}{\lambda} I_0^*(\lambda) + \frac{i}{\lambda} I_1(\lambda),$$

where

$$I_0^*(\lambda) = \int_{-\infty}^{\infty} \tfrac{1}{2} H_0^{(1)}(0, v) e^{-i\lambda v^2}\, dv,$$

$$I_1(\lambda) = \int_0^{\infty} \int_{-\infty}^{\infty} G_1(u, v) e^{i\lambda(u^2 - v^2)}\, dv\, du,$$

with $G_1(u, v) = \tfrac{1}{2} \nabla \cdot \mathbf{H}_0$. Note that $I_0^*(\lambda)$ is a one-dimensional integral, and $I_1(\lambda)$ is exactly of the same form as $I_0(\lambda)$. By repeating this process, show that

$$I(\lambda) \sim \sum_{s=0}^{\infty} c_s \lambda^{-1-(s/2)}, \qquad \text{as } \lambda \to +\infty,$$

where $c_0 = \pi G(0, 0)/2$.

10. Use the procedure outlined in Ex. 9 or the method of §6 and Ex. 8 to derive an asymptotic expansion for the integral

$$I(\lambda) = \int_0^{\infty} \int_0^{\infty} G(u, v) e^{i\lambda(u^2 - v^2)}\, du\, dv,$$

where G is a C^{∞}-function with compact support, and its support contains the origin. Note that the saddle point $(0, 0)$ is at a corner on the boundary of the domain.

11. Show that the double integral

$$I(\lambda) = \iint_D \exp[i\lambda x(1 - y - x^2)^2]\, dx\, dy,$$

where $D = \{(x, y): (x - 1)^2 + y^2 \le 1\}$, has the asymptotic formula

$$I(\lambda) \sim 2 \sqrt{\frac{\pi}{\lambda}}\, e^{i\pi/4} (\sqrt{b} - \sqrt{a}),$$

where a and b are the x-coordinates of the points of intersection of the curves $x^2 - 2x + y^2 = 0$ and $y = 1 - x^2$.

12. Formulate the two one-sided versions of equations (9.1) and (9.2) when the boundary Γ in condition (C_3) of §9 is allowed to have corners at A and B.

13. Let $\mathbf{M}: (s, t) \to (x, y)$ denote the transformation defined by (9.6). Show that for each point $(s, 0)$ with $-\varepsilon \le s \le L + \varepsilon$, $\varepsilon > 0$, there

exists a neighborhood $N_s = \{(\sigma, t): |\sigma - s| < \varepsilon_s, |t| < \delta_s\}$ in which M is one-to-one. From this, deduce that there exists a positive δ such that the line segment $[-\varepsilon, L + \varepsilon] \times \{0\}$ can be covered by a finite number, say n, of neighborhoods of the form $W_k = \{(\sigma, t): |\sigma - s_k| < \frac{1}{2}\varepsilon_{s_k}, |t| < \delta\}$. Show that if M is not one-to-one in $W = \bigcup_{k=1}^{n} W_k$, then there exist $(s, t) \in W_j$ and $(\sigma, \tau) \in W_k$, $j \neq k$, such that

$$(*) \qquad \begin{cases} \xi(s) - \xi(\sigma) = t\eta'(s) - \tau\eta'(\sigma) \\ \eta(s) - \eta(\sigma) = \tau\xi'(\sigma) - t\xi'(s) \end{cases}$$

and $|\sigma - s| \geq \frac{1}{2} \max\{\varepsilon_{s_j}, \varepsilon_{s_k}\}$.

Let $\bar{\varepsilon} = \min\{\xi_{s_1}, \ldots, \xi_{s_n}\}$, and define the distance function

$$d(\alpha, \beta) = [\xi(\alpha) - \xi(\beta)]^2 + [\eta(\alpha) - n(\beta)]^2$$

on the compact set $K = [-\varepsilon, L + \varepsilon]^2 \setminus \{(\alpha, \beta): |\alpha - \beta| < \frac{1}{2}\bar{\varepsilon}\}$. Show that, on one hand, there exists a positive number ρ such that $d(\alpha, \beta) \geq 2\rho^2$ for all $(\alpha, \beta) \in K$; consequently, one of the two inequalities, $|\xi(s) - \xi(\sigma)| \geq \rho$, $|\eta(s) - \eta(\sigma)| \geq \rho$, must hold. On the other hand, we may suppose

$$\delta < \frac{\rho}{2 \max\{\|\xi'\|_\infty, \|\eta'\|_\infty\}},$$

where $\|\cdot\|_\infty$ denotes the maximum norm on $[-\varepsilon, L + \varepsilon]$. Deduce from $(*)$ that $|\xi(s) - \xi(\sigma)| < \rho$ and $|\eta(s) - \eta(\sigma)| < \rho$, thus obtaining a contradiction.

14. Consider the integral

$$\Phi(z) = \int_{\alpha(z^2)}^{\beta(z^2)} G(s, t) \frac{2z}{F_t(s, t)}\, dw$$

given in (9.24). For notational convenience, write $u = \alpha(z^2)$ and $v = \beta(z^2)$. Show that from (9.15), we have

$$\frac{\partial t}{\partial w} = -\frac{F_s}{F_t} \qquad \text{and} \qquad \frac{\partial t}{\partial z} = \frac{2z}{F_t},$$

and hence

$$\Phi'(z) = G(v, t(v, z)) \frac{2z}{F_t(v, t(v, z))} \frac{dv}{dz} - G(u, t(u, z)) \frac{2z}{F_t(u, t(u, z))} \frac{du}{dz}$$

$$- \int_u^v (G_t t_z^2 + G t_{zz})\, dw.$$

Show also that

$$t_{zz} = \frac{2}{F_t} - \frac{2zF_{tt}t_z}{F_t^2} = \frac{2F_t^2 - 4FF_{tt}}{F_t^3},$$

and consequently, by (9.11),

$$t_{zz}(w, 0) = -\frac{2F_{ttt}(w, 0)}{3F_{tt}(w, 0)^2}.$$

Finally, prove that from (9.13), we have

$$t_{zz}(w, 0) = \frac{2[f_{xxx}(\eta')^3 - 3f_{xxy}(\eta')^2\xi' + 3f_{xyy}\eta'(\xi')^2 - f_{yyy}(\xi')^3]}{3(f_{xx} + f_{yy})^2},$$

where ξ', η' are evaluated at w, and f_{xx}, etc., at $(\xi(w), \eta(w))$.

15.   The lower limit of integration $w = \alpha(z^2)$ in (9.24) is determined by
the equation $H(w, t(w, z)) = 0$; see the last paragraph of §9. Thus,
in terms of notations in Ex. 14, we have $H(u, t(u, z)) = 0$. Show that

$$u'(z) = -\frac{H_t(u, t)t_z(u, z)}{H_w(u, t) + H_t(u, t)t_w(u, z)},$$

$$u'(0^+) = -\frac{H_t(0, 0^+)}{H_w(0, 0^+)}\sqrt{\frac{2}{F_{tt}(0, 0)}}.$$

Also show that, since $H(s, t) = h(x, y)$, we have

$$u'(0^+) = \frac{h_x(A+)\eta'(0) - h_y(A+)\xi'(0)}{h_x(A+)\xi'(0) + h_y(A+)\eta'(0)}\sqrt{\frac{2}{f_{xx}(A) + f_{yy}(A)}},$$

where

$$h_x(A+) = \lim_{t \to 0^+} \{h_x(x, y): (x, y) = \mathbf{M}(s, t), s = a(t)\};$$

cf. Figure 8.14. Similarly, derive the result

$$v'(0^+) = \frac{k_x(B+)\eta'(L) - k_y(B+)\xi'(L)}{k_x(B+)\xi'(L) + k_y(B+)\eta'(L)}\sqrt{\frac{2}{f_{xx}(B) + f_{yy}(B)}}.$$

16. Recalling $G(s, t) = g(x, y)\, \partial(x, y)/\partial(s, t)$, show that Exercises 14 and 15 together give

$$\Phi'(0^\pm) = g(B)\, \frac{2}{f_{xx}(B) + f_{yy}(B)}\, \frac{k_x(B\pm)\eta'(L) - k_y(B\pm)\xi'(L)}{k_x(B\pm)\xi'(L) + k_y(B\pm)\eta'(L)}$$

$$- g(A)\, \frac{2}{f_{xx}(A) + f_{yy}(A)}\, \frac{h_x(A\pm)\eta'(0) - h_y(A\pm)\xi'(0)}{h_x(A\pm)\xi'(0) + h_y(A\pm)\eta'(0)}$$

$$+ \int_0^L \left\{ \frac{2}{f_{xx} + f_{yy}}\, [-g_x\eta' + g_y\xi' + g(\xi''\eta' - \xi'\eta'')] \right.$$

$$\left. + \frac{2g[f_{xxx}(\eta')^3 - 3f_{xxy}(\eta')^2\xi' + 3f_{xyy}\eta'(\xi')^2 - f_{yyy}(\xi')^3]}{3(f_{xx} + f_{yy})^2} \right\} dw,$$

where functions g, f_{xx}, etc., occurring in the integrand, are evaluated at a typical point $(\xi(w), \eta(w))$ on the stationary curve γ.

17. Show that as $n \to \infty$,

$$\iint_D [\cos u \cos v \sin(u + v)]^{2n}\, du\, dv \sim \left(\frac{\sqrt{3}}{2}\right)^{6n + 1} \frac{\pi}{n},$$

where D is the square $|u| \le \pi/2$ and $|v| \le \pi/2$.

18. Derive the asymptotic formula given in (11.4)–(11.5).

Supplementary Notes

§1. For applications of the oscillatory integral (1.1) to diffraction theory, see Wolf (1951), Bremmer (1955), Berghuis (1955), Chako (1965, §§6(a) and 8), and Jones (1977). In addition to the papers already mentioned in the first paragraph of this section, we also call attention to those by Fedoryuk (1970), deKok (1971), and Cooke (1982). (Cooke's treatment is not rigorous.)

§2. Classification of critical points, similar to that presented here, can be found also in a survey article by Erdélyi (1959).

§3–5. The major difference between our presentation and the original one given by Jones and Kline (1958) is that we map the phase function $f(x, y)$ into its canonical form $F(u, v)$, whereas they work directly with $f(x, y)$. One of the advantages of our method is that from the simple nature of the function $F(u, v)$, it is evident that the level curves of $F(u, v)$ satisfy the conditions required in the method of resolution of multiple integrals (Theorem 9, Chapter V). Another advantage is that we have avoided the use of a Taylor-type expansion of the Dirac δ-function (Ex. 28, Chapter V). Neither of these procedures are properly justified in the paper of Jones and Kline.

§6. The transformation (6.3) is due to Focke (1954), but it requires the condition $d \neq 0$. For an alternative approach with a weaker condition, see McClure and Wong (1991). The last reference also contains a treatment of a stationary point at a corner.

§9. The material in this section is taken from McClure and Wong (1987b). The problem of higher-dimensional integrals with a curve of stationary points is mentioned in the paper of Servadio (1988, §5 and Appendix Bb), but the treatment there is only formal.

§10. Formula (10.10) was derived first by Hsu (1948a), but the presentation given here is based on McClure and Wong (1983).

§11. An alternative derivation of formula (11.4) can be found in the book of Bleistein and Handelsman (1975a, pp. 339–340).

Exercises.   Exercises 3, 4, and 5 are based on an example in Cooke (1982). The procedure of repeated application of the divergence theorem in Ex. 9 is given in Bleistein and Handelsman (1975a, pp. 327–328). For solutions to Exs. 11 and 12, see McClure and Wong (1987b). The material in Exs. 14, 15, and 16 also is taken from the last reference. The result in Ex. 17 is given in deBruijn (1970, p. 72).

IX

Higher Dimensional Integrals

1. Introduction

The methods in the previous chapter can easily be carried over to the higher dimensional integrals

$$I(\lambda) = \int_D g(x)e^{i\lambda f(x)} \, dx \qquad (1.1)$$

and

$$J(\lambda) = \int_D g(x)e^{-\lambda f(x)} \, dx, \qquad (1.2)$$

where, as before, λ is a large positive parameter, but now $x = (x_1, \ldots, x_n)$. In (1.1) D is a bounded domain in $\mathbb{R}^n$, whereas in (1.2) D is allowed to be unbounded. In both cases, $f(x)$ is a real-valued function on D. The analysis here would of course be more complicated than that of double integrals. There are, however, several other elegant methods in the literature, and each has its own advantages. For this reason, we shall present methods which are different from those given in Chapter VIII.

The n-dimensional analogue of the stationary phase approximation is given by

$$I(\lambda) \sim g(x_0)|\det A|^{-1/2} \exp\left\{i\lambda f(x_0) + \frac{i\pi\sigma}{4}\right\}\left(\frac{2\pi}{\lambda}\right)^{n/2}, \qquad (1.3)$$

where A is the Hessian matrix of f and σ is the signature of the matrix A (see §2 below). This formula is proved in §2 by two different methods. There we also show that if x_0 is a boundary stationary point, then the asymptotic approximation of $I(\lambda)$ is exactly half of that given by (1.3). Section 3 is devoted to the contributions from boundary points, which are not stationary points, but at which a surface $f(x) = $ constant is tangent to the boundary ∂D. Some degenerate cases are considered in §4. The Laplace approximation of the integral $J(\lambda)$ in (1.2) is derived in §5, where contributions from boundary extrema are also discussed. The last section (§6) contains a derivation of an asymptotic expansion of the multidimensional Fourier transform

$$\hat{f}(t) = \frac{1}{(2\pi)^{n/2}} \int_{\mathbb{R}^n} f(x)e^{it\cdot x}\, dx \qquad \text{as } |t| \to \infty. \qquad (1.4)$$

2. Stationary Points

By a *stationary* (or *critical*) point of $f(x)$, we again mean a point x_0 at which the gradient of f vanishes, i.e., $(\nabla f)(x_0) = 0$, or equivalently

$$\frac{\partial f}{\partial x_1}(x_0) = \cdots = \frac{\partial f}{\partial x_n}(x_0) = 0.$$

A stationary point x_0 is said to be *non-degenerate* if the Hessian matrix

$$A = \left(\frac{\partial^2 f}{\partial x_i\, \partial x_j}\right)\bigg|_{x = x_0} \qquad (2.1)$$

is nonsingular, i.e.,

$$\det A \neq 0. \qquad (2.2)$$

We shall assume that the support of the amplitude function $g(x)$ in (1.1) is contained in D and can be made as small as necessary. If the phase function $f(x)$ in (1.1) has no stationary point in the support of g

then, as in §2 of Chapter VIII (see also (3.4) below), the integral (1.1) satisfies

$$I(\lambda) = O(\lambda^{-N}), \qquad \text{as } \lambda \to +\infty. \qquad \text{for any } N \geq 0, \qquad (2.3)$$

thus demonstrating the fact that the essential contribution to the asymptotic expansion of $I(\lambda)$ must come from stationary points of $f(x)$. We shall assume that $f(x)$ has exactly one non-degenerate stationary point x_0 in the support of g. A result of Morse shows that near a non-degenerate stationary point the function f can be put into a simple canonical form by a change of coordinates. First, we need the following definition:

Let U, V be open sets in $\mathbb{R}^n$. Let $h: U \to \mathbb{R}^n$ and suppose that $h(U) = V$. Then h is called a *diffeomorphism* provided: (a) h is differentiable; (b) h has an inverse function $h^{-1}: V \to \mathbb{R}^n$ such that $h \circ h^{-1} = 1_V$, $h^{-1} \circ h = 1_U$; and (c) h^{-1} is differentiable.

Morse's Lemma. *Let $f(x)$ be a real-valued C^∞-function in a neighborhood of the nondegenerate critical point x_0. Then there exist neighborhoods U, V of the points $y = 0$, $x = x_0$ and a diffeomorphism $h: U \to V$ of class C^∞ such that*

$$(f \circ h)(y) = f(x_0) + \tfrac{1}{2}\langle Ay, y \rangle, \qquad (2.4)$$

where A is the Hessian matrix given in (2.1). Furthermore, the Jacobian of the transformation satisfies

$$\frac{\partial(x_1, \ldots, x_n)}{\partial(y_1, \ldots, y_n)}\bigg|_{y=0} = 1. \qquad (2.5)$$

The usual version of Morse's lemma gives only a diffeomorphism $\bar{h}: U \to \mathbb{R}^n$ such that $\bar{h}(0) = x_0$ and

$$(f \circ \bar{h})(z) = f(x_0) - z_1^2 - \cdots - z_l^2 + z_{l+1}^2 + \cdots + z_n^2;$$

see Poston and Stewart (1978, p. 54). By rescaling the variables z_j, $j = 1, \ldots, n$, we may without loss of generality write

$$(f \circ \bar{h})(z) = f(x_0) + \frac{1}{2} \sum_{j=1}^{n} \mu_j z_j^2, \qquad (2.6)$$

where the μ_j are the eigenvalues of A. From (2.6) we also have

$$\frac{\partial(x_1, \ldots, x_n)}{\partial(z_1, \ldots, z_n)}\bigg|_{z=0} = 1. \qquad (2.7)$$

To see this, we let Q denote the Hessian matrix of $(f \circ \bar{h})(z)$ at $z = 0$. From (2.6) it is easily seen that Q is the diagonal matrix whose entries are $\mu_1, \ldots, \mu_n$. On the other hand, by the chain rule, we have

$$Q = \bar{h}'(0)^{\mathrm{T}} A \bar{h}'(0),$$

where $\bar{h}'(z) = (\partial x_i / \partial z_j)$. Hence

$$\det Q = [\det \bar{h}'(0)]^2 \det A.$$

Since $\det Q = \det A = \mu_1 \cdots \mu_n$, $\det \bar{h}'(0) = \pm 1$. The $+$ sign can be ensured by interchanging the role of two of the variables $z_1, \ldots, z_n$ in (2.6), if necessary, and (2.7) is proved. Now, since A is real symmetric, if P is an orthogonal (unitary) matrix whose columns are normalized eigenvectors of A, and if $z = Py$, then

$$\langle Ay, y \rangle = \sum_{j=1}^{n} \mu_j z_j^2; \tag{2.8}$$

see Noble (1969, p. 388). Coupling (2.6) and (2.8), we have the identity (2.4). The result (2.5) follows from (2.7) and the fact that

$$\frac{\partial(z_1, \ldots, z_n)}{\partial(y_1, \ldots, y_n)} = \det P = 1. \tag{2.9}$$

Before stating the main result of this section, we first recall the meaning of *signature* of a real symmetric matrix A, which is denoted by sgn A, and is defined to be $v_+ - v_-$, where v_+ and v_- are the number of positive and negative eigenvalues of A, respectively.

Theorem 1. *If g is supported in D and if f has exactly one non-degenerate stationary point x_0 in the support of g, then the oscillatory integral (1.1) has the asymptotic expansion*

$$I(\lambda) \sim e^{i\lambda f(x_0)} \sum_{j=0}^{\infty} a_j(f, g) \lambda^{-j-n/2}, \qquad as \ \lambda \to +\infty, \tag{2.10}$$

where the coefficients $a_j(f, g)$ can be expressed in terms of the derivatives of f and g at x_0. In particular,

$$a_0(f, g) = (2\pi)^{n/2} |\det A|^{-1/2} g(x_0) e^{i\pi\sigma/4}, \tag{2.11}$$

where A is the Hessian matrix of f at x_0 and $\sigma = $ sgn A.

Proof. Inserting (2.4) in (1.1) gives

$$I(\lambda) = \exp[i\lambda f(x_0)] \int_{D'} \varphi(y) \exp\left(i\frac{\lambda}{2}\langle Ay, y\rangle\right) dy, \qquad (2.12)$$

where D' is the image of D and φ is the product of g and the Jacobian of the transformation involved. Put $\psi(y) = \exp(-\frac{1}{2}i\lambda\langle Ay, y\rangle)$, and consider the last integral as the inner product of φ and ψ. The Parseval formula suggests that

$$\int \varphi\bar{\psi}\, dy = \int \hat{\varphi}\bar{\hat{\psi}}\, d\eta, \qquad (2.13)$$

where $\hat{\varphi}$ and $\hat{\psi}$ are respectively the Fourier transforms of φ and ψ, that is,

$$\hat{\varphi}(\eta) = \frac{1}{(2\pi)^{n/2}} \int_{\mathbb{R}^n} \varphi(y)e^{i\eta\cdot y}\, dy. \qquad (2.14)$$

The inverse Fourier transform is defined by

$$\varphi(y) = \frac{1}{(2\pi)^{n/2}} \int_{\mathbb{R}^n} \hat{\varphi}(\eta)e^{-iy\cdot\eta}\, d\eta. \qquad (2.15)$$

Since φ is a C^∞-function with compact support, $\hat{\varphi}$ is a rapidly decreasing function; cf. Theorem 7 of Chapter V. Furthermore, since $(\bar{\hat{\psi}})^\wedge = \bar{\psi}$, formula (2.13) can be proved by an interchange of the order of integration. We shall now show that

$$\hat{\psi}(\eta) = \left(\frac{1}{\lambda}\right)^{n/2} |\det A|^{-1/2} \exp\left(-\frac{i\sigma\pi}{4} + \frac{i}{2\lambda}\langle A^{-1}\eta, \eta\rangle\right). \qquad (2.16)$$

First we recall that if z is real then the Fourier transform of $e^{-\mu z^2/2}$ where $\operatorname{Re}\mu \geq 0$ and $\mu \neq 0$ is $e^{-n^2/2\mu}/\sqrt{\mu}$ with $\sqrt{\mu}$ being defined for $\operatorname{Re}\mu \geq 0$ so that it is equal to 1 when $\mu = 1$. Thus the Fourier transform of $\exp(-i\lambda\sum\mu_j z_j^2/2)$, where μ_j are real and $\neq 0$, is

$$\left(\frac{1}{\lambda}\right)^{n/2} \left|\prod_{j=1}^{n} \mu_j\right|^{-1/2} \exp\left(-\frac{i\pi}{4}\sum \operatorname{sgn}\mu_j + \frac{i}{2\lambda}\sum_{j=1}^{n} \mu_j^{-1}\eta_j^2\right).$$

Since $\Pi\mu_j = \det A$ and $\sum \operatorname{sgn}\mu_j = \sigma$, by using (2.8) we have

$$\hat{\psi}(\eta) = \left(\frac{1}{\lambda}\right)^{n/2} |\det A|^{-1/2} \exp\left(-\frac{i\pi\sigma}{4} + \frac{i}{2\lambda}\sum_{j=1}^{n} \mu_j^{-1}(P^{\mathrm{T}}\eta)_j^2\right),$$

where $(P^{\mathrm{T}}\eta)_j$ denotes the jth component of $P^{\mathrm{T}}\eta$. The result (2.16) now follows from the fact that

$$\langle A^{-1}\eta, \eta\rangle = \sum_{j=1}^{n} \mu_j^{-1}(P^{\mathrm{T}}\eta)_j^2.$$

Returning to (2.12) and (2.13), we obtain

$$I(\lambda) = \left(\frac{1}{\lambda}\right)^{n/2} |\det A|^{-1/2} \exp\left[i\lambda f(x_0) + \frac{i\pi\sigma}{4} \right]$$

$$\times \int \hat{\varphi}(\eta) \exp\left\{ -\frac{i}{2\lambda} \langle A^{-1}\eta, \eta\rangle \right\} d\eta. \qquad (2.17)$$

By Taylor's formula

$$\exp\left(-\frac{i}{2\lambda} \langle A^{-1}\eta, \eta\rangle \right) = \sum_{|\alpha| < 2p} c_\alpha \lambda^{-|\alpha|/2}\eta^\alpha + R_{2p},$$

where

$$c_\alpha = \frac{1}{\alpha!} \frac{\partial^{|\alpha|}}{\partial \zeta^\alpha} \exp\left(-\frac{i}{2} \langle A^{-1}\zeta, \zeta\rangle \right)\Big|_{\zeta = 0}$$

and $\eta = \sqrt{\lambda}\zeta$. The remainder is given by

$$R_{2p} = \frac{1}{\lambda^p} \sum_{|\alpha| = 2p} \frac{1}{\alpha!} \eta^\alpha \frac{\partial^{|\alpha|}}{\partial \zeta^\alpha} \exp\left(-\frac{i}{2} \langle A^{-1}\zeta, \zeta\rangle \right)\Big|_{\zeta = \bar{\zeta}},$$

$\bar{\zeta}$ being on the line segment joining 0 to ζ. Note that the coefficients c_α vanish for all odd $|\alpha|$, and that, in view of Lemma 4 of Chapter V, we can regard $\int \eta^\alpha \hat{\varphi}(\eta)\, d\eta$ as the inverse Fourier transform of $(2\pi)^{n/2}(-1)^{|\alpha|}(D^\alpha\varphi)^\wedge$ evaluated at $y = 0$. Thus

$$\int \hat{\varphi}(\eta) \exp\left(-\frac{i}{2\lambda} \langle A^{-1}\eta, \eta\rangle \right) d\eta$$

$$= (2\pi)^{n/2} \sum_{|\alpha| < 2p} (-1)^{|\alpha|} c_\alpha D^\alpha\varphi(0)\lambda^{-|\alpha|/2} + O(\lambda^{-p}). \quad (2.18)$$

The final result (2.10) now follows from (2.17) and (2.18). ∎

The above elegant proof is due to Hörmander (1971). A drawback of this proof is that it does not lead to formulas for the coefficients in the expansion explicitly in terms of the original functions $f(x)$ and $g(x)$.

This is due to the fact that the diffeomorphism h in the Morse lemma is not explicitly known. To overcome this difficulty, we present the following modification given by Fedoryuk (1971, p. 76); see also Hörmander (1983, p. 220).

Let L be the differential operator

$$L = \frac{i}{2} \langle A^{-1} \nabla_x, \nabla_x \rangle, \qquad (2.19)$$

where A is again the Hessian matrix given in (2.1) and ∇_x is the gradient $(\partial/\partial x_1, \ldots, \partial/\partial x_n)$. Define

$$F_0(x) = f(x) - f(x_0) - \tfrac{1}{2} \langle A(x - x_0), x - x_0 \rangle \qquad (2.20)$$

and

$$G_\lambda(x) = g(x) \exp[i\lambda F_0(x)]. \qquad (2.21)$$

Theorem 2. *Under the conditions of Theorem 1, we have*

$$I(\lambda) \sim \left(\frac{2\pi}{\lambda}\right)^{n/2} \Omega(f) \exp[i\lambda f(x_0)] \sum_{j=0}^{\infty} \frac{\lambda^{-j}}{j!} (L^j G_\lambda)(x_0), \qquad (2.22)$$

where L is the operator defined in (2.19),

$$\Omega(f) = |\det A|^{-1/2} \exp\left(\frac{i\pi\sigma}{4}\right) \qquad (2.23)$$

and $(L^j G_\lambda)(x_0)$ is a polynomial in λ of degree $\leq [2j/3]$.

Proof. Inserting (2.20) in (1.1) gives

$$I(\lambda) = \exp[i\lambda f(x_0)] \int G_\lambda(x) \exp\left[\frac{i\lambda}{2} \langle A(x - x_0), x - x_0 \rangle\right] dx.$$

The Fourier transform of $\exp[-i\lambda\langle A(x - x_0), x - x_0 \rangle/2]$ is

$$\lambda^{-n/2} |\det A|^{-1/2} \exp\left(-\frac{i\sigma\pi}{4} + \frac{i}{2\lambda} \langle A^{-1}\eta, \eta \rangle + i\eta \cdot x_0\right).$$

By the Parseval formula (2.13),

$$\int G_\lambda(x) \exp\left[\frac{i\lambda}{2} \langle A(x - x_0), x - x_0 \rangle\right] dx$$

$$= \lambda^{-n/2} \Omega(f) \int \hat{G}_\lambda(\eta) e^{-ix_0 \cdot \eta} \exp\left(-\frac{i}{2\lambda} \langle A^{-1}\eta, \eta \rangle\right) d\eta.$$

Expanding the exponential in a Taylor series, we obtain

$$\int G_\lambda(x) \exp\left\{\frac{i\lambda}{2} \langle A(x - x_0), x - x_0\rangle\right\} dx$$

$$\sim \lambda^{-n/2}\Omega(f) \sum_{j=0}^{\infty} \frac{(-1)^j}{j!} \left(\frac{i}{2\lambda}\right)^j \int \langle A^{-1}\eta, \eta\rangle^j \hat{G}_\lambda(\eta) e^{-ix_0 \cdot \eta} \, d\eta.$$

Note that

$$\int \langle A^{-1}\eta, \eta\rangle^j \hat{G}_\lambda(\eta) e^{-ix_0 \cdot \eta} \, d\eta$$

$$= (-1)^j \int [\langle A^{-1}\nabla, \nabla\rangle^j G_\lambda]^\wedge \, e^{-i(x_0 - x)\cdot\eta} \, d\eta \bigg|_{x=0}$$

$$= (2\pi)^{n/2}(-1)^j \{\langle A^{-1}\nabla, \nabla\rangle^j G_\lambda(x_0 - x)\} \bigg|_{x=0}.$$

Expansion (2.22) now follows. To show that $(L^j G_\lambda)(x_0)$ is a polynomial in λ of degree $[2j/3]$, we first observe that the Taylor expansion of the function $F_0(x)$ in (2.20) begins with terms of order greater than or equal to three. We also note that since L is a linear differential operator of second order, it suffices to prove that if $f(z)$ and $g(z)$ are analytic functions of a single complex variable z in a neighborhood of $z = 0$ and $h(z) = z^3 g(z)$, then $\psi_j(\lambda) = (d/dz)^j f(z) \exp(i\lambda h(z))|_{z=0}$ is a polynomial of degree $\leq [j/3]$. Indeed, for small $\varepsilon > 0$, we have

$$\psi_j(\lambda) = \frac{j!}{2\pi i} \int_{|z|=\varepsilon} z^{-j-1} f(z) \exp[i\lambda h(z)] \, dz$$

$$= \frac{j!}{2\pi i} \sum_{k=0}^{\infty} \frac{(i\lambda)^k}{k!} \int_{|z|=\varepsilon} z^{3k-j-1} f(z)[g(z)]^k \, dz,$$

and all the terms in the last sum with $k \geq \frac{1}{3}(j + 1)$ vanish.   ∎

In view of the fact that $(L^j G_\lambda)(x_0)$ is a polynomial of degree at most $[2j/3]$, the terms in (2.22) can obviously be rearranged to yield an asymptotic expansion in descending powers of λ. Thus Theorem 1 can be deduced from Theorem 2.

We next consider the case where the stationary point x_0 is on the boundary of D. The following argument is an extension of what is given in §6 of Chapter VIII, and is taken from Jones (1982, p. 386). Let the boundary of D be given by $h(x) = 0$ with $\nabla h(x) \neq 0$, and suppose that near x_0,

$$f(x) = f(x_0) + \tfrac{1}{2}(x^{\mathrm{T}} - x_0^{\mathrm{T}})A(x - x_0) + \cdots,$$

$$h(x) = b^{\mathrm{T}}(x - x_0) + \tfrac{1}{2}(x^{\mathrm{T}} - x_0^{\mathrm{T}})B(x - x_0) + \cdots,$$

where b denotes the gradient of h evaluated at x_0, and A and B are respectively the Hessian matrix of f and h at x_0. Let P be the orthogonal matrix which diagonalizes the quadratic form $(x^{\mathrm{T}} - x_0^{\mathrm{T}})A(x - x_0)$, i.e., if $y = P(x - x_0)$, then

$$(x^{\mathrm{T}} - x_0^{\mathrm{T}})A(x - x_0) = \sum_{j=1}^{n} \mu_j y_j^2,$$

where the μ_j's are the eigenvalues of A; cf. (2.8). In terms of y, we have

$$f(x) = f(x_0) + \frac{1}{2} \sum_{j=1}^{n} \mu_j y_j^2 + \cdots, \tag{2.24}$$

$$h(x) = b^{\mathrm{T}}P^{\mathrm{T}}y + \frac{1}{2} y^{\mathrm{T}}PBP^{\mathrm{T}}y + \cdots. \tag{2.25}$$

We make one further transformation $y = Qu$ so that $b^{\mathrm{T}}P^{\mathrm{T}}y = c_0 u_1$, where c_0 is a constant and $u = (u_1, \ldots, u_n)$. We choose c_0 so that the interior of D corresponds to $u_1 > 0$. The objective of this transformation is to eliminate all linear terms in (2.25) except the first one, and in the meantime to retain the form of $f(x)$. Thus, the transformation $y = Qu$ gives

$$h(x) = c_0 u_1 + \frac{1}{2} u^{\mathrm{T}}Q^{\mathrm{T}}PBP^{\mathrm{T}}Qu + \cdots, \tag{2.26}$$

$$f(x) = f(x_0) + \frac{1}{2} \sum_{j=1}^{n} \mu_j u_j^2 + \cdots. \tag{2.27}$$

The matrix Q satisfies $\det Q = \pm 1$, and a necessary and sufficient condition for the existence of such a matrix is

$$\sum_{j=1}^{n} \frac{b_j^2}{\mu_j} \neq 0;$$

cf. Eq. (6.3) and Ex. 7 in Chapter VIII. For the existence of the matrix Q we refer to McClure and Wong (1990).

Returning to (2.26), we define X by $c_0 X_1 = h(x)$ and $X_j = u_j$ $(j \neq 1)$. Solving for the u_j's in terms of X_j's, we obtain

$$u_1 = X_1 + \text{quadratic terms} + \cdots \qquad (2.28)$$

$$f(x) = f(x_0) + \frac{1}{2} \sum_{j=1}^{n} \mu_j X_j^2 + \cdots. \qquad (2.29)$$

Let $F(X)$ denote the transform of $f(x)$ and define

$$G(X) = g(x) \left| \frac{\partial(x_1, \dots, x_n)}{\partial(X_1, \dots, X_n)} \right|.$$

Note that since the matrix P is unitary, and $\det Q = \pm 1$, the Jacobians of the transformations $y = P(x - x_0)$ and $y = Qu$ are equal to ± 1. Furthermore, from (2.28), it is easily seen that the value of the Jacobian of the transformation $u \to X$ is one. Thus

$$\frac{\partial(x_1, \dots, x_n)}{\partial(X_1, \dots, X_n)} \bigg|_{X=0} = 1,$$

and $G(0) = g(x_0)$ cf. the argument for (2.7). The integral (1.1) now becomes

$$I(\lambda) = \int G(X) e^{i\lambda F(X)} \, dX, \qquad (2.30)$$

where integration is over the half-space $X_1 > 0$. From (2.29), we have

$$F(X) = f(x_0) + \frac{1}{2} \sum_{j=1}^{n} \mu_j X_j^2 + \cdots.$$

The asymptotic expansion of the integral in (2.30) can be derived in several different ways, and probably the simplest one is to first make a change of variable $X \to z$, corresponding to that given in equation (3.5) of Chapter VIII, so that

$$F(X) = f(x_0) + \frac{1}{2} \sum_{j=1}^{n} \mu_j z_j^2,$$

and then write

$$I(\lambda) = e^{i\lambda f(x_0)} \int \varphi(z) \exp\left(\frac{i\lambda}{2} \sum_{j=1}^{n} \mu_j z_j^2 \right) dz. \qquad (2.31)$$

Here $\varphi(z)$ denotes the product of $G(X)$ and the Jacobian of the transformation, and the integration is again over a half-space, namely, $z_1 > 0$. Note that

$$\left.\frac{\partial(X_1, \ldots, X_n)}{\partial(z_1, \ldots, z_n)}\right|_{z=0} = 1.$$

Hence, $\varphi(0) = G(0) = g(x_0)$. To each dz_j-integral in (2.31) we can apply the one-dimensional stationary-phase approximation. Since the range of integration with respect to z_1 is from 0 to $+\infty$, whereas all other variables range from $-\infty$ to $+\infty$, we obtain, to the leading order,

$$\begin{aligned} I(\lambda) &\sim \frac{1}{2}\left(\frac{2\pi}{\lambda}\right)^{n/2} g(x_0)|\mu_1 \cdots \mu_n|^{-1/2} \exp\left[i\lambda f(x_0) + \frac{i\pi}{4}\sum_{j=1}^{n} \operatorname{sgn}\mu_j\right] \\ &= \frac{1}{2}\left(\frac{2\pi}{\lambda}\right)^{n/2} g(x_0)|\det A|^{-1/2} \exp\left[i\lambda f(x_0) + \frac{i\pi\sigma}{4}\right], \end{aligned} \tag{2.32}$$

which is exactly half of the approximation given in (1.3).

3. Points of Tangential Contact

These are classified as critical points of the second kind in the two-dimensional case discussed in Chapter VIII. They are boundary points of the domain D, which are not stationary points of f, but at which a surface $f(x) = $ constant is tangent to the boundary ∂D. We shall assume that $\nabla f \neq 0$ in $\overline{D}$. Define

$$\mathbf{u}_0 = \frac{\nabla f}{|\nabla f|^2}g, \tag{3.1}$$

and

$$g_{s+1} = \nabla \cdot \mathbf{u}_s, \qquad \mathbf{u}_{s+1} = \frac{\nabla f}{|\nabla f|^2}g_{s+1} \tag{3.2}$$

for $s = 0, 1, 2, \ldots$. Repeated application of the divergence theorem (Edwards 1973, p. 380), as illustrated in §1 of Chapter VIII, gives

$$I(\lambda) = -\sum_{s=0}^{n-1}\left(\frac{i}{\lambda}\right)^{s+1}\int_{\Sigma}\mathbf{u}_s\cdot\mathbf{n}\,e^{i\lambda f}\,dA + \left(\frac{i}{\lambda}\right)^n\int_{D}g_n e^{i\lambda f}\,dx, \tag{3.3}$$

where Σ is the oriented boundary of D, dA is the surface element and $\mathbf{n}$ is the unit outward normal to Σ. If the support of g is contained in the interior of D, then the terms under the summation sign in (3.3) all vanish. Thus, for any $n \geq 1$, we have

$$I(\lambda) = O(\lambda^{-n}), \qquad \text{as } \lambda \to +\infty. \tag{3.4}$$

If g is not identically zero on the boundary of D, then each integral on the right-hand side of (3.3) contributes, and the leading term gives

$$I(\lambda) \sim -\frac{i}{\lambda} \int_{\Sigma} \mathbf{u}_0 \cdot \mathbf{n}\, e^{i\lambda f}\, dA. \tag{3.5}$$

Let x_0 be a boundary point of D, at which a surface $f(x) = \text{constant}$ is tangent to ∂D, and assume that the $(n-1)$-dimensional surface Σ is parametrically represented by

$$x = x(u), \qquad u = (u_1, \ldots, u_{n-1}) \in \Omega,$$

so that $x_0 = x(u_0)$ for some u_0 in the interior of Ω, Ω being the parameter domain. Let J denote the $n \times (n-1)$ matrix

$$J = \left(\frac{\partial x_i}{\partial u_j}\right), \tag{3.6}$$

and let

$$\det J_i = \frac{\partial(x_1, \ldots, x_{i-1}, x_{i+1}, \ldots, x_n)}{\partial(u_1, \ldots, u_{n-1})}. \tag{3.7}$$

Furthermore, put

$$D(u) = [\det J^{T}J]^{1/2} = [\textstyle\sum (\det J_i)^2]^{1/2}. \tag{3.8}$$

(The second equality is a well-known result in linear algebra; see Edwards (1973, p. 327).) The components of $\mathbf{n}$ are then given by

$$n_i(x) = \frac{(-1)^{i-1}}{D(u)} \det J_i, \qquad i = 1, \ldots, n. \tag{3.9}$$

From the definition of a surface integral, we have

$$\int_{\Sigma} \mathbf{u}_0 \cdot \mathbf{n}\, e^{i\lambda f}\, dA = \int_{\Omega} q(u)e^{i\lambda p(u)}\, du, \tag{3.10}$$

where

$$p(u) = f(x(u)) \tag{3.11}$$

and

$$q(u) = (\mathbf{u}_0 \cdot \mathbf{n})(x)D(u); \tag{3.12}$$

see Ex. 7. The last integral is of dimension $n - 1$. Note that

$$\frac{\partial}{\partial u_i} p(u) = \sum_{l=1}^{n} \frac{\partial f}{\partial x_l} \frac{\partial x_l}{\partial u_i} = \nabla f \cdot \frac{\partial x}{\partial u_i}. \tag{3.13}$$

Since the surface $f(x) = $ constant is tangent to ∂D at x_0, ∇f is normal to Σ at x_0 and

$$\nabla f(x_0) \cdot \frac{\partial x}{\partial u_i}(u_0) = 0; \tag{3.14}$$

see Smith (1971, pp. 263–264). The last two equations together imply

$$\frac{\partial p}{\partial u_i}(u_0) = 0, \qquad i = 1, \ldots, n - 1, \tag{3.15}$$

i.e., u_0 is a stationary point of $p(u)$ in the interior of Ω. Thus, to the integral on the right-hand side of (3.10), we can apply the stationary phase approximation (1.3) with n replaced by $n - 1$. The result is

$$\int_{\Sigma} \mathbf{u}_0 \cdot \mathbf{n} \, e^{i\lambda f} \, dA \sim q(u_0) |\det B|^{-1/2} \left(\frac{2\pi}{\lambda}\right)^{(n-1)/2} \exp\left[i\lambda f(x_0) + \frac{i\pi\sigma}{4}\right], \tag{3.16}$$

where

$$B = \left(\frac{\partial^2 p}{\partial u_i \, \partial u_j}\right)\Big|_{u = u_0} \tag{3.17}$$

and σ is the signature of B. Formula (3.16) holds only if the $(n - 1) \times (n - 1)$ matrix B is nonsingular.

To express (3.16) in terms of the original data, namely, f and g, we first observe that since both ∇f and $\mathbf{n}$ are normal to the surface Σ, there exists a constant α such that $\nabla f(x_0) = \alpha \mathbf{n}(x_0)$. Taking the norm of each side of the equation gives $\alpha = \pm |\nabla f(x_0)|$. Hence

$$\nabla f(x_0) = \pm |\nabla f(x_0)| \mathbf{n}(x_0). \tag{3.18}$$

This together with (3.1) and (3.12) yields

$$q(u_0) = \pm \frac{g(x_0)}{|\nabla f(x_0)|} D(u_0). \tag{3.19}$$

To derive an alternative expression for $\partial^2 p/\partial u_i\, \partial u_j$, we suppose that the boundary Σ of the domain D is also represented by $h(x) = 0$ in a neighborhood of x_0. A unit normal to Σ at x_0 is then given by $\nabla h(x_0)/|\nabla h(x_0)|$. On account of (3.18), there is a constant K such that

$$\nabla f(x_0) = K\, \nabla h(x_0). \tag{3.20}$$

From (3.13) it follows that

$$\frac{\partial^2 p}{\partial u_i\, \partial u_j} = \sum_{l=1}^{n} \sum_{k=1}^{n} \frac{\partial^2 f}{\partial x_l\, \partial x_k} \frac{\partial x_k}{\partial u_j} \frac{\partial x_l}{\partial u_i} + \sum_{l=1}^{n} \frac{\partial f}{\partial x_l} \frac{\partial^2 x_l}{\partial u_i\, \partial u_j}. \tag{3.21}$$

Note that Σ is parametrically represented by $x = x(u)$, $u = (u_1, \ldots, u_{n-1})$, and is also given by $h(x) = 0$. Hence, $h(x(u)) = 0$ for $u \in \Omega$. Upon differentiation, we have

$$\sum_{k=1}^{n} \frac{\partial h}{\partial x_k} \frac{\partial x_k}{\partial u_j} = 0$$

and

$$\sum_{k=1}^{n} \sum_{l=1}^{n} \frac{\partial^2 h}{\partial x_l\, \partial x_k} \frac{\partial x_l}{\partial u_i} \frac{\partial x_k}{\partial u_j} + \sum_{k=1}^{n} \frac{\partial h}{\partial x_k} \frac{\partial^2 x_k}{\partial u_j\, \partial u_i} = 0.$$

In view of (3.20), we further obtain

$$\sum_{k=1}^{n} \sum_{l=1}^{n} \frac{\partial^2 h}{\partial x_l\, \partial x_k} \frac{\partial x_l}{\partial u_i} \frac{\partial x_k}{\partial u_j}(u_0) = -\frac{1}{K} \sum_{k=1}^{n} \frac{\partial f}{\partial x_k} \frac{\partial^2 x_k}{\partial u_j\, \partial u_i}(u_0). \tag{3.22}$$

Coupling (3.21) and (3.22) yields

$$\frac{\partial^2 p}{\partial u_i\, \partial u_j}(u_0) = \left[\sum_{k=1}^{n} \sum_{l=1}^{n} \left(\frac{\partial^2 f}{\partial x_l\, \partial x_k} - K \frac{\partial^2 h}{\partial x_l\, \partial x_k} \right) \frac{\partial x_l}{\partial u_i} \frac{\partial x_k}{\partial u_j} \right](u_0). \tag{3.23}$$

Let the partial derivatives now be denoted by subscripts. Then, from a well-known result in linear algebra (Edwards 1973, p. 326), we have

$$\det\left(\frac{\partial^2 p}{\partial u_i\, \partial u_j}(u_0) \right) = \frac{D^2(u_0)}{|\nabla f(x_0)|^2} \sum_{p=1}^{n} \sum_{q=1}^{n} f_{x_p} f_{x_q}\, \mathrm{cof}[f_{x_p x_q} - K h_{x_p x_q}], \tag{3.24}$$

where the symbol $\mathrm{cof}[a_{pq}]$ denotes the cofactor of the element a_{pq} in the matrix $[a_{pq}]$; see Ex. 9. Thus a combination of (3.16), (3.19), and (3.24) gives

$$\int_{\Sigma} \mathbf{u}_0 \cdot \mathbf{n} e^{i\lambda f}\, dA \sim \pm \frac{g(x_0)}{\sqrt{|J|}} \exp\left\{ i\lambda f(x_0) + \frac{i\pi\sigma}{4} \right\} \left(\frac{2\pi}{\lambda} \right)^{(n-1)/2}, \tag{3.25}$$

where

$$\mathbf{J} = \sum_{p=1}^{n} \sum_{q=1}^{n} f_{x_p} f_{x_q} \operatorname{cof}[f_{x_p x_q} - K h_{x_p x_q}]. \tag{3.26}$$

From (3.25) and (3.5), we conclude that the contribution from a point of tangential contact is given by

$$I(\lambda) \sim \mp \frac{i g(x_0)}{2\pi \sqrt{|\mathbf{J}|}} \exp\left[i\lambda f(x_0) + \frac{i\pi\sigma}{4} \right] \left(\frac{2\pi}{\lambda} \right)^{(n+1)/2}, \tag{3.27}$$

where the ambiguous sign is the opposite of that in (3.18).

4. Degenerate Stationary Point

A stationary point x_0 is degenerate if the Hessian matrix A in (2.1) is singular, which is equivalent to stating that the rank of A is less than n. In this case, Morse's lemma does not apply, but we have the following powerful result (Poston and Stewart 1978, p. 61).

Splitting Lemma. *Let* $f: \mathbb{R}^n \to \mathbb{R}$ *be a* C^∞-*function, and let* x_0 *be a stationary point of* f. *If the Hessian matrix of* f *at* x_0 *has rank* r *(and corank* $n - r$), *then there exist neighborhoods* U, V *of the points* $u = 0$, $x = x_0$ *and a diffeomorphism* $h: U \to V$ *such that*

$$f(h(u)) = \pm u_1^2 \pm \cdots \pm u_r^2 + p(u_{r+1}, \ldots, u_n), \tag{4.1}$$

where $p: \mathbb{R}^{n-r} \to \mathbb{R}$ *is* C^∞.

This result essentially allows us to split the function into two pieces, a Morse piece on one set of variables and a degenerate piece on a different set, whose number is equal to the corank. We shall consider only the simplest case, namely, when the rank of the Hessian matrix A in (2.1) is $n - 1$. Returning to (1.1), we shall assume that the domain of integration is the whole space so there are no boundary points to investigate, and that $g(x)$ has compact support close to the stationary point x_0. Thus

$$I(\lambda) = \int_{\mathbb{R}^n} g(x) e^{i\lambda f(x)} \, dx. \tag{4.2}$$

Inserting (4.1) in (4.2) gives

$$I(\lambda) = \int_{\mathbb{R}^n} G(u) \exp\left[i\lambda \sum_{i=1}^{n-1} \pm u_i^2 + i\lambda p(u_n) \right] du,$$

where $G(u)$ is the product of $g(x)$ and the Jacobian of the transformation. An asymptotic expansion of the integral with respect to $(u_1, \ldots, u_{n-1})$ can be obtained from Theorem 1 (cf. Ex. 12), and we are left with only a one-dimensional integral with respect to u_n. Thus, after changing notation, our problem is reduced to the study of the behavior of

$$\int_{-\infty}^{\infty} q(u) e^{i\lambda p(u)} \, du,$$

where q is a C^∞-function with compact support close to 0 and $p'(0) = p''(0) = 0$. A detailed discussion of this integral is given in Chapter II, Section 3.

It should be noted that, although powerful, the above procedure does not apply to simple cases such as

$$f(x_1, x_2) = x_1 x_2^2 \qquad \text{and} \qquad f(x_1, x_2, x_3) = x_1 x_2 x_3,$$

since the Hessian matrix of each of these functions at the origin has rank 0. It is therefore important to consider cases where the phase function is of the form

$$f(x) = \left(\prod_{i=1}^n x_i^{\alpha_i} \right) \varepsilon(x), \tag{4.3}$$

where $\varepsilon(x)$ is an invertible real analytic function (i.e., $\varepsilon(0) \neq 0$) and each α_i is a nonnegative integer. Any real analytic function $F(\not\equiv 0)$, defined in a neighborhood of $0 \in \mathbb{R}^n$, can be locally put in this form by the theorem of Hironaka on the resolution of singularities; see Atiyah (1970). In what follows, we shall present a brief outline of a procedure by which one can derive the asymptotic expansion of $I(\lambda)$ in (4.2) when the phase function is given by (4.3). Without loss of generality we may assume that $\varepsilon(x) = 1$ and the support of g is contained in $[-1, 1] \times \cdots \times [-1, 1]$.

First we note that for $0 < c < \frac{1}{2}$, we have

$$\frac{1}{2\pi i} \int_{c-i\infty}^{c+i\infty} \mu^{-z} \Gamma(z) e^{i\pi z/2} \, dz = e^{i\mu}, \tag{4.4}$$

whether μ is positive or negative; see Ex. 13. When μ is negative, the principal value of μ^{-z} in (4.4) must be taken. Now let $Q_1 = [0, 1] \times \cdots \times [0, 1]$ and consider

$$I_1(\lambda) = \int_{Q_1} g(x)e^{i\lambda x^\alpha}\, dx. \tag{4.5}$$

Take c_0 such that $0 < c_0 < \min(1/2, 1/\alpha_1, \ldots, 1/\alpha_n)$. From (4.4), it follows that

$$I_1(\lambda) = \int_{Q_1} g(x)\, \frac{1}{2\pi i} \int_{c_0 - i\infty}^{c_0 + i\infty} (\lambda x^\alpha)^{-z}\Gamma(z)e^{i\pi z/2}\, dz\, dx.$$

Interchanging the order of integration gives

$$I_1(\lambda) = \frac{1}{2\pi i} \int_{c_0 - i\infty}^{c_0 + i\infty} \lambda^{-z}\Gamma(z)e^{i\pi z/2}G(z)\, dz, \tag{4.6}$$

where

$$G(z) = \int_{Q_1} (x^\alpha)^{-z} g(x)\, dx;$$

see Ex. 14. Note that G is analytic and bounded in $\operatorname{Re} z < \delta$, for $\delta < \min\{1/\alpha_1, \ldots, 1/\alpha_n\}$. Furthermore, since g vanishes on $x_1 = 1, \ldots, x_n = 1$, by partial integration we obtain

$$G(z) = \frac{1}{\alpha_1 z - 1} \int_{Q_1} x_1^{-\alpha_1 z + 1} x_2^{-\alpha_2 z} \cdots x_n^{-\alpha_n z} \frac{\partial g}{\partial x_1}(x)\, dx_1 \cdots dx_n, \tag{4.7}$$

and hence by repetition

$$G(z) = \prod_{j=1}^{n} \left(\prod_{i=1}^{k_j} \frac{1}{\alpha_j z - i} \right) \int_{Q_1} x^{-\alpha z + k} D^k g(x)\, dx$$

for any multi-index $k = (k_1, \ldots, k_n)$. From this it is easily seen that the poles of $G(z)$ are located at the points i/α_j, $j = 1, \ldots, n$, $i = 1, 2, \ldots$. The asymptotic expansion of $I_1(\lambda)$ is then obtained by translating the contour in (4.6) to the right, the terms in the expansion appearing as residues; cf. Chapter III, Section 7.

Suppose that $1/\alpha_1 < 1/\alpha_j$ for $j > 1$. Then $G(z)$ has a simple pole at $z = 1/\alpha_1$. The residue of G at that pole is

$$
\begin{aligned}
r_{11} &= \frac{1}{\alpha_1} \int_{Q_1} x_2^{-\alpha_2/\alpha_1} \cdots x_n^{-\alpha_n/\alpha_1} \frac{\partial g}{\partial x_1}(x)\, dx \\
&= -\frac{1}{\alpha_1} \int_0^1 \cdots \int_0^1 x_2^{-\alpha_2/\alpha_1} \cdots x_n^{-\alpha_n/\alpha_1} g(0, x_2, \ldots, x_n)\, dx_2 \cdots dx_n.
\end{aligned}
\tag{4.8}
$$

Note that r_{11} defines a distribution supported on the coordinate plane $x_1 = 0$. From (4.6), we obtain

$$
I_1(\lambda) = -\lambda^{-1/\alpha_1} \Gamma\!\left(\frac{1}{\alpha_1}\right) e^{i\pi/2\alpha_1} r_{11} + \frac{1}{2\pi i} \int_{c_1 - i\infty}^{c_1 + i\infty} \lambda^{-z} \Gamma(z) e^{i\pi z/2} G(z)\, dz, \tag{4.9}
$$

where $1/\alpha_1 < c_1 < 1/\alpha_j$, $j > 1$. It is evident that the last integral is $O(\lambda^{-c_1}) = o(\lambda^{-1/\alpha_1})$. Thus, (4.9) gives the leading term of the asymptotic expansion of $I_1(\lambda)$. Higher terms in the expansion can be obtained in a similar manner.

The above analysis applies not only to the unit cube Q_1, but, in fact, can be carried out with any similar cube $I_1 \times \cdots \times I_n$, where each $I_j = [0, 1]$ or $[-1, 0]$. The final result is then obtained by summing over all such cubes.

5. Laplace's Approximation in $\mathbb{R}^n$

We now turn to the consideration of the Laplace-type integral

$$
J(\lambda) = \int_D g(x) e^{-\lambda f(x)}\, dx, \tag{5.1}
$$

where λ is again a large positive parameter and D is a possibly unbounded domain in $\mathbb{R}^n$. As before, we assume that both f and g are infinitely differentiable in D. Furthermore, we assume that the following conditions hold:

(i)   The integral $J(\lambda)$ converges absolutely for all $\lambda \geq \lambda_0$.

(ii) For every $\varepsilon > 0$, $\rho(\varepsilon) > 0$ where

$$
\rho(\varepsilon) = \inf\{f(x) - f(x_0): x \in D \quad \text{and} \quad |x - x_0| \geq \varepsilon\}.
$$

(iii)   The Hessian matrix

$$A = \left(\frac{\partial^2 f}{\partial x_i \, \partial x_j}\right)\Bigg|_{x = x_0}$$

is positive definite.

Condition (ii) implies that the function $f(x)$ has a minimum at, and only at, the point x_0. If x_0 is an interior point, this will in turn imply that x_0 is a critical point of f, i.e., $\nabla f(x_0) = 0$. Condition (iii) is equivalent to the statement that all eigenvalues of A are positive.

Theorem 3.   *If x_0 is an interior point of D, and if conditions* (i), (ii), *and* (iii) *hold, then the integral* (5.1) *has an asymptotic expansion of the form*

$$J(\lambda) \sim e^{-\lambda f(x_0)} \sum_{k=0}^{\infty} c_k \lambda^{-n/2 - k}, \qquad as \ \lambda \to \infty, \tag{5.2}$$

where the c_k are constants. In particular, we have the Laplace approximation

$$J(\lambda) \sim \left(\frac{2\pi}{\lambda}\right)^{n/2} g(x_0)(\det A)^{-1/2} \exp[-\lambda f(x_0)]. \tag{5.3}$$

Proof.   For any subdomain D_0 of D containing x_0 (Figure 9.1), we have

$$J(\lambda) = \int_{D_0} + \int_{D \setminus D_0} \equiv J_1(\lambda) + J_2(\lambda).$$

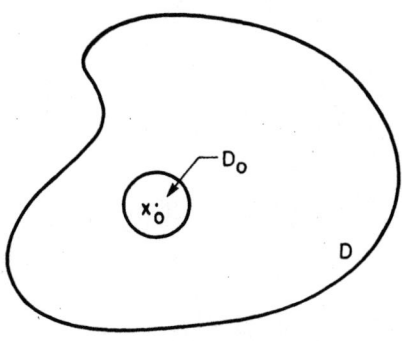

Fig. 9.1

By condition (ii), there exists a positive number c such that $f(x) \geq c + f(x_0)$ for all $x \in D \setminus D_0$. Therefore

$$|J_2(\lambda)| \leq K e^{-\lambda[f(x_0) + c]} \tag{5.4}$$

for some constant $K > 0$; cf. (10.3) and (10.4) in Chapter VIII. Now choose the domain D_0 so that the Morse lemma applies, i.e., there exists a diffeomorphism $h: \Omega \to D_0$, $y = 0 \in \Omega$, such that the substitution $x = h(y)$ gives

$$J_1(\lambda) = e^{-\lambda f(x_0)} \int_\Omega G(y) \exp\left(-\frac{\lambda}{2} \sum_{j=1}^n \mu_j y_j^2\right) dy, \tag{5.5}$$

where $\Omega = h^{-1}(D_0)$ and $G(y) = g(h(y)) \det h'(y)$. (Note that instead of the version of Morse's lemma given in (2.4), we have used its equivalent (2.6).) We also assume that D_0 has been chosen sufficiently small so that $\det h'(y) > 0$; cf. (2.7). By Taylor's theorem,

$$G(y) = \sum_{|\alpha| < p} \frac{1}{\alpha!} D^\alpha G(0) y^\alpha + R_p,$$

where

$$R_p = \sum_{|\alpha| = p} \frac{1}{\alpha!} D^\alpha G(\xi) y^\alpha$$

for some point $\xi \in \Omega$. Termwise integration then gives

$$J_1(\lambda) = e^{-\lambda f(x_0)} \left[\sum_{|\alpha| < p} \frac{1}{\alpha!} D^\alpha G(0) \int_\Omega y^\alpha \exp\left(-\frac{\lambda}{2} \sum_{j=1}^n \mu_j y_j^2\right) dy + \delta_p(\lambda) \right] \tag{5.6}$$

with

$$\delta_p(\lambda) = \sum_{|\alpha| = p} \frac{1}{\alpha!} \int_\Omega D^\alpha G(\xi) y^\alpha \exp\left\{-\frac{\lambda}{2} \sum_{j=1}^n \mu_j y_j^2\right\} dy. \tag{5.7}$$

To complete the proof, we recall the identity

$$\int_{-\infty}^\infty t^m e^{-vt^2} \, dt = \begin{cases} 0 & \text{if } m \text{ is odd,} \\[2ex] \Gamma\!\left(\dfrac{m+1}{2}\right) v^{-(m+1)/2} & \text{if } m \text{ is even,} \end{cases} \tag{5.8}$$

from which it follows that

$$\int_{\mathbb{R}^n} y^\alpha \exp\left(-\frac{\lambda}{2} \sum_{j=1}^n \mu_j y_j^2\right) dy = d_\alpha \lambda^{-(n+|\alpha|)/2}, \tag{5.9}$$

where d_α is zero if one of the α_i in $\alpha = (\alpha_1, \ldots, \alpha_n)$ is odd but otherwise is given by

$$d_\alpha = \left(\frac{2}{\mu}\right)^{(\alpha+1)/2} \Gamma\left(\frac{\alpha+1}{2}\right), \qquad \mu = (\mu_1, \ldots, \mu_n). \tag{5.10}$$

Here we have used the notation $\Gamma(v) = \Gamma(v_1) \cdots \Gamma(v_n)$ for any multi-index $v = (v_1, \ldots, v_n)$. Also, we recall the estimate

$$\int_\delta^\infty t^m e^{-vt^2} dt \le K_\delta e^{-v\delta^2}$$

for any $\delta > 0$ and $v > 1$, K_δ being independent of v. Thus, for any domain Ω containing 0, there exists a positive number ε such that

$$\int_\Omega y^\alpha \exp\left(-\frac{\lambda}{2} \sum_{j=1}^n \mu_j y_j^2\right) dy = d_\alpha \lambda^{-(n+|\alpha|)/2} + O(e^{-\varepsilon\lambda}). \tag{5.11}$$

Since $D^\alpha G$ is bounded on Ω, a simple estimation shows that the remainder $\delta_p(\lambda)$ in (5.7) satisfies

$$\delta_p(\lambda) = O(\lambda^{-(n+p)/2}). \tag{5.12}$$

Now insert (5.11) and (5.12) in (5.6), and replace p by $2p$. The result is

$$J_1(\lambda) = e^{-\lambda f(x_0)}\left[\sum_{k=0}^{p-1} c_k \lambda^{-n/2-k} + O(\lambda^{-n/2-p})\right],$$

which is equivalent to the relation (5.2). The constants c_k are given by

$$c_k = \sum_{|\alpha|=k} \frac{d_\alpha}{\alpha!} D^\alpha G(0). \tag{5.13}$$

In particular, from (2.7) and (5.10),

$$c_0 = d_0 G(0) = (2\pi)^{n/2}(\mu_1 \cdots \mu_n)^{-1/2} g(x_0) = (2\pi)^{n/2}(\det A)^{-1/2} g(x_0);$$

compare (5.3). This completes the proof of Theorem 3. ∎

    The Laplace approximation (5.3) is actually valid under much weaker smoothness conditions. Instead of $f(x)$ and $g(x)$ being C^∞, it is sufficient to assume that $g(x)$ is continuous and $f(x)$ has continuous second-order partial derivatives in a neighborhood of the critical point x_0. A proof of this assertation can be found in Ex. 5.

If the critical point x_0 is on the boundary of D, then we may proceed as in the last part of §2 (compare the discussion following the proof of Theorem 2). After a sequence of transformations $x \to y \to u \to X \to z$ as explained there, the integral $J(\lambda)$ can be expressed in the form

$$J(\lambda) = e^{-\lambda f(x_0)} \int \varphi(z) \exp\left(-\frac{\lambda}{2} \sum_{j=1}^{n} \mu_j z_j^2\right) dz, \qquad (5.14)$$

where $\varphi(0) = g(x_0)$ and the integration is over the half-space $z_1 > 0$; see (2.31). To each dz_j-integral, we can apply Watson's lemma. Since the range of integration with respect to z_1 is from 0 to ∞ and for all other variables it is from $-\infty$ to $+\infty$, the dominant contribution is given by

$$J(\lambda) \sim \frac{1}{2}\left(\frac{2\pi}{\lambda}\right)^{n/2} g(x_0)(\det A)^{-1/2} \exp[-\lambda f(x_0)], \qquad (5.15)$$

which is one-half of the approximation (5.3).

Finally we consider the case where the function $f(x)$ attains its minimum at a boundary point x_0, which is not a critical point (i.e., $\nabla f(x_0) \neq 0$). The analysis here parallels that given in §3. By using an estimate similar to (5.4), we may assume without loss of generality that D is bounded with a smooth boundary, and that $\nabla f \neq 0$ in $\bar{D}$. With the sequences $\{u_s\}$ and $\{g_s\}$ defined as in (3.1)–(3.2), we have, by repeated application of the divergence theorem,

$$J(\lambda) = -\sum_{s=0}^{n-1} \frac{1}{\lambda^{s+1}} \int_{\Sigma} \mathbf{u}_s \cdot \mathbf{n}\, e^{-\lambda f}\, dA + \frac{1}{\lambda^n} \int_{D} g_n e^{-\lambda f}\, dx;$$

cf. (3.3). Here Σ is the oriented boundary of D, dA is the surface element and $\mathbf{n}$ is the unit outward normal to Σ. The leading term of this expansion gives

$$J(\lambda) \sim -\frac{1}{\lambda} \int_{\Sigma} \mathbf{u}_0 \cdot \mathbf{n}\, e^{-\lambda f}\, dA. \qquad (5.16)$$

We shall assume, as in §3, that the $(n-1)$-dimensional surface Σ is parametrically represented by

$$x = x(u), \qquad u = (u_1, \ldots, u_{n-1}) \in \Omega,$$

and that $x_0 = x(u_0)$ for some u_0 in the interior of Ω. By the definition of a surface integral, we have

$$\int_\Sigma \mathbf{u}_0 \cdot \mathbf{n}\, e^{-\lambda f}\, dA = \int_\Omega q(u)e^{-\lambda p(u)}\, du, \qquad (5.17)$$

where

$$p(u) = f(x(u)) \qquad (5.18)$$

and

$$q(u) = (\mathbf{u}_0 \cdot \mathbf{n})(x)D(u); \qquad (5.19)$$

cf. (3.10)–(3.12). Since $f(x)$ has its absolute minimum at $x_0 \in \Sigma$, $p(u_0)$ is the minimum value of $p(u)$ in Ω. Since u_0 is an interior point of Ω, we see that

$$\frac{\partial p}{\partial u_i}(u_0) = 0, \qquad i = 1, \dots, n-1.$$

Thus, by Laplace's approximation (5.3) with n replaced by $n-1$, we have

$$\int_\Sigma \mathbf{u}_0 \cdot \mathbf{n}\, e^{-\lambda f}\, dA \sim q(u_0)|\det B|^{-1/2}\left(\frac{2\pi}{\lambda}\right)^{(n-1)/2} \exp[-\lambda f(x_0)], \quad (5.20)$$

provided the Hessian matrix

$$B = \left(\frac{\partial^2 p}{\partial u_i\, \partial u_j}\right)\Bigg|_{u=u_0} \qquad (5.21)$$

is nonsingular. To express this formula in terms of the original functions f and g, we recall from (3.19) and (3.24) that

$$q(u_0) = \pm\frac{g(x_0)}{|\nabla f(x_0)|}D(u_0) \qquad (5.22)$$

and

$$\det B = \frac{D^2(u_0)}{|\nabla f(x_0)|^2}\mathbf{J}, \qquad (5.23)$$

$D(u)$ and $\mathbf{J}$ being as given in (3.8) and (3.26) respectively. Since $f(x_0)$ is the absolute minimum of $f(x)$ in $\overline{D}$, $\nabla f(x_0)$ points into D. Hence $\nabla f(x_0) = -|\nabla f(x_0)|\mathbf{n}(x_0)$, and the "$-$" sign must be taken in (3.19) or equivalently (5.22). The combination of (5.16), (5.20), (5.22), and (5.23) gives the required result

$$J(\lambda) \sim \frac{g(x_0)}{2\pi\sqrt{|J|}} e^{-\lambda f(x_0)} \left(\frac{2\pi}{\lambda}\right)^{(n+1)/2}. \tag{5.24}$$

6. Multiple Fourier Transforms

We conclude this chapter with a discussion of the behavior of the n-dimensional Fourier transform

$$\hat{f}(t) = \frac{1}{(2\pi)^{n/2}} \int_{\mathbb{R}^n} f(x)e^{it\cdot x}\, dx, \tag{6.1}$$

where $t = (t_1, \ldots, t_n)$, $t\cdot x = t_1 x_1 + \cdots + t_n x_n$, and f is a locally integrable function in $\mathbb{R}^n$. In (6.1) it is assumed only that the Cauchy limit

$$\lim_{v \to \infty} \int_{|x| < v} f(x)e^{it\cdot x}\, dx$$

exists for every $t \in \mathbb{R}^n$. Thus the function f need not belong to $L^1(\mathbb{R}^n)$.

The notation to be used in this section will be the same as that given in §9 of Chapter V. If $p = (p_1, \ldots, p_n)$ is a multi-index of nonnegative integers, then we define the differential operator

$$D_p = (i)^{-|p|} D^p = \left(\frac{1}{i}\frac{\partial}{\partial x_1}\right)^{p_1} \cdots \left(\frac{1}{i}\frac{\partial}{\partial x_n}\right)^{p_n};$$

cf. (9.2) in Chapter V. For any open set $\Omega \subset \mathbb{R}^n$ and for each nonnegative integer m, the set $C^m(\Omega)$ consists of all complex functions f in Ω whose derivative $D^p f$ exists and is a continuous function in Ω for each multi-index p with $|p| \le m$.

In what follows, we shall assume that the function $f(x)$ in (6.1) satisfies the conditions (C$_1$)–(C$_3$) below.

(C$_1$) $f \in C^{2m}(\mathbb{R}^n \backslash \{0\})$, m being a nonnegative integer. Also, $f(x)$ is expressible in the form

$$f(x) = \sum c_{pq} x^p r^q + \varphi(x), \tag{6.2}$$

where the q's are real numbers, the p's are multi-indices, $r = \sqrt{x_1^2 + \cdots + x_n^2}$ and the sum is finite. For all p and q under the summation, we require $q + |p| > -n$, and if $Q = \max\{q + |p|\}$, then $2m - 1 \le Q + n < 2m + 1$.

(C_2)   *As $r \to 0^+$, the remainder $\varphi(x)$ in (6.2) satisfies*

$$(\Delta^j \varphi)(x) = O(r^{Q-2j+1}), \qquad j = 0, 1, \dots, m, \qquad (6.3)$$

if $Q + n \neq 2m - 1$, and

$$(\Delta^j \varphi)(x) = O(r^{Q-2j+2}), \qquad j = 0, 1, \dots, m, \qquad (6.4)$$

if $Q + n = 2m - 1$, Δ being the Laplacian operator.
(C_3)   *For some $\rho > 0$, each of the integrals*

$$\int_{|x| \geq \rho} (\Delta^j f)(x) e^{it \cdot x}\, dx, \qquad j = 0, 1, \dots, m,$$

converges uniformly for all sufficiently large $|t|$.

The above conditions ensure that the Fourier transform (6.1) converges uniformly for all sufficiently large values of $t \in \mathbb{R}^n$. Moreover, they imply that the Fourier transform

$$(\Delta^m \varphi)^\wedge (t) = \frac{1}{(2\pi)^{n/2}} \int_{\mathbb{R}^n} (\Delta^m \varphi)(x) e^{it \cdot x}\, dx \qquad (6.5)$$

also exists uniformly for all sufficiently large values of $t \in \mathbb{R}^n$. We shall demonstrate this assertion only for the case $Q + n \neq 2m - 1$. The case $Q + n = 2m - 1$ can be handled similarly. First we note that by (6.3)

$$(\Delta^m \varphi)(x) = O(r^{Q-2m+1}), \qquad \text{as } r \to 0^+.$$

Since $Q + n > 2m - 1$, near $x = 0$ the improper integral in (6.5) converges absolutely and uniformly for all values of t. Next we observe that in polar coordinates $x = (r, \theta_1, \dots, \theta_{n-1}) = (r, \theta)$, the Laplacian is given by

$$\Delta = \frac{\partial^2}{\partial r^2} + \frac{n-1}{r} \frac{\partial}{\partial r} + \frac{1}{r^2} L(\theta, D_\theta),$$

where $L(\theta, D_\theta)$ is a second-order linear differential operator with the independent variables $\theta_1, \dots, \theta_{n-1}$. Thus, from (6.2), it follows that

$$(\Delta^m \varphi)(x) = (\Delta^m f)(x) - \sum \psi_{pq}(\theta) r^{|p|+q-2m}, \qquad (6.6)$$

where the ψ_{pq}'s are linear combinations of products of the sine and cosine functions of $\theta = (\theta_1, \dots, \theta_{n-1})$. Since $Q + n < 2m + 1$, the powers of r in the last sum are all less than $1 - n$. Hence, by condition (C_3) with $j = m$, the integral in (6.5) also exists at infinity, uniformly for all large values of $t \in \mathbb{R}^n$. This establishes our assertion.

If $f(x)$ is a function of r only, say $f(x) = g(r)$, then it is well known that $\hat{f}(t)$ depends only on $|t|$. More specifically, we have

$$\hat{f}(t) = |t|^{(2-n)/2} \int_0^\infty r^{n/2} g(r) J_{(n-2)/2}(|t|r)\, dr, \qquad (6.7)$$

where $J_\nu(r)$ is the Bessel function of the first kind; see Schwartz (1966b, p. 203). The integral on the right-hand side of (6.7) is the Hankel transform of $g(r)$. Asymptotic behavior of this transform is discussed in Chapter IV, §3.

We now recall some of the basic properties of distributions studied in Chapter V. As usual, we let $\mathscr{S}$ denote the space of rapidly decreasing functions and $\mathscr{S}'$ the space of tempered distributions. (For simplicity, we have dropped the subscript n in $\mathscr{S}_n$.) If $\eta \in \mathscr{S}$ and $\Lambda \in \mathscr{S}'$ then we write $\langle \Lambda, \eta \rangle$ for the action of Λ on η. Any locally integrable function $f(x)$ on $\mathbb{R}^n$ of finite algebraic growth at infinity defines a distribution $f \in \mathscr{S}'$ by

$$\langle f, \eta \rangle = \int_{\mathbb{R}^n} f(x)\eta(x)\, dx; \qquad (6.8)$$

see (9.13), Chapter V. In particular, the function r^λ, $\operatorname{Re} \lambda > -n$, defines the tempered distribution

$$\langle r^\lambda, \eta \rangle = \int_{\mathbb{R}^n} r^\lambda \eta(x)\, dx.$$

If p is a multi-index and $u \in \mathscr{S}'$, then the formula

$$\langle D^p u, \eta \rangle = (-1)^{|p|} \langle u, D^p \eta \rangle$$

defines a distribution $D^p u \in \mathscr{S}'$. If f and $D^p f$ are both locally integrable in $\mathbb{R}^n$ with at most finite algebraic growth at infinity, then, under appropriate smoothness conditions, we have by integration by parts

$$\int_{\mathbb{R}^n} (D^p f)(x)\eta(x)\, dx = (-1)^{|p|} \int_{\mathbb{R}^n} f(x)(D^p \eta)(x)\, dx \qquad (6.9),$$

for every $\eta \in \mathscr{S}$; see the remark following (9.14) in Chapter V. The validity of (6.9) shows that the distributional derivative of f and the distribution defined by the ordinary derivative are equal (in the sense of distributions). A variant of (6.9) is the identity

$$\int_{\mathbb{R}^n} f(x)(\Delta \eta)(x)\, dx = \int_{\mathbb{R}^n} (\Delta f)(x)\eta(x)\, dx \qquad (6.10)$$

for every $\eta \in \mathscr{S}$; see (9.17) in Chapter V.

If $\eta \in \mathscr{S}$ then $\hat{\eta}$, as defined in (6.1), exists and belongs to $\mathscr{S}$ by Theorem 7 in Chapter V. If $u \in \mathscr{S}'$ then the Fourier transform $\hat{u}$ is defined by the equation

$$\langle \hat{u}, \eta \rangle = \langle u, \hat{\eta} \rangle, \qquad \eta \in \mathscr{S}. \tag{6.11}$$

By Theorem 8 of Chapter V, $\hat{u} \in \mathscr{S}'$ and the identities

$$(P(D)u)^\wedge = P(-t)\hat{u}, \qquad (Pu)^\wedge = P(D)\hat{u}, \tag{6.12}$$

hold for any polynomial $P(x) = \sum c_p x^p$ (Recall: $P(D) \equiv \sum c_p D_p$.)

We also recall that if f is locally integrable in $\mathbb{R}^n$ and increases no more rapidly than an algebraic function, and if the Cauchy limit in the integral (6.1) exists uniformly in $t \in \mathbb{R}^n$, then the distributional Fourier transform of f given in (6.11) and the ordinary Fourier transform of f given in (6.1) are equal; see the remark following (12.7) in Chapter V. The following result is crucial to our main theorem of this section.

Lemma. *Let $|p| + q + n > 0$. If $q + n \neq -2l$, $l = 0, 1, 2, \ldots$, then*

$$(x^p r^q)^\wedge = L(q) D_p(|t|^{-q-n}) \qquad in \ \mathscr{S}', \tag{6.13}$$

where $L(q) = 2^{q+n/2} \Gamma[(q+n)/2]/\Gamma(-q/2)$. If $q + n = -2l$, $l = 0, 1, 2, \ldots$, then

$$(x^p r^q)^\wedge = L^*(q) D_p(|t|^{2l} \log|t|) \qquad in \ \mathscr{S}' \tag{6.14}$$

where $L^*(q) = (-1)^{l+1}/2^{n/2+2l-1} l! \ \Gamma(\tfrac{1}{2}n + l)$.

Proof. From (12.12) and (12.19) in Chapter V, we have respectively

$$(r^q)^\wedge = L(q)|t|^{-q-n} \qquad in \ \mathscr{S}'$$

if $q + n \neq -2l$, $l = 0, 1, 2, \ldots$, and

$$(r^q)^\wedge = L^*(q)|t|^{2l} \log|t| + K|t|^{2l} \qquad in \ \mathscr{S}'$$

if $q + n = -2l$, $l = 0, 1, 2, \ldots$, where K is some constant. Thus (6.13) follows immediately from the second equation in (6.12). Since $D_p(|t|^{2l}) = 0$ for $|p| > 2l$, (6.14) also follows immediately from the second equation in (6.12), thus proving the lemma. ∎

Theorem 4. *Under the conditions (C_1), (C_2), and (C_3), we have*

$$\hat{f}(t) = {\sum}' c_{pq} L(q) D_p(|t|^{-q-n}) + {\sum}'' c_{pq} L^*(q) D_p(|t|^{2l} \log|t|) + \delta_m(t), \tag{6.15}$$

where $\sum'$ excludes those q's for which $q + n$ is a negative even integer and $\sum''$ includes only those q's for which $q + n = -2l$, a negative even *integer. The remainder in (6.15) satisfies*

$$\delta_m(t) = \frac{(-1)^m}{|t|^{2m}} (\Delta^m \varphi)^{\wedge}(t). \tag{6.16}$$

Proof.   Returning to (6.2), we note that each function in this equation generates a distribution in $\mathscr{S}'$ by means of the definition (6.8), and that these distributions are again related via this equation, i.e.,

$$f = \sum c_{pq} x^p r^q + \varphi \qquad \text{in } \mathscr{S}'.$$

Taking (distributional) Fourier transforms on both sides, we have by the preceding Lemma

$$\hat{f} = \sum{}' c_{pq} L(q) D_p(|t|^{-q-n}) + \sum{}'' c_{pq} L^*(q) D_p(|t|^{2l} \log |t|) + \hat{\varphi} \tag{6.17}$$

in $\mathscr{S}'$, where the sums $\sum'$ and $\sum''$ are as given in (6.15). Applying the first equation in (6.12) m times gives

$$(\Delta^m \varphi)^{\wedge} = (-1)^m |t|^{2m} \hat{\varphi} \qquad \text{in } \mathscr{S}'. \tag{6.18}$$

Inserting (6.18) in (6.17), we have as functionals

$$|t|^{2m} \hat{f} = \sum{}' c_{pq} L(q) |t|^{2m} D_p(|t|^{-q-n})$$
$$+ \sum{}'' c_{pq} L^*(q) |t|^{2m} D_p(|t|^{2l} \log|t|) + (-1)^m (\Delta^m \varphi)^{\wedge}. \tag{6.19}$$

In view of the remark following condition (C$_3$) and the remark following (6.12) concerning the ordinary and distributional Fourier transforms, we see that each distribution in (6.19) is defined by a locally integrable function in $\mathbb{R}^n$, and hence that equation (6.19) holds not only as functionals but also pointwise almost everywhere in $\mathbb{R}^n$. Since the Fourier transforms $\hat{f}(t)$ and $(\Delta^m \varphi)^{\wedge}(t)$ defined in (6.1) and (6.5) exist uniformly for all large values of $|t|$, they are in fact continuous functions of $t \in \mathbb{R}^n$. From this it follows that all functions associated with the distributions in (6.19) are continuous, and hence that (6.19) is valid for every $t \in \mathbb{R}^n$. This completes the proof of the theorem. ∎

As remarked earlier, the Fourier transform $(\Delta^m \varphi)^{\wedge}(t)$ given in (6.5) exists uniformly for all sufficiently large values of $t \in \mathbb{R}^n$. Hence, by a generalization of the Riemann–Lebesgue lemma, we have

$$\delta_m(t) = o(|t|^{-2m}), \qquad \text{as } |t| \to \infty; \tag{6.20}$$

cf. Lemma 3, Chapter IV. This confirms the asymptotic nature of (6.15).

Exercises

1. Use the identity

    $$\exp(x^2) = \pi^{-1/2} \int_{-\infty}^{\infty} \exp[-(k^2 + 2kx)] \, dk$$

    to show that the integral

    $$I_n = \int_0^{\infty} \cdots \int_0^{\infty} \exp\left[-\sum_{j=1}^{n} x_j^2 + \frac{1}{2n}\left(\sum_{j=1}^{n} x_j\right)^2 \right] dx_1 \cdots dx_n$$

    can be reduced to the form

    $$I_n = \pi^{-1/2}\left(\frac{\pi^{1/2}}{2}\right)^n (2n)^{1/2}[F_+(n) + F_-(n)],$$

    where

    $$F_{\pm}(n) = \int_0^{\infty} e^{-nk^2}[1 \pm \mathrm{erf}(k)]^n \, dk,$$

    erf(k) being the error function (Olver 1974a, p. 43). Find the asymptotic behavior of $F_+(n)$ by using Laplace's method, and show that $F_-(n)$ is negligible for large n. Deduce from these results the asymptotic formula

    $$I_n \sim \left[\frac{-2}{h''(k_0)}\right]^{1/2} \exp[nh(k_0)], \qquad \text{as } n \to \infty,$$

    where $h(k) = -k^2 + \log[1 + \mathrm{erf}(k)]$ and k_0 is the unique solution of the equation $ke^{k^2}[1 + \mathrm{erf}(k)] = \pi^{-1/2}$. (Numerically, $k_0 = 0.3578345$, $h(k_0) = 0.0784462$, and $[-2/h''(k_0)]^{1/2} = 1.1500399$.)

2. Consider the integral

    $$I_m = \int \cdots \int_T [\cos \varphi_1 \cdots \cos \varphi_n \cos(\varphi_1 + \cdots + \varphi_n)]^m \, d\varphi_1 \cdots d\varphi_n,$$

    where T is the hypercube $-\pi/2 \leq \varphi_j \leq \pi/2$ in $\mathbb{R}^n$. Let S be the set $\{(\varphi_1, \ldots, \varphi_n): |\varphi_1 + \cdots + \varphi_n| < \pi/2\}$. Show that the integrand is bounded by $(\cos 2\pi/n)^m$ outside S, and that the expression in brackets is positive in S and achieves its maximum 1 only at the origin. Let

    $$f(\varphi_1, \ldots, \varphi_n) = -\log[\cos \varphi_1 \cdots \cos \varphi_n \cos(\varphi_1 + \cdots + \varphi_n)].$$

Show that the Hessian matrix of f at $\varphi_1 = \cdots = \varphi_n = 0$ is

$$A = \begin{pmatrix} 2 & 1 & \cdots & 1 \\ 1 & 2 & \cdots & 1 \\ \vdots & & \ddots & \vdots \\ 1 & 1 & \cdots & 2 \end{pmatrix}.$$

By using induction, prove that the eigenvalues of A are $\mu_1 = \cdots = \mu_{n-1} = 1$, $\mu_n = n + 1$. Conclude from this that

$$I_m \sim \frac{1}{\sqrt{n+1}} \left(\frac{2\pi}{m} \right)^{n/2} \qquad \text{as } m \to \infty.$$

3. Consider

$$I_m = \int \cdots \int_T [\cos \varphi_1 \cdots \cos \varphi_n \sin(\varphi_1 + \cdots + \varphi_n)]^{2m} \, d\varphi_1 \cdots d\varphi_n,$$

where T is, as in Ex. 2, the hypercube $-\pi/2 \le \varphi_j \le \pi/2$ in $\mathbb{R}^n$. Show that

$$I_m = 2 \int \cdots \int_{T'} [\cos \varphi_1 \cdots \cos \varphi_n \sin(\varphi_1 + \cdots + \varphi_n)]^{2m} \, d\varphi_1 \cdots d\varphi_n,$$

where T' is that half of T which satisfies $\varphi_1 + \cdots + \varphi_n > 0$. Let

$$f(\varphi_1, \ldots, \varphi_n) = -\log \cos \varphi_1 - \cdots - \log \cos \varphi_n$$
$$- \log \sin(\varphi_1 + \cdots + \varphi_n),$$

and note that on the boundary of T', $f = +\infty$. Thus, there is no contribution from boundary points. Show that the critical point of f occurs at $(\varphi_1, \ldots, \varphi_n)$, where $\tan \varphi_1 = \cdots = \tan \varphi_n$, and deduce from this that $\varphi_1 = \cdots = \varphi_n = \alpha$, where $\alpha = (2k + 1)\pi/(2n + 2)$, $0 < \alpha < \pi/2$. Furthermore, show that the absolute minimum of f is $-(n + 1) \log \cos[(\pi/(2n + 2))]$ and occurs at $\varphi_1 = \cdots = \varphi_n = \pi/(2n + 2)$. Finally, prove that Hessian matrix of f at this point is given by

$$A = \frac{1}{\cos^2 \left(\dfrac{\pi}{2n+2} \right)} \begin{pmatrix} 2 & 1 & \cdots & 1 \\ 1 & 2 & \cdots & 1 \\ \vdots & & \ddots & \vdots \\ 1 & 1 & \cdots & 2 \end{pmatrix},$$

and derive the asymptotic formula

$$I_m \sim \frac{2}{\sqrt{n+1}} \left(\frac{\pi}{m}\right)^{n/2} \left[\cos\left(\frac{\pi}{2n+2}\right)\right]^{2m(n+1)+n}, \qquad \text{as } m \to \infty.$$

4. Let A be a real, symmetric and positive definite matrix. Show that

$$\int_{\mathbb{R}^n} \exp(-x^T A x)\, dx = \frac{\pi^{n/2}}{(\det A)^{1/2}}.$$

5. Consider the Laplace integral $J(\lambda)$ in (5.1), and assume that conditions (i), (ii), and (iii) in §5 hold. Assume also that f is twice-continuously differentiable and g is continuous in a neighborhood of x_0. (Note that this assumption is weaker than the C^∞-condition imposed in §5.)

   Show that without loss of generality, one may consider only the normalized case where $g(x_0) = 1$, $f(x_0) = 0$, and $x_0 = 0$. Let $D_\varepsilon = \{x: |x| < \varepsilon\} \subset D$. Show that the integral

$$J_2(\lambda) = \int_{D \setminus D_\varepsilon} g(x) e^{-\lambda f(x)}\, dx$$

is $O[\exp\{-(\lambda - \lambda_0)\rho(\varepsilon)\}]$ for all $\lambda \geq \lambda_0$ and for all ε. Write

$$2f(x) = x^T A x + \eta(x) x^T x,$$

where A is the Hessian matrix of f at 0, and put

$$\eta(\varepsilon) = \sup_{|x| < \varepsilon} |\eta(x)|.$$

Show that $\eta(\varepsilon) \to 0$ as $\varepsilon \to 0$. Furthermore, put

$$g^+(\varepsilon) = \sup_{|x| \leq \varepsilon} g(x) \qquad \text{and} \qquad g^-(\varepsilon) = \inf_{|x| \leq \varepsilon} g(x),$$

and show that $g^\pm(\varepsilon) \to 1$ as $\varepsilon \to 0$. Now prove that for sufficiently small ε,

$$J_1(\lambda) = \int_{D_\varepsilon} g(x) e^{-\lambda f(x)}\, dx$$

is bounded between

$$g^-(\varepsilon) \left\{ [\det(A + \eta(\varepsilon)I)]^{-1/2} \left(\frac{2\pi}{\lambda}\right)^{n/2} - J_3(\lambda) \right\},$$

and

$$g^+(\varepsilon)[\det (A - \eta(\varepsilon)I)]^{-1/2}\left(\frac{2\pi}{\lambda}\right)^{n/2},$$

where

$$J_3(\lambda) = \int_{\mathbb{R}^n\backslash D_\varepsilon} \exp\left[-\frac{\lambda}{2}(x^\mathsf{T}Ax + \eta(\varepsilon)x^\mathsf{T}x)\right] dx.$$

Let $\mu_0 > 0$ be the smallest eigenvalue of $A + \eta(\varepsilon)I$. Show that

$$J_3(\lambda) \leq \int_{\mathbb{R}^n\backslash D_\varepsilon} \exp\left(-\frac{\lambda}{2}\mu_0 x^\mathsf{T}x\right) dx = O\left(\lambda^{-1}\exp\left[-\frac{\lambda}{2}\mu_0\varepsilon^2\right]\right).$$

From the above information, derive the Laplace approximation

$$J(\lambda) \sim (\det A)^{-1/2}\left(\frac{2\pi}{\lambda}\right)^{n/2}, \qquad \lambda \to \infty,$$

for the normalized case.

6. Consider the integral

$$J(\lambda) = \int_{\mathbb{R}^n} g(x)e^{-\lambda f(x)}\,dx,$$

where λ is a large positive parameter, $g(x)$ is a bounded measurable function, and $f(x)$ is real-valued and continuous. Assume that $f(x)$ attains its absolute minimum m at, and only at, x_0, and also that $g(x) \geq \rho > 0$ in some neighborhood of x_0. Let $S_\varepsilon = \{x \in \mathbb{R}^n: f(x) \leq m + \varepsilon\}$, and put

$$J_\varepsilon^*(\lambda) = \int_{S_\varepsilon} g(x)e^{-\lambda[f(x)-m]}\,dx \qquad \text{and} \qquad \Phi_\varepsilon(\lambda) = e^{\lambda\varepsilon}J_{\varepsilon/2}^*(\lambda).$$

Show that there exists a positive number ε_0 such that $\Phi_\varepsilon(\lambda)$ satisfies the differential inequality $u'(\lambda) \geq \frac{1}{2}\varepsilon u(\lambda)$ for all $0 < \varepsilon \leq \varepsilon_0$, and deduce from this that $\Phi_\varepsilon(\lambda) \geq Ae^{\varepsilon\lambda/2}$, where $A = \Phi_\varepsilon(0) > 0$. Furthermore, show that

$$\frac{\log A}{\lambda} - \frac{\varepsilon}{2} \leq \frac{\log J_{\varepsilon_0}^*(\lambda)}{\lambda} \leq \frac{\log B}{\lambda}, \qquad \text{where} \qquad B = \int_{S_{\varepsilon_0}} g(x)\,dx,$$

and hence

$$\lim_{\lambda \to \infty} \frac{\log J_{\varepsilon_0}^*(\lambda)}{\lambda} = 0.$$

Finally, prove that $J(\lambda) = e^{-\lambda m}[J^*_{\varepsilon_0}(\lambda) + O(e^{-\lambda \varepsilon_0})]$, $\lambda \to \infty$, and conclude from this that

$$\lim_{\lambda \to \infty} \frac{\log J(\lambda)}{\lambda} = -m.$$

Note that no smoothness condition is imposed on $f(x)$ and $g(x)$.

7. Let Σ be an oriented smooth $(n-1)$-dimensional surface in $\mathbb{R}^n$ with the parametric representation

$$x = x(u), \qquad u = (u_1, \ldots, u_{n-1}) \in \Omega,$$

Ω being the parameter domain, and let $f(x)$ be a smooth function defined on Σ. Show that

$$\int_\Sigma f(x)\, dA = \int_\Omega f(x(u)) D(u)\, du,$$

where dA is the surface element, $D(u) = [\sum (\det J_i)^2]^{1/2}$, and

$$\det J_i = \frac{\partial(x_1, \ldots, x_{i-1}, x_{i+1}, \ldots, x_n)}{\partial(u_1, \ldots, u_{n-1})};$$

cf. (3.7) and (3.8). (Hint: the surface area form of Σ is given by

$$dA = \sum (-1)^i n_i(x)\, dx_1 \wedge \cdots \wedge dx_{i-1} \wedge dx_{i+1} \wedge \cdots \wedge dx_n,$$

where the n_i are the components (given in (3.9)) of the unit outward normal $\mathbf{n}$ to Σ; see Edwards (1973, p. 380). Apply the definition of the integral of differential forms; see Rudin (1976, p. 254) or Edwards (1973, p. 353, lines 3 and 4).)

8. Let Σ denote the upper half of the ellipsoid $x^2/a^2 + y^2/b^2 + z^2/c^2 = 1$, $z > 0$, where $a > b > 0$. Let d_0 and d_1 denote respectively the distances from $(0, 0, c_0)$ and $(0, 0, c_1)$ to a point on Σ, where $c_0 > c > 0$ and $c_1 < 0$. Show that the surface integral

$$I(\lambda) = \int_\Sigma e^{i\lambda(d_0 + d_1)}\, dA$$

has the asymptotic formula

$$I(\lambda) \sim i \frac{2\pi(c_0 - c)(c - c_1)}{\lambda(c_0 - c_1)} e^{i\lambda(c_0 - c_1)}, \qquad \text{as } \lambda \to +\infty.$$

9. To prove (3.24), we let P denote the $(n-1) \times (n-1)$ matrix whose elements p_{ij} are given by

$$p_{ij} = \frac{\partial^2 p}{\partial u_i \, \partial u_j} (u_0),$$

and let C denote the $n \times n$ matrix whose elements c_{lk} are given by

$$c_{lk} = \frac{\partial^2 f}{\partial x_l \, \partial x_k} - K \frac{\partial^2 h}{\partial x_l \, \partial x_k}.$$

Furthermore, let J be the $n \times (n-1)$ matrix

$$J = \begin{pmatrix} \dfrac{\partial x_1}{\partial u_1} & \cdots & \dfrac{\partial x_1}{\partial u_{n-1}} \\ \vdots & \ddots & \vdots \\ \dfrac{\partial x_n}{\partial u_1} & \cdots & \dfrac{\partial x_n}{\partial u_{n-1}} \end{pmatrix}.$$

Using (3.23), show that $P = J^T C J$. Now recall the *Binet-Cauchy product formula* (Edwards (1973, p. 326) or Noble (1969, p. 226)):

$$\det AB = \sum_P \det A_p \det B_p,$$

where A is a $k \times n$ matrix, B is an $n \times k$ matrix, $k \leq n$, A_p is the square matrix obtained by choosing any k columns of A, B_p is the matrix obtained by choosing the corresponding rows of B, and the sum is taken over all such possible choices. Apply this formula twice to conclude that

$$\det P = \sum_{p=1}^{n} \sum_{q=1}^{n} \det J_p \det C_{pq} \det J_q,$$

where J_p is the $(n-1) \times (n-1)$ matrix obtained from the matrix J above by deleting the pth row, and C_{pq} is the submatrix of C obtained by deleting the pth row and qth column. From this and (3.9), deduce equation (3.24).

10. For an arc of the form $y = \varphi(x)$, the curvature κ of the arc is given by

$$\kappa(x) = \frac{\varphi''(x)}{\{1 + [\varphi'(x)]^2\}^{3/2}}.$$

Let $n = 2$ in (3.26), and let κ and κ_1 be the curvatures of the two curves $f(x, y) = f(x_0, y_0)$ and $h(x, y) = 0$ at (x_0, y_0). Show that in this case, (3.26) reduces to

$$\mathbf{J} = (\kappa_1 - \kappa)|\nabla f|^3.$$

This demonstrates the fact that if the two curves have the same curvature (including sign) at (x_0, y_0), then the asymptotic formula (3.27) will not hold.

11. Let $f(x) = c$, $x = (x_1, \ldots, x_n)$, be an $(n-1)$-dimensional surface with $\nabla f(x_0) \neq 0$, $x_0 \in \mathbb{R}^n$, and suppose that this surface is parametrically represented by $x = x(u)$, $u = (u_1, \ldots, u_{n-1})$. Let $J = (\partial x_i / \partial u_j)$ and define $D(u) = [\det J^T J]^{1/2}$; cf. (3.6) and (3.7). The second fundamental form of differential geometry at x_0 is defined as the quadratic differential form $\sum \alpha_{kl} \, du_k \, du_l$, where the coefficients α_{kl} are given by

$$\alpha_{kl} = \frac{1}{D(u_0)} \begin{vmatrix} \dfrac{\partial^2 x_1}{\partial u_k \, \partial u_l} & \cdots & \dfrac{\partial^2 x_n}{\partial u_k \, \partial u_l} \\ \dfrac{\partial x_1}{\partial u_1} & \cdots & \dfrac{\partial x_n}{\partial u_1} \\ \vdots & \ddots & \vdots \\ \dfrac{\partial x_1}{\partial u_{n-1}} & \cdots & \dfrac{\partial x_n}{\partial u_{n-1}} \end{vmatrix} (0);$$

see Struik (1961, p. 75) or Thorpe (1979, p. 92). Let β_{kl} denote the coefficients in the second fundamental form at $x = x_0$ for the surface $h(x) = 0$. Prove or disprove that the quantity $\mathbf{J}$ in (3.26) satisfies

$$|\mathbf{J}| = \frac{|\nabla f|^{n+1}}{D^2(u_0)} |\det(\alpha_{kl} - \beta_{kl})|.$$

Geometrically, the second fundamental form gives information about the shape of the surface near the point x_0. If the above result is true and if the two surfaces $f(x) = f(x_0)$ and $h(x) = 0$ have the same curvature ($\alpha_{kl} = \beta_{kl}$), then $|\mathbf{J}| = 0$.

12.   Let $f\colon \mathbb{R}^n \to \mathbb{R}$ be a smooth function with $\nabla f(0) = 0$. Suppose that
the Hessian matrix of f at 0 is of the form

$$\begin{bmatrix} 1 & & & & & & \\ & \ddots & & 0 & & & \\ & & 1 & & & 0 & \\ & & & -1 & & & \\ & 0 & & & \ddots & & \\ & & & & & 1 & \\ \hline & 0 & & & & 0 & \end{bmatrix} \Big\} r$$

By the splitting lemma, we have

$$f(x) = \pm u_1^2 \pm \cdots \pm u_r^2 + p(u_{r+1}, \ldots, u_n)$$

for some smooth function $p\colon \mathbb{R}^{n-r} \to \mathbb{R}$. By retracing the argument
given in the proof of this lemma (see Poston and Stewart (1978, pp.
61–62)), it can be shown that the Jacobian of the transformation
$x \to u$ at 0 is 1, i.e.

$$\left.\frac{\partial(x_1, \ldots, x_n)}{\partial(u_1, \ldots, u_n)}\right|_{u=0} = 1.$$

Show that in general if the Hessian matrix A of f at 0 has rank r,
then

$$\left.\frac{\partial(x_1, \ldots, x_n)}{\partial(u_1, \ldots, u_n)}\right|_{u=0} = \frac{1}{\sqrt{|\mu_1 \cdots \mu_r|}},$$

where the μ_i's are the nonzero eigenvalues of A.

13.   The Mellin transform of e^{ix}, as defined by equation (2.1) in Chapter
III, is

$$\Gamma(z)e^{i\pi z/2} \qquad \text{for } 0 < \mathrm{Re}\, z < 1.$$

Show that the inversion formula

$$\frac{1}{2\pi i} \int_{c-i\infty}^{c+i\infty} x^{-z}\Gamma(z)e^{i\pi z/2}\, dz = e^{ix}, \qquad x \text{ real}, \quad 0 < c < \tfrac{1}{2},$$

does not follow directly from Theorem 2 in that chapter. Prove the
above identity by using the fact that the Mellin transforms of the

sine integral and cosine integral

$$\text{Si}(x) = \frac{\pi}{2} - \int_x^\infty \frac{\sin t}{t} \, dt, \qquad \text{Ci}(x) = -\int_x^\infty \frac{\cos t}{t} \, dt$$

are given by, respectively,

$$-\sin\left(\frac{\pi z}{2}\right) \frac{\Gamma(z)}{z}, \ -1 < \text{Re } z < 0, \text{ and } -\cos\left(\frac{\pi z}{2}\right) \frac{\Gamma(z)}{z}, \ 0 < \text{Re } z < 1.$$

14. By using the asymptotic formula for $\Gamma(x + iy)$ as $y \to \pm\infty$, and by shifting the vertical line of integration back and forth, justify the interchange of the order of integration in (4.6).

15. Consider the two-dimensional Fourier transform

$$\hat{f}(t) = \frac{1}{2\pi} \iint_{\mathbb{R}^2} f(x) e^{it \cdot x} \, dx,$$

where

$$f(x) = \frac{1}{r^{3/2}[1 + (x_1 + x_2)^2]}$$

and $r = \sqrt{x_1^2 + x_2^2}$. Show that the function $\varphi(x)$ defined by $f(x) = r^{-3/2} + \varphi(x)$ is continuous at $x = 0$, and its Fourier transform

$$\hat{\varphi}(t) = \frac{1}{2\pi} \iint_{\mathbb{R}^2} \varphi(x) e^{it \cdot x} \, dx$$

exists uniformly for all large values of $|t| = \sqrt{t_1^2 + t_2^2}$. Furthermore, prove that

$$|\Delta\varphi| \le \frac{52}{r^{3/2}[1 + (x_1 + x_2)^2]}.$$

Deduce from above the formula

$$\hat{f}(t) = \frac{\Gamma^2(\frac{1}{4})}{2\pi |t|^{1/2}} + \hat{\varphi}(t),$$

where $\hat{\varphi}(t) = o(1)$ as $|t| \to \infty$. Show also that

$$\hat{\varphi}(t) = -\frac{1}{|t|^2} (\Delta\varphi)^\wedge(t),$$

and hence conclude that

$$|\hat{\varphi}(t)| \le \frac{52\sqrt{\pi}\,\Gamma^2(\tfrac{1}{4})}{2^{1/4}|t|^2}.$$

16.   Consider the three-dimensional Fourier transform

$$\hat{f}(t) = \frac{1}{(2\pi)^{3/2}} \iiint\limits_{\mathbb{R}^3} f(x)e^{it\cdot x}\,dx, \qquad t \in \mathbb{R}^3,$$

where

$$f(x) = \frac{x_1 x_2 x_3}{(1+r)^6}, \qquad r = \sqrt{x_1^2 + x_2^2 + x_3^2}.$$

Note that $f \notin L^1(\mathbb{R}^3)$. Put

$$f(x) = x_1 x_2 x_3(1 - 6r + 21r^2) + \varphi(x),$$

and show that

$$\hat{f}(t) = 12\sqrt{\frac{2}{\pi}}\, D_p(|t|^{-4}) + \hat{\varphi}(t),$$

where $p = (1, 1, 1)$ and

$$\hat{\varphi}(t) = \frac{1}{|t|^8}\,(\Delta^4 \varphi)^{\wedge}(t).$$

Furthermore, put $\varphi(x) = x_1 x_2 x_3 F(r)$ and show that

$$\Delta^4 \varphi(x) = x_1 x_2 x_3 \left(\frac{d}{dr^2} + \frac{8}{r}\frac{d}{dr}\right)^4 F(r).$$

From this deduce the estimates

$$|\Delta^4 \varphi| \le \frac{C}{r^2(1+r)^2} \qquad \text{and} \qquad |\hat{\varphi}(t)| \le \sqrt{\frac{\pi}{2}}\,\frac{C}{|t|^8},$$

C being a constant.

Supplementary Notes

§1.   Books which contain an extensive discussion of the asymptotic behavior of n-dimensional integrals include Hsu (1958), Bleistein and Handelsman (1975a), Fedoryuk (1977), and Jones (1982, pp. 379–386). (Jones' treatment is formal.)

§2. An alternative derivation of the n-dimensional stationary phase approximation has been given by Bleistein and Handelsman (1975b); see also Chako (1965, pp. 407–414).

§3. The argument used here to deduce the asymptotic formula (3.27) from (3.16) seems to be simpler than that of Jones as outlined in Bleistein and Handelsman (1975a, pp. 339–340 and Exs. 8.7–8.10).

§4. Consideration here is rather brief; a more detailed discussion is needed for degenerate stationary points.

§5. The n-dimensional Laplace approximation (5.3) and the corresponding formula (5.15) for the case of boundary critical points seems to have been given first by Hsu (1948b). For a derivation of (5.2) without making any smoothness assumptions, see the article by Fulks and Sather (1961).

§6. The problem of finding the asymptotic behavior of n-dimensional Fourier transforms, when $n = 3$, was first considered by Duffin (1953). The case $n = 2$ was subsequently treated by Duffin and Shaffer (1960). (According to them, the latter case is more difficult.) The general result stated in Theorem 4 was proved first by Shivakumar and Wong (1979) by using the summability method (Chapter IV). The material in this section is taken, however, from Wong (1980d).

Exercises. Exercise 1 is essentially the solution by Grzesik (1983) to Problem 82-20 in *SIAM Review*. Exercise 2 is based on an example in Henrici (1974, pp. 414–415), whereas the material in Ex. 3 is taken from deBruijn (1970, pp. 73–75). The proof of the multidimensional Laplace approximation outlined in Ex. 5 is given in Henrici (1974, pp. 409–413). The result in Ex. 6 is due to Maslov and Fedoryuk (1981). Exercise 8 is taken from Jones (1982, p. 385, Ex. 40), whereas Exs. 10 and 11 are taken from Bleistein and Handelsman (1975a, p. 363, Exs. 8.12 and 8.13). For solutions to Exs. 15 and 16, see Shivakumar and Wong (1979).

Bibliography

Abramowitz, A. and Stegun, I. A. (1964). *Handbook of Mathematical Functions*. NBS Appl. Math. Series 55, Washington, D.C.

Armstrong, J. A. and Bleistein, N. (1980). Asymptotic expansions of integrals with oscillatory kernels and logarithmic singularities. *SIAM J. Math. Anal.*, **11**, 300–307.

Atiyah, M. F. (1970). Resolution of singularities and division of distributions. *Comm. Pure Appl. Math.*, **23**, 145–150.

Bakhoom, N. G. (1933). Asymptotic expansions of the function $F_k(x) = \int_0^\infty e^{-u^k + xu} \, du$. *Proc. London Math. Soc.* (2), **35**, 83–100.

Barnes, E. W. (1906). The asymptotic expansion of integral functions defined by Taylor's series. *Phil. Trans. Roy. Soc. (London)*, **A206**, 249–297.

Bender, E. A. (1974). Asymptotic methods in enumeration. *SIAM Review*, **16**, 485–515.

Bender, C. M. and Orszag, S. A. (1978). *Advanced Mathematical Methods for Scientists and Engineers*. McGraw–Hill, New York.

Berg, L. (1968). *Asymptotische Darstellungen und Entwicklungen*. VEB Deutscher Verlag der Wissenschaften, Berlin.

Berghuis, J. (1955). The method of critical regions for two-dimensional integrals and its application to a problem of antenna theory. Ph.D. Thesis, University of Delft.

Bleistein, N. (1966). Uniform asymptotic expansions of integrals with stationary points near algebraic singularity. *Comm. Pure Appl. Math.*, **19**, 353–370.

Bleistein, N. (1967). Uniform asymptotic expansions of integrals with many nearby stationary points and algebraic singularities. *J. Math. Mech.*, **17**, 533–559.

Bleistein, N. (1977). Asymptotic expansions of integral transforms of functions with logarithmic singularities. *SIAM J. Math. Anal.*, **8**, 655–672.

Bleistein, N. and Handelsman, R. A. (1975a). *Asymptotic Expansions of Integrals*. Holt, Rinehart & Winston, New York.

Bleistein, N. and Handelsman, R. A. (1975b). Multidimensional stationary phase. An alternative derivation. *SIAM J. Math. Anal.*, **6**, 480–487.

Bremermann, H. J. (1965). *Distributions, Complex Variables, and Fourier Transforms*. Addison–Wesley, Reading, Massachusetts.

Bremmer, H. (1955). Diffraction problems of microwave optics. *I.R.E. Trans. on Antennas and Propagation*, **V.AP-3**, 222–228.

Brüning, J. (1984). On the asymptotic expansion of some integrals. *Arch. Math.*, **42**, 253–259.

Carlson, B. C. (1977). *Special Functions of Applied Mathematics*. Academic Press, New York.

Carlson, B. C. and Gustafson, J. L. (1985). Asymptotic expansion of the first elliptic integral. *SIAM J. Math. Anal.*, **16**, 1072–1092.

Chako, N. (1965). Asymptotic expansions of double and multiple integrals arising in diffraction theory. *J. Inst. Math. Appl.*, **1**, 372–422.

Chester, C., Friedman, B., and Ursell, F. (1957). An extension of the method of steepest descents. *Proc. Cambridge Philos. Soc.*, **53**, 599–611.

Connor, J. N. L. and Farrelly, D. (1981). Molecular collisions and cusp catastrophes: Three methods for the calculation of Pearcey's integral and its derivatives. *Chem. Phys. Letters*, **81**, 306–310.

Connor, J. N. L., Curtis, P. R., and Farrelly, D. (1984). The uniform asymptotic swallowtail approximation: Practical methods for oscillating integrals with four coalescing saddle points. *J. Phys. A: Math. Gen.*, **17**, 283–310.

Cooke, J. C. (1982). Stationary phase in two dimensions. *IMA J. Appl. Math.*, **29**, 25–37.

Copson, E. T. (1935). *Theory of Functions of a Complex Variable.* Oxford University Press, London.

Copson, E. T. (1965). *Asymptotic Expansions.* Cambridge Tracts in Math. and Math. Phys. No. 55, Cambridge University Press, London.

Courant, R. and John, F. (1974). *Introduction to Calculus and Analysis*, Vol. 2. John Wiley and Sons, New York.

Davis, P. J. and Rabinowitz, P. (1975). *Methods of Numerical Integration.* Academic Press, New York.

de Bruijn, N. G. (1970). *Asymptotic Methods in Analysis.* North-Holland, Amsterdam.

Debye, P. (1909). Näherungsformelm für die Zylinderfunktionen für grosse Werte des Arguments und unbeschränkt veränderliche Werte des Index. *Math. Ann.*, **67**, 535–558.

de Jager, E. M. (1970). Theory of Distributions, In *Mathematics Applied to Physics* (E. Roubine, ed.). Springer-Verlag, New York, 52–110.

De Kok, F. (1971). On the method of stationary phase for multiple integrals. *SIAM J. Math. Anal.*, **2**, 76–104.

Dieudonné, J. (1968). *Calcul infinitésimal.* Hermann, Paris.

Dingle, R. B. (1973). *Asymptotic Expansions: Their Derivation and Interpretation.* Academic Press, New York.

Doetsch, G. (1944). *Theorie und Anwendung der Laplace Transformation.* Dover Publications, New York.

Doetsch, G. (1955). *Handbuch der Laplace-Transformation*, Vol. II. Birkhäuser, Basel.

Duffin, R. J. (1953). Discrete potential theory. *Duke Math. J.*, **20**, 233–251.

Duffin, R. J. and Shaffer, D. H. (1960). Asymptotic expansion of double Fourier transforms. *Duke Math. J.*, **27**, 581–596.

Durbin, P. (1979). Asymptotic expansion of Laplace transforms about the origin using generalized functions. *J. Inst. Maths. Appl.*, **23**, 181–192.

Edwards, C. H., Jr. (1973). *Advanced Calculus of Several Variables.* Academic Press, New York.

Erdélyi, A. (1955). Asymptotic representations of Fourier integrals and the method of stationary phase. *J. Soc. Indust. Appl. Math.*, **3**, 17–27.

Erdélyi, A. (1956). *Asymptotic Expansions*. Dover, New York.

Erdélyi, A. (1959). On the principle of stationary phase. In *Proceedings of the Fourth Canadian Mathematical Congress* (M. S. Macphail, ed.). University of Toronto Press, Toronto, 137–146.

Erdélyi, A. (1960). Asymptotic forms for Laguerre polynomials. *J. Indian Math. Soc.*, Golden Jubilee Commemorative Volume, **24**, 235–250.

Erdélyi, A. (1961). General asymptotic expansions of Laplace integrals. *Arch. Rational Mech. Anal.*, **7**, 1–20.

Erdélyi, A. (1970). Uniform asymptotic expansion of integrals. In *Analytic Methods in Mathematical Physics* (R. P. Gilbert and R. G. Newton, eds.). Gordon and Breach, New York, 149–168.

Erdélyi, A. (1974). Asymptotic evaluation of integrals involving a fractional derivative. *SIAM J. Math. Anal.*, **5**, 159–171.

Erdélyi, A. and Wyman, M. (1963). The asymptotic evaluation of certain integrals. *Arch. Rational Mech. Anal.*, **14**, 217–260.

Erdélyi, A., Magnus, W., Oberhettinger, F., and Tricomi, F. (1953a). *Higher Transcendental Functions*, Vol. 1. McGraw–Hill, New York.

Erdélyi, A., Magnus, W., Oberhettinger, F., and Tricomi, F. (1953b). *Higher Transcendental Functions*, Vol. 2. McGraw–Hill, New York.

Erdélyi, A., Magnus, W., Oberhettinger, F., and Tricomi, F. (1954). *Tables of Integral Transforms*, Vol. 1. McGraw–Hill, New York.

Evgrafov, M. A. (1961). *Asymptotic Estimates and Entire Functions* (translated by A. L. Shields). Gordon and Breach, New York.

Evgrafov, M. A. (1966). *Analytic Functions*. Saunders, Philadelphia.

Fedoryuk, M. V. (1970). The stationary phase method in the multidimensional case. Contribution from the region boundary. *Zh. vychisl. Mat. mat. Fiz.*, **10**, 286–299. [English transl.: *U.S.S.R. Computational Math. and Math. Phys.*, **10** (1970), No. 2, 4–23.]

Fedoryuk, M. V. (1971). The stationary phase method and pseudo-differential operators. *Russian Math. Surveys*, **26**, 65–115.

Fedoryuk, M. V. (1977). *The Saddle-Point Method* (Russian). Nauka, Moscow.

Feller, W. (1966). *An Introduction to Probability Theory and Its Applications*, Vol. II. Wiley, New York.

Fields, J. L. (1966). A note on the asymptotic expansion of a ratio of gamma functions. *Proc. Edinburgh Math. Soc.*, **15**, 43–45.

Fields, J. L. (1968). A uniform treatment of Darboux's method. *Arch. Rational Mech. Anal.*, **27**, 289–305.

Fleming, W. H. (1965). *Functions of Several Variables*. Addison–Wesley, Reading, Massachusetts.

Focke, J. (1954). Asymptotische Entwicklungen mittels der Methode der stationären Phase. *Berichte Über die Verhandlungen der sächsischen Akademie der Wissenschaften zu Leipzig*, **101**, Heft 3, 1–48.

Franklin, J. and Friedman, B. (1957). A convergent asymptotic representation for integrals. *Proc. Cambridge Philos. Soc.*, **53**, 612–619.

Frenzen, C. L. (1987). Error bounds for asymptotic expansions of the ratio of two Gamma functions. *SIAM J. Math. Anal.*, **18**, 890–896.

Frenzen, C. L. and Wong, R. (1985). A note on asymptotic evaluation of some Hankel transforms. *Math. Comp.*, **45**, 537–548.

Frenzen, C. L. and Wong, R. (1988). Uniform asymptotic expansions of Laguerre polynomials. *SIAM J. Math. Anal.*, **19**, 1232–1248.

Friedman, B. (1959). Stationary phase with neighboring critical points. *SIAM Journal*, **7**, 280–289.

Fu, J. C. and Wong, R. (1980). An asymptotic expansion of a Beta-type integral and its application to probabilities of large deviations. *Proc. Amer. Math. Soc.*, **79**, 410–414.

Fulks, W. (1960). Asymptotics I: A note on Laplace's method. *Amer. Math. Monthly*, **67**, 880–882.

Fulks, W. and Sather, J. O. (1961). Asymptotics II: Laplace's method for multiple integrals. *Pacific J. Math.*, **11**, 185–192.

Gabutti, B. (1979). On high precision methods for computing integrals involving Bessel functions. *Math. Comp.*, **33**, 1049–1057.

Gabutti, B. (1985). An asymptotic approximation for a class of oscillatory infinite integrals. *SIAM J. Numer. Anal.*, **22**, 1191–1199.

Gabutti, B. and Lepora, P. (1987). A novel approach for the determination of asymptotic expansions of certain oscillatory integrals. *J. Comput. Appl. Math.*, **19**, 189–206.

Gabutti, B. and Minetti, B. (1981). A new application of the discrete Laguerre polynomials in the numerical evaluation of the Hankel transform of a strongly decreasing even function. *J. Comput. Phys.*, **42**, 277–287.

Gautschi, W. (1982). On generating orthogonal polynomials. *SIAM J. Sci. Stat. Comput.*, **3**, 289–317.

Gelfand, I. M. and Shilov, G. E. (1964). *Generalized Functions*, Vol. 1. Academic Press, New York.

Gilmore, R. (1981). *Catastrophe Theory for Scientists and Engineers.* John Wiley and Sons, New York.

Glauber, R. J. (1959). *Lectures in Theoretical Physics*, Vol. 1. Interscience, New York.

Gradshteyn, I. S. and Ryzhik, I. M. (1980). *Table of Integrals, Series and Products.* Academic Press, New York.

Greene, D. H. and Knuth, D. E. (1981). *Mathematics for the Analysis of Algorithms.* Birkhäuser, Boston.

Greenhill, A. G. (1886). Solutions of the cubic and quartic equations by means of Weierstrass's elliptic functions. *Proc. Lond. Math. Soc.*, **18**, 262–287.

Greenhill, A. G. (1892). *The Applications of Elliptic Functions.* MacMillan, London.

Grosjean, C. C. (1965). On the series expansion of certain types of Fourier integrals in the neighborhood of the origin. *Bull. Soc. Math. Belgigue*, **17**, 251–418.

Grosswald, E. (1966). Generalization of a formula of Hayman and its application to the study of Riemann's Zeta function. *Illinois J. Math.*, **10**, 9–23.

Grosswald, E. (1969). Correction and completion to the paper "Generalization of a formula of Hayman". *Illinois J. Math.*, **13**, 276–280.

Grzesik, J. (1983). Asymptotic behavior of an n-fold integral. *SIAM Review* (problem sect.), **25**, 576–577.

Handelsman, R. A. and Bleistein, N. (1973). Asymptotic expansion of integral transforms with oscillatory kernels: A generalization of the method of stationary phase. *SIAM J. Math. Anal.*, **4**, 519–535.

Handelsman, R. A. and Lew, J. S. (1969). Asymptotic expansion of a class of integral transforms via Mellin transforms. *Arch. Rational Mech. Anal.*, **35**, 382–396.

Handelsman, R. A. and Lew, J. S. (1970). Asymptotic expansion of Laplace transforms near the origin. *SIAM J. Math. Anal.*, **1**, 118–129.

Handelsman, R. A. and Lew, J. S. (1971). Asymptotic expansion of a class of integral transforms with algebraically dominated kernels. *J. Math. Anal. Appl.*, **35**, 405–433.

Hardy, G. H. (1942). Note on Lebesgue's constants in the theory of Fourier series. *J. London Math. Soc.*, **17**, 4–13.

Hardy, G. H. (1949). *Divergent Series*. Oxford University Press (Clarendon), London.

Harris, B. and Schoenfeld, L. (1968). Asymptotic expansions for the coefficients of analytic functions. *Illinois J. Math.*, **12**, 264–277.

Hayman, W. K. (1956). A generalization of Stirling's formula. *J. Reine Angew. Math.*, **196**, 67–95.

Henrici, P. (1974). *Applied and Computational Complex Analysis*, Vol. 2. Wiley, New York.

Hildebrand, F. B. (1974). *Introduction to Numerical Analysis, Second Edition*. McGraw-Hill, New York.

Hörmander, L. (1971). Fourier integral operators I. *Acta Math.*, **127**, 79–183.

Hörmander, L. (1983). *The Analysis of Linear Partial Differential Operators* 1. Springer-Verlag, Berlin.

Hsu, L. C. (1948a). Approximations to a class of double integrals of functions of large numbers. *Amer. J. Math.*, **70**, 698–708.

Hsu, L. C. (1948b). A theorem on the asymptotic behavior of a multiple integral. *Duke Math. J.*, **15**, 623–632.

Hsu, L. C. (1958). *Asymptotic Integration and Integral Approximation* (Chinese). Science Press, Beijing.

Jeanquartier, P. (1970). Développement asymptotique de la distribution de Dirac. *C.R. Acad. Sci. Paris. Ser. A.*, **271**, 1159–1161.

Jeffreys, H. (1962). *Asymptotic Approximations*. Oxford University Press, London.

Jones, D. S. (1966). *Generalized Functions*. McGraw-Hill, London.

Jones, D. S. (1969). Generalized transforms and their asymptotic behaviour. *Philos. Trans. Roy. Soc. London Ser. A*, **265**, 1–43.

Jones, D. S. (1972). Asymptotic behavior of integrals. *SIAM Review*, **14**, 286–317.

Jones, D. S. (1977). The mathematical theory of noise shielding. *Prog. Aerospace Sci.*, **17**, 149–229.

Jones, D. S. (1982). *The Theory of Generalized Functions*. Cambridge University Press, Cambridge.

Jones, D. S. and Kline, M. (1958). Asymptotic expansions of multiple integrals and the method of stationary phase. *J. Math. Phys.*, **37**, 1–28.

Kaminski, D. (1987). *Asymptotic expansions of some canonical diffraction integrals.* Ph.D. Thesis, University of Manitoba.

Kaminski, D. (1989). Asymptotic expansion of the Pearcey integral near the caustic. *SIAM J. Math. Anal.,* to appear.

Kaper, H. G. (1977). Asymptotic evaluation of two families of integrals. *J. Math. Anal. Appl.,* **59**, 415–422.

Kelvin (Lord). (1887). On the waves produced by a single impulse in water of any depth, or in a dispersive medium. *Philos. Mag.* [5], **23**, 252–255. [Reprinted in *Mathematical and Physical Papers*, Vol. 4 (1910). Cambridge University Press, London, 303–306.]

Knuth, D. E. (1980). Asymptotic behavior of a sequence. *SIAM Review* (problem sect.), **22**, 101–102.

Kontorovich, M. I., Karatygin, V. A., and Rozov, V. A. (1970). Asymptotic evaluation of a surface integral for the case of a stationary line. *Zh. vychisl. Mat. mat. Fiz.,* **10**, 811–817. [English transl.: *U.S.S.R. Computational Math. and Math. Phys.*]

Lam, P. F. (1976). Nesting property for Liapunov functions associated with an equilibrium point. *Rend. Circ. Matem. di Palermo*, **25**, 79–82.

LaSalle, J. P. and Lefschetz, S. (1961). *Stability by Liapunov's Direct Method with Applications.* Academic Press, New York.

Lauwerier, H. A. (1974). *Asymptotic Analysis, Second Edition.* Mathematisch Centrum, Amsterdam.

Lebedev, N. N. (1965). *Special Functions and Their Applications.* Prentice–Hall, Englewood Cliffs, New Jersey.

Leighton, W. (1976). *An Introduction to the Theory of Ordinary Differential Equations.* Wadsworth, Belmont, California.

Levey, H. C. and Mahony, J. J. (1968). Series representations of Fourier integrals. *Q. Appl. Math.,* **26**, 101–109.

Levinson, N. (1961). Transformation of an analytic function of several variables to a canonical form. *Duke Math. J.,* **28**, 345–353.

Lewis, R. M. (1967). Asymptotic theory of transients. In *Electromagnetic Wave Theory*, Part 2 (J. Brown, ed.). Pergamon Press, New York, 864–869.

Lighthill, M. J. (1958). *Fourier Analysis and Generalized Functions.* Cambridge University Press, Cambridge.

Lorch, L. (1966). Comparison of two formulations of Sonin's theorem and of their respective applications to Bessel functions. *Studia Scientiarum Mathematicarum Hungarica*, **1**, 141–145.

Luke, Y. L. (1968). An asymptotic expansion. *SIAM Review* (problem sect.), **10**, 229–232.

Malgrange, B. (1966). *Ideals of Differentiable Functions.* Oxford University Press, Bombay.

Malgrange, B. (1974). Intégrales asymptotiques et monodromie. *Ann. Sci. Ecole Norm Sup.* (4), **7**, 405–430.

Marichev, O. I. (1983). *Handbook of Integral Transforms of Higher Transcendental Functions, Theory and Algorithmic Tables.* Ellis Horwood, West Sussex, England.

Markushevich, A. I. (1977). *Theory of Functions of a Complex Variable, Second Edition.* Chelsea, New York.

Maslov, V. P. and Fedoryuk, M. V. (1981). Logarithmic asymptotic of the Laplace integrals. *Mat. Zametki,* **30**, 763–768. [English transl.: *Mathematical Notes of the Academy of Sciences of the USSR,* **30** (1981), 880–883.]

McClure, J. P. and Wong, R. (1978). Explicit error terms for asymptotic expansions of Stieltjes transforms. *J. Inst. Math. Appl.,* **22**, 129–145.

McClure, J. P. and Wong, R. (1979). Exact remainders for asymptotic expansions of fractional integrals. *J. Inst. Math. Appl.,* **24**, 139–147.

McClure, J. P. and Wong, R. (1983). Error bounds for multidimensional Laplace approximation. *J. Approximation Theory,* **37**, 372–390.

McClure, J. P. and Wong, R. (1986). Asymptotic approximation of an integral involving the normal distribution. *Can. Math. Bull.,* **29** (2), 167–176.

McClure, J. P. and Wong, R. (1987a). Asymptotic expansion of a multiple integral. *SIAM J. Math. Anal.,* **18**, 1630–1637.

McClure, J. P. and Wong, R. (1987b). Asymptotic expansion of a double integral with a curve of stationary points. *IMA J. Appl. Math.,* **38**, 49–59.

McClure, J. P. and Wong, R. (1990). Multidimensional stationary phase approximation: boundary stationary point. *J. Comp. Appl. Math.,* **30**, 213–225.

McClure, J. P. and Wong, R. (1991). Two-dimensional stationary phase approximation: stationary point at a corner. *SIAM J. Math. Anal.,* **22**, 500–523.

Meinardus, G. (1985). Remark on a lemma by R. Wong and J. P. McClure. *Math. Comp.,* **45**, 197-198.

Meyer, R. E. (1980). Exponential asymptotics. *SIAM Review,* **22**, 213-224.

Miller, J. C. P. (1946). *The Airy Integral*. British Assoc. Adv. Sci. Mathematical Tables, Part-Vol. B., Cambridge University Press, Cambridge.

Moser, L. and Wyman, M. (1955a). On solutions of $x^d = 1$ in symmetric groups. *Can. J. Math.*, **7**, 159–168.

Moser, L. and Wyman, M. (1955b). An asymptotic formula for the Bell numbers. *Trans. Roy. Soc. Can.*, *Series III*, **49**, Sec. III, 49–54.

Moser, L. and Wyman, M. (1956). Asymptotic expansions. *Can. J. Math.*, **8**, 225–233.

Moser, L. and Wyman, M. (1957). Asymptotic expansions II. *Can. J. Math.*, **9**, 194–207.

Moser, L. and Wyman, M. (1958). Stirling numbers of the second kind. *Duke Math. J.*, **25**, 29–44.

Murray, J. D. (1984). *Asymptotic Analysis*. Springer-Verlag, New York.

Noble, B. (1969). *Applied Linear Algebra*. Prentice–Hall, Englewood Cliffs, New Jersey.

Oberhettinger, F. (1959). On a modification of Watson's lemma. *J. Res. Nat. Bur. Standards, Section B*, **63**, 15–17.

Oberhettinger, F. (1972). *Tables of Bessel Transforms*. Springer-Verlag, New York.

Oberhettinger, F. (1974). *Tables of Mellin Transforms*. Springer-Verlag, New York.

Olver, F. W. J. (1968). Error bounds for the Laplace approximation for definite integrals. *J. Approximation Theory*, **1**, 293–313.

Olver, F. W. J. (1970a). Why steepest descents? *SIAM Review*, **12**, 228–247.

Olver, F. W. J. (1970b). A paradox in asymptotics. *SIAM J. Math. Anal.*, **1**, 533–534.

Olver, F. W. J. (1974a). *Asymptotics and Special Functions*. Academic Press, New York.

Olver, F. W. J. (1974b). Error bounds for stationary phase approximations. *SIAM J. Math. Anal.*, **5**, 19–29.

Olver, F. W. J. (1975). Unsolved problems in the asymptotic estimation of special functions. In *Theory and Application of Special Functions*. (R. Askey, ed.). Academic Press, New York, 99–142.

Olver, F. W. J. (1980). Asymptotic approximations and error bounds. *SIAM Review*, **22**, 188–203.

Osborn, T. A. and Wong, R. (1983). Schrödinger spectral kernels: Higher order asymptotic expansions. *J. Math. Phys.* **24**, 1487–1501.

Ott, H. (1943). Die Sattelpunktsmethode in der Umgeburg eines Poles mit Anwendungen auf die Wellenoptik und Akustik. *Ann. Physik*, **43**, 393–403.

Otter, R. (1948). The number of trees. *Ann. of Math.* (2), **49**, 583–599.

Pearcey, T. (1946). The structure of an electromagnetic field in the neighbourhood of a cusp of a caustic. *Phil. Mag.*, **37**, 311–317.

Pederson, R. N. (1965). Laplace's method for two parameters. *Pacific J. Math.*, **15**, 585–596.

Perron, O. (1917). Über die näherungsweise Berechnung von Funktionen großber Zahlen. *Sitzungsber. Bayr. Akad. Wissensch.* (*Münch. Ber.*), 191–219.

Peters, A. S. (1949). A new treatment of the ship wave problem. *Commun. Pure Appl. Math.*, **2**, 123–148.

Pólya, G. and Szegö, G. (1972). *Problems and Theorems in Analysis*, Vol. I. Springer-Verlag, Berlin.

Poston, T. and Stewart, I. (1978). *Catastrophe Theory and Its Applications*. Pitman, Boston.

Pourahmadi, M. (1984). Taylor expansion of $\exp(\sum_{k=0}^{\infty} a_k z^k)$ and some applications. *Amer. Math. Monthly*, **91**, 303–308.

Rabinowitz, P. and Weiss, G. (1959). Tables of abscissas and weights for numerical evaluation of integrals of the form $\int_0^{\infty} e^{-x} x^n f(x)\, dx$. *MTAC*, **13**, 285–294.

Ray, B. K. (1970). On the absolute summability of some series related to a Fourier series. *Proc. Camb. Phil. Soc.*, **67**, 29–45.

Riekstiņš, E. (1966). On the use of neutrices for asymptotic representation of some integrals (Russian). *Latvian Math. Yearbook*, 5–21.

Riekstiņš, E. (1974). Asymptotic expansions for some type of integrals involving logarithms. *Latvian Math. Yearbook*, **15**, 113–130.

Riekstiņš, E. (1974, 1977, 1981). *Asymptotic Expansions of Integrals* (Russian). Vols. 1, 2, 3. Zinatne, Riga, U.S.S.R.

Robinson, R. (1951). A new absolute geometric constant? *Amer. Math. Monthly*, **58**, 462–469.

Rosser, J. B. (1955). Explicit remainder terms for some asymptotic series. *J. Rational Mech. Anal.*, **4**, 595–626.

Rudin, W. (1973). *Functional Analysis*. McGraw–Hill, New York.

Rudin, W. (1974). *Real and Complex Analysis, Second Edition*. McGraw–Hill, New York.

Rudin, W. (1976). *Principles of Mathematical Analysis, Third Edition*. McGraw–Hill, New York.

Schmidt, H. (1937). Beiträge zu einer Theorie der allgemeinen asymptotischen Darstellungen. *Math. Ann.*, **113**, 629–656.

Schwartz, L. (1966a). *Théorie des distributions, Nouvelle édition (augmentée)*. Hermann, Paris.

Schwartz, L. (1966b). *Mathematics for Physical Sciences*. Addison–Wesley, Reading, Massachusetts.

Servadio, S. (1982). Calculation of $O(1)$ interferences of double-scattering waves in 3–3 collision. *Nuovo Cimento*, **69**, 1–22.

Servadio, S. (1988). Unitarity constraint on the "truly 3 body" scattering. *Nuova Cimento B*, **100**, 565 and 587.

Shao, T. S., Chen, T. C., and Frank, R. M. (1964). Tables of zeros and Gaussian weights of certain associated Laguerre polynomials and the related generalized Hermite polynomials. *Math. Comput.*, **18**, 598–616.

Shivakumar, P. N. and Wong, R. (1979). Asymptotic expansion of multiple Fourier transforms. *SIAM J. Math. Anal.*, **10**, 1095–1104.

Shivakumar, P. N. and Wong, R. (1982). Asymptotic expansion of the Lebesgue constants associated with polynomial interpolation. *Math. Comp.*, **39**, 195–200.

Shivakumar, P. N. and Wong, R. (1988). Error bounds for a uniform asymptotic expansion of the Legendre function $P_n^{-m}(\cosh z)$. *Quart. Appl. Math.*, **46**, 473–488.

Sirovich, L. (1971). *Techniques of Asymptotic Analysis*. Springer-Verlag, New York.

Smith, K. T. (1971). *Primer of Modern Analysis*. Bogden & Quigley, Tarrytown-on-Hudson, New York.

Soni, K. (1978). A note on asymptotic expansions. *Amer. Math. Monthly*, **85**, 268–269.

Soni, K. (1980). Exact error terms in the asymptotic expansion of a class of integral transforms I (oscillatory kernels). *SIAM J. Math. Anal.*, **11**, 828–841.

Soni, K. (1982). Asymptotic expansion of the Hankel transform with explicit remainder terms. *Quart. Appl. Math.*, **50**, 1–14.

Soni, K. (1983). A note on uniform asymptotic expansion of incomplete Laplace integrals. *SIAM J. Math. Anal.*, **14**, 1015–1018.

Soni, K. and Soni, R. P. (1981). A note on uniform asymptotic expansions of finite K_ν and related transforms with explicit remainder. *J. Math. Anal. Appl.*, **79**, 163–177.

Soni, K. and Soni, R. P. (1985). A note on summability and asymptotics. *SIAM J. Math. Anal.*, **16**, 392–404.

Steffensen, J. F. (1950). *Interpolation, Second Edition*. Chelsea, New York.

Stenger, F. (1981). Numerical methods based on Whittaker cardinal, or sinc functions. *SIAM Review*, **23**, 165–224.

Stoyanov, B. J. and Farrell, R. A. (1987). On the asymptotic evaluation of $\int_0^{\pi/2} J_0^2(\lambda \sin x)\, dx$. *Math. Comp.*, **49**, 275–279.

Struik, D. J. (1961). *Lectures on Classical Differential Geometry, Second Edition*. Addison–Wesley, Reading, Massachusetts.

Szegö, G. (1921). Über die Lebesgueschen Konstanten bei den Fourierschen Reihen. *Math. Zeitschrift*, **9**, 163–166.

Szegö, G. (1932). Über einige asymptotische Entwicklungen der Legendreschen Funktionen. *Proc. London Math. Soc.* (2), **36**, 427–450.

Szegö, G. (1967). *Orthogonal polynomials*. Colloquium Publications, Vol. 23, Third Edition, Amer. Math. Soc., Providence, Rhode Island.

Temme, N. M. (1975). Uniform asymptotic expansions of the incomplete gamma functions and the incomplete beta function. *Math. Comp.*, **29**, 1109–1114.

Temme, N. M. (1976). Remarks on a paper of A. Erdélyi. *SIAM J. Math. Anal.*, **7**, 767–770.

Temme, N. M. (1979). The asymptotic expansion of the incomplete gamma functions. *SIAM J. Math. Anal.*, **10**, 757–766.

Temme, N. M. (1985). *Special functions as approximants in uniform asymptotic expansions of integrals: A survey*. Special Functions: Theory and Computation, Rendiconti del Seminario Matematico, Torino.

Thorpe, J. A. (1979). *Elementary Topics in Differential Geometry*. Springer-Verlag, New York.

Titchmarsh, E. C. (1939). *The Theory of Functions, Second Edition.* Oxford University Press, London.

Titchmarsh, E. C. (1959). *Introduction to the Theory of Fourier Integrals, Second Edition.* Oxford University Press, London.

Tolstov, G. P. (1962). *Fourier Series.* Prentice–Hall, Englewood Cliffs, New Jersey.

Tricomi, F. (1933). Determinazione del valore asintotico di un certo integrale. *Rend. Lincei* (6), **17**, 116–119.

Tricomi, F. G. and Erdélyi, A. (1951). The asymptotic expansion of a ratio of gamma functions. *Pacific. J. Math.*, **1**, 133–142.

Ursell, F. (1970). Integrals with a large parameter: Paths and descent and conformal mapping. *Proc. Cambridge Philos. Soc.*, **67**, 371–381.

Ursell, F. (1972). Techniques of asymptotic expansion. Integrals with a large parameter. Unpublished Notes. University of Manchester, 26 pp.

Ursell, F. (1972). Integrals with a large parameter. Several nearly coincident saddle-points. *Proc. Camb. Phil. Soc.*, **72**, 49–65.

Ursell, F. (1983). Integrals with a large parameter: Hilbert transforms. *Math. Proc. Camb. Phil. Soc.*, **93**, 141–149.

Ursell, F. (1984). Integrals with a large parameter: Legendre functions of large degree and fixed order. *Math. Proc. Camb. Phil. Soc.*, **95**, 367–380.

Uspensky, J. V. (1948). *Theory of Equations.* McGraw–Hill, New York.

van der Corput, J. G. (1934). Zur Methode der stationären Phase I. *Compositio Math.*, **1**, 15–38.

van der Corput, J. G. (1936). Zur Methode der stationären Phase II. *Compositio Math.*, **3**, 328–372.

van der Corput, J. G. (1948). On the method of critical points, I. *Proc. Nederl. Akad. Wetensch.*, **51**, 650–658.

van der Waerden, B. L. (1951). On the method of saddle points. *Appl. Sci. Res.*, **B2**, 33–45.

van Kampen, N. G. (1949, 1950). An asymptotic treatment of diffraction problems. *Physica*, **XIV**, 575–589 and **XVI**, 817–821.

Watson, E. J. (1981). *Laplace Transforms and Applications.* van Nostrand Reinhold, New York.

Watson, G. N. (1918). Harmonic functions associated with the parabolic cylinder. *Proc. Lond. Math. Soc.* (2), **17**, 116–148.

Watson, G. N. (1930). The constants of Landau and Lebesgue. *Quart. J. Math.*, **1**, 310–318.

Watson, G. N. (1944). *A Treatise on the Theory of Bessel Functions, Second Edition.* Cambridge University Press, London.

Widder, D. V. (1941). *The Laplace Transform.* Princeton University Press, Princeton, New Jersey.

Wimp, J. (1980). Uniform scale functions and the asymptotic expansion of integrals. In *Ordinary and Partial Differential Equations* (Proc. Fifth Conf. Univ. Dundee, Dundee, 1978). Springer Lecture Notes in Math., No. 827, Berlin, 251–271.

Wolf, E. (1951). The diffraction theory of aberrations. *Reports on Progress in Physics*, **XIV**, 95–120. The Physical Society, London.

Wong, R. (1970). On a Laplace integral involving logarithms. *SIAM J. Math. Anal.*, **1**, 360–364.

Wong, R. (1973). On uniform asymptotic expansion of definite integrals. *J. Approximation Theory*, **7**, 76–86.

Wong, R. (1976). Error bounds for asymptotic expansions of Hankel transforms. *SIAM J. Math. Anal.*, **7**, 799–808.

Wong, R. (1977). Asymptotic expansions of Hankel transforms of functions with logarithmic singularities. *Comput. Math. Appl.*, **3**, 271–286.

Wong, R. (1978). Asymptotic expansions of fractional integrals involving logarithms. *SIAM J. Math. Anal.*, **9**, 835–842.

Wong, R. (1979). Explicit error terms for asymptotic expansions of Mellin convolutions. *J. Math. Anal. Appl.*, **72**, 740–756.

Wong, R. (1980a). Asymptotic expansion of the Hilbert transform. *SIAM. J. Math. Anal.*, **11**, 92–99.

Wong, R. (1980b). On a uniform asymptotic expansion of a Fourier-type integral. *Quart. Appl. Math.*, **38**, 225–234.

Wong, R. (1980c). Error bounds for asymptotic expansions of integrals. *SIAM Review*, **22**, 401–435.

Wong, R. (1980d). Distributional derivation of an asymptotic expansion. *Proc. Amer. Math. Soc.*, **80**, 266–270.

Wong, R. (1981). Asymptotic expansions of the Kontorovich–Lebedev transform. *Applicable Analysis*, **12**, 161–172.

Wong, R. (1983). Applications of some recent results in asymptotic expansions. Proc. 12th Winnipeg Conference on Numerical Methods of Computing. *Congress. Numer.*, **37**, 145–182.

Wong, R. (1988). Asymptotic expansion of $\int_0^{\pi/2} J_\nu^2(\lambda \cos\theta)\,d\theta$. *Math. Comp.*, **50**, 229–234.

Wong, R. and Lin, J. F. (1978). Asymptotic expansions of Fourier transforms of functions with logarithmic singularities. *J. Math. Anal. Appl.*, **64**, 173–180.

Wong, R. and McClure, J. P. (1981). On a method of asymptotic evaluation of multiple integrals. *Math. Comp.*, **37**, 509–521.

Wong, R. and McClure, J. P. (1984). Generalized Mellin convolutions and their asymptotic expansions. *Can. J. Math.*, **36**, 924–960.

Wong, R. and Wyman, M. (1972). A generalization of Watson's lemma. *Can. J. Math.*, **24**, 185–208.

Wong, R. and Wyman, M. (1974). The method of Darboux. *J. Approximation Theory*, **10**, 159–171.

Wright, E. M. (1934). The asymptotic expansion of the generalized Bessel function. *Proc. London Math. Soc.* (2), **38**, 257–270.

Wyman, M. (1959). The asymptotic behavior of the Laurent coefficients. *Can. J. Math.*, **11**, 534–555.

Wyman, M. (1960). Asymptotics. Unpublished notes. University of Alberta, 84 pp.

Wyman, M. (1963). The asymptotic behaviour of the Hermite polynomials. *Can. J. Math.*, **15**, 332–349.

Wyman, M. (1964). The method of Laplace. *Trans. Roy. Soc. Canada*, **2**, 227–256.

Wyman, M. (1965). Asymptotic Analysis. Unpublished notes. University of Alberta, 216 pp.

Wyman, M. and Wong, R. (1969). The asymptotic behaviour of $\mu(z, \beta, \alpha)$. *Can. J. Math.*, **21**, 1013–1023.

Zauderer, E. (1983). *Partial Differential Equations of Applied Mathematics*. John Wiley and Sons, New York.

Zayed, A. I. (1982). Asymptotic expansions of some integral transforms by using generalized functions. *Trans. Amer. Math. Soc.*, **272**, 785–802.

Symbol Index

$I_n(x)$	modified Bessel function, 128
$J_v(x)$	Bessel function of the first kind, 24, 97
$\mathbf{J}_v(x)$	Anger's function, 82
K^+	space of all distributions which are the derivatives of some locally integrable functions, 255
$K_v(x)$	modified Bessel function, 225
$L_{\mathrm{loc}}(I)$	collection of integrable functions on I, 250
$L_{\mathrm{loc}}^+(\mathbb{R})$	class of functions which are locally integrable on $\mathbb{R}$ and which vanish on $(-\infty, 0]$, 254
L_n	Lebesgue constants, 1
$L_n^{(\mu)}(t)$	Laguerre polynomial, 222, 372
$\mathrm{li}(z)$	logarithmic integral, 43
$M[f; z]$	Mellin transform, 147
O, o	order symbols, 5
$P_n(x)$	Legendre polynomial, 141
$P_n^{-m}(z)$	Legendre function, 407
Prin	principal part, 183
$Q_v^{-\mu}(z)$	Legendre function of the second kind, 47
Res	residue, 148
$\mathcal{S}$	space of rapidly decreasing functions, 261
$\mathcal{S}'$	space of tempered distributions, 262
$\mathcal{S}_n$	space of rapidly decreasing functions in $\mathbb{R}^n$, 266
$\mathcal{S}_n'$	space of tempered distributions in $\mathbb{R}^n$, 267
sgn A	signature of a matrix A, 480
$\mathrm{Si}(x)$	sine integral, 513
$U(a, t)$	parabolic cylinder function (J. C. P. Miller's notation), 185
$W_{k,m}(z)$	Whittaker function, 410
$Y_v(x)$	Bessel function of the second kind, 171
$(\alpha)_n$	Pochhammer notation, 47
$\Gamma(\alpha, x)$	complementary incomplete Gamma function, 28
γ	Euler (-Mascheroni) constant, 2, 29, 39
∇	gradient, 478
$\delta(x)$ (or δ)	Dirac delta function, 243–245
$\delta(x - a)$ (or δ_a)	translated delta function, 244
$\delta(P)$	delta function concentrated on the surface $P = 0$, 282
Λ_f	distribution generated by $f(x)$, 243
$\langle \Lambda, \phi \rangle$	action of the distribution Λ on the test function φ, 243
$\chi_{[a,b]}$	characteristic function, 335
$\psi(\alpha, \gamma, z)$	confluent hypergeometric function, 46
$\psi(z)$	logarithmic derivative of the Gamma function, 186
$\sim$	asymptotic equality, 1, 5
$\approx$	asymptotic equality with respect to an asymptotic sequence, 11
$[x]$	integer part, 34
$t \cdot x = t_1 x_1 + \cdots + t_n x_n$	inner product in $\mathbb{R}^n$, 274
$\langle x, y \rangle = x_1 y_1 + \cdots + x_n y_n$	inner product in $\mathbb{R}^n$, 480

Author Index

535

Subject Index

A

Abel limit, 198
Abel summability, 198
Abscissas, 222
Airy function, 46, 51, 90, 133, 370
  asymptotic expansion of, 47, 93
  differential equation for, 91
  generalized, 386
  integral of, 46, 91
  Maclaurin expansion of, 91
  of negative argument, 220
    asymptotic expansion of, 220
    relation to Bessel functions, 220
Airy integral, *see* Airy function,
Amplitude function, 429
Anger function $\mathbf{A}_\nu(x)$, 418
  uniform asymptotic expansion of, 418
Anger's function $\mathbf{J}_\nu(x)$, 82
Asymptotic
  equalities, 2
  expansions, 5
    compound, 11
    failure of, 18
    generalized, 10

    of Poincaré type, 5, 11
    of power series form, 11
    uniform, 353, 356
  formulas, *see* asymptotic equalities
  sequences, 10
Asymptotics, 1
  paradox in, 146

B

Barnes' lemma, 48, 229, 355
Bernoulli numbers, 2, 32
Bernoulli polynomials, 32
  generalized, 47
Bessel function, 24, 45
  asymptotic formulas, 205, 206
  integral representation of, 24, 97, 112, 134, 417
  modified, 50, 225
    asymptotic expansion of, 227
  of large argument, 24–26, 112–116, 134
  of large order, 98, 100, 102–103
  of the second kind, 236
  relation to Hankel functions, 97, 211, 227

539